IC Converter Cookbook

by

Walter G. Jung

Howard W. Sams & Co., Inc.
4300 WEST 62ND ST. INDIANAPOLIS, INDIANA 46268 USA

FIRST EDITION
FIRST PRINTING—1978

International Standard Book Number: 0-672-21527-6
Library of Congress Catalog Card Number: 78-54453

Printed in the United States of America.

Preface

It is an obvious reality to virtually anyone acquainted with the field of electronics that there has been spectacular growth over the past few years in digital technology. The computer, at various levels of sophistication, has become a dominant central force of electronic systems. More recently, the low cost microcomputer or microprocessor has accelerated this pervasiveness to the extent that even the smallest systems often find them in use. Along with this large scale application of digital techniques has come an accompanying increase in hardware to interface digital circuits to the real world environment.

It is the analog to digital (a/d) and digital to analog (d/a) converters, each a specific interface device, which enable the fast and accurate transfer of information between the analog and digital domains. Quite like computer hardware, a/d and d/a converters have proliferated to the point where low cost but sophisticated single-chip devices are now performing the tasks once relegated to large and expensive racks of equipment. And, quite naturally, this phase of development has (and is) affecting the use of these devices to the point where their application is more and more widespread. This is due to the fact that they may be regarded as components as well as systems.

Potential users of a/d and d/a converters include virtually all who use electronic circuitry, and on all levels of experience. It is this audience to which this book is addressed—the problem solving practicing engineer, technician, hobbyist, or the student. Like the author's previous two "cookbooks" *IC Op-Amp Cookbook* and *IC Timer Cookbook,* this book is also intended to be a working guide to the understanding and use of IC a/d and d/a converters.

The book is divided into three sections. Part I is concerned with the basic principles and theories of data conversion, with particular emphasis on how they are applied in the form of representative IC devices. Along with the basic theory, this section includes definition of terms and a brief discussion of the operation of the devices used in the book, including peripheral conversion devices such as sample holds, instrumentation amps, and analog switches.

Part II of the book is devoted to converter applications, and includes discussion of various a/d and d/a circuits, with many application examples in every major category and class of converter. As in the author's previous books, the circuits are complete working entities, with specified part numbers and component values. As part of the application discussion, performance specifications are given along with appropriate suggestions for modifications.

The book concludes with Part III, an appendix of manufacturers catalog sheets, parts number cross references, and list of manufacturers.

WALTER G. JUNG

To the memory of my father.

Acknowledgments

In the preparation of this book the author was fortunate to be able to draw on a number of valuable sources of technical information. One significant factor towards the completion of the book was critical comment and informational inputs by a number of industry experts. For this, thanks must go to the following individuals.

Special thanks for a countless number of helpful suggestions go to Jeff Riskin of Analog Devices. Sincere thanks also to Russ Apfel and John Schoef of Advanced Micro Devices; to Lew Counts and Dan Sheingold of Analog Devices; to Joe Santen of Burr-Brown; to Don Jones of Harris Semiconductor; to Bill O'Neil of Intersil; to Lothar Stern of Motorola; to Bob Dobkin, Carl Nelson, and Bob Pease of National Semiconductor; to Donn Soderquist of Precision Monolithics (and to Will Ritmanich, formerly of Precision Monolithics); to Tom Cate and Jim Schmoock of Raytheon; to Steve Bolger, Jim Graham, and Barry Harvey of Siliconix; and to Michael Paiva of Teledyne Semiconductor.

General thanks is extended to the following firms for the permission to use portions of their technical literature.

Advanced Micro Devices
Analog Devices, Inc.
Burr-Brown Corporation
Fairchild Semiconductor
Harris Semiconductor
Intech/Function Modules
Intersil, Inc.
Motorola Semiconductor Products
National Semiconductor Corporation

Precision Monolithics, Inc.
Raytheon Semiconductor
RCA Solid State Division
Siliconix, Inc.
Teledyne Semiconductor

Also, the following publications were helpful in allowing use of copyrighted material, and their assistance is appreciated: *Electronic Design, EDN* and *Journal of the AES.*

Finally, to my wife Anne there goes again a special note of thanks for the typing of the manuscript. Thanks also go to my research assistants, Jeannie and Mark.

Contents

CHAPTER 4

CHAPTER 5

PART II

IC Converter Applications

CHAPTER 6

CHAPTER 7

CHAPTER 8

PART III

Appendices

APPENDIX A

APPENDIX B

Introduction

The data conversion and processing function is one which has until recently been largely dominated by the modular "black box." Now in the late seventies, this function is beginning to be penetrated by the IC converter, from a multitude of points.

Operational amplifiers experienced a parallel form of growth to maturity, beginning in the late sixties. Today, IC op amps are available at low cost which satisfy all but the most esoteric uses. Ultimately IC data converters will follow this same path of progress, but the road will be a bit more bumpy, and the trip perhaps longer.

One reason for the absence of an overnight transition to a high performance, low cost IC converter is the magnitude of the technical task. Another is the diversity of converter types, that in turn spreads the market concentration widely and tends to both slow the learning curve and hold costs up. Simply stated, there is no such thing as a "741" converter, and there may never be such a thing, that is, a d/a or a/d converter with a comparable pervasiveness.

Nevertheless, the attributes which all ICs have, namely high performance/cost ratio, small size and weight, low power consumption, and minimization of system complexity, make IC converters highly desirable. Development will continue with all types of data converters until they are efficiently and satisfactorily realized in IC forms to fulfill a high percentage of needs.

If the preceding discussion appears to reflect that great progress is yet to be seen in IC converter development, this is the intent. However, it should not be interpreted to mean that useful IC converter realizations are not presently available, as they are. This book discusses IC data converters. with examples of devices for all major d/a and a/d converter types. Many of these are available through multiple sources at low cost; this already is evidence of maturity in some regards.

The IC converter is a highly useful and efficient circuit element today, and this book describes how to work with them and exploit their advantages. The field is a highly active one and future developments are anticipated which will be interesting and exciting to a broad spectrum of users.

PART I

Introducing the IC Converter

1

Data Conversion Fundamentals

The general data converter function, be it either the a/d or d/a process, is to interface between the digital and analog worlds. A prerequisite of the proper understanding of data converters is therefore an understanding of the terms of both the analog and digital domains.

In the digital world, data is most often in the binary, or two state, form. Accordingly, a/d and d/a converters most often generate (or respond) to binary-coded data. Another form of digital data often used is binary-coded decimal, or bcd. This section reviews briefly these two number systems, and how they relate to converter operation in practice.

THE BINARY NUMBER SYSTEM

Computers and digital logic system are by far predominantly binary in operation. Data bits are represented by a binary "1" for the *on* (or true) state, and a binary "0" for the *off* (or false) state. Thus a single data bit can communicate two possible states, on or off. In positive logic systems this is defined as a 1 or (high) for on and a 0 (or low) for off.

In general, such a system can communicate for "n" bits, 2^n possible states. Thus for example a 2-bit system has four possible states, 3 bits, eight, and so on. This pattern can be continued indefinitely to accommodate any conceivable number of states, and the range is doubled for each additional bit added.

Fig. 1-1 illustrates the binary number system with a simple progression through three bits, where $2^3 = 8$ and also the general limit of 2^n. Using this system any decimal number such as we are accustomed to dealing with may be expressed in binary notation, given a sufficient number of bits.

$$2^0 = 1$$
$$2^1 = 2$$
$$2^2 = 4$$
$$2^3 = 8$$
$$\bullet$$
$$\bullet$$
$$\bullet$$
$$2^{n-1} = 2^{n-1}$$
$$2^n = 2^n$$

$$\text{Decimal Number} = a_0\,2^0 + a_1\,2^1 + a_2\,2^2 + a_3\,2^3 \ldots a_n\,2^{n-1}$$

where $a_0 \ldots a_n$ are coefficients, either 1 or 0

Example:

$$13 = (1)(1) + (0)(2) + (1)(4) + (1)(8)$$
$$= 1 + 0 + 4 + 8$$

Written in binary notation (4 bits)

$$13 = 1101$$

LSB (coefficient of 2^0 term)

MSB (coefficient of 2^{n-1} term)

Fig. 1-1. Binary number system.

A decimal number can be expressed as a sum of the succeeding powers of two, using all those terms necessary to yield the number. Each term of the series has a coefficient of the form a_n, where n denotes the order. This coefficient is either 1 or 0; 1 denoting the presence (or *on* state) of the term, 0 denoting the absence (or *off* state) of the term. The terms 2^0, 2^1 . . . 2^n are the "bits" of the system, and the highest order bit, 2^n, is termed the *resolution* of the system.

In the example shown, the number 13 is expressed as a sum of 2^0, 2^2, and 2^8, or 1, 4, and 8. Note that this uses three *active* bits of the 4-bit system.

To write this number in binary notation however, *all* coefficients are used. For the number 13 the coefficients of 2^0, 2^1, 2^2, 2^4 are 1, 0, 1, and 1 respectively.

Conventional practices of binary notation always list the most significant bit (MSB) of a binary number first. Here the MSB is 2^8; so the number 13 is written in binary is 1101. Note that this is just the reverse of the order of the terms as written in the general expression.

A further lilustration is given in Fig. 1-2, a complete listing of the numbers 0 through 15, and their binary equivalents. These numbers are shown in the 1, 2, 4, 8 component form, then the list of coefficients, and finally, the binary notation form. This table could of course be extended to any arbitrary length.

Although it is not absolutely necessary to do so, it is sometimes convenient to list binary numbers in groups of 4 bits each, as is shown in

		2^0		2^1		2^2		2^3		Coefficients of: 2^0	2^1	2^2	2^3		As written in binary notation: 2^3	2^2	2^1	2^0
0	=	0	+	0	+	0	+	0	=	0	0	0	0		0	0	0	0
1		1	+	0	+	0	+	0		1	0	0	0		0	0	0	1
2		0	+	2	+	0	+	0		0	1	0	0		0	0	1	0
3		1	+	2	+	0	+	0		1	1	0	0		0	0	1	1
4		0	+	0	+	4	+	0		0	0	1	0		0	1	0	0
5		1	+	0	+	4	+	0		1	0	1	0		0	1	0	1
6		0	+	2	+	4	+	0		0	1	1	0		0	1	1	0
7		1	+	2	+	4	+	0		1	1	1	0	➡	0	1	1	1
8		0	+	0	+	0	+	8		0	0	0	1		1	0	0	0
9		1	+	0	+	0	+	8		1	0	0	1		1	0	0	1
10		0	+	2	+	0	+	8		0	1	0	1		1	0	1	0
11		1	+	2	+	0	+	8		1	1	0	1		1	0	1	1
12		0	+	0	+	4	+	8		0	0	1	1		1	1	0	0
13		1	+	0	+	4	+	8		1	0	1	1		1	1	0	1
14		0	+	2	+	4	+	8		0	1	1	1		1	1	1	0
15	=	1	+	2	+	4	+	8	=	1	1	1	1		1	1	1	1
															MSB			LSB

Fig. 1-2. Binary notation table (4 bits).

Fig. 1-3. Written as such, the number 147 is written 1001 0011 rather than 10010011. Technically however, either form is correct for binary notation.

Fig. 1-3 also shows the general pattern of bit weighting for the two different examples. The MSB carries a weight of 2^{n-1}; in general, the lesser bits are progressively smaller in weight as indicated. Thus for example, the MSB of a 4-bit word has a weight of 2^{4-1}, or $2^3 = 8$.

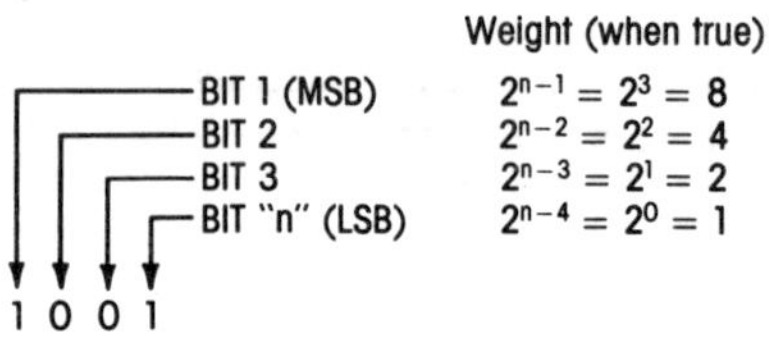

(A) 4-bit example.

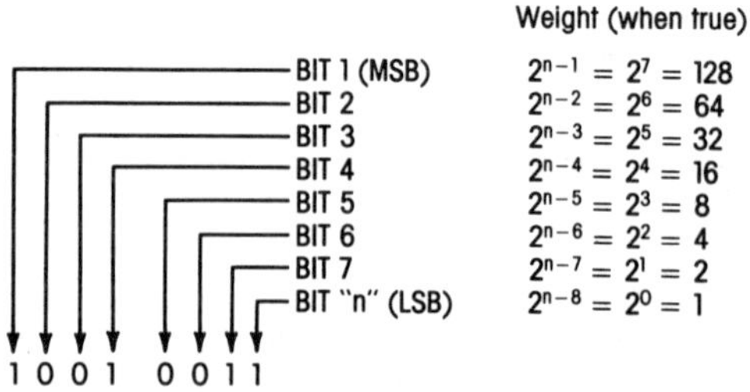

(B) 8-bit example.

Fig. 1-3. Binary notation (general).

Table 1-1. Base 2 Table

2^0	1	2^{11}	2048
2^1	2	2^{12}	4096
2^2	4	2^{13}	8192
2^3	8	2^{14}	16384
2^4	16	2^{15}	32768
2^5	32	2^{16}	65536
2^6	64	2^{17}	131072
2^7	128	2^{18}	262144
2^8	256	2^{19}	524288
2^9	512	2^{20}	1048576
2^{10}	1024		

The MSB of an 8-bit word would be $2^{8-1} = 2^7 = 128$. As may be noted, this pattern results in an LSB with a weight of $2^0 = 1$, for any number of bits. Table 1-1 illustrates the powers of 2 through $n = 20$, which may be used to calculate any binary word with an MSB weight of up to 1,048,576.

The preceding discussion has been oriented toward the *general* definition of binary notation. In data converters, the binary notation is used to define *fractional* numbers. The general concept is similar, except for fractional numbers the arrangement and weights are different.

The binary fractional number system is illustrated in Fig. 1-4 and as may be noted, it is generally similar to that previously described. The first term in the general expression is the MSB and since it is a 2^{-1} term, its weight is always ½. Succeeding terms progress as 2^{-2}, 2^{-3}, etc., and their weights are always ½ that of the preceding bit. The coefficients a_1-a_n are either 1 or 0, depending upon whether a particular bit is on (1) or off (0). For 13/16, the example shown, the fractional binary code is .1101. The preceding decimal point is technically complete, but it is usually dropped in converter use, as all codes are fractional.

Fractional Numbers $= a_1 2^{-1} + a_2 2^{-2} + a_3 2^{-3} \ldots a_n 2^{-n}$

where $a_1 \ldots a_n$ are coefficients either 1 or 0

Example:

$$13/16 = (1)(1/2) + (1)(1/4) + (0)(1/8) + (1)(1/16)$$
$$= 8/16 + 4/16 + 0/16 + 1/16$$

Written in fractional binary notation (4 bits)

13/16 = .1101

DECIMAL* POINT — MSB — LSB

*Usually dropped in all-fractional systems

Fig. 1-4. Binary fractional number system.

One thing which may be noted from this system is that the binary number can *approach* unity, but will never actually reach it. For any number of bit the highest number will be 11 . . . 11, which will result in a fraction of $2^{n-1}/2^n$. The inherent difference from unity in this relation is $1/2^n$, or 1 LSB.

Fig. 1-5 shows 4- and 8-bit examples of fractional binary numbers. In each case as can be noted the MSB is ½ and the LSB is 2^{-n}. The preceding decimal point is shown in these two examples to emphasize that they are fractional numbers. In the remainder of the book the decimal point will be dropped, as it is in conventional practice.

A more complete summary of the key fractional binary relationships is contained in Table 1-2. Here are shown various forms of information corresponding to bit lengths of up to 20. In essence, this table summarizes the value of an LSB (2^{-n} or $1/2^n$) for the various numbers of bits, making it convenient for reference. The data is given in terms of the decimal fraction, its equivalent percentage, its equivalent in parts per million (ppm) and also the decibel equivalent.

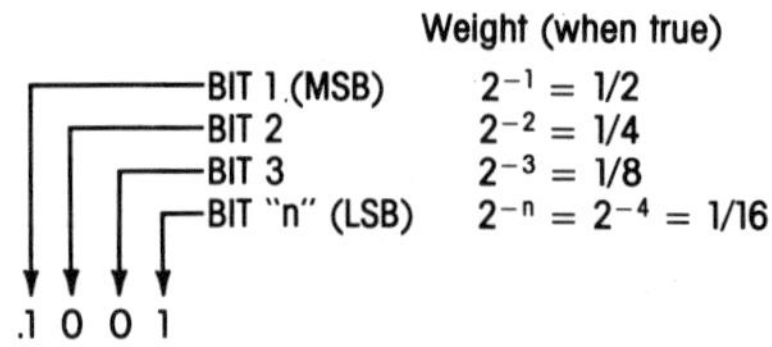

(A) 4-bit example (N = 9/16).

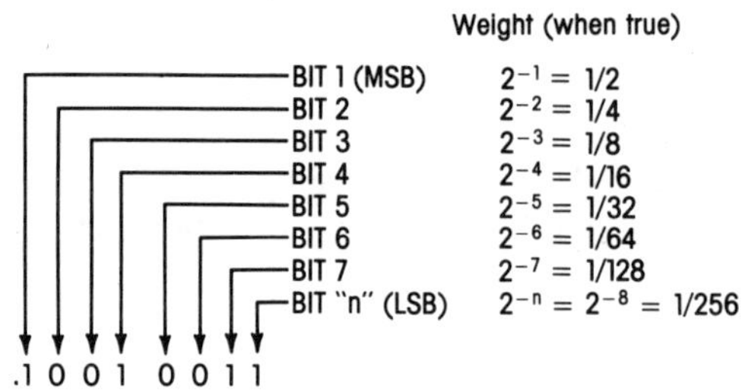

(B) 8-bit example (N = 147/256).

Fig. 1-5. Binary fractional notation.

THE BCD SYSTEM

A number system used quite often with data converters is the binary-coded decimal (bcd) system. This system is explained in Fig. 1-6, and is shown in fractional form, as would be used with converters.

Like the binary fractional system, the bcd fractional system is limited inherently to numerical values less than unity. Its major advantage is in the ease of interfacing between man and machine, via either thumbwheel

Table 1-2. Fractional-Binary Relationships

Bits	2^{-n}	$1/2^n$	Decimal	Percent	PPM	dB**
Full Scale	2^0	1/1	1.0	100	1.0×10^6	0
MSB 1	2^{-1}	1/2	.5	50	5.0×10^5	−6.02
2	2^{-2}	1/4	.25	25	2.5×10^5	−12.04
3	2^{-3}	1/8	.125	12.5	1.25×10^5	−18.06
4	2^{-4}	1/16	.0625	6.25	6.25×10^4	−24.08
5	2^{-5}	1/32	.03125	3.125	3.125×10^4	−30.10
6	2^{-6}	1/64	.015625	1.15625	1.5625×10^4	−36.12
7	2^{-7}	1/128	7.8125×10^{-3}	0.78125	7.8125×10^4	−42.14
8	2^{-8}	1/256	3.9063×10^{-3}*	0.39063*	3.9063×10^3*	−48.16
9	2^{-9}	1/512	1.9531×10^{-3}*	0.19531*	1.9531×10^3*	−54.19
10	2^{-10}	1/1024	9.7656×10^{-4}*	9.7656×10^{-2}*	977.56*	−60.21
11	2^{-11}	1/2048	4.8828×10^{-4}*	4.8828×10^{-2}*	488.28*	−66.23
12	2^{-12}	1/4096	2.4414×10^{-4}*	2.4414×10^{-2}*	244.14*	−72.25
13	2^{-13}	1/8192	1.2207×10^{-4}*	1.2207×10^{-2}*	122.07*	−78.27
14	2^{-14}	1/16384	6.1035×10^{-5}*	6.1035×10^{-3}*	61.035*	−84.29
15	2^{-15}	1/32768	3.0518×10^{-5}*	3.0518×10^{-3}*	30.518*	−90.31
16	2^{-16}	1/65536	1.5259×10^{-5}*	1.5259×10^{-3}*	15.259*	−96.33
17	2^{-17}	1/131072	7.6294×10^{-6}*	7.6294×10^{-4}*	7.6294*	−102.35
18	2^{-18}	1/262144	3.8147×10^{-6}*	3.8147×10^{-4}*	3.8147*	−108.37
19	2^{-19}	1/524288	1.9073×10^{-6}*	1.9073×10^{-4}	1.9073*	−114.39
20	2^{-20}	1/1048576	9.5367×10^{-7}*	9.5367×10^{-5}*	0.9537*	−120.41

* Rounded data.

** $dB = 20 \log 2^{-n}$.

A bcd fractional number $= 10^{-1}[a_3 2^3 + a_2 2^2 + a_1 2^1 + a_0 2^0] + 10^{-2}[a_3 2^3 + a_2 2^2 + a_1 2^1 + a_0 2^0] \ldots + 10^{-m}[a_3 2^3 + a_2 2^2 + a_1 2^1 + a_0 2^0]$

where $a_0 \ldots a_3$ are coefficients, either 1 or 0, and m is the number of decades.

Example:

$$0.12 = 10^{-1}[(0)(8) + (0)(4) + (0)(2) + (1)(1)] + 10^{-2}[(0)(8) + (0)(4) + (1)(2) + (0)(1)]$$

$$= \frac{0+0+0+1}{10} + \frac{0+0+2+0}{100}$$

$$= 1/10 \quad = \quad 2/100$$

Written in bcd fractional notation (2 decades)

$$0.12 = \underbrace{\overset{\text{MSB}}{0}\ 0\ 0\ \overset{\text{LSB}}{1}}_{\text{MSD}}\quad \underbrace{\overset{\text{MSB}}{0}\ 0\ 1\ \overset{\text{LSB}}{0}}_{\text{LSD}}$$

Fig. 1-6. The bcd fractional system.

switches or displays coded with a range of digits from 0 through 9. Each set of such digits comprises a *bcd decade,* and 1, 2, 3 or more decades are used depending upon system resolution/accuracy requirements. Since human beings tend naturally to think in terms of 0 to 9, 10 to 90, 100 to 900, and so forth in counting, bcd conversion schemes are most often used in instruments such as digital voltmeters.

In Fig. 1-6 it may be noted that a bcd fractional number is composed of a series of decades which are coded in binary form, and each being one decade smaller in value (weight). The first decade has an overall coefficient of 10^{-1} and is known as the tenths decade; it is also the MSD. Succeeding smaller decades, when used, are 10^{-2} for hundredths, 10^{-3} for thousandths, and so forth. The specific number within each decade is described by the coefficients a_0 to a_3, which signify the binary code.

In the example shown, the number 0.12 requires two bcd decades for representation, a tenths and hundredths decade. The code for the tenths decade is 0001, corresponding to 1/10, while the hundredths decade code is 0010, corresponding to 2/100. These two numbers added equal 0.12; the bcd code is 0001 0010.

In the notation, the MSD is always listed first, followed by succeeding decades. To clarify the point that a bcd code is represented, the decades are separated into groups of four binary quads, as shown. Although the code is fractional, the preceding decimal point is usually dropped for clarity; this will be the practice followed in this book.

Fig. 1-7 shows further examples of the bcd code, for one and two decades. In Fig. 1-7A, the entire tenths decade is outlined, from 000 through 1001, corresponding to 0.0 through 0.9. Note that no codes are allowed higher than 1001 in bcd, although the binary numbers are capable of higher range. This is a basic tradeoff of bcd coding—the inherent inefficiency of this mode use of the numbering system. This dis-

NUMBER	CODE
0.0	0 0 0 0
0.1	0 0 0 1
0.2	0 0 1 0
0.3	0 0 1 1
0.4	0 1 0 0
0.5	0 1 0 1
0.6	0 1 1 0
0.7	0 1 1 1
0.8	1 0 0 0
0.9	1 0 0 1

Note: Codes above 1001 are not allowed in bcd (for any decade).

(A) Single decade.

NUMBER	CODE	
0.00	0 0 0 0	0 0 0 0
0.01	0 0 0 0	0 0 0 1
••	•	•
••	•	•
0.09	0 0 0 0	1 0 0 1
0.10	0 0 0 1	0 0 0 0
0.11	0 0 0 1	0 0 0 1
0.12	0 0 0 1	0 0 1 0
•	•	•
•	•	•
0.19	0 0 0 1	1 0 0 1
0.20	0 0 1 0	0 0 0 0
0.21	0 0 1 0	0 0 0 1
•	•	•
•	•	•
0.97	1 0 0 1	0 1 1 1
0.98	1 0 0 1	1 0 0 0
0.99	1 0 0 1	1 0 0 1
	TENTHS DECADE (MSD)	HUNDREDTHS DECADE (LSD)

(B) Two decades.

Fig. 1-7. Some bcd notation tables.

advantage is offset by the increased ease of use of bcd in interfaces, as described previously.

Fig. 1-7B shows how a two-decade bcd-system advances from 0000 0000 through 1001 1001, or from 0.00 through .99. For further resolution three (or four) decades could be used, which would range to 0.999 (or 0.9999). In any bcd system the resolution is defined by the LSB of the LSD; as shown here, this would be one hundredth, or 0.01.

Table 1-3 shows by comparison the relative performance capabilities of binary and bcd fractional systems. It is relatively easy to see from this table that a straight binary system always has greater resolution and dynamic range, versus a bcd system for the same number of bits.

In some bcd conversion systems this inherent disadvantage of dynamic range/resolution limitation is mitigated to a degree by the use of an "overrange" bit. This practice is most often seen in digital voltmeters integrating a/d converters which have "3½ or 4½" decade

Table 1-3. Comparative Resolution/Dynamic Range, Binary/BCD

		LSB		Dynamic Range	
Bits	**BCD Decades**	**Binary**	**BCD**	**BCD**	**Binary**
4	1	0.0625	0.1	9	15
8	2	3.9063×10^{-3}	0.01	99	255
12	3	2.4414×10^{-4}	0.001	999	4,096
16	4	1.5259×10^{-5}	0.0001	9999	65,535
20	5	9.5367×10^{-7}	0.00001	99999	1,048,575

ranges. The fractional decade in these cases is an overrange bit which allows the instrument to range beyond its nominal full scale. An example would be a 1.000 volt full-scale meter which can read (accurately) to 1.200 volts rather than 0.999 volt, through the use of a 1-volt "overrange" bit. This operation is described in greater detail within the applications section on integrating a/d converters.

THE DATA CONVERTER FUNCTION

The basic function of a data converter, whether it is an a/d or d/a, is to accept data in one information format and convert it accurately into the other. As their names obviously imply, a/d converters convert analog information into a digital equivalent; d/a converters convert digital information to analog equivalents.

This basic process for either type of converter can be visualized by the input-output transfer characteristic, as shown in Fig. 1-8. Here, in the case of a d/a converter for instance, various input digital words result in corresponding analog outputs, ranging from zero to full scale. In a case of a *unipolar* analog output d/a converter, the output ranges from zero at a digit input of zero (00 . . . 00), to full scale with a full scale digital input (11 . . . 11). Full scale analog output is some maximum voltage (or current), here shown as +10 volts. (It could just as easily be −10 volts; and this would in no way change the concept.)

Thus, for increasing binary inputs to a d/a converter, the analog output increases in magnitude. For a unipolar d/a converter, at a digital half-scale input of 10 . . . 00 the analog output would be 5 volts

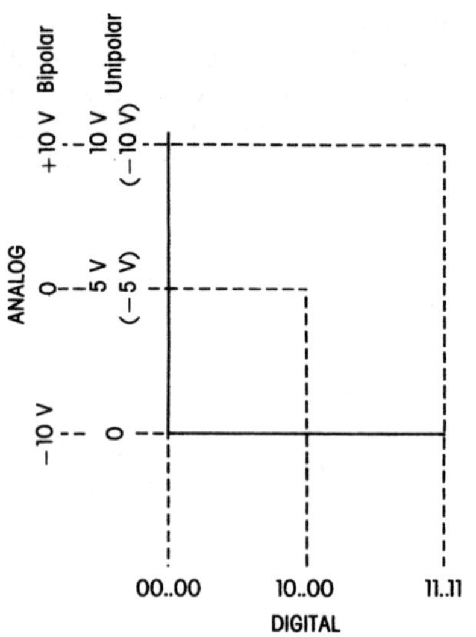

Fig. 1-8. Data converter input-output transfer characteristics.

(or −5 volts), as shown. Bipolar d/a converters have an offset analog output. They will be more completely described later.

If the preceding described process is reversed, it becomes the transfer function of an a/d converter. A unipolar input a/d converter, for instance, accepts analog inputs in the range of zero to full scale of 10 volts (or −10 volts), and produces digital outputs ranging from zero (00 . . . 00) to full scale (11 . . . 11). Thus the two conversion processes are seen to be complementary.

Functional diagrams of d/a and a/d converters shown in Fig. 1-9 may make their respective conversion processes more clear. The d/a converter accepts digital data inputs of "n" bits and produces analog outputs. It may use all of the power supplies as shown, and may have separate connections for digital and analog grounds. Particularly note the symbol for analog ground; it is often used to denote the analog reference point.

The preceding discussion is a very simplistic description of the two conversion functional processes and omits many details. It is, however, conceptually complete, and sufficient for the time being.

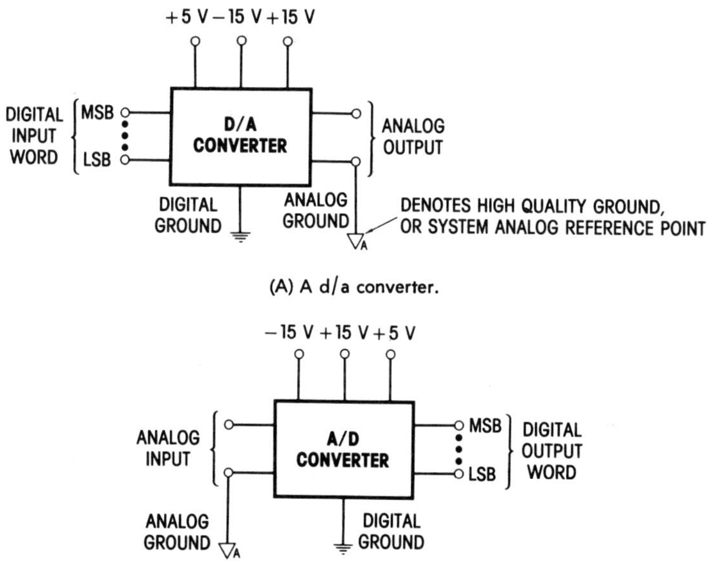

(B) A a/d converter.

Fig. 1-9. Data converter functional diagrams.

CONVERTER CODING

All data converters use specific input/output *coding* relationships, regardless of operating speed, number of bits, or the type or class of converter. The coding of a data converter is simply a concise definition of its input-output digital/analog (or analog/digital) conversion relationship. Coding is necessary for effective specification and use of a converter.

Coding arrangements take on many different forms, but a few of them evolve as most popular. They can be divided into two major categories; unipolar codes, which imply a unipolar analog scale, and bipolar codes which imply a bipolar analog scale. Within these two categories there are further subdivisions which deal mainly with the digital aspects.

The codes as described below are applicable to both d/a and a/d converters; one has to merely interchange the input/output roles. Thus the information is presented in "digital" and "analog" terms, which could represent either an output or input.

Unipolar Codes

The most popular coding arrangement of all is the *unipolar binary code,* which is quite straight forward. It is outlined in Table 1-4.

In the unipolar binary code, all zeros correspond to a zero analog scale, while all ones correspond to full scale minus 1 LSB. The example illustrated is an 8-bit one, scale for 10 volts full scale. One LSB is then

Table 1-4. Unipolar Converter Codes

BINARY						
Digital				Analog		
Binary		Complementary Binary		General		10 V (8 bits)
1111	1111	0000	0000	FS — 1 LSB	$2^N - 1/2^N$	9.961
1000	0000	0111	1111	FS/2	$2^N - 1/2^N$	5
0111	1111	1000	0000	FS/2 — 1 LSB	$(2^{N-1} - 1)/2^N$	4.961
0000	0001	1111	1110	1 LSB	$1/2^N$	.039
0000	0000	1111	1111	0	0	0
BCD						
Digital				Analog		
BCD		Complementary BCD				
1/10 decade	1/100 decade	1/10 decade	1/100 decade	General		10 V (2 decades)
1001	1001	0110	0110	FS — 1 LSB	$10^M - 1/10^M$	9.9
0101	0000	1010	1111	FS/2	$\frac{10^{(M-1)+\log_{10}5}}{10^M}$	5.0
0100	1001	1011	0110			
0000	0001	1111	1110	1 LSB	$1/10^M$	0.1
0000	0000	1111	1111	0	0	0

39 millivolts, and full scale −1 LSB is 9.961 volts. For other full scale analog levels, the fractional number represented by the binary word is simply multiplied by the full scale level to yield the actual analog level. Other code levels along the scale are illustrated in the table.

Also used is a *complementary binary unipolar code.* This is simply a complete interchange of 1s and 0s from the straight binary code used to obtain equivalent analog levels.

The *bcd unipolar code* is also illustrated in Table 1-4. Again the example chosen is 8 bits, which in this case yields two bcd decades. The full scale analog level here is 10 volts. In general, like the binary unipolar system, the actual analog voltage is arrived at by multiplying the full scale voltage times the bcd fractional numbers. Here the maximum voltage is 9.9 volts. Other code levels are also shown for the purpose of illustration.

Like the binary unipolar code, bcd unipolar codes are also sometimes operated in complementary fashion, as shown.

Bipolar Codes

There are a great variety of coding arrangements used with bipolar converters. These are diagrammed in Table 1-5, with examples chosen for 8 bits and an analog scale of ±10 volts.

Table 1-5. Bipolar Binary Codes

DIGITAL				ANALOG	
Offset Binary	2s Complement	1s Complement	Sign-Magnitude Binary	General	±10 V
1111 1111	0111 1111	0111 1111	11111 1111	FS − 1 LSB	+9.922
1100 0000	0100 0000	0100 0000	11000 0000	FS/2	+5
1000 0001	0000 0001	0000 0001	10000 0001	+1 LSB	+0.078
1000 0000	0000 0000	0000 0000	10000 0000	0 (+)	0*
------	------	1111 1111	00000 0000	0 (−)	0*
0111 1111	1111 1111	111 1110	00000 0001	−1 LSB	−0.078
0100 0000	1100 0000	1011 1111	01000 0000	−FS/2	−5
0000 0001	1000 0001	1000 0000	01111 1111	−FS + 1 LSB	−9.922
0000 0000	1000 0000	------	------ SIGN BIT	−FS	−10

* Both 1s complement and sign-magnitude have two zero code words, termed 0(+) and 0(−).

The simplest and most often used of these coding arrangements is *offset binary.* Offset binary simply transposes the entire analog scale, to align all zero digital code words with the negative analog FS limit, shown here as −10 volts.

Note that as compared to a unipolar code system, the analog range is doubled; for example, here it is ±10 volts versus simply 0 to 10 volts. The amount of the offset to accomplish the scale transposition is exactly equal to the MSB. This can be noted from the code; at the MSB code value of 10 . . . 00, the analog level is 0.

It should be noted also that this binary code doubles the value of the LSB, simply because of the doubling of the analog range.

The *twos complement* code is also often seen used, and is very close to the offset binary code. The only difference between the two is that the MSB is complemented. This may be noted by comparing codes for two like analog levels, between columns one and two. In practice, the twos complement code can be easily derived from the offset binary (or vice-versa), by inverting the MSB. In a/d converters for instance, both signs of the MSB are often made available, thus implementing either code.

The *ones complement code* is one of the least used codes. It is formed by complementing each positive value to obtain the negative set. Unfortunately, this includes zero also, and the process by nature yields two code values for zero called 0 (+) and 0 (−).

The ones complement code is symmetrical about the zero analog value and has a maximum range of ±(FS − 1 LSB). Note that the −FS level in offset binary and twos complement is *not present* in this code, because of the extra zero code.

The *sign-magnitude binary code* is one in which a code similar to the unipolar binary code is used to denote magnitude, and an additional sign bit used to denote polarity. As can be noted in the table, for "n" bit resolution this results in n+1 total bits to signify sign + magnitude. As shown here the 8-bit code (exclusive of the sign bit) yields analog levels equivalent to the other 8-bit codes.

Like the ones complement code, this code has two code values of zero, 0 (+) and 0 (−). It is also symmetrical about zero, with a maximum range of ±(FS − 1LSB).

Code Conversion

It may sometimes be necessary to convert a coding system from one of the above formats to another. This can be accomplished through appropriate logic, which is often simple. The general rules for conversion

Table 1-6. Bipolar Code Conversion Relations

To Convert	Rule
offset binary ⇄ 2s comp.	Complement MSB
offset binary ⇄ 1s comp.	Complement MSB—(if new MSB = 1, add 11 . . . 11)
offset binary ⇄ sign-mag. binary	Complement MSB—(if new MSB = 1, complement bits and add 1 LSB)
2s comp ⇄ 1s comp.	If MSB is a 1, add 11 . . . 11
2s comp ⇄ sign-mag. binary	If MSB is a 1, complement remaining bits and add 1 LSB
1s comp ⇄ sign-mag. binary	If MSB is a 1, complement remaining bits

among the four bipolar codes are described in Table 1-6. These conversion rules are bilateral; that is they can convert codes in either direction.

Other Codes

A form of sign-magnitude coding, sign magnitude bcd is often used in digital voltmeters, and other types of instrumentation. Not shown here, sign magnitude bcd is simply the basic bcd code already described, plus a sign bit to indicate polarity.

It should be obvious of course that many possible modifications exist for the above described codes. All of them can be seen in complemented form (and often are), a practice which often simplifies the hardware design. Details of other more specific modifications may be found in the references. Generally however, the codes described will be those most often encountered in practice. The above discussion of codes has been most appropriate in regarding the operation of the d/a converter. The a/d converters operate similarly with analog-input, digital-output coding. A subtle but important difference, however, is the fact that many a/d scales are *offset* by the value of the LSB/2. The reasons behind this are discussed in greater detail in Chapter 6.

REFERENCES

1. Allan, R. "A/D and D/A Converters." *EDN,* February 5, 1973.
2. "A/D, D/A Design Reference Guide" (wallchart). Analogic Corp., Wakefield, Massachusetts.
3. Bruck, D. B. "Data Conversion Handbook." Hybrid Systems Corp. Burlington, Massachusetts, 1974.
4. Daley, F. D. Jr. "Analog-to-Digital Conversion Techniques." *Electro-Technology,* May, 1967.
5. Hoeschele, D. F. Jr. *Analog-to-Digital/Digital-to-Analog Conversion Techniques.* John Wiley & Sons, Inc., New York, 1968.
6. Schmid, H. *Electronic Analog/Digital Conversions.* Van Nostrand Reinhold, New York, 1970.
7. ———. "Practical Guide to A/D Conversion." Part 1, *Electronic Design,* December 5, 1968. Part 2, *Electronic Design,* December 19, 1968.
8. ———. "Practical Guide to D/A Conversion." *Electronic Design,* October 24, 1968.
9. Sheingold, D. *Analog Digital Conversion Handbook.* Analog Devices, Inc., Norwood, Massachusetts, 1972.
10. ———. *Analog-Digital Conversion Notes.* Analog Devices, Inc., Norwood, Massachusetts, 1977.
11. Zuch, E. "Know Your Converter Codes." *Electronic Design,* October 25, 1974.

2

Data Converter Types and Specifications

Although there are many different ways in which d/a and a/d converters can be implemented with circuitry, only a few of the possible techniques tend to be widely used, particularly so within IC converters. Accordingly, it is these types of converter designs which will be discussed here. Information on the various other less popular solutions may be found described within the references listed at the end of the chapter.

A logical discussion of data converter design approaches begins with the d/a converter, which is also often a component part of certain a/d converter types.

D/A CONVERTERS

A basic block diagram for a d/a converter is shown in Fig. 2-1. Within a given d/a device all, or portions of, the various components shown here may be included.

At the heart of a d/a converter is a set of logic-operated switches. This switch array consists of a single switch point for each bit of resolution; that is, n switches for an "n" bit converter. These switches either connect or disconnect a reference voltage fed to the legs of a resistor network. Again, for "n" bits of resolution, the resistor network will have "n" legs, or inputs. The arrangement of resistor values within this network, or its "weighting," determines the scaling of the converter. Most often the scaling is either binary or bcd, although there are exceptions to these general rules, such as special nonlinear scales.

The two functional components just described make up the essential minimum complement that can really be called a d/a converter. There are examples of ICs configured as such, as will be seen subsequently. The complete function of a d/a converter includes both the reference voltage, and an output buffer amplifier. The reference voltage source is a stable voltage supply which is used to drive the resistor network, and is a major element of consideration in the d/a converter's overall accuracy. It is usually a zener diode circuit configured for high

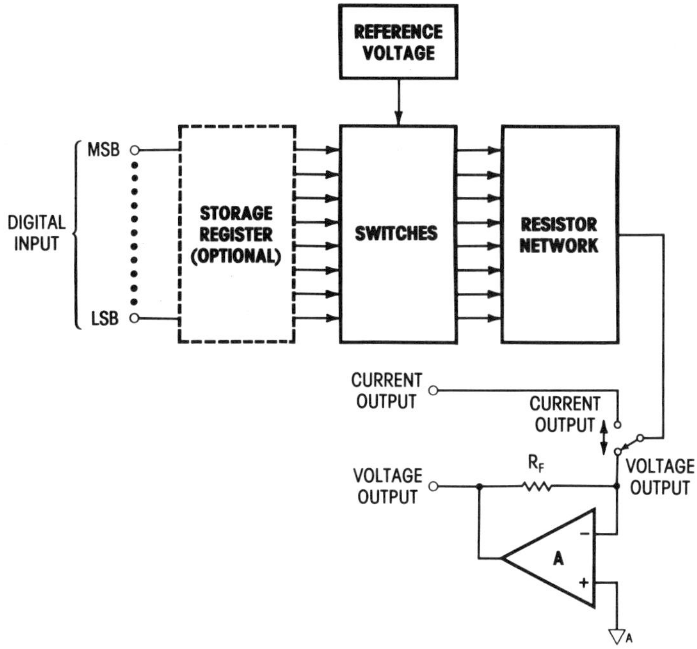

Fig. 2-1. Block diagram of a basic d/a converter.

stability and accuracy, with a typical output in the range of 6 to 10 volts. The output amplifier is an op amp which is used to buffer the d/a output, so as to deliver an output voltage which is accurate, and independent of loading.

Conceptually, many d/a converters may be used in either a current or voltage output mode, depending on the application. A current output d/a converter will use the output of the converter directly; a voltage output device will use an op amp in a current-to-voltage converter mode, connected as shown in Fig. 2-1.

Some d/a converters may include a digital storage register, shown here as an optional block. This is not universal in IC types, but this function is often desirable or necessary on an overall system basis.

Binary Weighted Ladder D/A Converter

A simple example which further illustrates the working process is the binary weighted d/a converter shown in Fig. 2-2. For simplicity, only those components necessary to illustrate the key points are shown here.

Here switches S1-Sn connected to switch legs of the resistor work between the reference voltage and ground. As can be noted, beginning with the MSD leg switched by S1, the resistor values are related as R 2R, 4R . . . (2^{n-1})R. In our 4-bit example, this yields resistor values of 10K, 20K, 40K and 80K. These are typically used values, as seen in actual circuits.

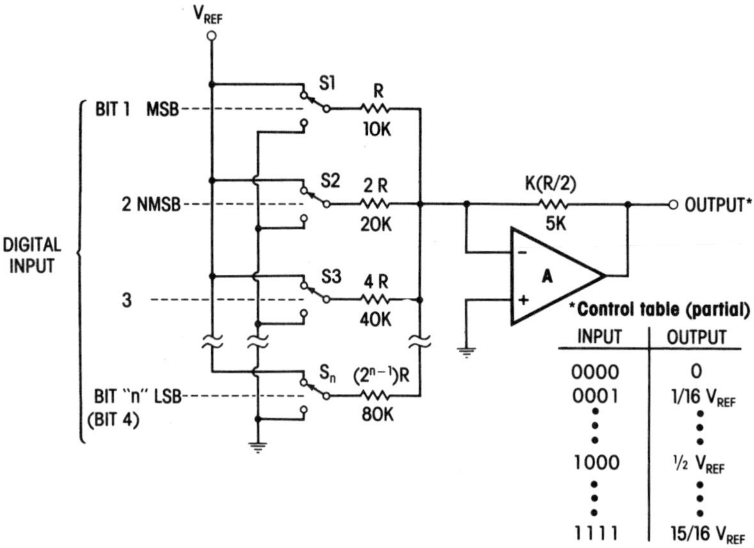

Fig. 2-2. Binary weighted ladder d/a converter.

It can be seen that the weighting of the resistor values produces a current into the op-amp summing point which decreases by a factor of 2 in each leg. For a 10-volt reference, it would here be 1, 0.5, 0.25 and 0.125 mA, respectively. This is in accordance with the fractional binary coding relation described above. The op amp, functioning as a voltage-to-current converter, produces an output voltage which is simply the reference voltage multiplied by the fraction represented by the digital word input.

A partial control table is shown in the figure's inset, as appropriate to 4-bit operation. Note that in general the output is zero with a 00 . . . 00 digital input, ½ V_{REF} with the MSB on (10 . . . 00), and full scale or V_{REF}−1 LSB with all 1s (11 . . . 11). The reader is encouraged to fill in this table to completion, and to verify the general relationship for a greater number of binary bits.

Binary Weighted Quad Groups D/A Converter

In theory, the just described principles can be extended to any number of bits, but there are serious practical reasons why it is seldom done. This is because it becomes extremely difficult to manufacture, calibrate, and manitain stability for a network of resistors with a very large span of values. An 8-bit converter, for instance, would require the 8th leg to be 128 times the value of the first. Instead, what is done is to arrange the bits in groups of four (quads) with *interquad attenuators,* to give the proper quad-to-quad attenuation.

In Fig. 2-3 this approach is shown with two quad 1-2-4-8R arrays, separated by an attenuator resistance. It is the function of this resistor

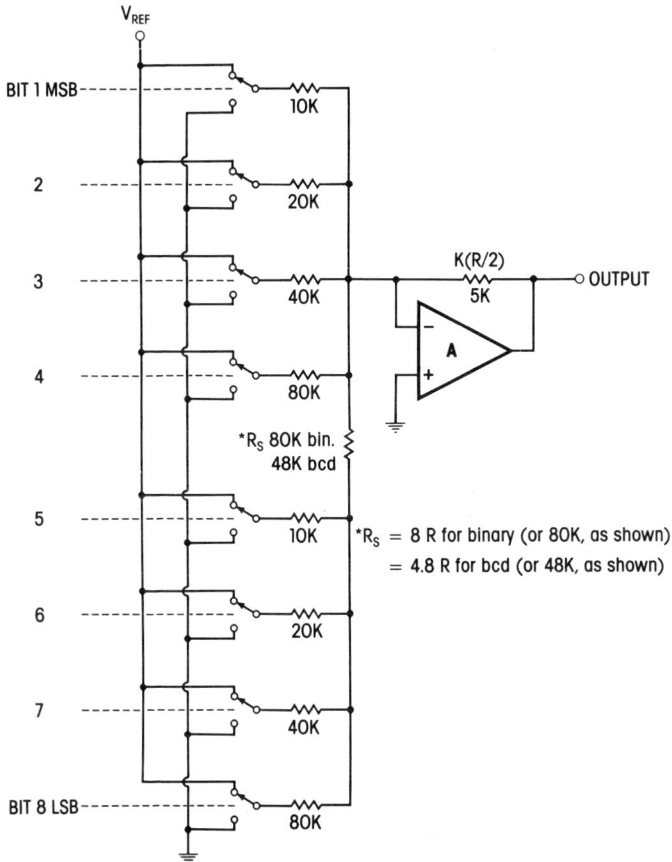

Fig. 2-3. Binary weighted quad groups d/a converter.

(or resistors if a pair is used, as is often) to attenuate the current from the second quad by a factor of 16/1 (or 10/1 for bcd).

It can be noted from this arrangement that the span of resistor values is kept to only 8/1, which is quite tolerable from a production standpoint. This scheme can be further extended to 12 (or 16) bits, if required, by adding further 16/1 (or 10/1 for bcd) attenuators between quads.

R/2R Ladder D/A Converter

In IC designs, it is desirable to keep the span of resistance values to an absolute minimum, and even the 8/1 span just described is generally unattractive. A scheme which allows binary scaled operation is the R/2R ladder scheme, shown in Fig. 2-4.

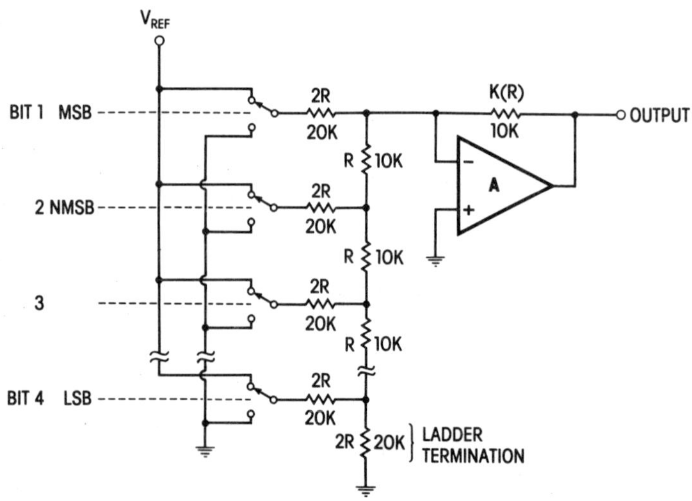

Fig. 2-4. An R/2R ladder d/a converter.

In this circuit the switches either apply the reference voltage to, or ground, the 2R resistance legs of a ladder network. It can be seen that each switch input resistor (2R) looks into a pair of R valued resistors. Between each point on the ladder and an adjacent one, there is a current attenuation of 2/1, which makes its operation inherently binary. The ladder is terminated at its lower tap, the LSB, by a 2R resistance to ground to maintain proper operation.

This circuit behaves just as the other two insofar as binary operation by generating an output which is the product of V_{REF} and the binary fraction. Output voltage can be scaled by the feedback resistor; it is scaled for K = unity, as shown in Fig. 2-4.

The important advantage of this approach is that it is not "bit-limited" as R/2R sections can be added almost indefinitely without regard to high value ratios. Thus the scheme is a popular one, being used in IC d/a converters of from 6- to 12-bit operation.

Inverted R/2R Ladder D/A Converter

The R/2R ladder d/a converter just described is sometimes referred to as a "voltage switched" R/2R ladder as the switches actually switch the reference potential. An alternate form of R/2R ladder d/a converter is shown in Fig. 2-5. This circuit is known as an "inverted" R/2R ladder, as it applies the reference to the end of the ladder, and switches the legs onto, or from, a common output bus. The circuit is also referred to as a R/2R "current switched" ladder, since the switches are actually switching a defined current at a small voltage level.

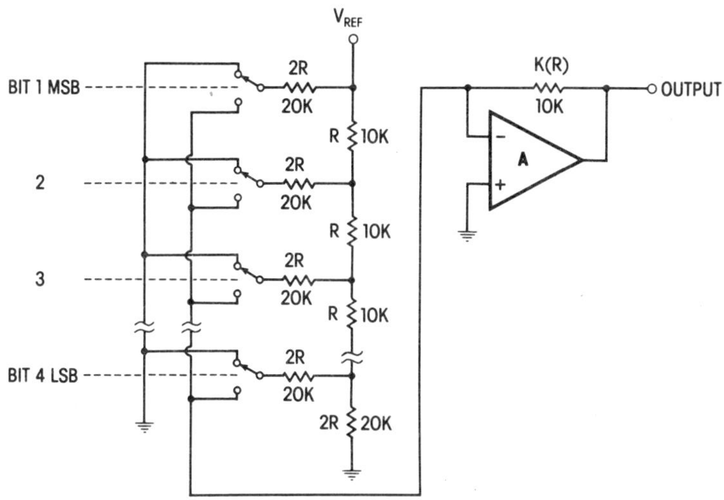

Fig. 2-5. Inverted R/2R ladder d/a converter.

The operation of the circuit is identical to the standard R/2R ladder d/a converter, and it has its same advantages as to ease of expansion. This circuit finds wide use within IC d/a converters, because of the fact that it eases some of the switch specification requirements.

A/D CONVERTERS

Like the d/a form of converter, there are a great number of a/d converter types. However only a few of them tend to be useful, particularly in IC form. These types are discussed in the following text and additional background information is available in the references at the end of the chapter.

Counter Type A/D Converter

The counter type a/d converter shown in Fig. 2-6 is a feedback form of converter which uses a d/a converter as part of its feedback loop. This is one of the most simple a/d types, and is readily implemented with common components.

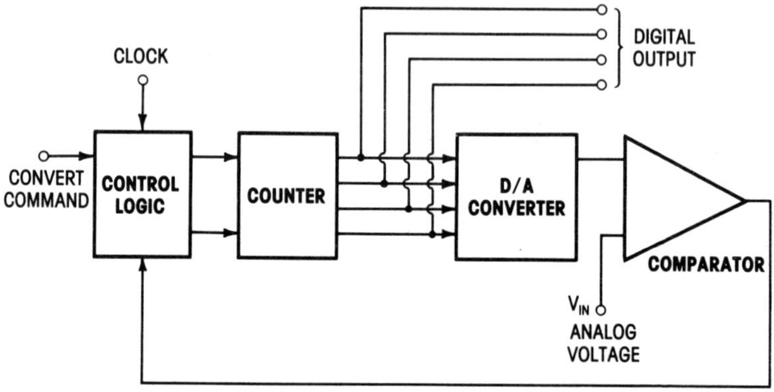

(A) Block diagram.

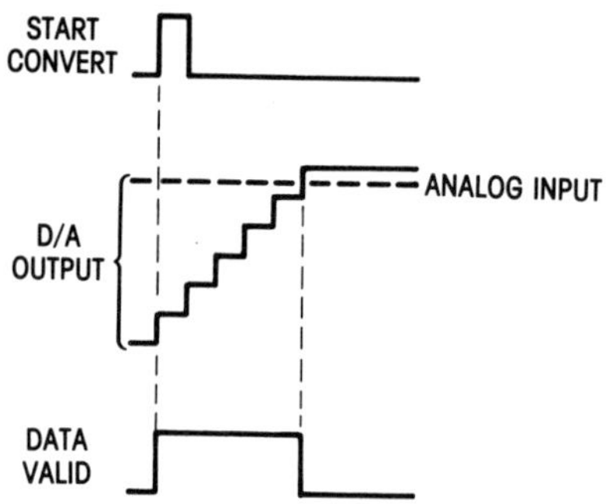

(B) Timing diagram.

Fig. 2-6. Counter type a/d converter.

The basis of this a/d converter operation is comparison of an increasing d/a output voltage with the unknown analog input, V_{in}. This is a *driven* a/d system, requiring an external start command to begin a conversion.

Upon receipt of a convert command pulse, the control logic resets the counter to all zeros, which also drives the d/a converter output to zero. The clock is then gated into the counter which starts counting up towards full scale. Similarly the d/a output rises with each clock pulse, until it is equal to the analog input (Fig. 2-6B). At this point the comparator changes state which signals the control logic to stop the counter. The digital number in the counter is then equal to the input voltage (within 1 LSB). The logic will hold this number indefinitely, or until a new convert command is received.

As may be appreciated from the timing diagram, this system is inherently slow in operation, as it must increment the d/a output from zero, up to the unknown level. For a full-scale input, this will take 2^n clock periods—that is, 1024 cycles for a 10-bit conversion, 4096 for 12-bit, etc.

This relatively slow speed is the counter type a/d converter's main disadvantage. Its advantages are relative simplicity and low cost. Accuracy is that of the d/a converter used.

Tracking (Servo) A/D Converter

A very useful modification of the basic counter type a/d converter is use of an up/down counter, rather than a conventional one. With the addition of some simple logic, this allows a counter a/d converter to *track* a varying analog input.

The disadvantage of a long time to acquire lock remains in the tracking type, if the analog input goes through large step changes or changes faster than the loop can follow. For input voltages which are below the tracking rate limit however, the output data continuously (and accurately) follows the analog input. In general, this form of counter is more useful than the ramping type, if the above mentioned precautions are taken in its use.

Integrating A/D Converter

The integrating a/d converter, as its name implies, uses an op-amp integrator as part of its basic operation. The functioning of this type of converter is best appreciated by regarding its block diagram and timing in Fig. 2-7.

An operating cycle of this system is made up of two basic time intervals, t1 and t2. Interval t1 is a fixed length time interval as determined by design. During this interval the input signal is connected by the switch S1, which causes the integrator's output voltage V_{int} to go positive. Dependent upon the level of V_{in}, the voltage V_{int} will reach some positive level ($V_{int\ a}$ or $V_{int\ b}$) at the end of t1.

During time t2 the input is disconnected and the opposite polarity reference voltage is connected, causing V_{int} to ramp downward at a fixed slope. When the ramp voltage reaches zero and trips the comparator, t2 is ended.

Times t1 and t2 are time intervals which can be measured by the counter. It can be shown that the time t2 is a measure of the unknown input voltage V_{in}, as $V_{in} = V_{REF}\ (t2/t1)$.

With V_{REF} and t1 constant, the time t2 is a direct measure of V_{in}. This time interval is counted by an appropriate counter arranged for binary or bcd output. The output of this counter is the digital output.

The advantages of this scheme are that the components R and C are not critical as to accuracy, to a large degree. Further, the clock timing is relatively noncritical as long as it does not change appreciably during

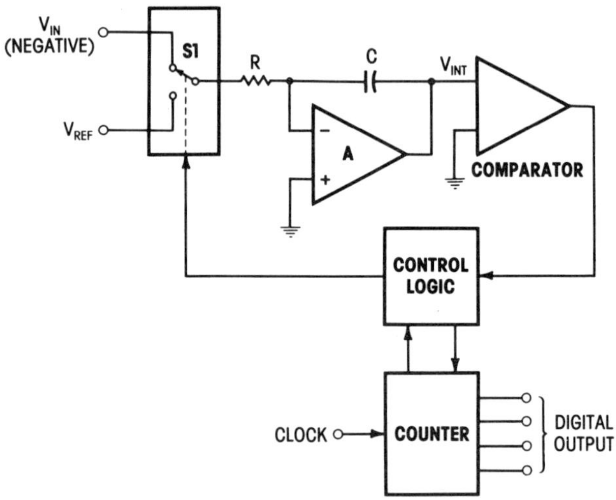

(A) Block diagram.

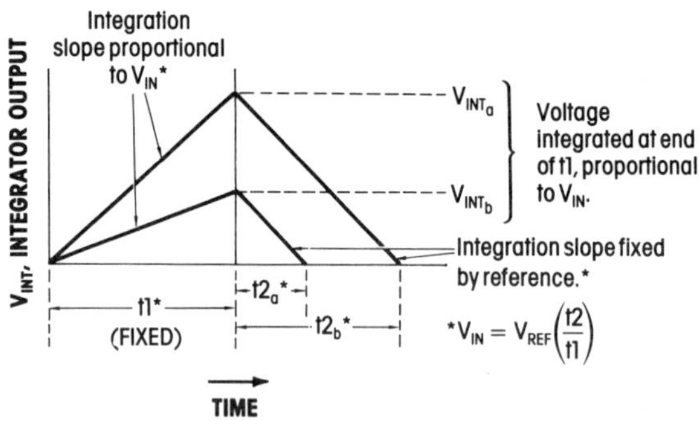

(B) Timing diagram.

Fig. 2-7. Integrating type a/d converter (dual slope).

a conversion cycle. Absolute accuracy is mainly dependent upon the reference used, and dual slope converters are easily capable of 12 binary bits, or 3 (or more) bcd digit operation. Finally, one of the most significant advantages of this technique is the inherent noise rejection. This is due to the integration of the input signal in part, but rejection at a specific frequency such as 50 or 60 Hz can be maximized by operating the conversion cycle in synchronism to provide rejection levels of 50 or 60 dB.

A major disadvantage of the integrating a/d converter is the conversion time, usually on the order of 10 milliseconds or more. Integrating converters find their main applications in instrumentation such as lab digital voltmeters, where the relatively long response time is not a disadvantage. More sophisticated variations of the basic concept exist which add further "self calibration" phases to the basic two-phase cycle. Examples are three-phase integrating a/d converters which correct for zero drift, and four-phase converters which correct for both zero and gain drift. Further details of devices which utilize these principles are found later in the book.

Successive Approximation A/D Converter

A successive approximation a/d converter is diagrammed in Fig. 2-8. This converter is conceptually simple, and like the counter and servo types, operates on a feedback principle.

From Fig. 2-8A it can be noted that the comparator compares the output of the d/a converter to the analog input, and drives the successive approximation register. This register is a specialized MSI logic element designed specifically for this function, which operates as follows.

Upon receipt of a convert command pulse, the system will sequentially enter each bit input to the d/a converter, to determine the relative weighting in relation to the input. This will be made clear by regarding the timing diagram.

Regarding the analog level 1 example first (Fig. 2-8B), the register first turns on the MSB (bit 1) and compares it to the input. If it is less than the input voltage (as shown) it is left on. If more, it is turned off. Then the next bit is examined to see if a summation with the first bit is more or less than the input. If more, it is turned off, if less (as shown), it is left on. This process is continued in sequence for all "n" bits until they all have been exercised.

Note that the d/a voltage builds up in successive steps for each bit until it is within 1 LSB of the input voltage. A second example of a lower analog voltage is shown as Analog Input 2. The analog input to a successive approximation converter must be stable during conversion, or change less than ½ LSB. At the end of the conversion cycle, the data in the register is then the converter digital word, available in parallel. Also, successive approximation registers usually make the data available in serial form.

This type of converter may either be driven (as shown), or continuously cycled (free-run). It performs an a/d conversion in n + 1 clock cycles—one cycle/bit in general, plus an additional clock cycle for register reset. It is one of the faster of the conversion processes, and speeds can go to well below 1 microsecond/bit in some units.

A successive approximation converter can be highly accurate, but it is only as good as the reference and d/a converter within it. If the d/a resistor network is not well matched or drifts, the a/d converter will

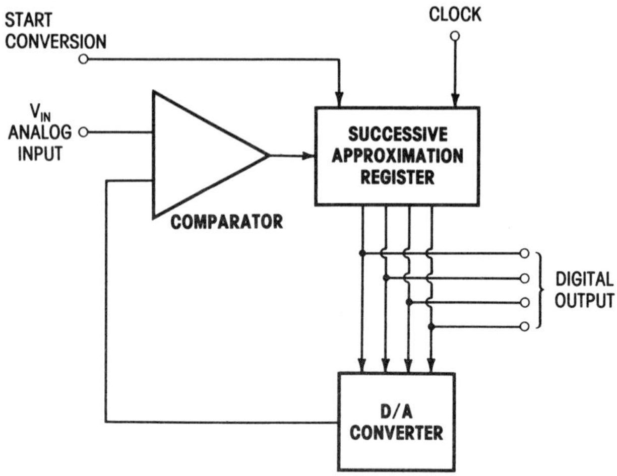

(A) Block diagram.

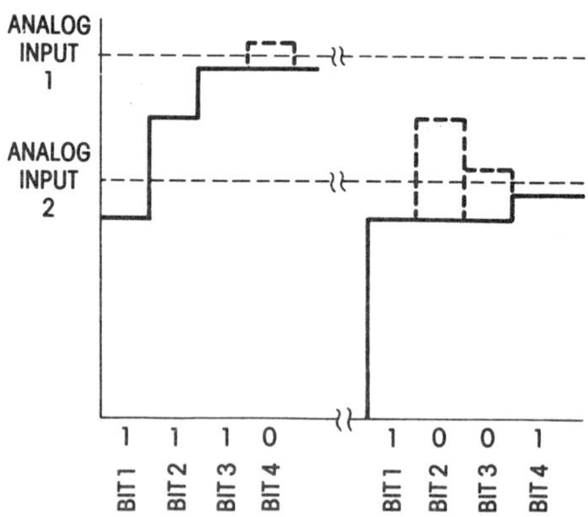

(B) Timing diagram.

Fig. 2-8. Successive approximation a/d converter.

be nonlinear and may also miss some codes and be unstable. Successive approximation type converters are currently available with up to 12 bits of resolution with 10 to 20 microsecond conversion times. All coding arrangements are available with this type converter.

Parallel A/D Converter

The parallel (sometimes called flash) type of a/d converter is illustrated in Fig. 2-9. As the name implies, this form of a/d converter operates by simultaneously comparing all analog code levels to the input. It is the fastest of all a/d types, as it is limited in speed to only the combined propagation delay of the comparators used, and the decoding logic.

In operation, a resistive divider chain biases one input of a bank of comparators to each code level, as appropriate to the required resolution. This technique requires $2^n - 1$ comparators for a "n" bit comparison; thus, it becomes very complicated in terms of hardware for resolutions beyond 5 or 6 bits. The output of the comparators themselves is not in binary form; thus a logic decoder is required.

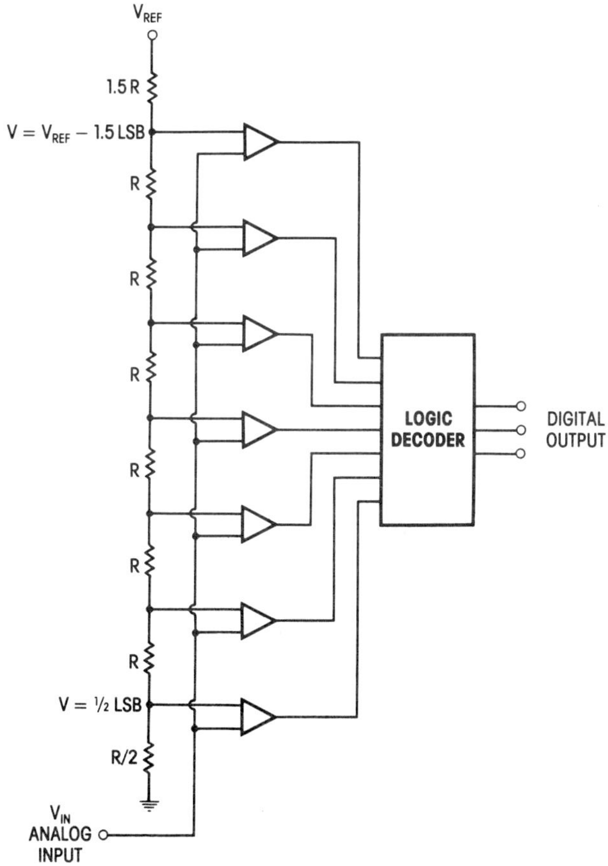

Fig. 2-9. Parallel a/d converter.

Another advantage of this technique is real time or continuous operation, in addition to the high inherent speed. It is a standard technique used for a/d conversion of high speed analog signals, such as video waveforms. Conversion rates are often in excess of 10 MHz.

DATA CONVERTER SPECIFICATIONS

Resolution

The resolution of a converter is a measure of the smallest basic increment the analog component (input or output) can be divided. Although it is often confused with accuracy, in truth it should not, as the two can be quite independent.

The basic resolution of a converter is the full scale output (d/a) or input (a/d), divided by 2^n for binary devices. For a 10-volt output 8-bit d/a converter for instance, resolution would be 10 V/2^8 = 0.039 V (to closest mV).

While the resolution tells us that we can have no finer gradation of analog level for a given number of bits, the accuracy can indeed be better than the resolution would imply. An 8-bit converter, for instance, could be accurate to much closer than 1/2^8 (39 mV). In fact, it might be accurate to 10 mV, or less.

Resolution is also stated (correctly) as the value of an LSB, which it is in general terms. For bcd coded devices, resolution is the LSB of the least significant digit. Resolution may sometimes also be seen stated in terms of percent or ppm.

Absolute Accuracy

Absolute accuracy of a converter defines a measure of the deviation from an ideal full scale output which is referred to an absolute (or traceable to absolute) measurement standard, such as the NBS volt. As such, it encompasses a great number of individual sources of error, including offset and gain error, nonlinearity, and the drift of all of these. In short, absolute accuracy is *any* deviation from the ideal with reference to a standard. In practice the major sources of converter error tend to be the three just listed.

Relative Accuracy

Relative accuracy is the deviation from an ideal transfer function, exclusive of offset or full-scale gain calibration. It is largely a function of the nonlinearity of the converter.

Converters are specified in terms of ½ LSB accuracy by convention, independent of resolution. This of course equates to differing percentages of accuracy, dependent upon the number of bits. An accuracy of ±½ LSB for a given number of bits may be expressed as +1 LSB for n+1 bits. Using Table 1-2 for n+1 bits will then yield the equivalent percent of accuracy, or ppm. For example, 12 bit ±½ LSB accuracy is roughly 0.012% (which is 13 ± 1 LSB accuracy).

"Accuracy" or "relative accuracy" is a misnomer in a sense, since the terms being discussed are errors or deviations, which are more correctly termed "inaccuracies." Industry convention however uses accuracy to describe these errors, with the connotation that lower errors yield better performance.

Linearity

Linearity is probably the single most important specification of a converter, as it usually represents the ultimate limit to performance. It is defined as the deviation from an ideal straight line transfer function, drawn from zero and full scale calibration points (after their calibration). An example of a 3-bit d/a transfer function with perfect linearity is shown in Fig. 2-10A. Each digital input level results in an ideal analog output, all of which are spaced ½n volts apart.

A converter transfer function with nonlinearity is shown in Fig. 2-10B. Although some nonlinearity is present it is less than ±½ LSB. When this is true, the converter will be *monotonic,* or its output will always increase (or remain the same) for an increasing digital input.

Nonmonotonicity is illustrated in Fig. 2-10C where it can be noted that the code transitions 001-010 and 101-110 result in a decrease in output. A study of the inset will reveal the source of this error.

If the linearity of a converter is ±½ LSB, any two adjacent codes can differ by no more than 1 LSB. For the example shown, the code 001 can be 1 LSB (ideally), or as high as 1.5 LSB. The next higher code, 010, should be 2 LSB (ideally), but can be as low as 1.5 LSB (2 − ½ LSB). These two adjacent codes could therefore actually result in the same output, 1.5 LSB.

If they are absolutely the same, the converter is still monotonic. It can be seen that if either code should exceed the ½ LSB linearity limit, the change from a lower to higher digital code will result in less, rather than more, (or the same) output.

Linearity is perhaps the single most important converter spec, as it is nonreducible in a given device. Many other errors (such as gain, offset) can be trimmed to zero. It is also subject to temperature drifts, as the d/a network components will not track perfectly over temperature changes. This may (often will) result in linearity errors greater than ½ LSB, that reduces the effective resolution and useful accuracy of the converter.

Differential Linearity

Differential linearity is the deviation of a converter step size from the ideal value of 1 LSB. For a perfectly linear, ideal converter, the differential linearity is zero, as all steps are 1 LSB. For a step of 1.5 LSB the differential linearity would be ½ LSB, as it would for a step of ½ LSB.

To guarantee monotonicity in terms of only differential linearity, it must be ±½ LSB, or less. This low a differential linearity is not necessary for monotonicity however, if the nonlinearity is held to ±½ LSB. In a monotonic converter, with ±½ LSB (or less) of nonlinearity, the differential linearity actually could be as high as 1 LSB.

From the foregoing it should be appreciated that the nonlinearity specification is the key in determining monotonicity. Differential linearity can be appreciably higher than nonlinearity in a monotonic converter, by as much as a factor of 2/1.

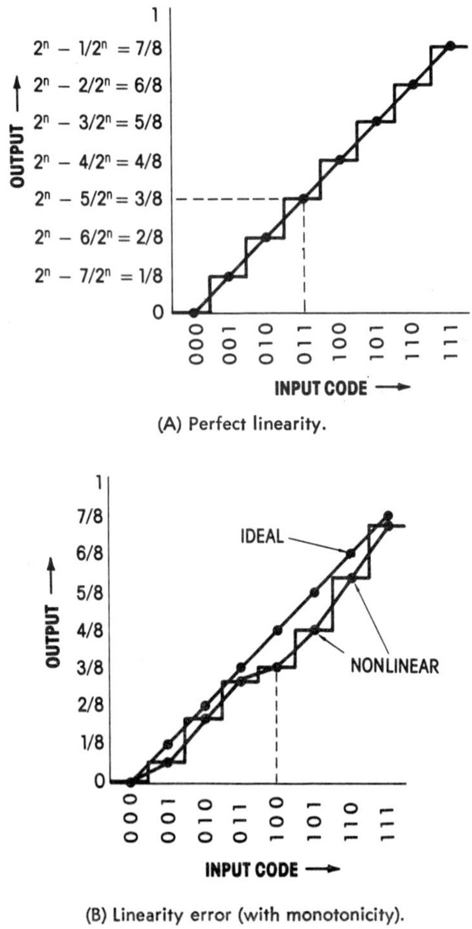

(A) Perfect linearity.

(B) Linearity error (with monotonicity).

Fig. 2-10. Examples of

Monotonicity

A monotonic d/a converter is one in which the output always increases (or remains the same) for an increasing digital count. In an a/d converter, monotonic behavior of the internal d/a converter is necessary to ensure that no digital output codes will be missed. This particularly applies to successive approximation and counter type a/d converters.

Monotonicity is absolutely essential to d/a converters which are part of a feedback system. In more general closed-loop systems, nonmonotonic operation can result in oscillation, as by definition it is a reversal of the input-output transfer phase relations.

Some types of a/d converters are inherently monotonic by the nature of their design. Examples of these are integrating a/d converters and V/F converters.

Missing Codes

It is possible for an a/d converter which uses a monotonic d/a converter to have certain analog input voltage ranges which will *not* result in successive digital output codes. The codes that are skipped due to this behavior are called "missing codes." They can result from a monotonic, but *nonincreasing* d/a output for increasing digital input—which in a feedback path causes codes to be skipped.

Offset

A shift of the entire transfer function of a converter by a fixed dc level is termed offset. It is illustrated in Fig. 2-11. Here the code 000, which should result in an output of zero, instead produces an output of 1 LSB. Succeeding codes also result in outputs 1 LSB high; thus, the

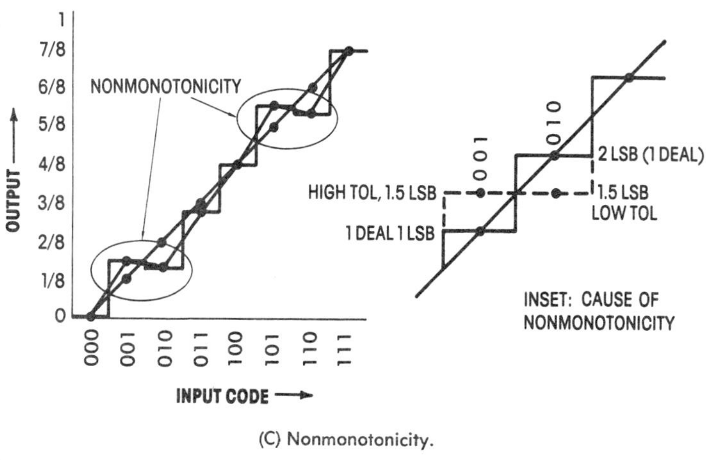

(C) Nonmonotonicity.

linearity errors.

entire transfer curve is offset. Although a positive offset is shown, it could be negative as well. In general, offset shifts the entire transfer curve by the fixed amount of the offset.

Offset error is adjustable to zero, which calibrates the all zeros digital code in d/a converters. A similar technique is used in a/d converters, although the zero code in an a/d converter is usually aligned to an analog input level of ½ LSB. This is done to split the inherent 1 LSB quantizing error to ±½ LSB.

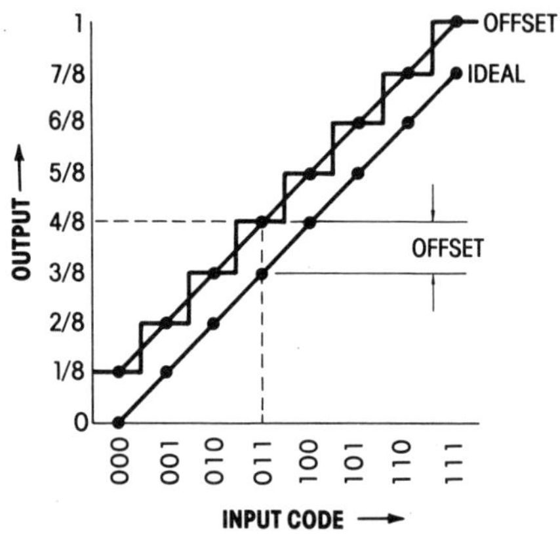

Fig. 2-11. Offset error.

Bipolar code converters are adjusted for offset to their negative full scale level (usually), rather than to zero. Although offset is adjustable to zero at room temperature, it will also have a temperature coefficient, or offset TC. This will typically be specified as a percentage or as an ppm/°C, and may be used to calculate offset drift, for a given temperature range.

Gain Error

The gain error of a converter is the deviation from the ideal full scale value. It is typically specified in percent, and is externally adjustable to zero. Gain error compensation adjustments usually alter the value of the converter's reference, thus directly changing the scaling.

Gain error is illustrated in Fig. 2-12, along with an ideal transfer curve for comparison. As can be noted, gain error can be either (+) or (−) resulting in either a high or low full-scale (FS) level.

Although gain error is adjustable to zero at room temperature, it will still have a temperature coefficient, or gain TC. This TC is largely due to the drift of the reference, but a significant part is also contributed by network and scaling-resistor drifts. Gain error TC is typically specified in ppm of FS/°C and may be used to calculate gain drift for a given temperature range.

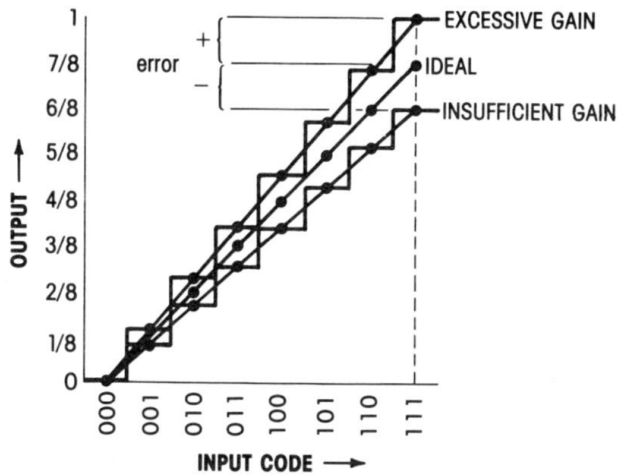

Fig. 2-12. Gain error.

Drift

Virtually all of the performance parameters of converters are affected by temperature changes, and almost always in a detrimental manner. However, the specifications which tend to be of major importance are few in number. They are the temperature coefficients of linearity, offset, and gain.

All of these parameters have individual temperature coefficients, usually given in ppm/°C. If given in other terms they may readily be converted to equivalent ppm/°C. The net drift of a converter can then be calculated by adding these three TCs and multiplying by the temperature change. A 100 ppm/°C net TC for a 10°C temperature change will result in a 1000 ppm drift, or error.

This drift figure may be used by itself to indicate resultant accuracy, which in this case would be 0.1%. In terms of ½ LSB error, this would be (nearly) a 9-bit accuracy.

Although drift can be readily calculated for cases such as the above, a general graph which illustrates the relationship of drift is very useful. Such a graph is contained in Fig. 2-13. This chart may be used to predict accuracy (to ½ LSB), given a combined TC in ppm/°C, and the temperature range. For the example cited, the 100 ppm and 10°C lines intersect just to the right of the vertical 9-bit reference line indicating the combination is just shy of 9 bit to ½ LSB accuracy.

Conversely, one can follow a given number of bits reference line upward to a given temperature difference, and read the required TC at the intersection. For instance, 11-bit accuracy over a 25°C temperature range requires a TC of just under 10 ppm/°C.

In using this graph, bear in mind that the accuracy being calculated here is for temperature alone; a practical case must also include other errors as well. Therefore, in practice, the error due to drift will usually

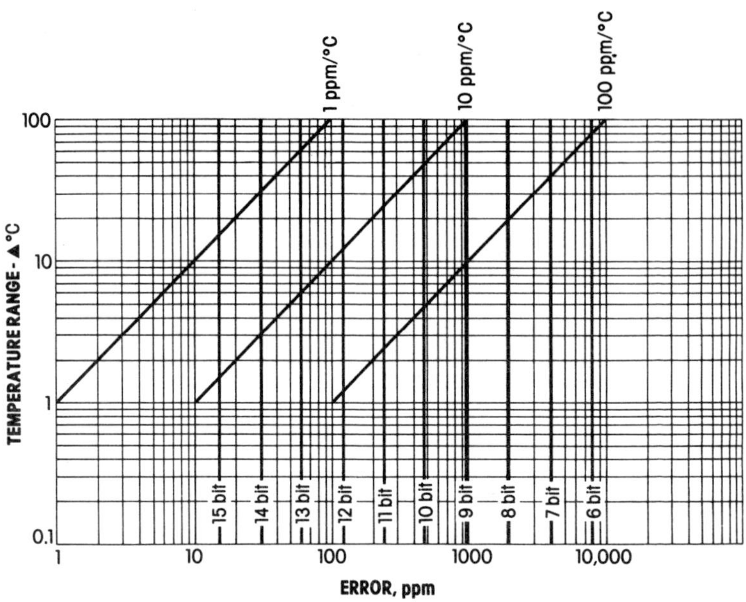

Fig. 2-13. Allowable drift/temperature range for various system accuracies.

want to be held to much less than ½ LSB. This can be easily extrapolated for ¼, or ⅛ LSB for instance, by dividing the required TC by 2 or 4 respectively (for a given temperature range).

From this chart it is easy to appreciate that highly accurate conversion over a wide temperature range necessitates low TC for the three major specifications of linearity, offset, and gain. In some instances the latter two may be correctable over temperature by automatic calibration computer routines. But linearity TC, like basic linearity, is largely non-correctable. It is best minimized by design and/or specification.

Quantizing Error

An error inherent to a/d converters is quantizing error. This is the error which results from the fact that the a/d converter by its nature of operation divides the analog input into a number of analog bands, each of which is a single increment, or *quantum*. A quantum in general is the full scale range divided by the number of resolving steps, or $FS/2^n$. This relation is very much akin to a d/a converter's resolution; in an a/d converter the quantizing error does in fact result from the finite resolution.

The quantizing error results from the fact that due to finite (and thus limited) resolution the a/d converter cannot recognize small level changes between two adjacent code levels. For the range of analog levels from 1 LSB to 2 LSB, for instance, an a/d converter will not produce a change in output codes; they will remain constant, due to quantizing error.

An a/d converter's basic quantizing error is 1 LSB. In general, this is basically nonreducible. In practice, quantizing error is reduced to tolerable levels by adding bits of resolution until the quantizing error is sufficiently low. It is also standard practice to bias an a/d converter's transfer function upward by ½ LSB, which splits the quantizing error to ±½ LSB.

Supply Sensitivity

Supply-voltage sensitivity is a measure of the degree to which converter performance parameters are affected by changes in power supply voltages. Typically, the parameters most sensitive are gain (or scale factor) and offset, although others can also be affected, dependent upon the particular design.

Supply sensitivity is typically specified as a percentage change in gain per volt of supply change (%/V), and as a microvolt per volt (μV/V) change in offset. For the typically used (+) and (−) 15-volt supplies, the supply rejection figures can be (and usually are) different.

Long Term Drift

Long term drift (as opposed to temperature drift) is a measure of the ability of a converter to maintain its performance over a period of time. It is not always specified, but is an important indicator of the necessary recalibration interval for a device.

Long term drift is predominantly associated with the aging of the reference diode (see reference chapter). Other potential sources of long term drift are resistor networks aging, and comparator and/or op amp offset drifts with time.

Dynamic Errors

Both d/a and a/d converters have dynamic or speed-related errors which are considered separately from the previously described static errors. These are discussed below.

Glitches

A problem which plagues high speed d/a conversion is the "glitch." A glitch is an undesired amplitude overshoot (or undershoot), caused by switch timing errors, either within the d/a converter or its associated logic. It is illustrated in Fig. 2-14.

A glitch is caused by switch-timing errors which result in a *momentarily* false analog output state. The error is seen as sharp positive or negative amplitudes which are a large percentage of full scale. Glitches are most often encountered at major code states, when a large number of bits are changing simultaneously. In Fig. 2-14A is shown an ideal "glitchless" output voltage change of 1 LSB, with the digital code changing from 01111111 to 10000000. In this situation all of the less significant bits are being replaced by the MSB, which is 1 LSB greater in amplitude.

Instead of the ideal step shown in Fig. 2-14A, glitched results are as shown in Figs. 2-14B and C. In Fig. 2-14B the MSB turns on slightly

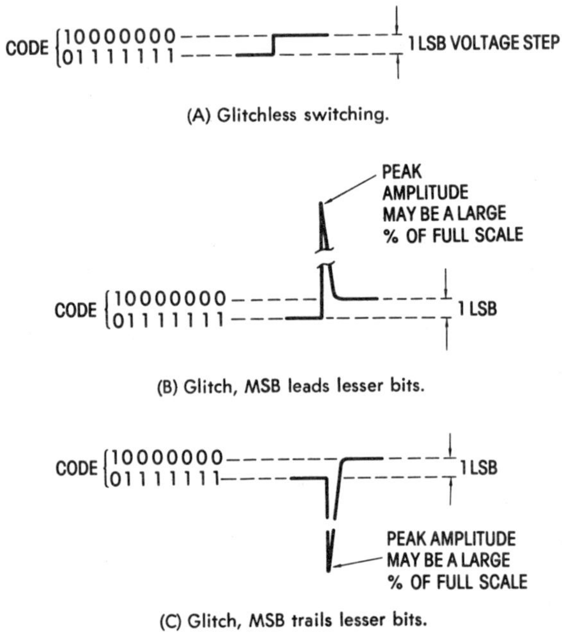

Fig. 2-14. Glitch errors.

faster than the lesser bits turn off, resulting in a large positive transient error, or glitch. After the glitch settles out, the output is at its correct amplitude. If the MSB turns on slower than the lesser bits turn off, there will be a momentary interruption in output, resulting in a negative transient, as illustrated in Fig. 2-14C.

Glitches are minimized in d/a designs by employing very fast switching elements, with matched transition times. This will minimize the glitching due to the d/a converter itself, but it should be appreciated that the problem can still occur due to time skewing within the input logic drive to the d/a converter.

Glitching is most troublesome in d/a applications where the output data is used continuously. One example is computer driven crt displays, where d/a glitches show up as ragged characters.

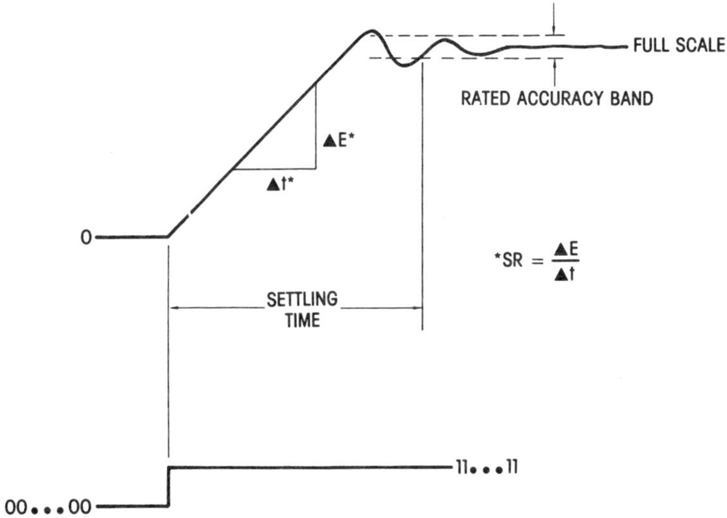

Fig. 2-15. Slew rate and settling time measurement.

Slew Rate

The large signal output rate of change of a d/a converter is governed by its *slew rate.* Slew rate is specified in V/μs for voltage output d/a converters, and in mA/μs for current output devices.

Slew rate is measured simply as the linear rate of rise of the output, usually from 0 to FS. It is a rough indicator of d/a converter speed; however, it is incomplete as the d/a converter must still *settle* to within rated accuracy, after a slew interval.

A waveform depicting these two parameters is shown in Fig. 2-15. With a full scale step input from 00 . . . 00 to 11 . . . 11, the d/a converter output must change from one analog limit to another, or from zero to FS. In doing so, it will traverse a slewing interval followed by a settling time region of smaller duration.

Slew rate is measured as the rate of change per unit of time, as shown. It is a measure of the slope of the initial (linear) region of the output waveform. The d/a slew rates are typically on the order of 1 or 2 V/μs; for high speed d/a converters, may be 10 or 20 V/μs.

Settling Time

The *settling time* of a d/a converter is a more complete description of the output voltage rate of change, since it defines the time required to *settle and remain* within a rated accuracy band around the final value.

As is shown in Fig. 2-15, the output may initially overshoot before settling within the rated accuracy; therefore, settling time must be measured to this final crossing to within the accuracy window.

Settling time is the most useful d/a converter speed figure, as it tells how fast the converter can be exercised at its rated accuracy. It must be specified for a full-scale output change, and to within an accuracy consistent with the d/a converter in use; otherwise, it is not meaningful. Typically, settling time is specified to an accuracy of ±½ LSB, for an appropriate number of bits.

The above discussion is a specific example of a device settling time measurement, and its interpretation. Often in conversion systems, even simple time constants, such as the single pole filter diagrammed in Fig. 2-16, must be considered in accuracy calculations. Although diagrammed as a simple resistor and capacitor combination, actual circuitry often finds R as the source resistance plus switch resistance in a multiplexer, and C as stray capacitance plus device shunt capacitance. Tables 2-1 and 2-2 show how the time related error of such a filter can be readily found. These tables are valid for single pole response such as a filter shown, or a single pole op amp.

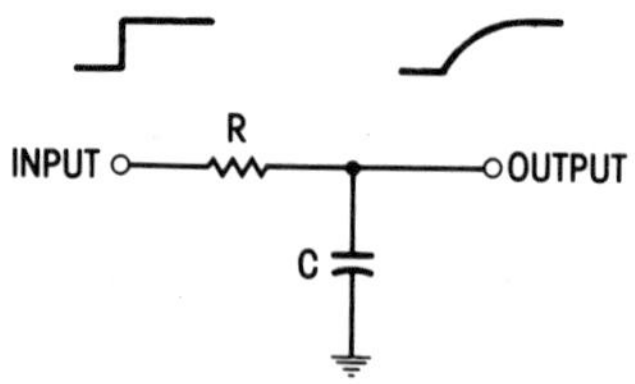

Fig. 2-16. Single pole RC filter.

Table 2-1. Settling Time Errors

Number of Time Constants	Output $1 - e^N$		Error e^N	
N	Fraction	%	Fraction	%
1	.63212	63.212	.36788	36.788
2	.86466	86.466	.13534	13.534
3	.95021	95.021	.04979	4.979
4	.98168	98.168	.01831	1.831
5	.99326	99.326	.00674	.674
6	.99752	99.752	.00248	.248
7	.99909	99.909	9.1×10^{-4}	.091
8	.99966	99.966	3.4×10^{-4}	.034
9	.99988	99.988	1.2×10^{-4}	.012
10	.99995	99.995	4.5×10^{-5}	.0045

Table 2-1 lists settling time errors as a function of the number of RC time constants, and Table 2-2 shows the number of time constants required for a given error.

Table 2-1 can be used as a rough guide to selecting RC combinations to yield a given error, but the format of Table 2-2 may be a bit more helpful in some cases.

Table 2-2. Time Constants for a Given Settling Time Error

Allowable Error Error (%)	Number of Time Constants $N = -\ln \frac{\text{Error (\%)}}{100}$
10	2.3025
1	4.6052
0.1	6.9078
0.01	9.2103
0.001	11.513

Since the settling time funciton being discussed is an exponential one, it may be very usefully graphed on semilog coordinates. This is shown in Fig. 2-17, where any desired error can be equated with the number of necessary time constants.

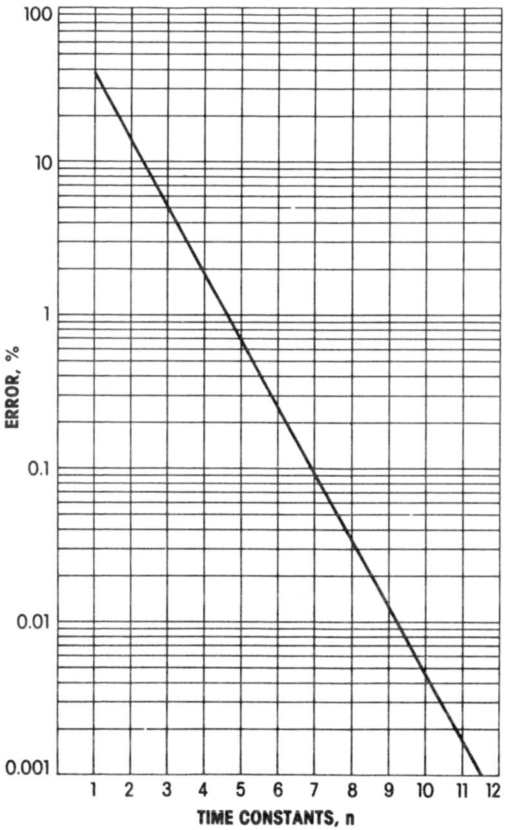

Fig. 2-17. Settling time errors due to single pole filter.

Conversion Time

Conversion time, in a general sense of the word, is simply the period of time which is necessary for a converter to execute its d/a or a/d conversion to rated accuracy. Conversion time places an uppermost limit on system operating speed. The d/a conversion time is specified simply as the device's settling time to ½ LSB accuracy. This settling time will be lowest (fastest conversion time) for current output d/a converters. Voltage output devices, which include an op amp, will necessarily be slower. It is not uncommon for the settling time of current versus voltage output d/a converters to differ by an order of magnitude, given a device which can operate in either mode.

The a/d conversion time is a very strong function of the a/d converter. Parallel a/d converters have the fastest conversion times, followed, in order, by successiver approximation converters, tracking-type converters, and integrating type a/d converters.

Aperture Error

A time-related error inherent to a/d converters is aperture error. Aperture error is the voltage error that will exist in a signal input variation during the conversion period. For virtually anything but a constant dc level, aperture error is a major factor for practical a/d conversion times.

The effect may be understood by considering a sine wave ac signal to be digitized during a conversion interval, tc. During time tc, it is desirable that the input not change more than ½ LSB so that the instantaneous voltage can be measured accurately. For a sine wave signal, the maximum rate of change or slew rate is:

$$SR = 2\pi V_p f$$

where,

V_p is the signal's *peak* voltage,
f is its frequency.

If this slew rate is equated to a ½ LSB maximum error voltage and the conversion time tc, we will have an expression which shows the relationship between aperture error and signal frequency. This expression is:

$$f = \frac{1}{tc\pi(2^{n+1})}$$

or in terms of tc,

$$tc = \frac{1}{f\pi(2^{n+1})}$$

The relationship is summarized in graphical form in Fig. 2-18 which shows data for converters of from 6 to 13 bits, for ½ LSB error.

Using this graph for instance, will show that a frequency of 1 kHz can be converted to 10-bit accuracy in a time of 155 ns (max).

The above, and other examples, easily demonstrate how restrictive finite conversion times can be on even moderately slow and accurate ac to digital conversions. The above example for instance would be very difficult to achieve in a direct conversion process.

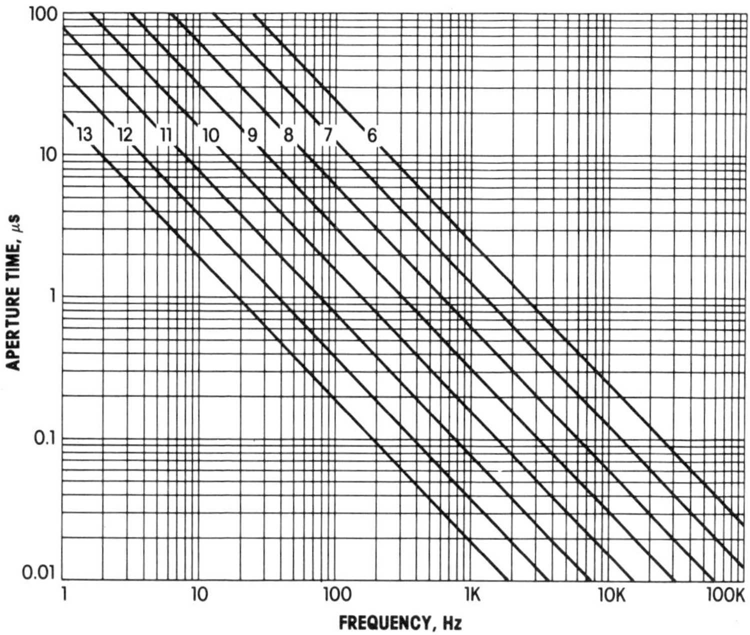

Fig. 2-18. Aperture time versus frequency for various system accuracies (½ LBS error).

Most practical a/d converter systems therefore use alternative means to solve the aperture error problem. This takes the form of a sample/hold, which freezes the ac input during the a/d conversion process. In this manner the burden of ultrahigh conversion speed is not placed upon the a/d converter. The *sample/hold* however still must have a switching related aperture time going from sample-to-hold which satisfies the relationship shown in Fig. 2-18. Sample/holds and their applications are dealt with in detail in a later chapter.

REFERENCES

1. Allan, R. "A/D and D/A Converters." *EDN*, February 5, 1973.
2. ———. "A/D and D/A Converters. Do You Know What the Parameters Mean" *EDN*, December 15, 1972.

3. "A/D, D/A Design Reference Guide" (wallchart). Analogic Corp., Wakefield, Massachusetts.

4. Brown, C. "Notes on Converter System Component Selection." *Analog Dialogue,* Vol. 6 No. 1, Spring, 1972.

5. Bruck, D. B. "Data Conversion Handbook." Hybrid Systems Corp., Burlington, Massachusetts, 1974.

6. Cate, T. "Speaks Out on Converter Specs." *EDN/EEE,* June 1, 1971.

7. Daley, F. D. Jr. "Analog-to-Digital Conversion Techniques." *Electro-Technology,* May. 1967.

8. Gadway, J. R. "A Data Converter Selection Guide." *EE Times,* November 5, 1973.

9. ———. "S/H or High Speed A/D Converters, How Do You Decide?" *EDN,* September 15, 1972.

10. Gordon, B. M. "The ABC's of A/D and D/A Converter Specifications." *EDN,* August 15, 1972.

11. ———. "What's Wrong With A/D Converter Specs." *EEE* Magazine (out of print), reprinted by Analogic Corp., Wakefield, Massachusetts.

12. Hoeschele, D. F. Jr. *Analog-to-Digital/Digital-to-Analog Conversion Techniques.* John Wiley & Sons, Inc., New York, 1968.

13. Karp, H. R. "D/A Converters: Trading Off Bits and Bucks." *Electronics,* March 13, 1972.

14. Miller, W. D. "Pick the Right DAC." *Electronic Design,* May 10, 1974.

15. Schmid, H. "Practical Guide to D/A Conversion." *Electronic Design,* October 24, 1968.

16. ———. "Practical Guide to A/D Conversion." Part 1. *Electronic Design,* December 5, 1968. Part 2. *Electronic Design,* December 19, 1968.

17. ———. *Electronic Analog/Digital Conversions.* Van Nostrand Reinhold, New York, 1970.

18. Sheingold, D. *Analog Digital Conversion Handbook.* Analog Devices Inc., Norwood, Massachusetts, 1972.

19. ———. *Analog Digital Conversion Notes.* Analog Devices, Inc., Norwood, Massachusetts, 1977.

20. Sherwin, J. *Specifying A/D and D/A Converters.* National Semiconductor Application Note AN-156. National Semiconductor Corp., Santa Clara, California.

21. Teeple, C. R. "Don't Forget D/A Converter Tempo!" *Electronic Design,* April 26, 1973.

22. Zuch, E. "Know Your Converter Codes." *Electronic Design,* October 25, 1974.

3

References

As was discussed in the previous chapters, one of the key elements of either a d/a or a/d converter (of any type) is the reference. Usually the reference of a converter takes the form of a voltage, but current references may also be used in certain instances.

Since the function of the reference circuit of a converter is of such basic importance, the quality of performance of the converter ultimately attained can be (and often is) largely dependent upon the accuracy of the reference. It is therefore important that the various methods of deriving reference voltages (or currents) be understood, so that their relative performances can be correctly assessed as part of the converter as a whole. This is true in general for any type of converter, but especially with many of the IC converters, which often do not contain a built-in reference to achieve the conversion function. So, in these cases the burden of system performance is left entirely to the user.

Reference circuits can assume a wide range of forms, dependent upon the performance required. In some cases it is relatively easy (and also economical) to design and build a reference from appropriate components. This can be either in the form of discrete components or by using voltage-reference ICs, which are readily available. In cases of very high performance requirements, such as drifts below 10 ppm, or comparable long term stability, it is usually a better choice to go to packaged hybrid IC references or aged certified zener diodes, which provide optimized system packaged performance, often approaching the state of the art.

A variety of reference circuits will be discussed in this chapter, with diversity ranging from the most simple to complex, and analogous performances. Devices currently available for reference use satisfy many different combinations of size, style, performance, and price, so there are many to choose from.

With the information gained in this chapter the reader will be able to more readily appreciate the overall converter system, and to better understand its performance specification limits. This is true whether a "complete" converter is being considered for purchase, or a reference circuit is being designed for use with one of the IC converters.

REFERENCE DEVICES

To begin, it is first appropriate to look in a broad sense at various devices which are appropriate for use as references. Generally, these consist of voltage regulator diodes (zeners), or complete integral regulator circuits which produce a reference voltage. Given a stable and accurate voltage, a current can always be derived, when necessary.

To give some perspective as to the general performance level being considered here for "reference" quality, it should be understood that this implies total regulation and/or stability to small fractions of a percent. A 12-bit converter for instance has a basic resolution of 0.025%. Some, if not a major portion of such a converter's inherent capability would be lost if the reference used in it were not capable of a performance better than the 0.025% figure; i.e., a lower percentage of error—this includes drift, regulation errors, and so on. The importance of this becomes even more apparent when temperature variations are considered, as TC (temperature coefficient) is the largest single source of reference error.

In light of the previous information, it is not at all unusual to speak of reference-source regulation (source and/or load induced) errors in the range of 0.001% to 0.01% (10 to 100 ppm), TCs of 1 to 20 ppm/°C, and long term drifts of 10-20 ppm per month, or even year. These specifications generally should be maintained even with load current variations, input voltage fluctuations, temperature variations, and time itself. A reference is, after all what its name implies, the quantity to which all input or output levels of an a/d or d/a converter are referred. If the reference is variable or uncertain, a converter can measure to an accuracy no better than that of the reference, regardless of the resolution, or *suggested* accuracy, of the converter. So, it is important that a converter's reference circuit be consistent with its intended use. If 12-bit accuracy is needed, you need a good reference. On the other hand, if only 8-bit performance is required, the reference burden is eased quite considerably.

Table 3-1 lists the key performance features of a number of devices which are suitable for use as converter references. Listed are such parameters as output voltages, current rating, output impedance, TC, input voltage and stability.

Table 3-1. Performance Features of Converter Reference Devices

Device	Classification*	Output Voltage (V)	Iz or Io (mA)	Rz or Ro (Ω)	TC (as noted)	Input Voltage (V)	Long Term Stability (as noted)
IN821-829 series	A	6.2 ± 5%	7.5 ± 0.01	15	5 ppm/°C (max) (IN829A)	↑	—
PRD series (CODI)	A	6.2 ± 5%	7.5 ± 0.01	15	1 ppm/°C (max) (25-45°C)		5 to 60 ppm/year
IN4611-4611C series	A	6.6 ± 5%	2 ± 0.2	75	5 ppm/°C (max) (IN4611C)	determined by user	—
LM336	B	2.49 ± 5%	0.3 to 10 mA range, 2 mA typ.	0.2	50 ppm/°C (max)		20 ppm (typ)
LM329B	B	6.9 ± 5%	0.6 to 16 mA max 1 mA typ.	0.8	20 ppm/°C (max)		20 ppm (typ)
LM3999	C	6.95 ± 5%	1 mA typ.	0.6	5 ppm/°C (max)		20 ppm/1000 hrs (typ)
LM399	C	6.95 ± 5%	1 mA typ.	0.5	2 ppm/°C (max)		20 ppm/1000 hrs (typ)
LM399A	C	6.95 ± 5%	1 mA typ.	0.5	1 ppm/°C (max)	↓	20 ppm/1000 hrs (typ)
AD580M	D	2.5 V ± 1%	10	1 (max)	10 ppm/°C (max)	4.5 to 30	100 ppm (typ)
MC1403A	D	2.5 V ± 1%	10	0.91 (max)	25 ppm/°C (max)	4.5 to 40	—
REF-01E	D	10 V ± 0.3%	10	0.8	8.5 ppm/°C (max)	12 to 40	—
REF-02E	D	5 V ± 0.3%	10	0.8	8.5 ppm/°C (max)	7 to 40	—
AD581L	D	10 V ± 0.05%	10	0.5	5 ppm/°C (max)	12 to 40	25 ppm/1000 hrs (typ)
LH0070-2	E	10 ± 0.10%	5	0.2	0.04% (max)	13 to 33	500 ppm/yr (max)
LH0071-2	E	10.24 ± 0.01%	5	0.2	over temp	13 to 33	500 ppm/yr (max)
AD2700L, V	E	+10 ± 0.025%	10	0.05 (max)	3 ppm/°C (max)	+13 to +18	100 ppm/1000 hrs (max)
AD2701L, V	E	−10 ± 0.025%	10	0.05 (max)	3 ppm/°C (max)	−13 to −18	100 ppm/1000 hrs (max)
AD2702L, V	E	±10 ± 0.025%	5	0.05 (max)	5 ppm/°C (max)	±13 to ±18	100 ppm/1000 hrs (max)

* A = Temperature compensated reference diode.
B = Active temperature compensated reference diode (IC).
C = Type B, with thermally stabilized substrate.
D = Three terminal monolithic IC.
E = Hybrid IC, actively trimmed to high accuracy.

In this table there are two broad categories—reference diodes and integral reference circuits. Classes A and B are diodes, as is class C in a sense. Class-C units, however, are reference diodes on a thermally stabilized substrate and as such, exhibit some of the lowest TCs of all those shown. Device classes D and E are integral reference circuits which can be simply plugged into a circuit to deliver a rated output voltage.

Further details of performance for the various devices are covered in their individual discussions which follow.

BASIC DEVICE HOOKUP AND DETAILS OF OPERATION

This section is a brief look at how each reference device appears schematically, what it takes to operate it, and how it is packaged. In some cases, as for instance that of the integral references, this information is all that is needed to apply the device to achieve the specified rated performance.

Classification A: Temperature Compensated Reference Diodes—1N821–1N829 and 1N4611–1N611C Series

These two series of devices are actually combinations of zener and conventional diodes, hooked up in series. This cancels the respective diode reverse and forward TCs, and thus achieves a nominal net TC of zero. Lowest maximum TC of the JEDEC registered "1N" devices is 5 ppm/°C, as typified by the 1N821 and 1N4611C. The TC of a reference diode is critically dependent upon its current, however, and this current must be closely regulated. This is particularly true in the case of 1N829 types, but somewhat less so in the 1N4611 series units. Very low TCs, on the order of 1 ppm/°C and very high long term stability achieved by the "PRD" series, proprietary units by CODI.

No schematic of these devices is shown, as they are represented by the familiar zener symbol. They are packaged in standard packages.

Classification B: Active Temperature-Compensated Reference Diodes

The LM329 is like other temperature-compensated diodes in that it has a low nominal TC. Three grades are available; these are 50, 30, and 15 ppm/°C typical drift, represented by suffixes "D," "C," and "B" respectively. However, a major difference in this type of device is the fact that the dynamic impedance is very low, typically less than one ohm. Also, the temperature coefficient is relatively independent of current. These features are in actuality achieved by an IC which does, in fact, comprise the "diode." The zener diode is a "subsurface" type—one which is inherently more stable than surface zeners.

The LM329 is packaged in a TO-92 outline as shown in Fig. 3-1. The schematic symbol used is that of a zener, although the IC circuit actually consists of many components.

(A) Schematic. (B) Connection diagram.

Fig. 3-1. LM329 zener diode (TO-92 package).

The LM336 (Fig. 3-2) is also an active zener type of reference, but is synthesized with a low voltage architecture, with a reference voltage of 2.5 volts, based on the bandgap voltage of silicon. The TC is somewhat higher than the subsurface zener type, but is still of general reference quality.

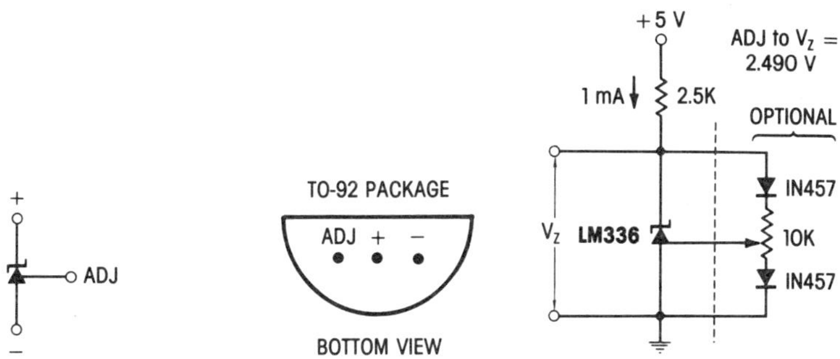

(A) Schematic. (B) Connection diagram. (C) Trimmed, low-temperature coefficient connection.

Fig. 3-2. LM336 zener diode (TO-92 package).

The LM336 is packaged in a TO-92 outline as shown in Fig. 3-2B. Note that the device has a third terminal, which is used for trim purposes. When trimmed as in Fig. 3-32C, for a voltage of 2.490 volts, the TC will be a minimum, typically 15 ppm/°C. Trimming is optional, and can be deleted. Another form of trim uses the pot without the diodes, and adjusts terminal voltage without change of TC.

Classification C: Active, Temperature-Stabilized Reference Diodes—LM3999, LM399, LM399A

The LM3999 is similar to the LM329 insofar as it is an active, temperature-compensated, subsurface zener reference diode. However, it goes a step further in that it also incorporates a temperature stabilizer (or heater) which keeps the internal substrate temperature at a constant 85°C. This minimizes sensitivity to external temperature changes, allowing a 5 ppm/°C drift to be achieved.

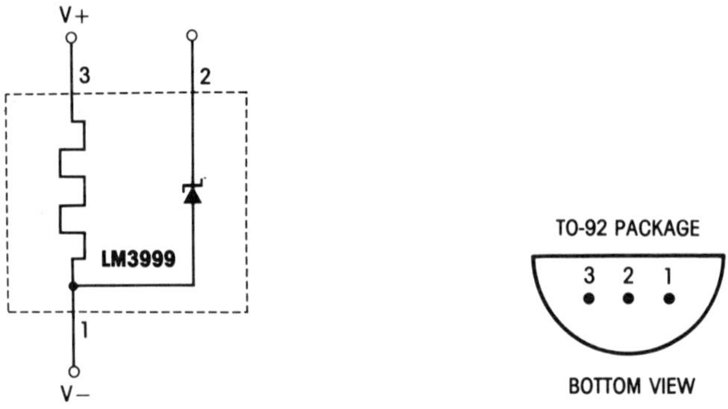

(A) Schematic. (B) Connection diagram.

Fig. 3-3. LM3999 zener diode (TO-92 package).

As indicated by the schematic (Fig. 3-3A), the diode and heater have a common terminal (1) and the device is contained in a TO-92 package. The heater operates from supplies from 9 to 36 volts, typically consuming 360 mW of power in 25°C (still) air. Initial heater surge current is typically 140 mA, and warmup time is 5 seconds.

The LM399 and LM399A (Fig. 3-4) are electrically similar to the LM3999, but feature improved performance and flexibility. These devices are packaged in an insulated TO-46 outline, using a four-lead configuration. This provides electrical isolation of heater and diode common terminals.

The LM399 provides a TC of 2 ppm/°C maximum, while the LM399A has a maximum TC of only 1 ppm/°C. Typically, both devices have a TC of 0.3 ppm/°C. Power consumption of both units is typically 255 mW at 25°C, and warmup time is 3 seconds.

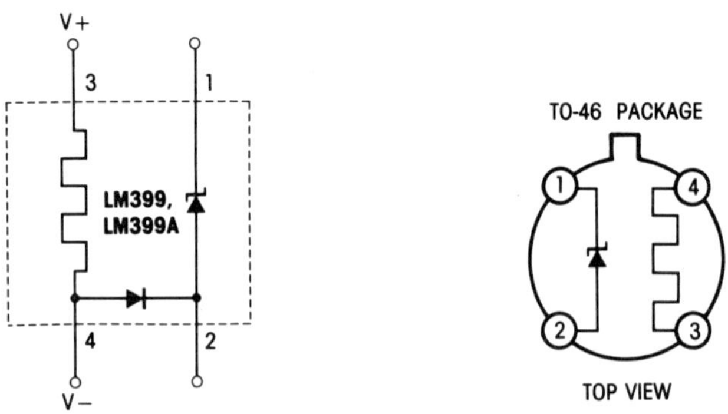

(A) Schematic. (B) Connection diagram.

Fig. 3-4. LM399, LM399A zener diodes (TO-46 package).

Classification D: Three Terminal "Bandgap" Reference Monolithic ICs

The AD580 is an IC reference device configured for three terminal voltage-in, voltage-out operation. It accepts an unregulated input voltage in the range of +4.5 volts to +30 volts, and produces a +2.5-volt ±1% reference voltage. Load current range is up to 10 mA, and the TC is 10 ppm/°C (max) (for the AD580M). Long term stability is 100 ppm/year, typically. This device like the others of this class, uses a voltage reference principle based on the bandgap voltage of silicon. A minimum component hookup for applying the AD580 is shown in Fig. 3-5A. It is supplied in a TO-52 package.

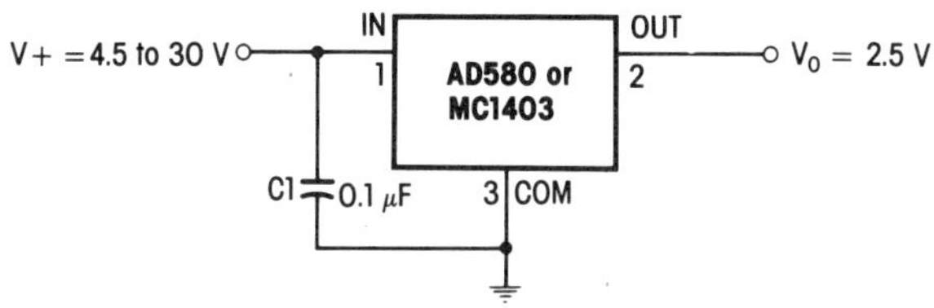

(A) Schematic.

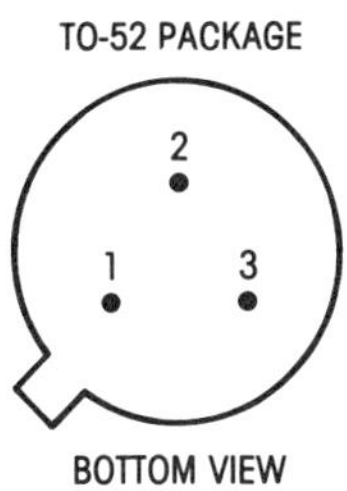

(B) Connection diagram, AD580.

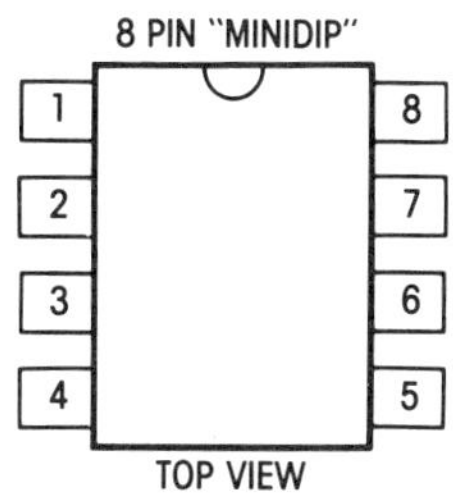

(C) Connection diagram, MC1403.

Fig. 3-5. AD580 and MC1403 monolithic ICs.

The MC1403 is a three-terminal IC reference with operation and specifications similar to the AD580. It is packaged in an 8 lead "MINI-DIP" package, and may also be applied as noted in Fig. 3-5A.

The REF-01 (Fig. 3-6A) is a three-terminal IC reference which operates somewhat similar to the AD580 in basic concept, but with some important differences. The device is packaged in an 8 pin TO-99 package.

The REF-01 is a 10-volt output device, with an input range of 12 to 40 volts. Initial output tolerance is ±0.3% for the REF-01E, which also has a 8.5 ppm/°C (max) TC.

As may be noted from the hookup in Fig. 3-6A, this device has an additional terminal, labeled TRIM (pin 5). If left open, the output voltage tolerance is as mentioned above. However at the user's option, a

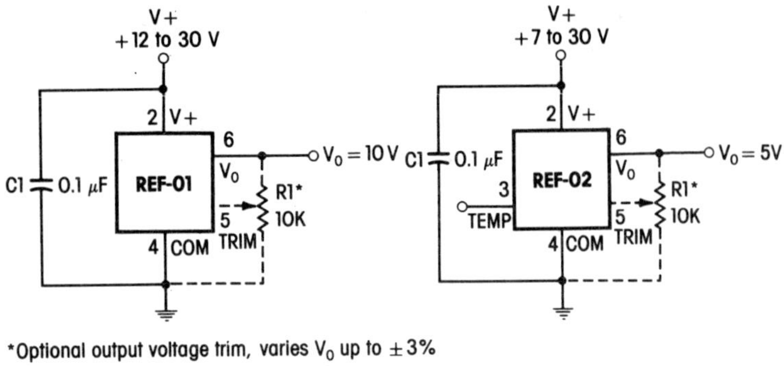

(A) REF-01. (B) REF-02.

Fig. 3-6. REF-01, REF-02 series (TO-99 package).

trim pot R1 may be used to set the output anywhere within the range of ±3% about nominal. Load current rating is 10 mA.

The REF-02 (Fig. 3-6B) is similar to the REF-01, except that its output voltage is a nominal 5 volts, and input range is 7 to 40 volts. It is trimmed (when desired) in a like manner, and has a similar load rating.

One additional pin is provided with the REF-02 however, termed TEMP (pin 3). This pin provides an output voltage proportional to temperature, allowing the device to be used for temperature sensing.

The AD581 is a three-terminal reference based on the bandgap principle, and is illustrated in Fig. 3-7. It has an output voltage of 10 volts, with tolerances as low as 0.05%, and a TC of 5 ppm/°C. It is packaged in a TO-5 package, and has a 10 mA load current rating.

An interesting feature of the AD581 is its ability to operate in a two terminal zener mode, as shown in Fig. 3-7C. With this feature, the device can be used for negative outputs, as well as positive. Device zener current should be maintained in the range of 1 to 5 mA, by selection of R_s.

Classification E: Actively Trimmed, Hybrid IC References

The LH0070 and AD2700 series are three-terminal references which operate in a *functional* sense similar to the three-terminal monolithic ICs. There are, however, significant advantages of this class of device in terms of performance. This is due to active circuit trimming (by laser, or similar means) that allows precise adjustment of final output voltage and a minimum TC to be achieved. This group of devices allows the greatest measure of "as delivered" accuracy, and a minimum of user design attention.

The LH0070 and LH0071 are devices which deliver 10 and 10.24 ±0.01% output voltages, respectively. Load current is 5 mA and TC is

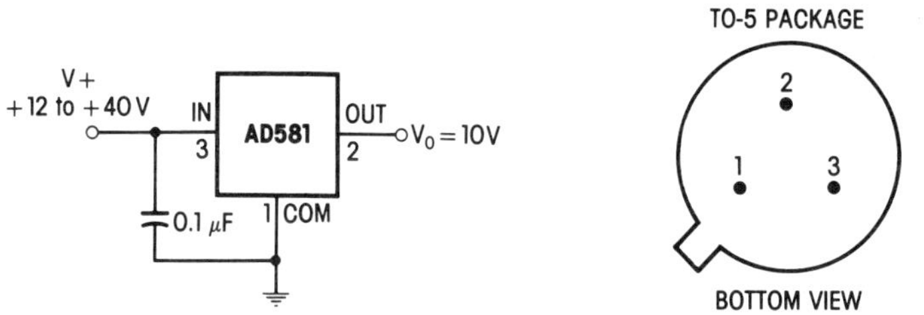

(A) Schematic.

(B) Connection diagram.

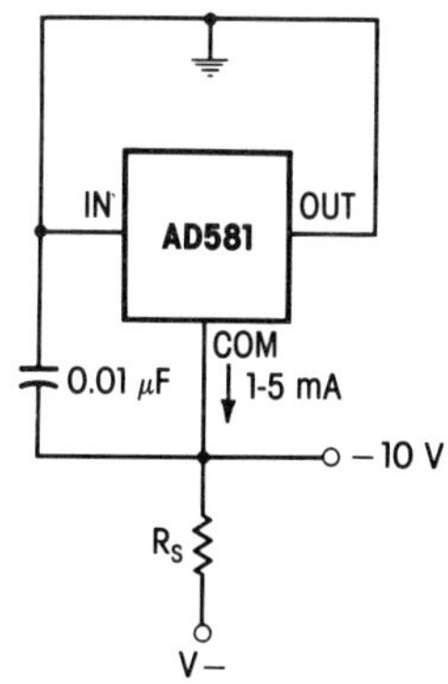

(C) Zener-mode operation.

Fig. 3-7. AD581 (TO-5 package).

specified as 0.04% (max) over the device temperature range (for the -2 suffix). Input voltage range is 13 to 33 volts and long term stability is specified as 500 ppm/year. They are packaged in a TO-39 package, and a minimum component hookup is diagrammed in Fig. 3-8A.

The AD2700 series of references delivers output voltages of +10 volts, −10 volts and +10 volts, in devices AD2700, AD2702 and AD2702 respectively. These units are packaged in a 14 pin DIP package.

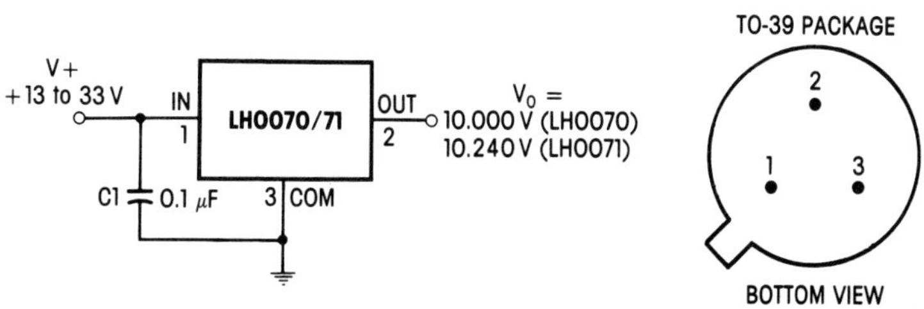

(A) Schematic.

(B) Connection diagram.

Fig. 3-8. LH0070,71 (TO-39 package).

Initial accuracy is 10 volts ±0.025%, and load current is 10 mA (5 mA for the AD2702). Input voltage range is 13 to 18 volts, and long term stability is 100 ppm/1000 hours.

Minimum component hookups are as noted in Fig. 3-9. Although the minimum hookup is three terminal, an optional output trim may be used with any device to adjust the output voltage ±30 mV, when desired.

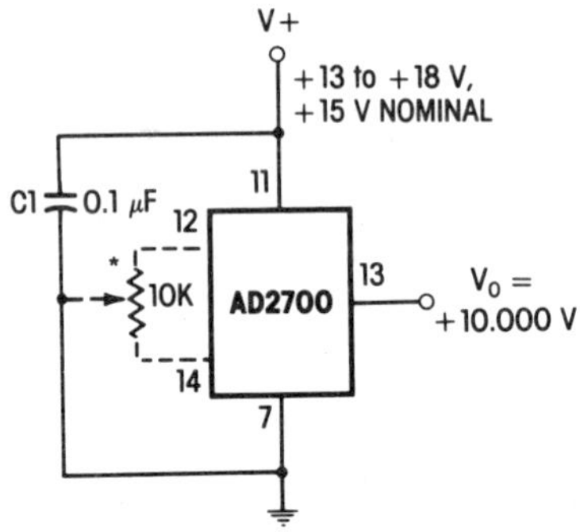

(A) Schematic, AD2700.

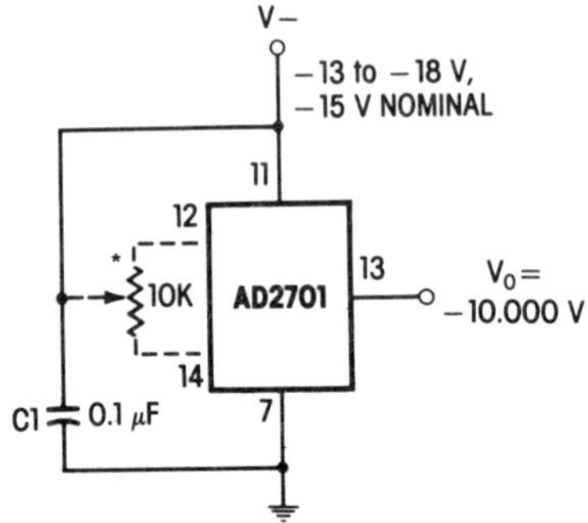

(B) Schematic, AD2701.

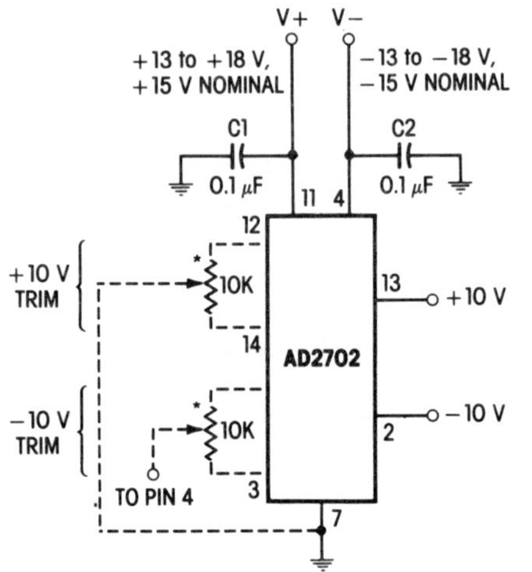

*Optional output voltage trim, varies V_0 up to ±30 mV.

(C) Schematic, AD2702.

Fig. 3-9. AD2700 series (14 pin DIP package).

APPLICATIONS OF REFERENCES

In this section the various devices just discussed are illustrated in a variety of reference applications, which serve to bring out the various points of consideration in optimum reference designs. This will begin with the most simple of circuits and progress through more advanced designs.

Simple Reference Sources

The most basic reference source has traditionally been the temperature-compensated diode. The circuit can in its most simple form consist of only an input resistor to the supply line, and the reference diode itself. The output voltage is taken from across the diode with suitable attention given to minimizing load current variations.

A modern and very practical form of such a circuit is shown in Fig. 3-10A using the LM329 reference diode. Because of its very low dynamic impedance of 0.8 ohm, the LM329 is attractive in such a circuit, as the low impedance minimizes errors due to input voltage variations, load variations, and feed resistor drift. The resulting major error of such a circuit will be the diode's basic TC, as will be seen from the following discussion of performance.

Included in Fig. 3-10A is a tabulation of the major worst case errors. The conditions assumed are a 15-volt ±1 volt supply, and a 0.5-mA load current which varies no more than 25%. The performance figures are tabulated in worst case fashion.

From the error budget it is fairly easy to see that the predominant error is due to TC which comprises about 90% of the total error. The total error in a 0-70°C temperature range (a change of 45°C from 25°C) is about 0.1%. This would restrict this circuit to converters of 8-bit, or less, accuracy, if the reference source error is to be held to ¼ LSB or less.

The circuit of Fig. 3-10B is similar in concept, but it uses the temperature stabilized LM3999 to achieve a lower TC. Its total error is 352 ppm or less (0.035%), which would be suitable for converter accuracies of 9 bits or less, over a 0-70°C range.

The circuit of Fig. 3-10C is optimized for high performance and uses the LM399A with its 1 ppm/°C (max) TC. Consistent with the high performance goal, the supply voltage is restricted to ±1% and a stable resistor type used for R1. With these measures this circuit is able to achieve a worst case error of less than 190 ppm over this range, which is suitable for 11-bit accuracy.

Errors which these analyses have not included are those due to initial voltage tolerance of the diodes, and the total effects of the load, R_L. In typical use, the initial tolerance would be trimmed out by using only a fraction of the output voltage, such as 5 or 6 volts for instance. The divider resistors thus used would have to be included in the overall error budget.

For current reference input converters, a resistor such as R_L can feed into an op amp summing point (a virtual ground) and the TC of R_L would then be a part of the error budget. Many popular IC converters such as the 1408 or DAC-08 types operate in this manner, which makes this simple type of reference quite attractive.

The "Bootstrapped-Diode" Reference Source

Although the simple reference circuits just described will suffice for many applications, quite often a prescribed voltage level is required (such as 10 volts), and at a current output level of several milliamperes. This is usually beyond the capabilities of simple unbuffered diode circuits so additional support circuitry is required. An example of this form of circuit is shown in Fig. 3-11.

This circuit has appeared in the literature in many forms (see references) and is quite popular as a voltage reference source. This is due to its high performance without undue complexity.

This circuit "bootstraps" the reference diode, because the diode is used not only to regulate its output voltage, but also to set and maintain its own current, a prerequisite for a low TC.

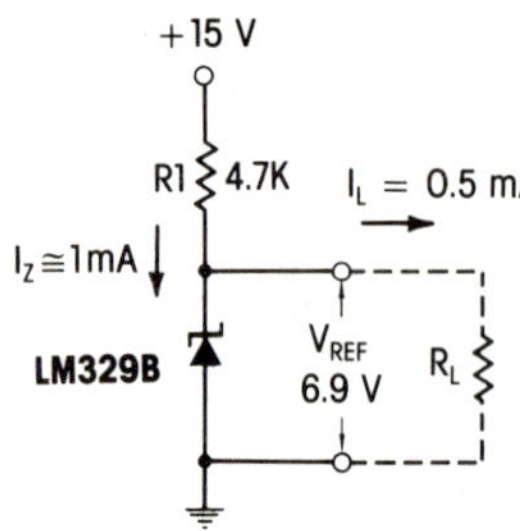

TYPICAL PERFORMANCE

Temp. Coefficient:	20 ppm/°C (MAX) x 45° C =	900 ppm
Input Regulation:	425 μV/V =	62 ppm
Load Regulation:	0.0035 % ($\Delta I_L \leq$ 2.5%)	35 ppm
Long Term Stability:	20 ppm	20 ppm
		1017 ppm
Total Error, 0-70° C:		≤ 1017 ppm
		or ≅ 0.1%
		suitable ≤ 8 bits

(A) Using the LM329.

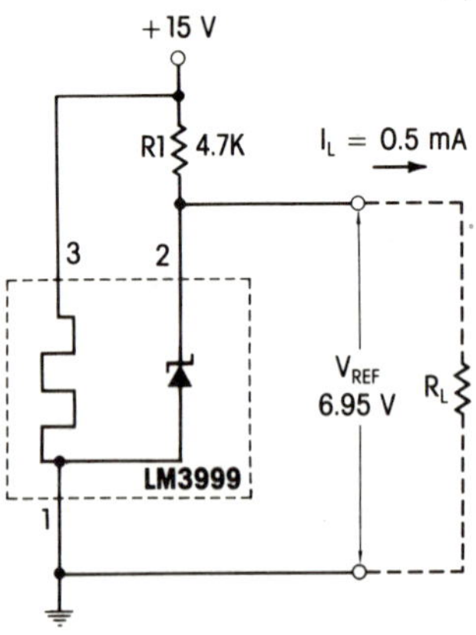

TYPICAL PERFORMANCE

Temp. Coefficient:	5 ppm/°C (MAX) x 45°C =	225 ppm
Input Regulation:	467 μV/V =	70 ppm
Load Regulation:	0.0037 % ($\Delta I_L \leq$2.5%)	37 ppm
Long Term Stability:	20 ppm	20 ppm
		352 ppm
Total Error, 0-70° C:		≤ 352 ppm
		or ≤ 0.035 %
		suitable ≤ 9 bits

(B) Using the LM3999.

Fig. 3-10. Simple

So that the output can assume only one polarity to ensure startup, the amplifier is operated from a single supply voltage. Once the diode has broken down, the negative feedback loop sets the output voltage as:

$$V_o = V_{REF}\left(\frac{R2 + R3}{R2}\right)$$

R1 sets the current in the diode, since the voltage drop across it is simply the difference between output and reference voltages. Since these two voltages are stable, this stabilizes the current.

The key to effective performance is the correct choice of components suitable to the required performance. The diode used should, of course, be consistent with the desired TC, and representative types are listed. R2 and R3 are most critical from the standpoint of their *tracking* TC, and their absolute value (or TC) is of lesser importance. R4 is selected to be equal to R2 ÷ R3, for minimum bias current error. The amplifier used should be capable of low drift, on the order of 5 μV/°C or less (this is equivalent to 1 ppm/°C or less for a 6-volt reference diode).

The selection of R1 is governed by the diode used. Thus, 3 values are shown, for use with the LM329, IN4611 or IN829 (or 7.5 mA) types respectively. This will yield nominal currents of 1, 2, and 7.5 mA respectively. For 7.5-mA use, addtiional current is supplied by R5 to ease the burden on the op amp.

For the IN4611, and in particular the IN829 type units, the value of R1 should be trimmed to optimize (minimize) the overall TC of the circuit. This has the effect of adjusting the TC of the reference diode about its nominal zero point, and will serve to minimize the *net* drift errors of the circuit, including the op amp and resistors R1, R2, and

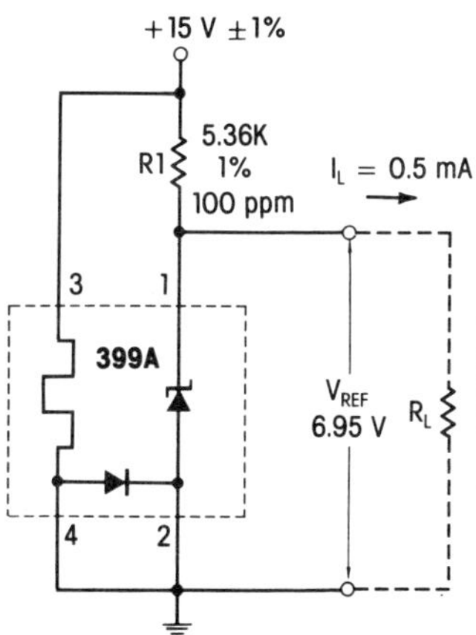

PERFORMANCE
(worst case)

Temp. Coefficient:	1 ppm/°C (MAX) x 45°C =	45 ppm
Input Regulation:	280 μV/V x 0.15 V	6 ppm
Load Regulation:	0.00087% ($\Delta I_L \leq$ 25%)	26 ppm
Long Term Stability:	20 ppm	20 ppm
		97 ppm
Total Error, 0-70°C:		≤ 97 ppm
		or ≤ 0.0097%
		suitable ≤ 11 bits

(C) Using the LM399A.

reference sources.

R3. R2 and R3 are trimmed for output voltage calibration. The trim process is interactive to an extent.

Because of the foregoing interrelated factors which influence the overall TC, it is difficult to give specific component recommendations for this circuit, particularly for the very low TCs. Below drifts of 10 ppm/°C the circuit must be considered as a complete thermal system and designed from both a mechanical and electrical standpoint accordingly. It is this type of approach, with this form of circuit, which does, in fact, result in the high performance of actively trimmed hybrid references sources such as the LH0070 and AD2700 series, and references of lab standard quality such as the CODI "Certavolt" (see references).

For those who desire to build and characterize a circuit of this type some general guidelines are given in the figure, but these should only be considered as starting points. It should be understood that only

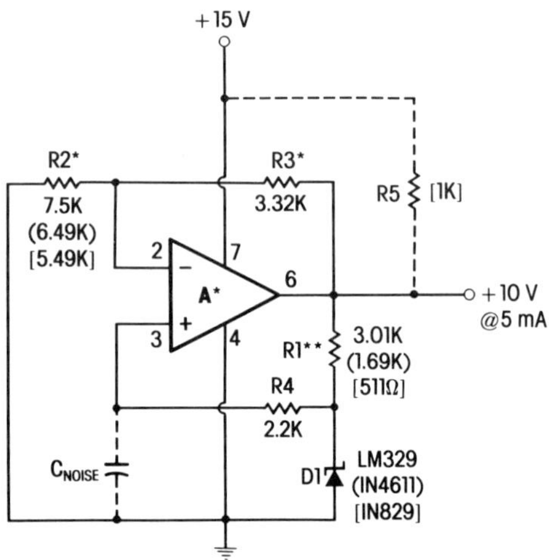

*Trim ratio to calibrate output voltage.
**Trim to minimize temp drift (IN4611, IN829).

SUGGESTED COMPONENT COMBINATIONS

Overall TC	D1	A1	R2/R3	R1
< 50 ppm	329B	741, 301A	10 ppm/°C tracking	100 ppm
< 25 ppm	IN4611B or IN827	741, 301A	5 ppm/°C tracking	↑
10-15 ppm	IN4611B or IN827	AD741L, OP-02E	2 ppm/°C tracking	optimize-
≅ 5 ppm (or less)	IN4611C or IN829 "PRD" or LM399A	AD741L, OP-02E	1 ppm/°C tracking or less	see text ↓

(A) Diode common.

Fig. 3-11. Basic bootstrapped

through extensive testing and trimming after component qualification, will the highest performance be reached.

For negative forms of this reference circuit, the diode is reversed, pin 7 of the op amp is grounded, and the unregulated negative input is applied to pin 4. Resistor R5 (when used) would then be returned to pin 4, instead of pin 7.

Various other reference circuit schemes have been devised to attack specific problems or sources of error which arise in varying applications. The following series of applications discussions illustrate some of these techniques.

Stacked References

Where unusually high variations in input voltage are encountered, the typically high line regulation of IC reference devices may prove to be inadequate. The "stacked" reference schemes illustrated in Fig. 3-12 can virtually eliminate errors due to input variations, reducing a full rated device input range swing to only several microvolts at the output.

Fig. 3-12A* shows a method using an AD580M (A1) as a 2.5-volt reference, whose input is regulated by A2. The common terminal of A2 is returned to A1's output, hence the term "stacked." R1 provides a bleed path for A2's common leg current. Note that only A1 need be a precision device in this circuit; the A2 function is relatively noncritical. A 5-volt "auxiliary" output is optionally available.

* W. G. Jung, "Applications of the 580," *Analog Dialogue,* Issue 9-2, 1975.

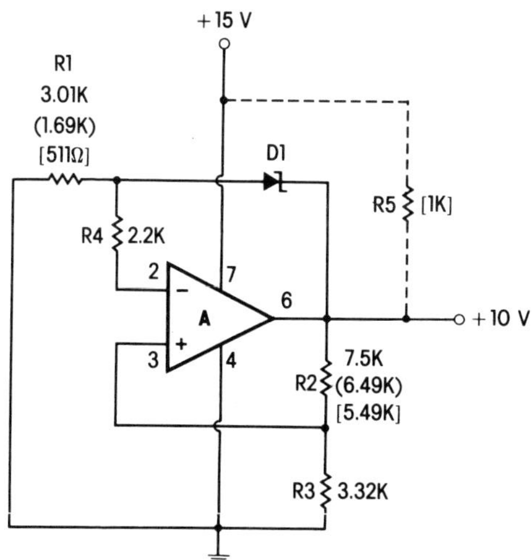

(B) Diode floating.

reference source.

Fig. 3-12B shows a similar scheme using REF-02 devices to generate a precision 5-volt reference output. Operation is generally similar, but here R1 serves not only as a bleeder for A2 (as in Fig. 3-12A) but

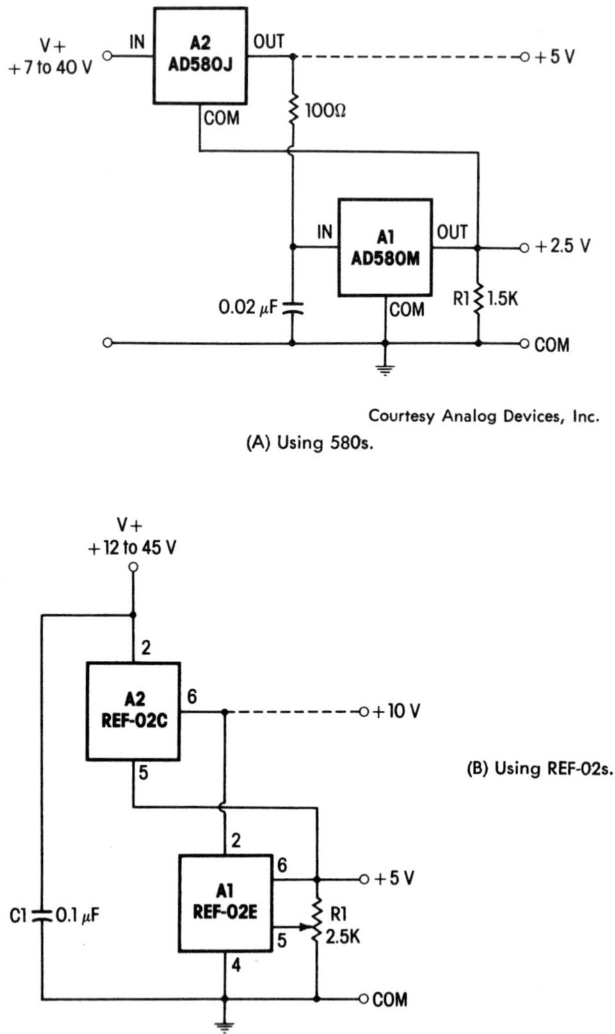

(A) Using 580s.

(B) Using REF-02s.

Fig. 3-12. Stacked references.

also trims the output voltage. If trim is not needed, a fixed resistor should be used for R1. Again the device used for A2 is relatively noncritical, and a bonus output is available (+10V).

Shunt Regulated Reference

With some additional support circuitry, IC references can operate as two terminal shunt regulators, as shown in Fig. 3-13†. Here a low-voltage reference, the AD580, is used in conjunction with the op amp A1 and LED1. The negative feedback loop formed around A1 programs output terminal voltage V_o to a multiple of the reference voltage.

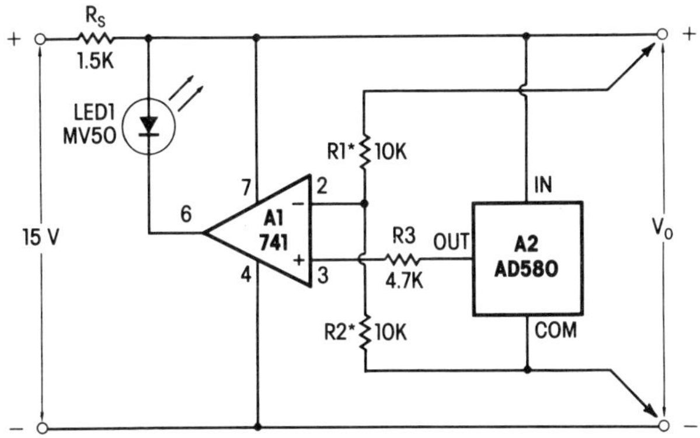

$$^*V_O = \frac{R1 + R2}{R2} V_{REF}$$

where V_{REF} is A2 output voltage
(V_O = 5V as shown)

Courtesy Electronic Design

Fig. 3-13. Shunt-regulated reference.

R_s is a series input resistor chosen to supply the necessary load current, plus the current in A1 and A2 (approximately 3 mA), and as shown is suitable for 3 mA loads.

Note that this circuit can "float," as it has no necessarily common point.

Precision Reference Sources

With a combination of circuit techniques, it is possible to design references where the total error is reduced to only the TC and long term drift of the reference diode itself. This is accomplished through combinations of bootstrapping, reference device buffering, and input preregulation.

In Fig. 3-14*, an IC reference is connected in a bootstrapped configuration which multiplies its output voltage by the ratio

† W. G. Jung, Programmable Voltage Reference Is Stable, Yet Simple," *EDN*, November 5, 1975.

* W. G. Jung, "Precision Reference Source Eliminates Power-Supply Regulation Errors," *Electronic Design*, September 1, 1975.

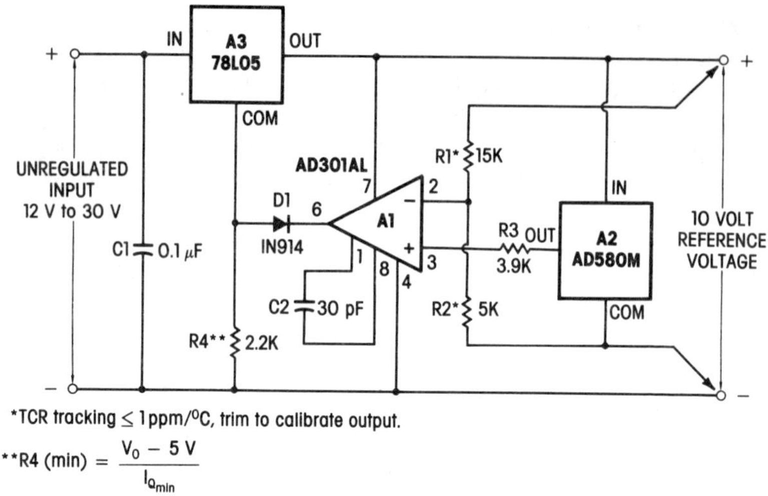

Courtesy Electronic Design

Fig. 3-14. Precision reference.

$$\frac{R1 + R2}{R2}$$

thus

$$V_o = V_{REF}\left(\frac{R1 + R2}{R2}\right)$$

A low-drift op amp is used, and its supply pins as well are bootstrapped, eliminating its supply voltage errors. Input voltage variations are absorbed by A3, a low cost 5-volt three-terminal regulator which can also supply up to 25 mA of output. R4 and D1 are used for start-up purposes.

Assuming that a low-tracking TC pair are used for R1 and R2, the TC will be essentially limited by the reference device used for A2. With the AD580M, this will be in the 10 ppm/°C region. A REF-02E will allow slightly better TC, due to its better specified TC, and also the fact that A1's offset drift will be a smaller percent of error. For best results, output voltage should be sensed directly at the load terminals, as shown.

Fig. 3-15* carries this approach of error minimization to perhaps its present state of the art, and uses a low TC LM399A as the reference diode. This circuit is functionally similar to Fig. 3-14, with the exception of the reference itself, and has the same features of low errors.

* W. G. Jung, "Precision Reference Source Features Minimum Errors," *EDN*, August 5, 1976.

Here, because the reference diode has such a low inherent TC, the scaling resistors R_A and R_B are especially important to a low overall TC. They should have tracking TCs of preferably 0.5 ppm/°C or less. This means they should be high stability wirewound units of Evanohm or Manganin alloys.

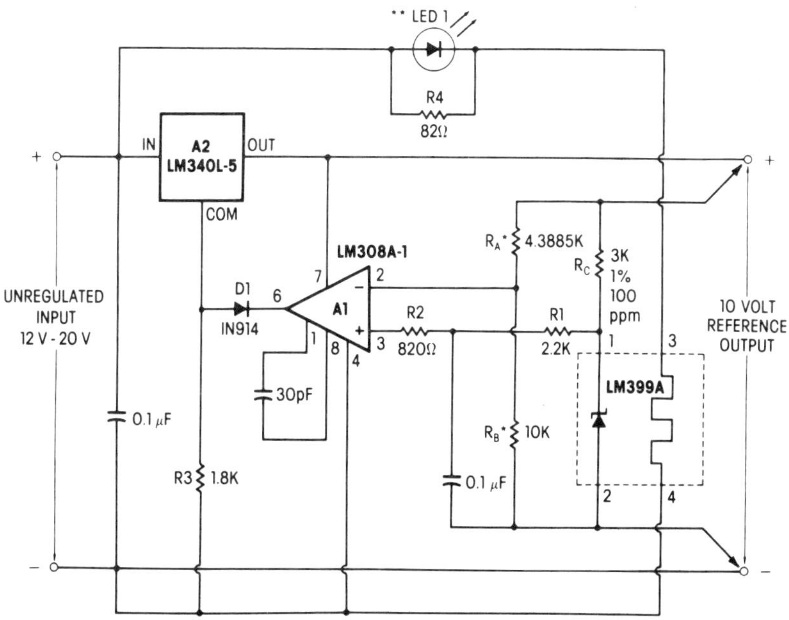

$*\ \dfrac{R_A + R_B}{R_B} = \dfrac{V_O}{6.95}$, trim ratio for exact calibration. R_A/R_B tracking TC $\leq$ 0.5 ppm/°C (see text).

**Out of tolerance indicator - oven stabilized when extinguished.

Courtesy Electronic Design

Fig. 3-15. Low TC precision reference.

A low drift op amp is used for A1, to minimize its TC contribution. The amplifier type specified has a drift of less than 0.2 ppm/°C. This, along with the remaining errors should yield overall drift of 2 ppm/°C or less if care is taken in construction. The ultimate performance limit of this circuit will then be the diode's basic long-term stability.

Negative References

Thus far the discussions have generally been in terms of positive reference sources, at least with those circuits involving the IC reference devices. Negative reference sources can be implemented with reversed zener diodes of course, as mentioned previously. The circuits in Fig. 3-

16* show how IC references, which are basically positive output devices, can be used to produce negative reference voltages with some attractive features.

In Fig. 3-16A is shown a basic negative source, using a three-terminal IC regulator (A1), in this case a REF-02E. Other types and voltage outputs can be used, as well.

---

* W. G. Jung, "Positive Reference Voltage IC Is Flipped Negative By Adding a Single Component," *Electronic Design,* February 15, 1976.

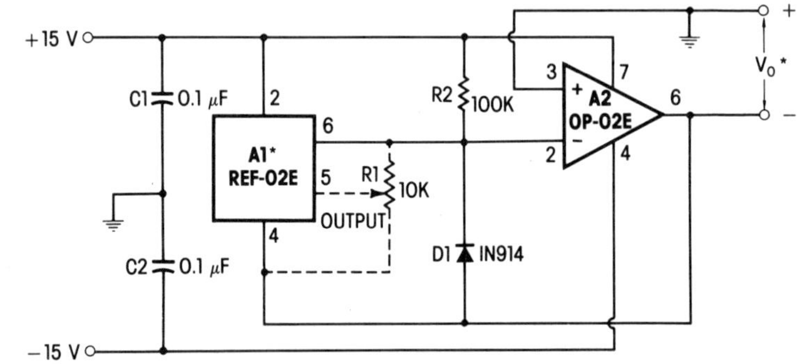

*$V_O = -V_{REF}$, where V_{REF} equals A1 output voltage (as shown, $V_O = -5V$)

(A) Basic circuit.

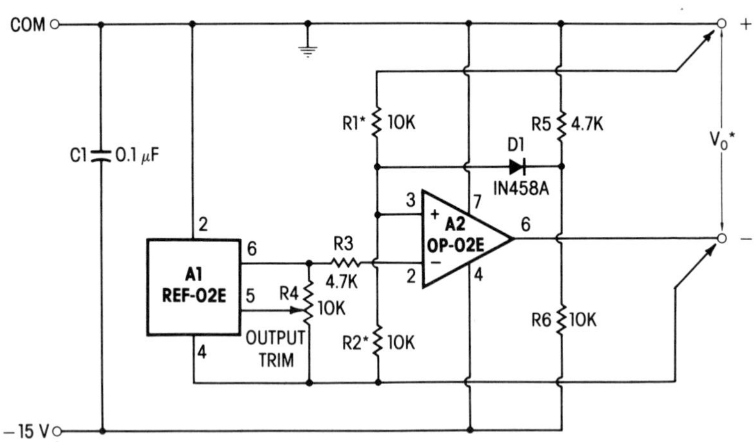

$$*V_O = -\frac{R1 + R2}{V_{REF}}$$

where V_{REF} is A2 output voltage ($V_O = -10$ V as shown)

(B) Scaled negative reference.

Fig. 3-16. Negative references. Courtesy Electronic Design

In this circuit the op amp supplies the output current. A1 biases the op amp input-to-output at a voltage equal to its output voltage, in this case 5 volts. With the (+) input of the op amp referred to ground, the output of the op amp is negative.

R1 serves its normal trim purpose (when desired) and the output current of the circuit is determined by the op amp. In this case the output current is 5 milliamperes. Overall TC of this circuit is less than 10 ppm/°C, or slightly higher than the TC of reference IC which is 8.5 ppm. R2 and D1 are used for startup and protection purposes.

Fig. 3-16B shows how the negative reference idea can be extended to scale the output voltage to a specific level, and eliminate some of the supply input errors.

Here A2 is connected with its input terminal grounded, but otherwise as previously. R1 and R2 are output scaling resistors which set the output voltage as shown. The positive supply pin of A1 is grounded, allowing the circuit to operate from a single negative supply voltage line.

This particular configuration has problems in starting, or assuming the proper output state, if precautions are not taken. Here, a startup network composed of R5, R6, and D1 is used to force the (+) input of A1 to −5 volts at switch on. This causes the output to go negative after which the feedback loop takes over, generating the proper output, a −10 volts in this case. Diode D1 is back-biased after startup; thus, it does not induce any error. For output voltages other than 10 volts, R5 should be selected for a nominal drop equal to, or less than, that across R1.

The TC of this circuit is about 10 ppm/°C, if R1 and R2 track to within 1 ppm/°C. The supply input rejection is 6 ppm/V (max) due to A1, so the −15-V supply should be desirably regulated to within a few percent.

An extension of the negative reference idea is illustrated in Fig. 3-17.* This circuit can be used to divide a given reference voltage into equal (or unequal) smaller voltages. The outputs are then bipolar tracking reference voltages.

With a given reference IC output voltage, such as 2.5 volts, from the AD580 shown here, the resistors R1 and R2 can be adjusted for any desired positive and negative output voltages, with an absolute value sum equal to the IC's output voltage. Positive (+) output current is determined by A2, while negative (−) output is set by A1.

One of the basic problems which often arises in the application of voltage references is a means of adjustment for calibration of system gain or scale factors. While generally any reference can be adjusted by a voltage divider, this usually gives rise to additional error factors

* W. G. Jung, "Simple Regulator Circuits Provide Tracking Bipolar Reference Voltages," *Electronic Design*, July 5, 1976.

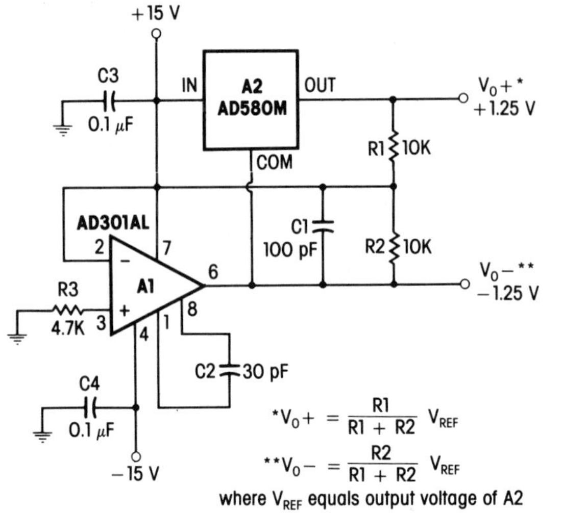

Courtesy Electronic Design

Fig. 3-17. Bipolar tracking reference source.

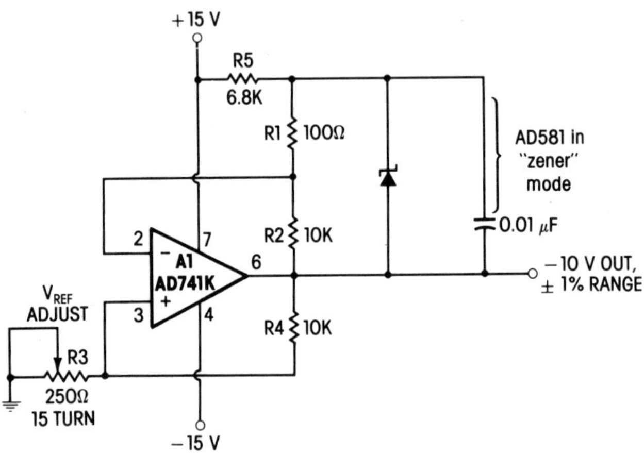

Fig. 3-18. Adjustable reference source.

due to divider resistances. Also, a voltage divider is unbuffered so it sacrifices output impedance and it cannot, of course, furnish a higher voltage, if this is desired.

The circuit of Fig. 3-18 illustrates a means of trimming a reference source an equal amount about its nominal level, with the output appearing in low impedance form. Here an AD581 (or other zener type reference) is operated within the negative-feedback path of A1, with the zener current set by R5. If the (+) input of A1 were grounded, the output of the circuit would simply be the voltage of the reference diode, a −10 volts in this case.

In the circuit here, R1 and R2 reduce the voltage from the AD581 by 1%, which produces a −1% trim limit. Positive feedback through R4 to R3 allows the op amp to bootstrap the output by the amount of voltage across R3. Resistor R3 can then be trimmed for exact calibration, with range limits of ±1% about the nominal voltage.

The circuit can generally be used with virtually any reference type of zener, and is applicable to positive outputs by reversing the diode (and feeding it from −15 volts).

Precision Current Sources

As mentioned initially, converter references are often required to take on the form of a regulated current, rather than a voltage. The following circuits show how voltage reference devices can be configured to precisely regulate currents.

The circuit of Fig. 3-19* uses a voltage reference IC to create a proportional regulated current. This regulated current is maintained in spite of variations in input voltage, and/or the load itself. It is suitable for applications where a current reference is necessary, and provides a source of current into a ground, or negative supply-referred load.

In operation, the voltage reference A2 provides a regulated voltage between A1's output and the OUT terminal of A2. This voltage, here

* W. G. Jung, "Applications of the 580," *Analog Dialogue,* Issue 9-2, 1975.

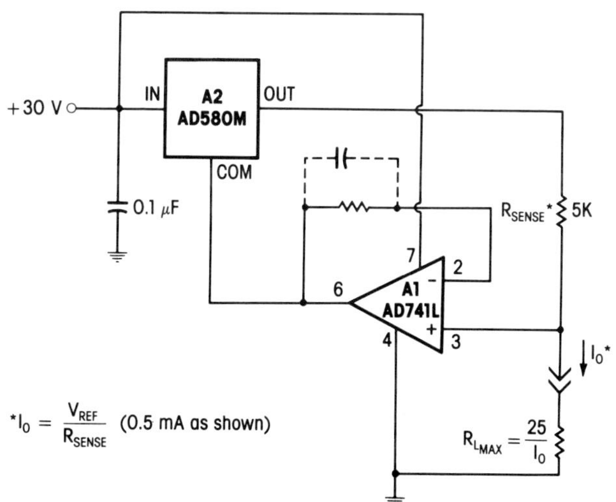

Courtesy Electronic Design

Fig. 3-19. A precision current source.

2.5 volts, is used to set up a current in the precision resistor, R_{sense}. Since A1 is connected as a voltage follower, the regulated voltage effectively appears across R_{sense}. While the voltage across R_{sense} is constant, the common mode voltage (that across R_L) need not be, as the output of A1 can freely assume any desired level (within design constraints of course). The voltage that A1-A2 can therefore supply to the load is the input-output difference minimum of A2, plus its output voltage. Here this is 5 volts, so the circuit can comply with up to 25 volts of drive to R_L.

A1 is a device chosen for specifications consistent with overall accuracy, and the current I_o. The AD741L type shown will operate well down to currents of about 10 microamperes, but for the lower levels a FET input type such as the AD506L type would be more appropriate.

High Compliance Precision Current Source

The circuit of Fig. 3-20 improves upon the precision current source of Fig. 3-19 in two major regards. It has a higher output voltage compliance, and better basic regulation. Another important feature is its ability to easily trim the output current without direct access to the sensing resistor.

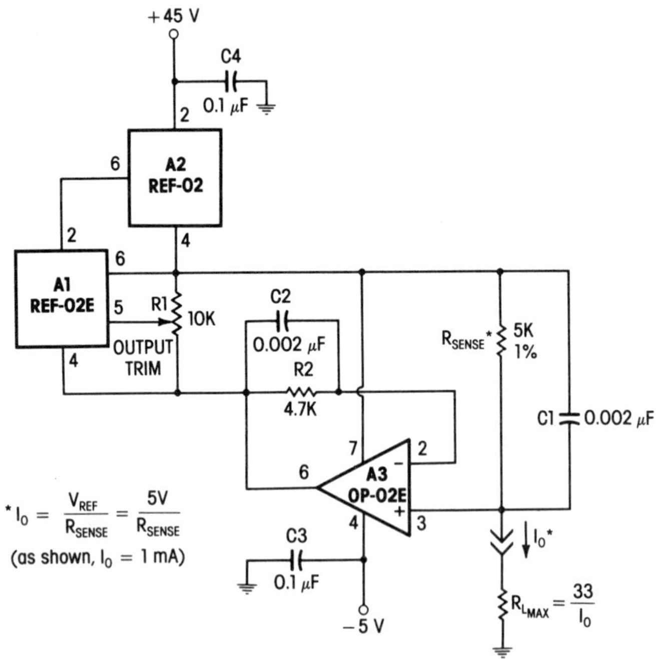

Courtesy Electronic Design

Fig. 3-20. High-compliance precision current source.

This circuit operates essentially the same as the previous one, but in this case the input voltage to the voltage reference device A1 is bootstrapped by A2. This arrangement maintains the dissipation of A1 constant (for a given load) which can minimize its temperature errors. It also eliminates input or load-induced voltage change errors.

Similarly, sensing op amp A3 is bootstrapped by tying its positive supply terminal to A1's output. This eliminates the positive voltage-induced errors of A3, leaving only negative supply-line errors to influence accuracy. With the typical 3 μV/V supply rejection of the OP-02E this results in only a 20 ppm current change in regulated current I_o.

The design, as shown, is suitable for currents from an upper limit of 10 millamperes, down to about 10 microamperes, at which point the bias current of the OP-02E becomes a significant error. When desired, lower currents can be accommodated by using an FET input amplifier, such as the 355A or OP-15, which will allow regulation down to about 10 nanoamperes, with low errors.

Precision Current Sink

Fig. 3-21 illustrates a form of current regulator which is closer to the classical method of supplying a constant current into a varying load resistance. This circuit uses an op amp to sense the voltage drop across a precision current sample resistor, R_{sense}. This voltage drop is maintained by the loop to be equal to the reference voltage; thus, the current is regulated.

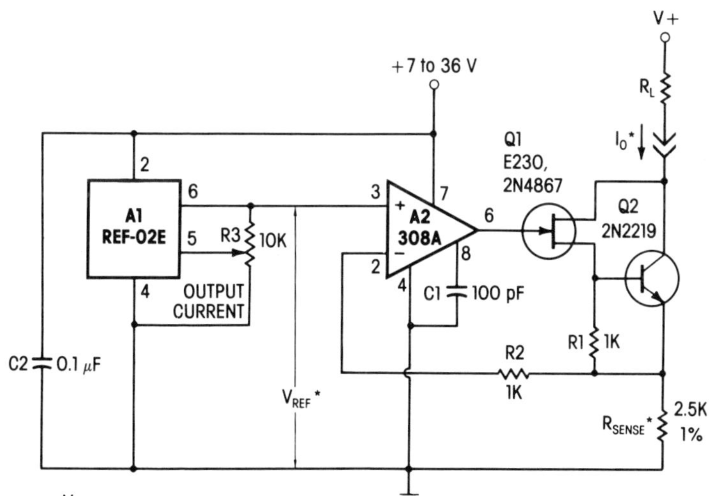

$$*I_O = \frac{V_{REF}}{R_{SENSE}},\ 1\ \mu A \le I_O \le 10\ mA\ \text{(as shown, } I_O = 2\ mA\text{)}.$$

Fig. 3-21. Precision current sink.

One of the keys to accuracy in this circuit is to ensure that the current delivered to the load, which flows in the collector/drain of Q2/Q1, is identical to that in the emitter leg (the regulated current). In practice, this means the transistor used for the Q1-Q2 function must have a very high current gain, preferably 10,000 or more. For this reason, an FET is used to drive Q2. Another possibility would be a single transistor, an integral npn Darlington with power/voltage rating comparable to the devices shown, or a monolithic power transistor, such as the LM395.

As diagrammed here, the circuit is suitable for currents up to about 10 mA. For currents lower than 0.5 mA, Q2 can be deleted and R1 shorted, and Q1 used alone. As shown, the circuit will drive loads referred to positive voltages. For loads referred to ground, the A1-A2 positive supply line should be grounded, and the bus shown as ground should be ungrounded and returned to a negative voltage of 36 volts or less.

Sample resistor R_{sense} should be a precision, low TC type, consistent with the desired performance. Output current can be trimmed ±3% by trimming the output voltage of A1 as shown. Trimming of the sense resistor should be avoided (preferably) because of reliability considerations. An alternate form of voltage reference which could be used with this general circuit would be one of the thermally stabilized diode types such as the LM399, with a voltage divider on the output to adjust the voltage applied to A2. This will allow TCs down to a few ppm/°C.

REFERENCES

1. Accardi, L. "Self-Stabilized Zener Insures Constant Current in Op-Amp Voltage Reference." *Electronic Design,* December 21, 1972.
2. ———. "Superstable Reference Voltage Source." *EDN,* October 1, 1971.
3. Brokaw, A. P. "A Simple, Three-Terminal IC Bandgap Reference." *IEEE Journal of Solid State Circuits* Vol. SC-9, No. 6, December 1974.
4. Dobkin, R. C. "Don't Forget A/D Reference Stability." *EDN,* June 20, 1977.
5. ———. *High Stability Regulators.* National Semiconductor LB-15, January 1971. National Semiconductor Corp., Santa Clara, California.
6. ———. *IC Voltage Reference Has 1 PPM Per Degree Drift.* National Semiconductor Application Note AN-161, June 1976. National Semiconductor Corp., Santa Clara, California.
7. ———. "On Chip Heater Helps To Stabilize Monolithic Reference Zener." *Electronics,* September 16, 1976.
8. Giles, J. N. *Fairchild Semiconductor LIC Handbook.* Fairchild Semiconductor, Mountain View, California, 1967.
9. Goldfarb, W. "Single-Supply Reference Source Uses Self-Regulated Zener." *Electronics,* June 7, 1973.

10. Gunn, J. "10 Volt (±1mV) References." *Analog Dialogue,* Vol. 10, No. 1, 1976.

11. Henderson, B., Glasser, S. "Constant Voltage Diodes Are Getting Sharper." *Electronic Products,* March 18, 1974.

12. Jung, W. G. "Applications of the AD580." *Analog Dialogue,* Vol. 9, No. 2, June, 1975.

13. ———. *IC Op Amp Cookbook.* Howard W. Sams & Co., Inc., Indianapolis, 1974.

14. ———. "IC Regulated Power." *The Audio Amateur,* April, 1974.

15. ———. "Precision Reference Source Eliminates Power Supply Regulation Errors." *Electronic Design,* September 1, 1975.

16. ———. "Precision Reference Source Features Minimum Errors." *EDN,* August 5, 1976.

17. ———. "Programmable Voltage Reference Is Stable, Yet Simple." *EDN,* November 5, 1975.

18. ———. "Simple Regulator Circuits Provide Tracking Bipolar Reference Voltages." *Electronic Design,* July 5, 1976.

19. Kennedy, E. J. "Temperature Stable Supply Uses IC as Error Amplifier." *Electronic Design,* November 8, 1969.

20. Kesner, D. "Choose Regulator Op Amps With Care." *Electronic Design,* June 21, 1969.

21. Koep, K. "Solid State Voltage Standards." *Instruments and Control Systems,* October, 1972.

22. Kuijk, K. E. "A Precision Reference Voltage Source." *IEEE Journal of Solid State Circuits,* Vol. SC-8, No. 3, June, 1973.

23. Pouliot, D. "Build Precision Reference Sources." *Electronic Design,* July 5, 1976.

24. Ritmanich, W. "Bandgap Voltage Reference Applications." *Proceedings of EE Times IC Applications Seminar,* 1976.

25. Routh, W. S. *Applications Guide for Operational Amplifiers.* National Semiconductor Application Note AN-20, National Semiconductor Corp., Santa Clara, California.

26. Shah, M. J. "Stable Voltage Reference Uses Single Power Supply." *Electronics,* March 13, 1972.

27. Shirley, F. R., Vanderlosk, L. "A Stable Voltage Reference Uses Only Six Components." *Electronic Design,* January 18, 1968.

28. Widlar, R. J. "New Development in IC Voltage Regulators." *IEEE Journal of Solid State Circuits,* Vol. SC-6, No. 1, February 1971.

29. Williams, J. "Don't Bypass the Voltage Reference That Best Suits Your Needs." *EDN,* October 5, 1977.

30. Product Catalog, PRD series precision reference diodes; CODI Corp., Fairlawn, New Jersey.

4

Converter Devices

This chapter examines a wide variety of different IC devices which are suitable for either d/a or a/d data conversion uses. In concept, most of them can be readily categorized into the general types previously described. There are, however, a number of details and features which make each IC device unique and it is the purpose of this chapter to fill in this detailed background, before we begin applying the devices in circuits.

Because of the sheer mass of information available on the various units it is not feasible to cover each of them in all facets of detail. What is presented is a highlight summary of device operation, including such key data as functional or block diagrams, pin configurations, truth tables, transfer characteristics, and other such necessary fundamental information. This material is supported by a discussion which describes the important advantages or features of a given unit. Selected data sheets are included in an appendix for the more popular units, as well as all manufacturers' addresses. Also, the references at the end of the chapter provide further background information, both on specific devices and on general device design concepts. Detailed application discussions are covered for all devices in Chapter 6.

D/A CONVERTER DEVICES

One of the keys to understanding the operation of IC converters is an appreciation of the elements of which they are composed. Since many a/d converters use d/a converters as integral components, a logical starting point for device discussions is with the IC d/a converter. We can go even further than this, and examine the basic circuit elements which either comprise IC d/a converters, or have given rise to similarly operating IC d/a devices.

D/A Quad Current Switches

A major fundamental element of IC d/a converters is the technique of current switching, which is used almost exclusively in a very high percentage of d/a applications.

A simplified diagram of a 4-bit d/a converter employing a current-switched cell is shown in Fig. 4-1. In this circuit transistors Q1 through Q4, when on, conduct currents of 1000, 500, 250 and 125 μA respectively, as determined by the scaled emitter resistors of 10K, 20K, 40K, and 80K. The collectors of these transistors feed a common bus line, which carries the output current, I_0. It can be seen that if Q1 through Q4 are considered as switches, they comprise a simple 4-bit binary d/a converter where transistor Q1 is the MSB leg, and transistor Q4 is the LSB leg.

Q1 through Q4 are switched either on or off by pnp transistors Q5 through Q8, respectively. Considering the MSB as typical, if the logic input is high, Q5 will be on, thus robbing Q1 of emitter current and turning off the MSB. Conversely, when the MSB logic input is low, Q5 is off, and Q1 and the MSB current are on. The common base bias line for Q5 through Q8 is chosen to be in the range of 1.2 to 1.4 volts, to provide a TTL compatible logic input. Note that the coding of this structure is complementary binary; that is, the respective bits are on (true) for a low logic input.

The output current of this converter can be used directly in current form to feed a load resistor (not shown) to develop a negative output voltage. Or, an op amp can be used (connection shown dotted) in the current-to-voltage mode, to develop a buffered voltage output.

The actual currents conducted by Q1 through Q4 must, of course, be predictable and stable for this structure to be effective. This is accomplished by the resistor network composed of binary weighted 1-2-4-8 legs, plus the additional leg in the emitter of transistor QR. Transistor QR is a reference transistor whose function is to allow duplication of the reference current I_R in Q4, and scaled currents of 2X, 4X, and 8X the reference current in Q3, Q2, and Q1. Q1 through Q4 are all monolithically matched, with the emitter areas of Q1, Q2, and Q3 scaled in proportion to their currents. Thus, all five transistors will have equal base-to-emitter voltages, resulting in equal voltage drops across the emitter resistances. This factor and the matched transistor betas allow precisely scaled resistors to yield precisely scaled currents.

The reference current in QR is developed by a control loop consisting of A1 and QR. An external reference voltage applied to input resistor R develops a reference current of 125 μA, which flows in the collector of QR. As all transistor bases are driven in parallel by A1,

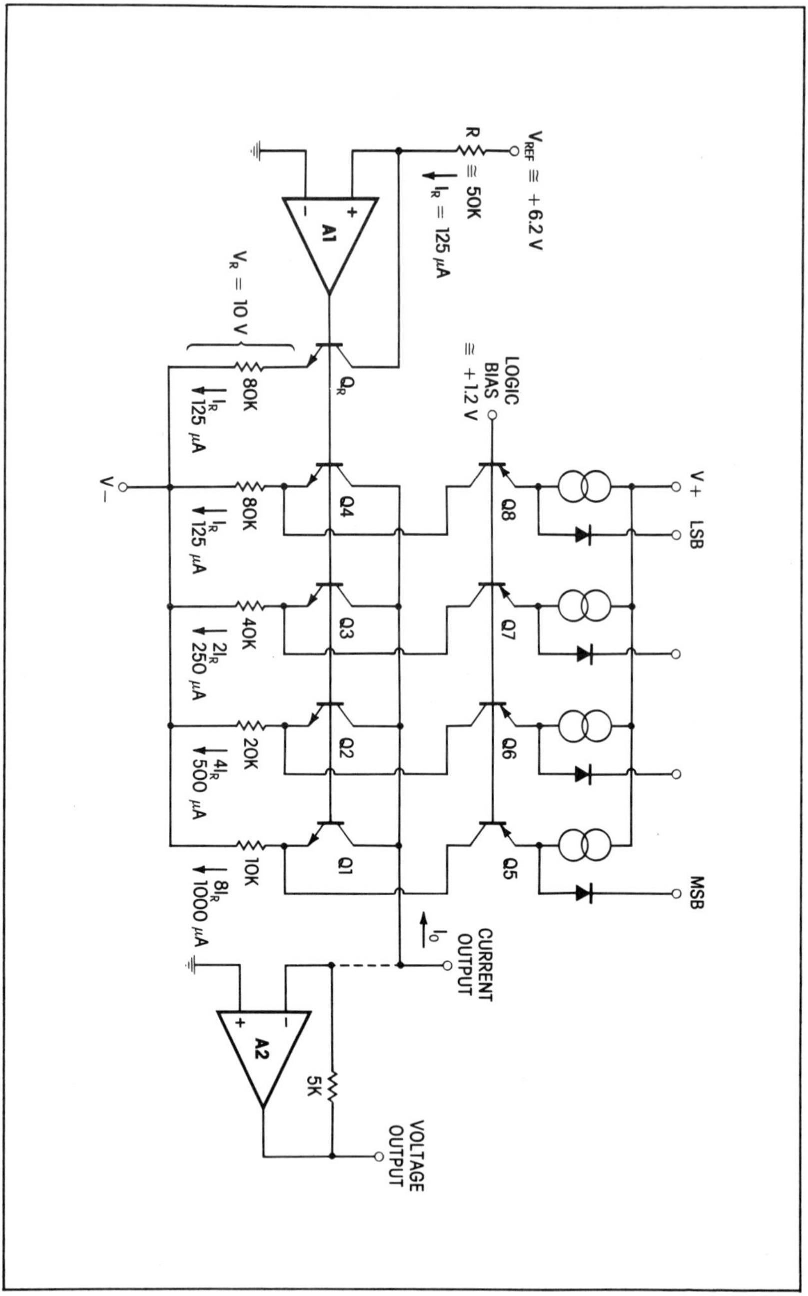

Fig. 4-1. A quad d/a current-switching cell.

the previously described matching technique forces proportional currents in the slave transistors Q1 through Q4. These currents are maintained uniformly in the presence of temperature changes by the control loop.

The two main ingredients which make this scheme work are the matched current switches and the matched resistor network. One of the earliest forms of IC based d/a circuits operates essentially as just described, and this technique is still in wide use today in hybrid IC d/a converters. Variations of current switching and reference current development are also widely used.

The AD550 and AD553

The Analog Devices AD550 and AD553 quad current switches operate on these principles, and are illustrated in Fig. 4-2. The schematic shows the scaling of emitter areas among the four current-switching transistors, and in this case Q31 is the reference transistor. The four control inputs can be noted, with diodes driving the pnp switch drivers. For a reference input current of 125 μA, the complete truth table is shown, progressing from 0 to 1.875 mA, in complementary binary fashion.

The AD550 and AD553 are identically pinned, similarly operating devices with the AD553 being faster in switching time, and thus settling time to rated accuracy. Rated supply voltages are +5 volts and −15 volts, and the logic control inputs are TTL compatible.

Multiple Quad Converters

To construct converters with resolutions of more than four bits, additional quad current switches are added, as shown in Fig. 4-3. This circuit uses three AD550 or AD553 quad switches, plus a single precision resistor network to implement a 12-bit d/a converter.

This type of converter uses interquad attenuation, as described previously in Chapter 2. All quad switches operate at the same dc current levels, but the outputs of the second and third quad are attenuated by factors of 1/16 and 1/256, respectively. I_1 is the first quad current output, I_2 the second, and I_3 the third. As can be noted, I_o is then composed of $I_1 + I_2/16 + I_3/256$.

The 937.5 ohm and 14,062.5 ohm resistors attenuate I_3 by 1/16; the attenuated I_3 plus I_2 are then further attenuated by the second stage current divider of 1000 ohms and 14,062.5 ohms.

All of the emitter resistances, the interquad attenuators, the reference input resistors, and the output scaling resistors are contained in a single resistor network (Fig. 4-4A). This is necessary to maintain overall accuracy, by holding the resistor tracking TCs to 1 ppm/°C. The table in Fig. 4-4C lists several manufacturers which supply pin equivalent versions of this network.

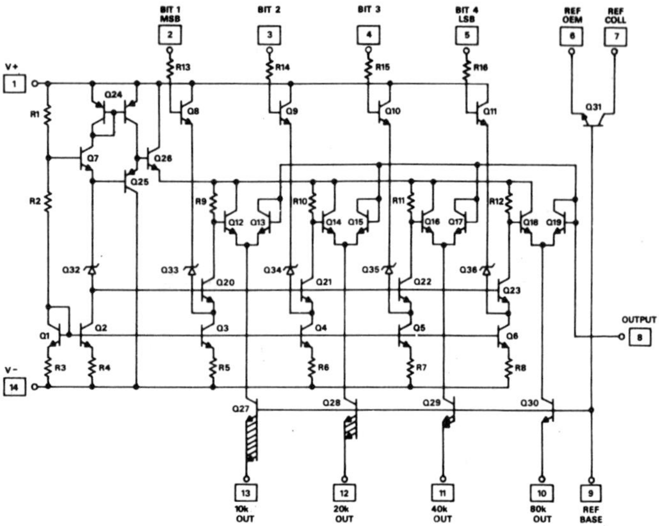

(A) Functional diagram.

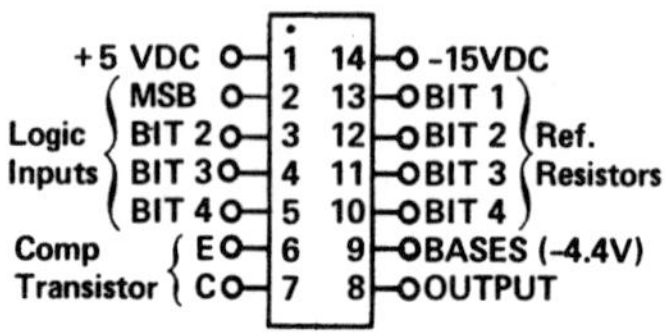

(B) Pin configuration.

Logic Input	Nominal Output Current (mA)
0 0 0 0	1.875
0 0 0 1	1.750
0 0 1 0	1.625
0 0 1 1	1.500
0 1 0 0	1.375
0 1 0 1	1.250
0 1 1 0	1.125
0 1 1 1	1.000
1 0 0 0	0.875
1 0 0 1	0.750
1 0 1 0	0.625
1 0 1 1	0.500
1 1 0 0	0.375
1 1 0 1	0.250
1 1 1 0	0.125
1 1 1 1	0.000

(C) Truth table.

	AD550	AD553
Switching Time	500 ns	60 ns to 90%
Settling Time (12 bits)	1.8 μs	200 ns
Reference Current	125 μA	125 μA
TC (typ) (L grade unit)	2ppm/°C	2 ppm/°C

(D) Key specifications.

Courtesy Analog Devices

Fig. 4-2. AD550, AD553 quad switches.

The circuit as shown in Fig. 4-3 may be used to construct 12-bit accurate converters, with suitable additional attention to details. As can be noted, the most significant quad switch unit is necessarily the highest accuracy grade (L), and the remaining two units can be of looser toler-

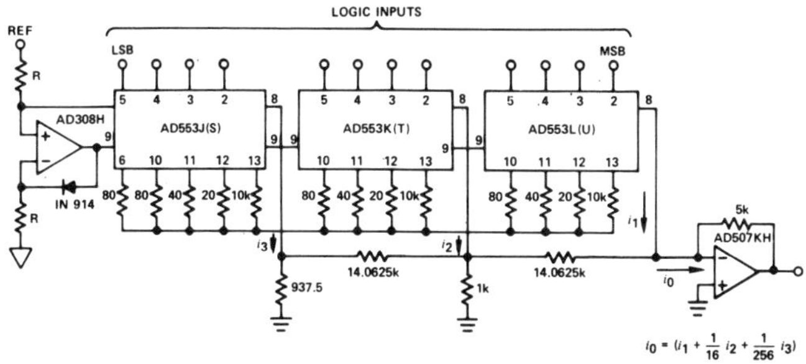

Courtesy Analog Devices

Fig. 4-3. AD553 (AD550) used in 12-bit converter.

ances (K and J). Not shown in this general circuit are power supply connections, which are +5 and −15V for the quad switches, and ±15V for the op amps. Also, appropriate frequency compensation must be used with the op amp(s) employed.

The ICL 8018A, 4550 Series Quad Switches

Another type of quad current switch which operates similarly to those just described is the Intersil ICL8018A and Burr-Brown 4550 series, shown in Fig. 4-5. The devices are actually very similar designs, as they were originally a joint development. As can be noted, these devices are pinned identically and have an identical truth table. The main performance difference between the AD550 and the ICL8018 or 4550 types is speed. The 8018A and 4550 switch in 40 ns, and have a settling time to 12 bits of 200 ns. This performance is essentially identical to the AD553.

These devices are also graded for accuracy, as required for 12, 8, or 4 bit uses. The 8018A or 4550L is 12-bit accurate; the 8019A or 4550K is 8-bit accurate; and the 8020A or 4550J is 4-bit accurate. A set of devices for a 12-bit converter would employ three of either of the two respective families.

A 12-bit d/a configuration using either 8018A or 4550 series devices is shown in Fig. 4-6. This circuit also uses an integral resistor network of 10K-20K-40K-80K construction, such as described previously. Again, this is a simplified diagram to illustrate the concept. Not shown are power connections and compensation for the output amplifier. Recommended devices are listed in the 8019 or 4550 data sheets.

A great many currently available hybrid and modular d/a converters are based on IC quad switches and this circuit topology, making such devices attractively priced due to the volume of use.

Further information on quad switches is contained in the chapter references. As will be seen from the device discussions which follow, many IC units employ similar arrangements, or variations on the basic scheme.

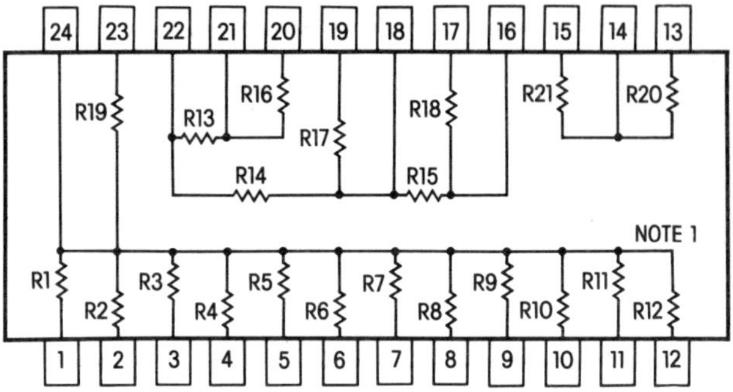

Note 1: Common bus resistor resistance shall be ≤150mΩ between pin 24 and the R11-R12 junction.

(A) Pinout (24 pin package).

R	Nominal Value (ohms)	REF. R	Max. Pwr. Dissipation (mW)	Accuracy
1	10K	-	20	±5.0% Abs.
2	20K	R1	10	0.015%
3	40K	R1	5	0.03%
4	80K	R1	2.5	0.05%
5	10K	R1	20	0.10%
6	20K	R1	10	0.20%
7	40K	R1	5	0.40%
8	80K	R1	2.5	0.80%
9	10K	R1	20	1.0%
10	20K	R1	10	1.0%
11	40K	R1	5	1.0%
12	80K	R1	2.5	1.0%
13	5K	R1	40	0.10%
14	14.0625K	R17	0.5	0.05%
15	14.0625K	R17	0.5	1.0%
16	5K	R1	40	0.10%
17	1K	-	8	±5.0% Abs.
18	937.5Ω	R17	8	1.0%
19	80K	R1	2.5	0.10%
20	6K	R1	12	0.10%
21	48.4K	R1	2	0.10%

(B) Specifications.

Analog Devices	AD1850	HyComp Box 250 Maynard, MA 01754	HC-130
Allen Bradley Milwaukee, WI 53204	FN226 + FN227 (2 packages)		
Hybrid Systems Burlington, MA 01803	HSR 4900 C	Micro Networks 324 Clark Street Worcester, MA 01606	MN142 + MN143 (2 packages)

(C) Manufacturers.

Fig. 4-4. A 12-bit "quad switch" resistor network.

Multiplying D/A Converters

A multiplying d/a converter is one in which the reference can be varied, and the d/a converter's output is a product of the reference voltage and the fractional digital input. This is a general form of definition, which defines the function performed by the multiplying d/a converter.

In the context used here, which is one generally applicable to IC d/a devices, we will extend this definition to specific forms of IC d/a types. There are a great many devices available for d/a use which consist of only a set of switches and a ladder network, plus in some cases a reference current amplifier. These devices omit the reference source and output amplifier, which would make up a complete d/a converter (if present).

While it can be said that such a device is incomplete in the strictest sense of the d/a function, there are many situations where it is desirable to the user to be able to adjust the reference to his exact needs, or to vary it on an ac basis. These are generally termed multiplying applications. And the absence of an output amplifier is not necessarily a disadvantage, in all cases.

This type of IC d/a converter is then called a "multiplying" type, as it uses an external reference, which can be variable. In most cases the reference is restricted to unipolar (or dc) voltage (currents), but in some devices the reference can be bipolar (or ac). For buffered voltage outputs, multiplying converters require an output op amp.

While the multiplying d/a converter offers a maximum of flexibility, it also requires more design time to interface the required external devices with the d/a unit itself to achieve a given function. Because of their relative simplicity, multiplying d/a converters tend to be low in cost.

There are a great number of multiplying IC d/a devices, in resolutions of 6 to 12 bits, and various accuracies. Many of them are readily available through multiple sources.

The MC1406/MC1506

The Motorola MC1406 (0-70°C) and MC1506 (−55 to +125°C) 6-bit multiplying d/a converter is illustrated in Fig. 4-7. This device series, one of the first IC d/a converters to be introduced, established an early pattern in IC converter development. Its general concept of operation is employed in several other similarly operating IC d/a devices in the Motorola line, as well as devices from other manufacturers.

The MC1406 consists of a reference current amplifier, a 6-bit R-2R ladder, and six current switches, along with associated support circuitry, as illustrated in Fig. 4-7A. In operation, the reference amplifier drives the R-2R ladder with a current equal to that applied externally to the V_{REF} (+) pin. The R-2R ladder scales this current into six binary-weighted levels, which are then switched on or off by the current switches.

Input logic coding is complementary binary, and is TTL logic compatible. For a reference current of 2 mA, the output full scale current is 2 mA. The output current is current sinking, with a small compliance

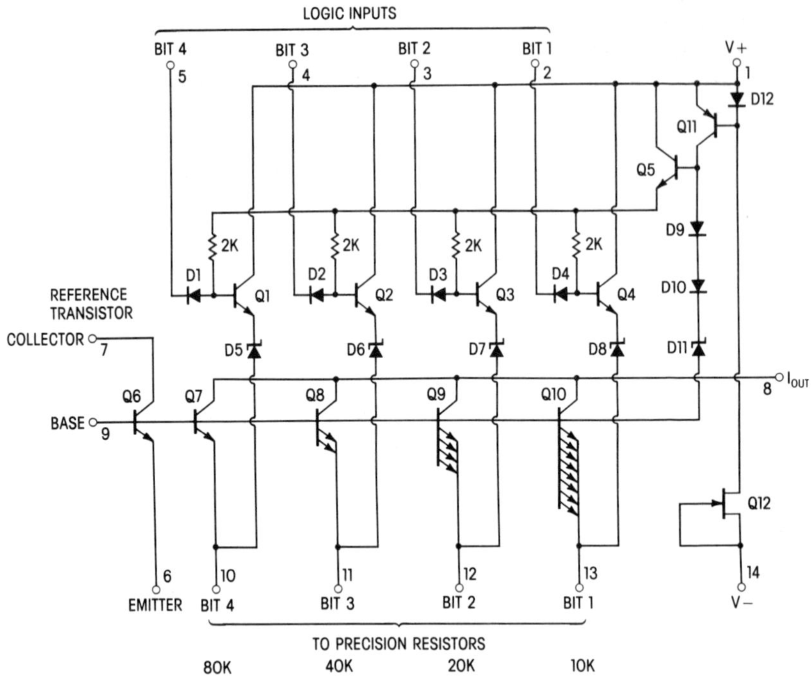

(A) Functional diagram.

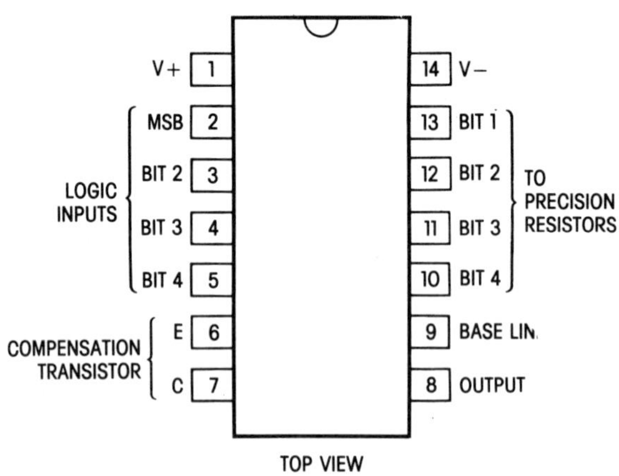

(B) Pin configuration.

Fig. 4-5. ICL8018A, 4550

voltage range. The device has a 150 ns settling time, and uses +5 and −15 volt supplies.

A summary of performance highlights is given in Fig. 4-7C. The MC1406 (MC1506) is supplied in a 14 pin DIP.

The MC1408/MC1508, MC3408

The Motorola MC1408 (0-70°C) MC1508 (−55 to +125°C) and MC3408 (0-70C) are 8-bit multiplying d/a converters, and are illustrated in Fig. 4-8. The MC1408 family of devices is very similar in concept of operation to the just described MC1406, with of course the increase of resolution to eight bits, and the attendant greater accuracy.

As can be noted from the block diagram of Fig. 4-8A, the MC1408 types contain a reference amplifier, an 8-bit R-2R ladder, and eight current switches. Input logic coding is binary coded, and TTL compatible. A reference current of 2 mA results in a full scale output current of 2 mA. The output is current sinking, with a +0.4 to −5 volt range. Settling time of this family is 300 ns, and it uses +5 and −15V supplies.

The MC1408 family is available in 8-, 7-, or 6-bit accuracy grades, as designated by an 8, 7, or 6 suffix to the basic part number (example:

Logic Input				Nominal Output Current (mA)
0	0	0	0	1.875
0	0	0	1	1.750
0	0	1	0	1.625
0	0	1	1	1.500
0	1	0	0	1.375
0	1	0	1	1.250
0	1	1	0	1.125
0	1	1	1	1.000
1	0	0	0	0.875
1	0	0	1	0.750
1	0	1	0	0.625
1	0	1	1	0.500
1	1	0	0	0.375
1	1	0	1	0.250
1	1	1	0	0.125
1	1	1	1	0.000

(C) Truth table.

Switching Time	40 ns	Reference Current	125 μA
Settling Time (12 bits)	200 ns	TC (typ) (ICL8018A or 4550L)	2 ppM/°C

(D) Key specifications.

series quad switches.

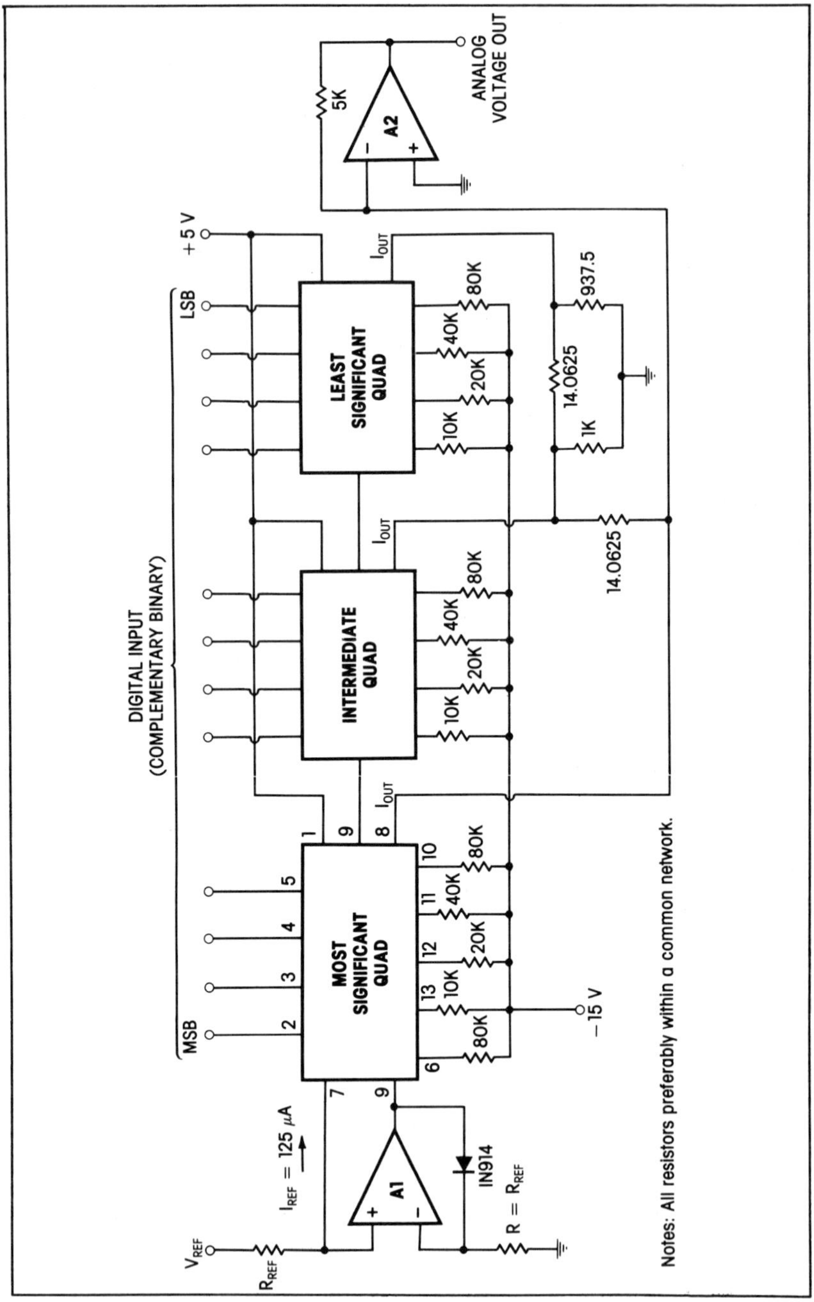

Fig. 4-6. The 8019A or 4550 series used in a 12-bit converter.

MC1408P8). The MC1508 is 8-bit accurate. The MC3408 is a lower cost device, with a 0.5% accuracy.

A summary of performance highlights is given in Fig. 4-8E. The devices are supplied in 16 pin DIPs. The MC1408 and MC1508 series are second sourced by a number of manufacturers. A device data sheet is included in the appendix.

The MC3410/MC3510

The Motorola MC3410 (0 to 70°C) and MC3510 (−55 to +125°C) are 10-bit multiplying d/a converters, and are illustrated in Fig. 4-9. These devices are generally similar in operation to the MC1406 and MC1408 series, but with 10-bit resolution and increased accuracy.

As can be noted from the block diagram of Fig. 4-9A, the MC3410 series contains a reference current amplifier, a 10-bit R-2R ladder, and ten current switches. Input logic is binary coded, and TTL compatible. In the MC3410/MC3510 devices there is a scaling factor of two between the reference current and full scale output; thus, a 2 mA reference current results in a 4 mA full scale current output. The output is current sinking, with a +0.2 to −2.5 volt range. Settling time for the devices is 250 ns, and they use +5 and −15 volt power supplies.

The MC3410 is available in accuracy grades of both 10 and 9 bits. A summary of performance highlights is given in Fig. 4-9E. The device is supplied in a 16 pin DIP.

As may be noted from the above discussions on these three types of devices, they are all quite similar in concept of operation. They all require an external 0 to 2 mA reference current for operation, which may be linearly varied for multiplying applications. All devices have a basic current output; a voltage output requires an external amplifier.

The DAC-08

The Precision Monolithics DAC-08 8-bit multiplying d/a converter series is illustrated in Fig. 4-10. This series is similar to the MC1408 series in block form in some regards, but does have some important major differences.

Similarities to the MC1408 series are multiplying operation, the reference current amplifier configuration, and the R-2R ladder driving current switches. In the DAC-08, however, the current switches operate differentially, providing both true and complementary current outputs. These outputs have a very wide voltage compliance, −10 to +18 volts. The settling time of the DAC-08 is quite fast, 85 ns. Its current reference range for monotonic operation is 100 μA to 4 mA (a 40/1 range). Power supply range is ±4.5 to ±18 volts.

An additional useful device feature is logic programmability, which enables the binary-coded input thresholds to be adjusted for various logic families.

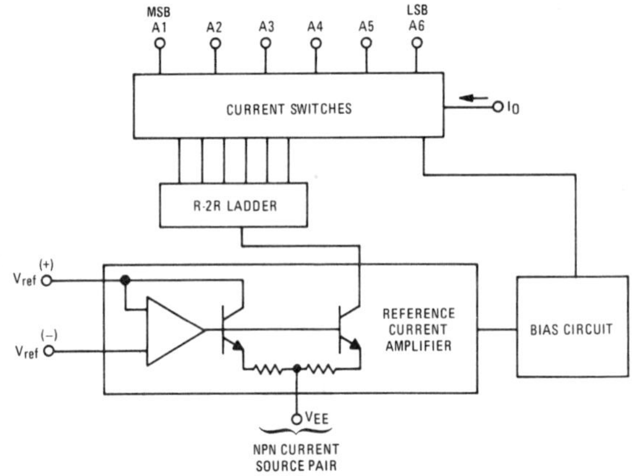

(A) Functional diagram.

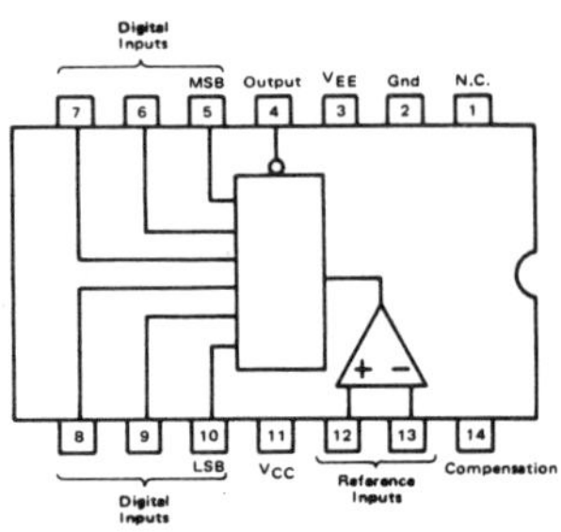

(B) Pin configuration.

Settling Time 150 ns
2 mA Output Current
+5, −15 V Supplies
Complementary Binary TTL Inputs
Multiplying Operation
Positive or Negative Reference
Wide Reference Bandwidth

(C) Key specifications and features.

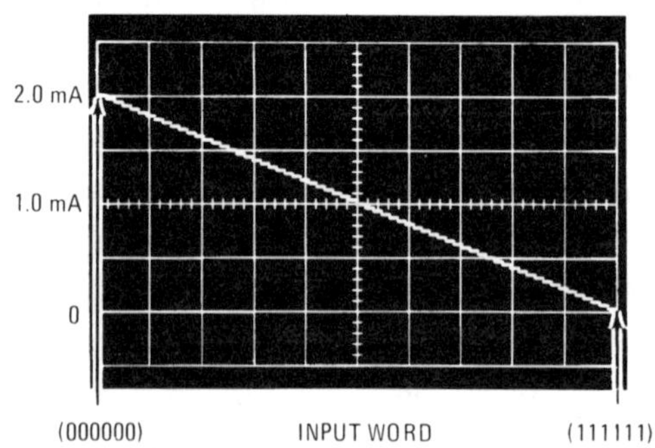

(D) D-to-A transfer characteristics.

Fig. 4-7. The MC1406/MC1506

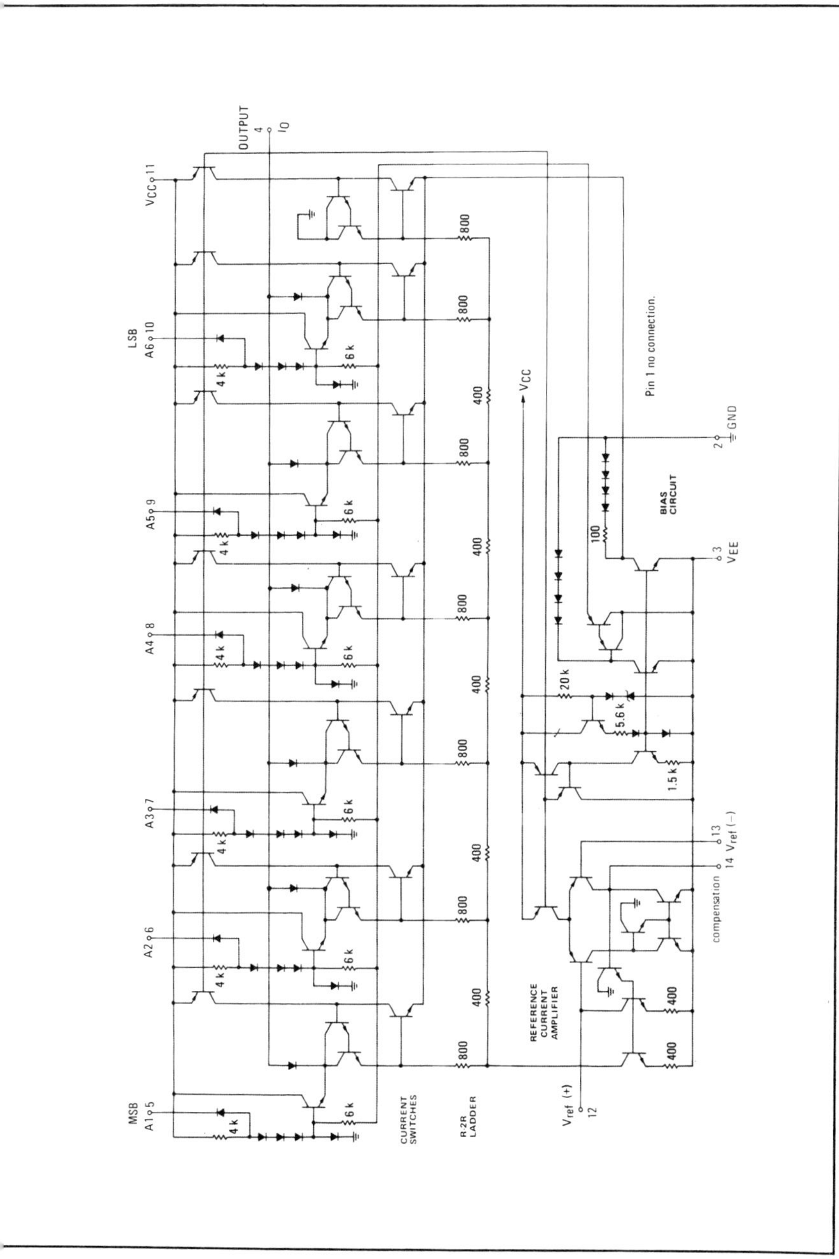

(E) Circuit schematic.

Courtesy Motorola Semiconductor Products, Inc.

6-bit multiplying d/a converters.

DAC-08 devices are available as DAC-08H, DAC-08E, and DAC-08C (0 to 70°C) with nonlinearities of +0.1, +0.19 and +0.39%, and the DAC-08A and DAC-08 (−55 to +125°C) with nonlinearities of 0.1 and 0.19%.

A summary of performance highlights is given in Fig. 4-10D. The DAC-08 series is supplied in a 16 pin DIP. The series is second sourced by a number of manufacturers. A device data sheet is included in the appendix.

The DAC-20

The Precision Monolithics DAC-20 2 digit bcd multiplying d/a converter is illustrated in Fig. 4-11. This device is similar in many regards to the just described DAC-08, with the exception that the DAC-20 is coded for 2 digit bcd operation instead of binary, as in the DAC-08. All

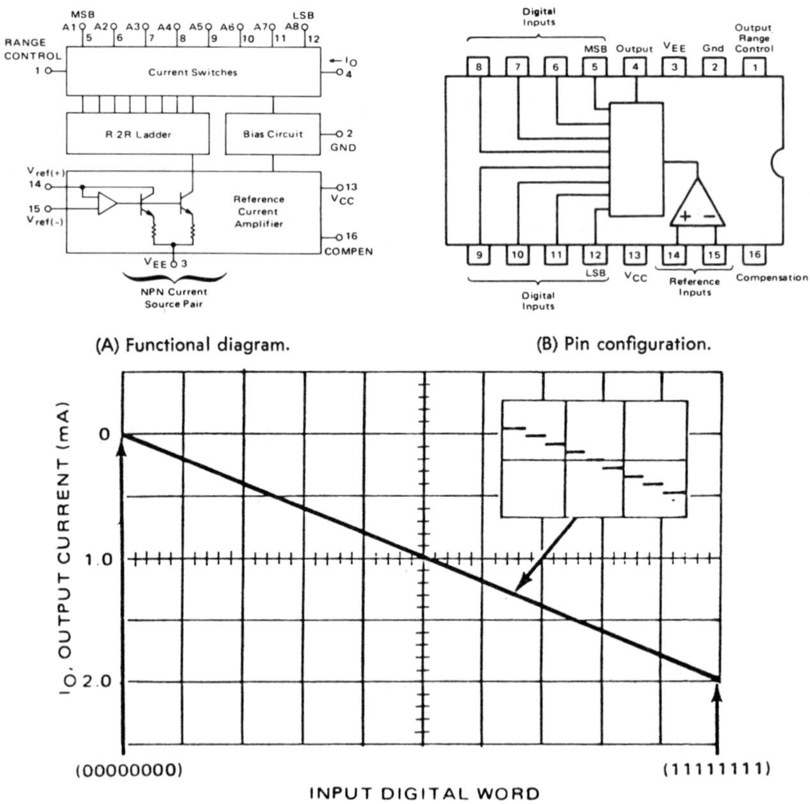

(A) Functional diagram.

(B) Pin configuration.

(C) D-to-A transfer characteristics.

Fig. 4-8. MC1408/MC1508, MC3408

specifications of the DAC-20 are equal to the DAC-08, except those related to coding and linearity.

There are four device grades available; the DAC-20C and DAC-20E are for 0 to 70°C operation with maximum nonlinearities of ±¼ and ±½LSB, and the DAC-20A and DAC-20 are for −55 to +125°C with similar (respective) linearities.

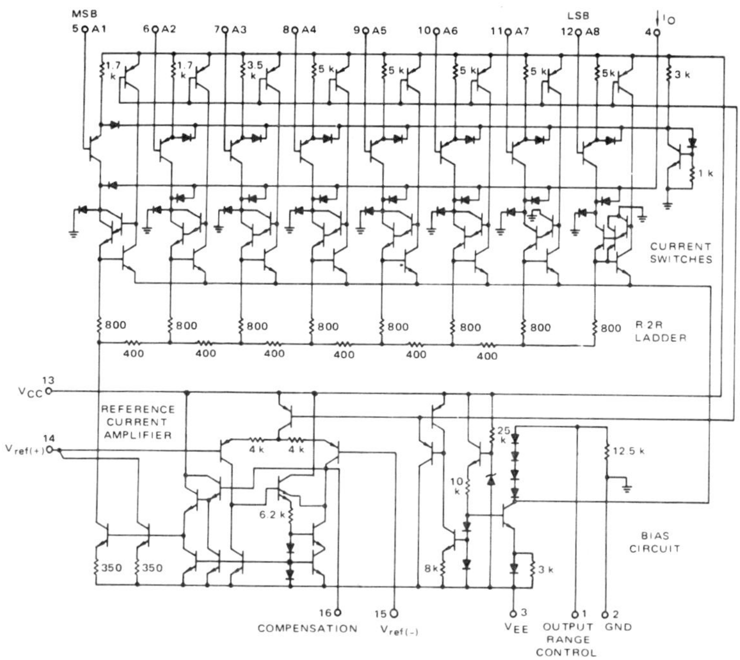

(D) Circuit schematic.

Settling Time – 300 ns
Output Voltage Compliance – +0.4 to −5 V
Various Accuracy Grades – 8, 7 and 6 bits
2 mA Output Current
+5, −15 V Supplies
Binary TTL Inputs
Multiplying Operation
Positive or Negative Reference
Wide Reference Bandwidth

(E) Key specifications and features.

Courtesy Motorola Semiconductor Products, Inc.

8-bit multiplying d/a converters.

A summary of performance highlights is given in Fig. 4-11D. The DAC-20 series is supplied in a 16 pin DIP.

The AD7520 Series

The Analog Devices AD7520 series of 10- and 12-bit multiplying d/a converters includes a number of devices, which are illustrated in Fig. 4-12.

The AD7520 (AD7521) consists of an inverted R-2R ladder of a 10 (or 12) bit length, and 10 (or 12) CMOS current switches. The logic switches direct the ladder current to either bus $I_{out\ 2}$ (ground) or $I_{out\ 1}$, the output bus. An external voltage reference is in the range of ±10V is applied to the ladder, and an external op amp is used in the

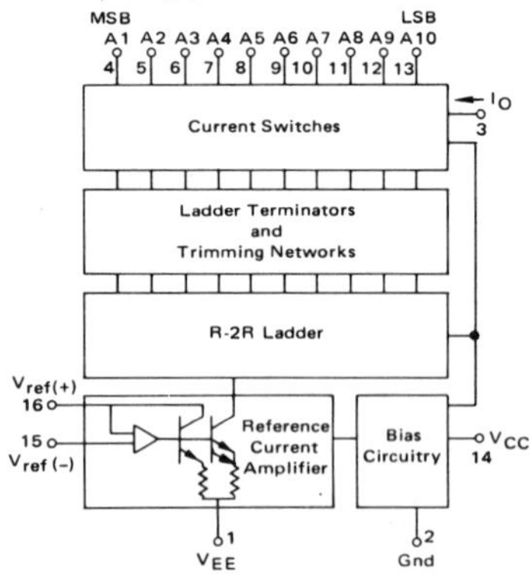

(A) Functional diagram.

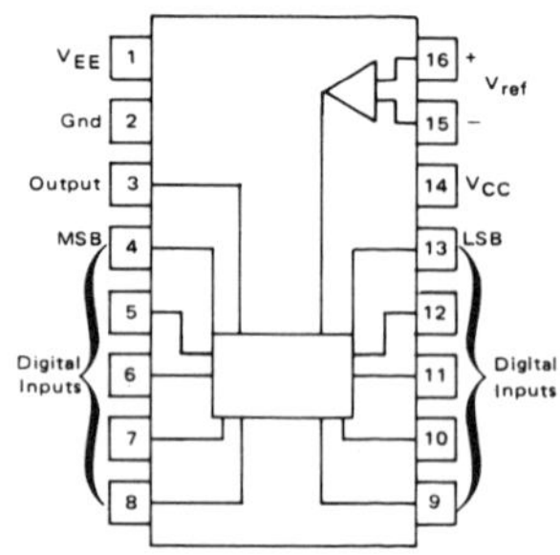

(B) Pin configuration.

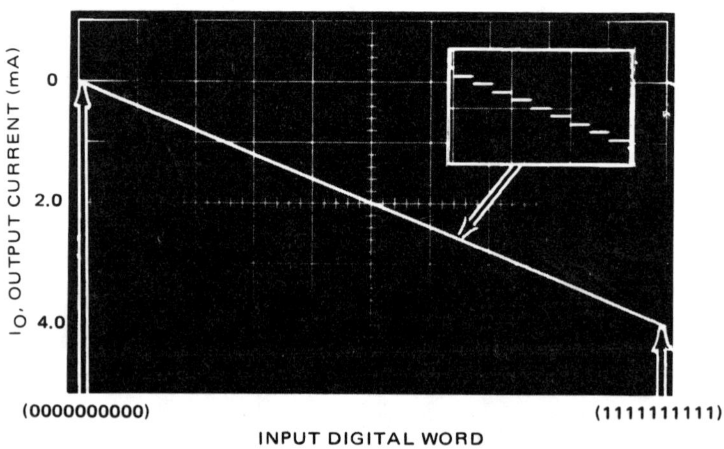

(C) D-to-A transfer characteristics.

Fig. 4-9. The MC3410/MC3510 10-bit

current-to-voltage mode at bus $I_{out\ 1}$. An on-chip 10K resistor serves as this amplifier's feedback path, allowing the basic operation of the device to be ratiometric.

The AD7520 was the forerunner of the device series and features 10-bit resolution with 10-, 9- and 8-bit nonlinearities, denoted by suffixes L, K, and J respectively for 0 to 75°C applications, and S, T, and U suffixes for −55 to 125°C. The AD7521 is a 12-bit resolution device with similar nonlinearity gradings. The AD7530 is a low-cost version of the AD7520 for 0 to 75°C temperature ranges, with similar linearity errors, but slightly higher feedthrough error. The AD7531 is a 12-bit resolution device, with specs otherwise similar to the AD7530.

Input coding for all devices is binary, and both TTL and CMOS compatible (for any supply voltage). The devices feature very low power

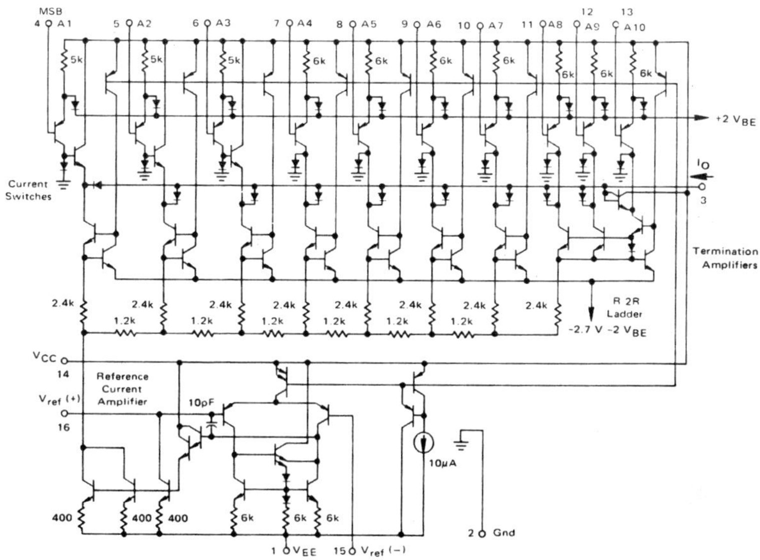

(D) Equivalent circuit schematic.

Settling Time – 250 ns
Output Voltage Compliance – +0.2 to −2.5 V
Various Accuracies – 10 or 9 bit
4 mA Output Current
+5, −15 V Supplies
Binary TTL Inputs
Multiplying Operation
Positive or Negative Reference
Wide Reference Bandwidth

(E) Key specifications and features.

Courtesy Motorola Semiconductor Products, Inc.

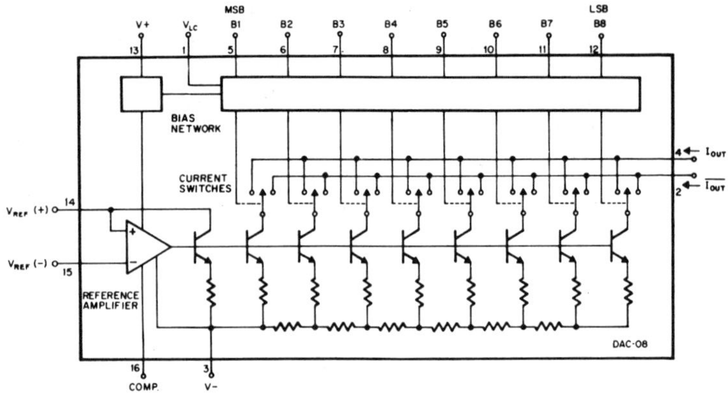

(A) Functional diagram.

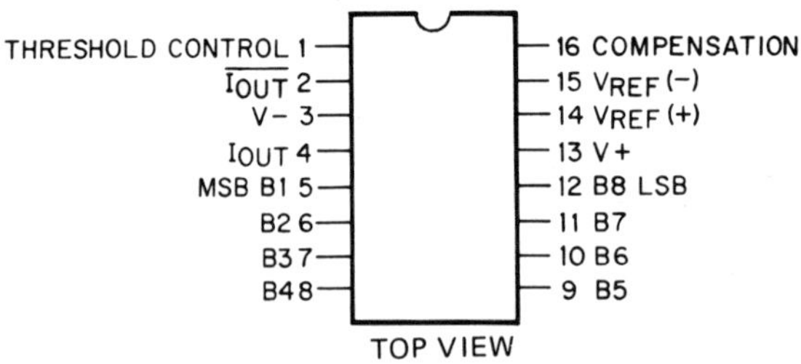

(B) Pin configuration.

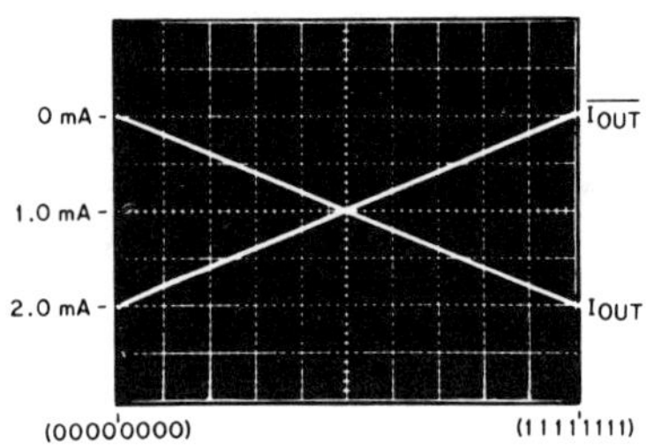

(C) Transfer characteristics for complementary outputs.

Nonlinearity - 0.1 to 0.39%
Settling Time - 85 ns
Gain TC-100 ppm/°C
Voltage Compliance - −10 V to +18 V
Monotonic Multiplying
Range - 100 μA to 4 mA
Differential Current Outputs
Binary Coding
Programmable Logic Threshold
Supply Range ±4.5 V to ±18 V

(D) Key specifications and features.

Courtesy Precision Monolithics, Inc.

Fig. 4-10. DAC-08 8-bit multiplying d/a converter.

operation, typically 20 mW. Both linearity and gain TC errors are quite low, being 2 and 10 ppm/°C, respectively. The devices have a settling time of 500 ns, and low feedthrough error.

The AD7520 series is designed to be used with external reference voltages, which can be either dc or ac. This unique feature makes the series well suited to programmable attenuation and gain stages. The external op amp, which must be used with the device, will largely determine its overall dynamic performance for the voltage output mode.

A summary of performance highlights is given in Fig. 4-12C. The devices are supplied in 16 pin (AD7520, AD7530) and 18 pin DIPs (AD7521, AD7531). The AD7520 device series is second sourced by other manufacturers, and a data sheet is included in the appendix. Recently announced devices to this family are the AD7523 and AD7541 series.

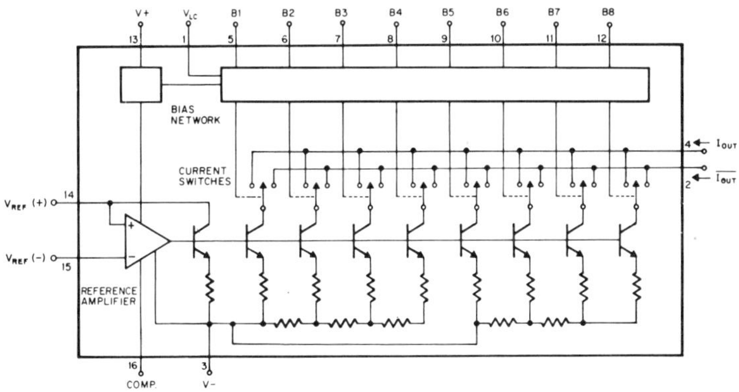

(A) Functional diagram.

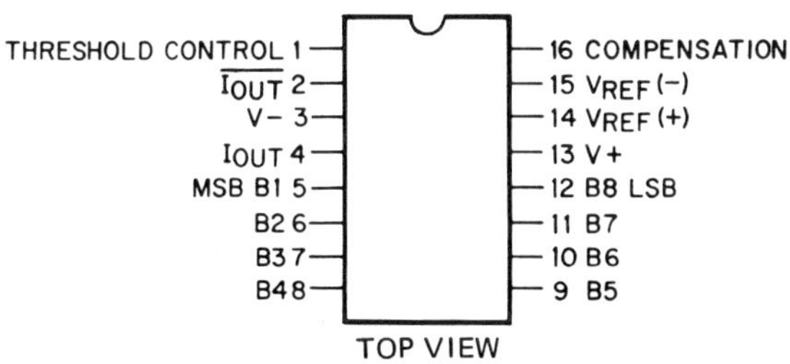

(B) Pin configuration.

(C) Key specifications and features.

Nonlinearity – ¼ to ½ LSB
Settling Time – 85 ns
Gain TC – 10 ppm/°C
Voltage Compliance – −10 V to +18 V
Monotonic Multiplying Range - 100 μA to 4 mA
Differential Current Outputs
Bcd Coding
Programmable Logic Threshold
Supply range ±4.5 to ±18 V

Courtesy Precision Monolithics, Inc.

Fig. 4-11. DAC-20 2 digit bcd multiplying d/a converter.

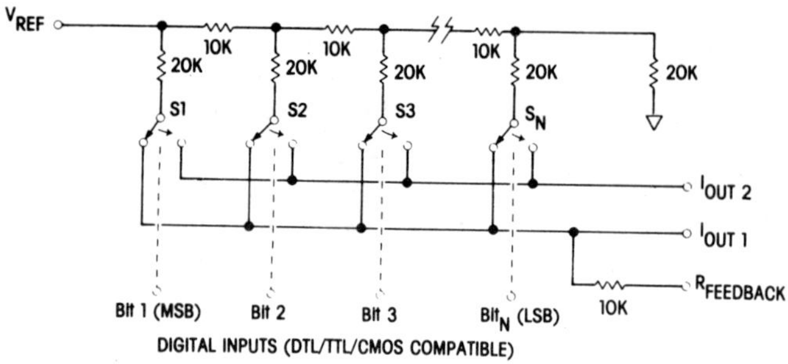

AD7520: N=10
AD7521: N=12
Logic: A switch is closed to $I_{OUT\ 1}$ for its digital input in a "HIGH" state.

(A) Functional diagram.

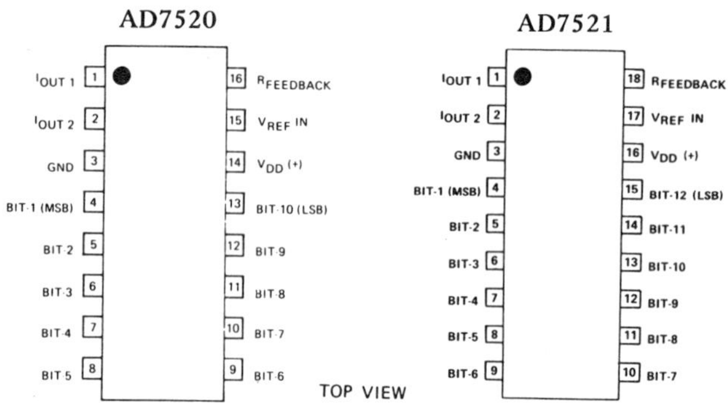

(B) Pin configuration.

Settling Time – 500 ns
TC - 2 ppm/°C Linearity, 10 ppm/°C Gain
Feedthrough – 0.05% @ 100 kHz
Various Accuracies and Resolutions
Single Supply Operation, +5 to +15 V
Wide Multiplying Range, Bipolar Reference (±10 V)
Ratiometric Operation
Low Current, TTL/CMOS Inputs

(C) Key specifications and features.

Courtesy Analog Devices

Fig. 4-12. AD7520, AD7521 (AD7530, AD7531) 10- and 12-bit multiplying d/a converters.

The AD7522

The Analog Devices AD7522 10-bit multiplying d/a converter is a device which is essentially like the previously described AD7520, but with the addition of double-buffered logic input registers. It is illustrated in Fig. 4-13.

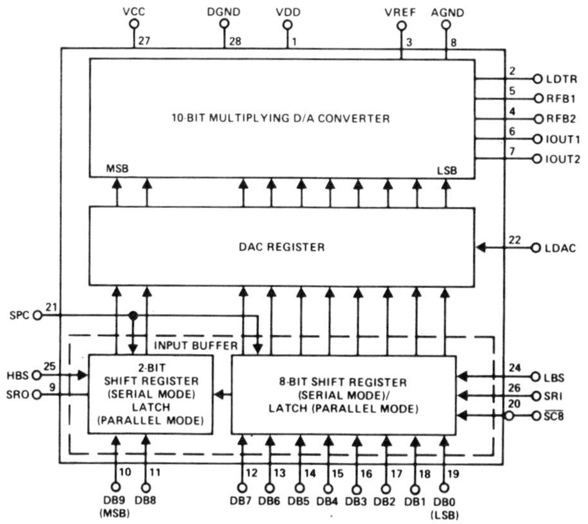

(A) Functional diagram.

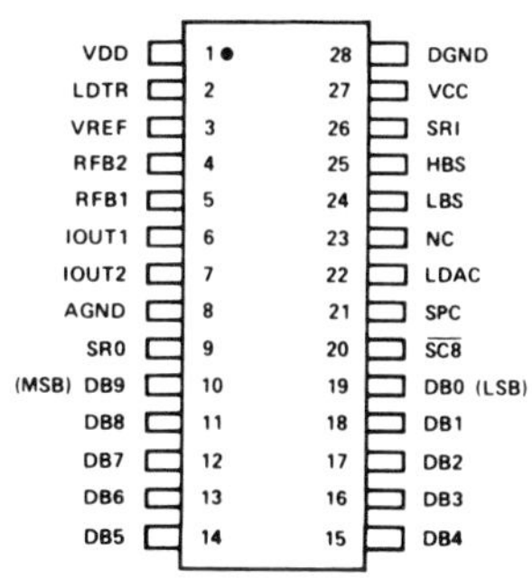

(B) Pin configuration.

Settling Time - 500 ns
TC - 2 ppm/°C Linearity, 10 ppm/°C Gain
Feedthrough - 0.05% @ 10 kHz
Various Accuracies Available; 8, 9, 10 bits
Single Supply Operation
Wide Multiplying Range, Bipolar Reference (±10 V)
Ratiometric Operation
Low Current, TTL/CMOS Inputs
Double Buffered Input Resistors
Serial or Parallel Leading

(C) Key specifications and features.

Courtesy Analog Devices

Fig. 4-13. AD7522 10-bit multiplying d/a converter.

The AD7522 consists of an inverted R-2R ladder with a resolution of 10 bits, like the basic AD7520. It does, however, have additional feedback taps for analog scaling flexibility. Analog operation is very much akin to the AD7520.

The main difference of the AD7522 is the addition of the double buffered input registers, which can be loaded in either serial or parallel fashion. Logic input load capability is in two bytes of 8 and 2 bits, for 8-bit bus capability. The device operates from a single supply, and is either TTL or CMOS compatible.

The AD7522 is available in three accuracy/linearity grades, and two basic temperature ranges. The AD7522 L, K, and J are 10-, 9-, and 8-bit devices for 0 to 75°C operation, with the AD7522 U, T, and S are analogous accuracy units for −55 to 125°C operation.

A summary of AD7522 performance highlights is given in Fig. 4-13C. The devices are supplied in 28 pin DIPs.

The AD562

The Analog Devices AD562 12-bit multiplying d/a converter is illustrated in Fig. 4-14. This device is one of the most accurate multiplying IC d/a converters available, with nonlinearities as low as ¼ LSB, and very low TCs for gain and linearity error.

As noted in Fig. 4-14A, the AD562 is constructed with a two-chip approach; one chip being composed of the active devices, the other including a thin film resistor network. The thin film resistor array is actively laser trimmed with the device in operation, a feature which allows the AD562 to achieve its high overall accuracy.

Fig. 4-14B shows the functional schematic diagram of the AD562. It is composed of a reference control amplifier, a set of twelve current mode switches with scaled currents, an output summing network, and the application resistors required for various output voltage/coding arrangements. Binary or bcd coding is achieved by distributing the weighting between the current switches and the interquad dividers. The device uses a positive supply in the range of 5 to 15 volts, and a 15-volt negative supply. Full scale output current is 2 mA for an input reference current of 0.5 mA.

Logic levels are programmed by a threshold adjust pin, and may accommodate either TTL or CMOS levels. Binary or bcd coding is available by specifying the appropriate part number suffix wanted, either BIN or BCD.

The AD562 is available in various temperature ranges and accuracy grades. The AD562K is a 0 to 70°C range device, with ½ LSB nonlinearity; the AD562A is for −25 to 85°C with identical nonlinearity. The AD5625 is used for a temperature range of −55 to 125°C, with ¼ LSB nonlinearity.

All devices have a 3 ppm/°C (max) gain TC. The AD562K and AD562A have a 2 ppm/°C typical differential nonlinearity TC, while in the AD562S it is held to 1 ppm/°C. Monotonicity is guaranteed for all devices over their respective temperature ranges.

Output voltage ranges are pin programmable, through access to the internal application resistors. Ranges available are 0 to +5 volts, 0 to +10 volts, ±2.5 volts, ±5 volts, and ±10 volts.

A summary of performance highlights is given in Fig. 4-14D. The AD562 series is supplied in a 24 pin DIP. The device is second sourced by other manufacturers.

The HI-562

The Harris Semiconductor HI-562 high speed 12-bit multiplying d/a converter is illustrated in Fig. 4-15. This device, as its model number suggests, is similar to the AD562 in terms of function. However, it is quite different in terms of actual specifications, as it is nearly an order of magnitude faster in speed.

Although the internal circuitry of the HI-562 is quite a bit different than the AD562, it is pinned similarly. The HI-562 uses a low impedance R-2R ladder technique, with a relatively high current of 2.5 mA per bit. This high current, the low impedance ladder, the low parasitics and high speed of the dielectrically isolated circuitry combine to allow a typical settling time of 200ns (12 bits).

The HI-562 is available in various temperature ranges and accuracy grades. The HI-562 is a 0 to 70°C device, with a ½ LSB nonlinearity; the HI-562-4 is for −25 to +85°C with the same nonlinearity. The HI-562-2 is for −55 to 125°C, with ½ LSB nonlinearity.

The HI-562-4 and -5 have a 3 ppm/°C (max) gain TC, and the HI-562-2 has a TC of 2 ppm/°C, respectively. Monotinicity is guaranteed for all devices, over their respective temperature ranges.

A summary of performance highlights is given in Fig. 4-15D. The HI-562 series is supplied in a 24 pin DIP.

The DAC-76

The precision Monolithics DAC-76 companding d/a converter is illustrated in Fig. 4-16. This device is a unique type of d/a converter, with a transfer characteristic which is nonlinear in nature. It is intended for telecommunications applications which employ a signal compression/expansion scheme (companding) for increased dynamic range.

The DAC-76 is composed of the elements illustrated by the block diagram in Fig. 4-16A. It is intended for both d/a and a/d applications, and thus has an encode/decode control input (E/D), and two corresponding output pairs I_{OE} and I_{OD}. These are current outputs, a single one (+ or −) of which is selected as active by the sign bit input (SB).

The exponential coding of the DAC-76 is generated by a nonlinear current output d/a converter. This d/a converter has 7-bit control inputs, which are designated as three chord select inputs (B1 to B3) and four step-select inputs (B4 to B7). The d/a transfer characteristic is then a segmented series of eight chords, each of which has a series of sixteen steps. This characteristic is shown graphically in Fig. 4-16C.

In Fig. 4-16C are shown both a/d and d/a conversion transfer characteristics. As can be noted, the composite combination of these two complementary curves will result in a straight-line (linear) characteristic. In the summary table is shown information relating to step size, and the equivalent binary d/a resolution.

The DAC-76 exponential 7-bit + sign control is equivalent to 12-bit + sign operation, in terms of total dynamic range, and resolution near zero. This dynamic range is 72 dB, defined as 20 log 1111111/0000001.

Although its coding is radically different than a conventional binary d/a converter, there are some related specifications. Montonicity is speci-

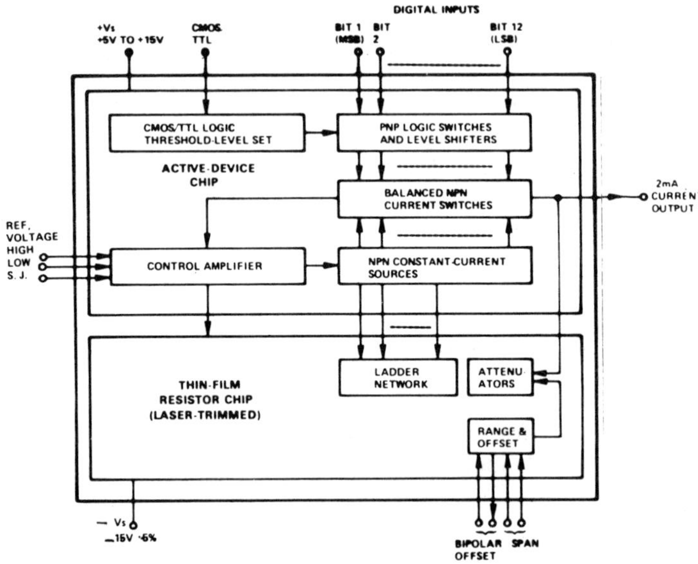

(A) Block diagram.

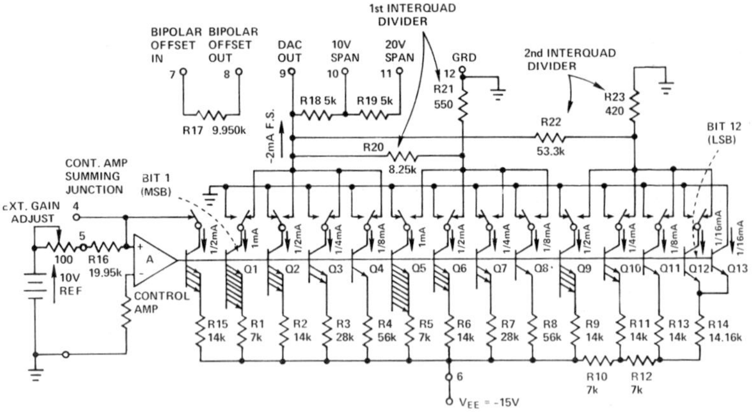

(B) Functional diagram.

Fig. 4-14. AD562 12-bit

fied over the complete range of 128 steps. Differential linearity is also specified, in terms of a fractional step. For the DAC-76B and DAC-76 (−55 to +125°C range) differential linearity is ±½ and ±1 step, respecively. For the DAC-76 and DAC-76C (0 to 70°C range) it is also ±½ and ±1 step, respectively. The full scale current output of the DAC-76 is 2 mA and is realized with a reference input current of 0.5 mA. Settling time is 500 ns to ½ step. Output voltage compliance is −5 to +18V. The device uses supplies of ±12 to ±18 volts.

A summary of performance highlights is given in Fig. 4-16D. The DAC-76 series is supplied in a 18 pin DIP. The device is second

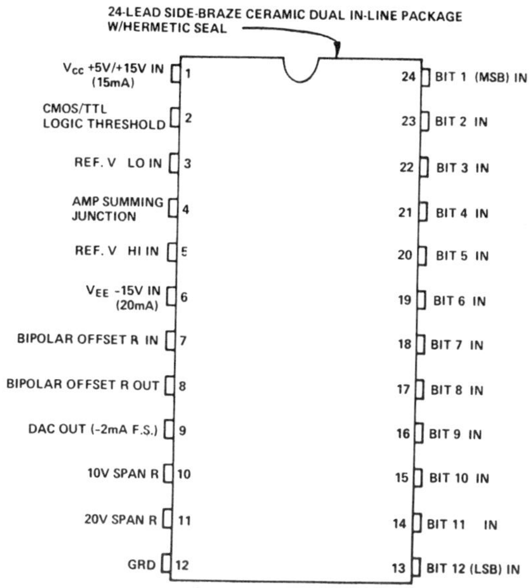

(C) Pin configuration.

Nonlinearity – 0.0048 to 0.012%
(¼ to ½ LSB)
Settling Time – 3.5 μs
TC - 3 ppm/°C Gain, 2 ppm/°C Linearity
Fully Trimmed Accuracy
Multiplying Operation
Ratiometric Operation
Programmable Ranges (Includes Application Resistor)
Binary or BCD Coding
Programmable Logic Threshold
2 mA I_o ($I_o = 4 I_R$)
Supplies +5 to +15 V, −15 V

(D) Key specifications and features.

Courtesy Analog Devices

multiplying d/a converter.

sourced by other manufacturers.

Internal Reference D/A Converters

An internal reference d/a converter is one which includes both a reference circuit and the d/a switches, plus their associated ladder.

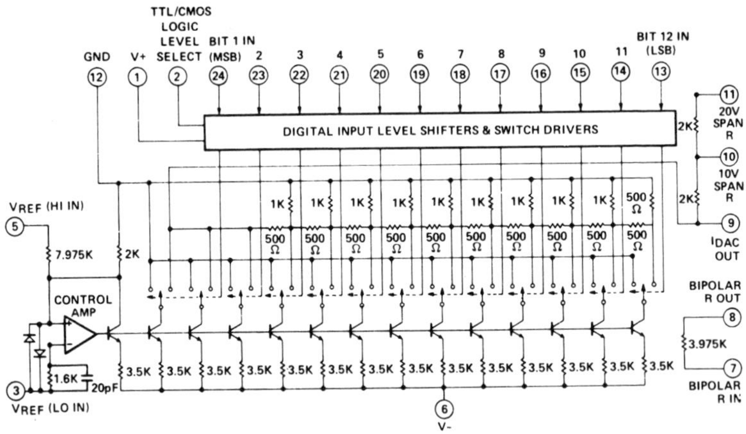

(A) Functional diagram.

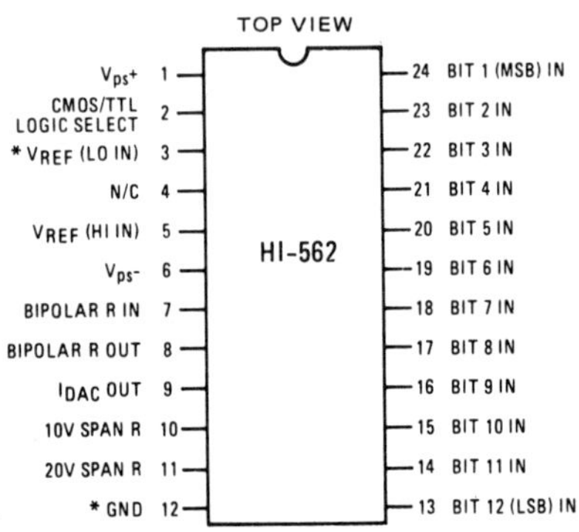

(B) Pin configuration.

* Pins 3 and 12 connected to top and bottom case for high frequency shielding.

Settling Time - 200 ns
TC - 3 ppm/°C Gain,
3 ppm/°C Linearity
Fully Trimmed Accuracy
Multiplying Operation
Ratiometric Operation

Programmable Ranges (Includes Application Resistors)
Binary Coding
Programmable Logic Threshold
5 mA I_o
Supplies +5, −15 V

(C) Key specifications and features.

Courtesy Harris Semiconductor

Fig. 4-15. HI-562 high-speed 12-bit multiplying d/a converter.

Although it does not include an output op amp, it may still be considered functionally complete for some applications. An internal reference d/a converter has an output in the form of a current. It can also include application resistors to use for output voltage scaling and offset purposes (in most cases should, for best accuracy).

Dependent upon the voltage compliance limits of the d/a converter the current output can often be used to directly produce an unbuffered voltage by means of a simple resistive load to ground. In general, this type of connection will result in the fastest settling time—that of the d/a converter itself. This settling time is often strongly dependent upon the load resistance value.

A good example of how an internal reference, current output d/a converter can be used to advantage is within feedback-type a/d converters. In such applications the output current can be compared against the analog input (in current form), and this scheme will yield the best or lowest conversion time, due to faster operation of the d/a converter for current outputs. Consequently this form of operation is almost always used in successive approximation a/d converters.

Alternately, the current output of the d/a converter can be applied to an external op amp for amplification to produce a voltage output. Most often this op amp is used in the current-to-voltage (inverting) mode, but in some cases it may be desirable to operate it in a noninverting mode. It should be noted that low voltage compliance d/a converters can limit flexibility in this regard as they can effectively constrain the d/a converter to inverting op amp only uses, due to limited voltage compliance.

Regardless of which buffer mode is to be used, it is highly advantageous to also have the gain scaling and offset resistors integral to the d/a converter. This setup will result in the lowest overall TCs, and thus highest accuracy. Typically, these resistors are part of the same network as the internal ladder, so they will provide optimum temperature tracking characteristics. External jumper options using these resistors can be used to produce the popular output voltage arrangements to arrive at several ranges, such as 0 to +5 volts, 0 to +10 volts, ±2.5 volts, ±5 volts and ±10 volts.

The DAC-100

The Precision Monolithics DAC-100 10-bit internal reference d/a converter is illustrated in Fig. 4-17. This device is available in a broad range of accuracies, TCs and output ranges, making it one of the more flexible d/a converters available.

The DAC-100 is composed of an internal reference, ten-bit switches, an R-2R ladder, and output scaling resistors, as shown in Fig. 4-17A. The DAC-100 current switches are arranged for complementary binary coding and are compatible with 5-volt logic, both TTL and CMOS. The current sinking output is a nominal 2 mA, but the output mode is designed to work into a virtual ground using an external current-to-voltage converter op amp. Output voltage is chosen as 10 volts or 5 volts by part number suffix (see Fig. 4-17A). Supply voltage range is ±6 to ±18 volts.

A wide range of basic nonlinearity and gain TC options are available with the DAC-100, as indicated by various part number suffixes. These are respectively; A = 0.05% and 15 ppm/°C; B = 0.1% and 30 ppm/°C; C = 0.2% and 60 ppm/°C; and D = 0.3% and 120 ppm/°C.

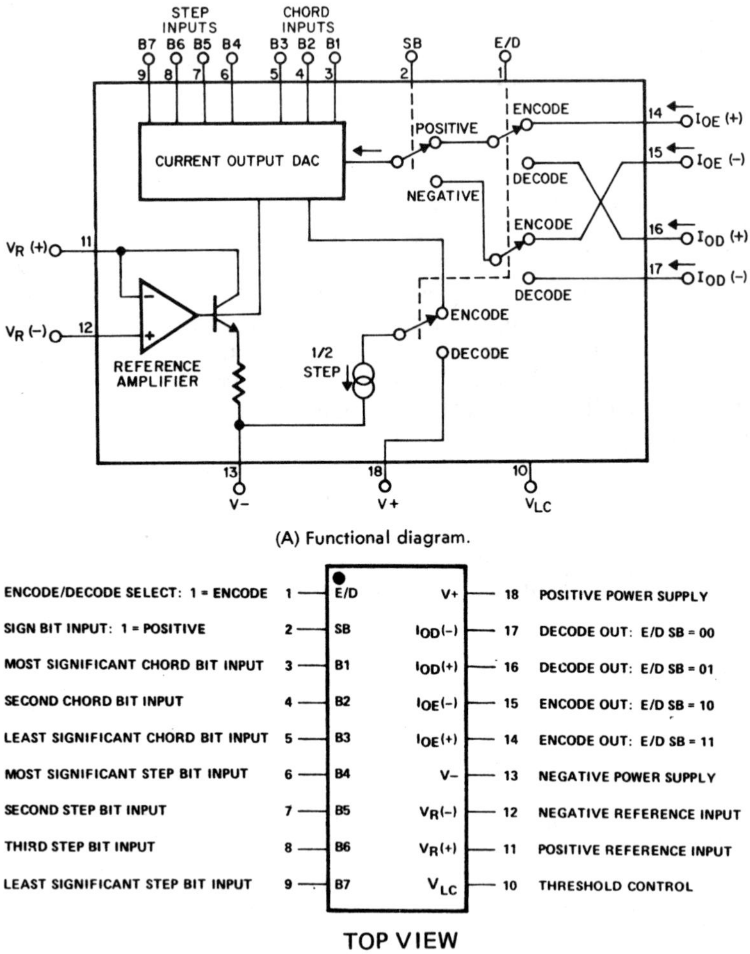

(A) Functional diagram.

(B) Pin configuration.

Fig. 4-16. DAC-76 companding

These suffixes may be mixed if desired; a DAC-100 AB for instance would be 0.05% nonlinearity and 30 ppm/°C TC. All devices have 10-bit resolution, but as may be noted from the nonlinearities, monotonicity is guaranteed to be consistent with 10, 9, 8, and 7 bit operation, for A, B, C, and D suffixes, respectively.

Three basic temperature range devices are available 0 to +70°C, −25 to +85°C, and −55 to +125°C. A summary of performance highlights is given in Fig. 4-17C. The DAC-100 series is supplied in a 16 pin DIP.

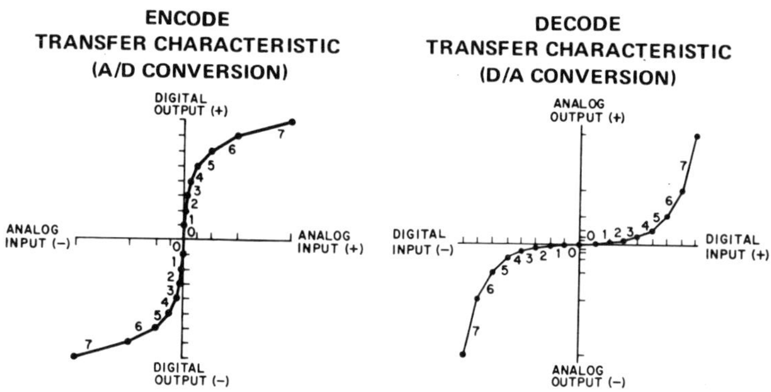

STEP SIZE SUMMARY TABLE DECODE OUTPUT (SIGN BIT EXCLUDED)

CHORD	STEP SIZE NORMALIZED TO FULL SCALE	STEP SIZE IN µA WITH 2007.75 µA F.S.	STEP SIZE AS A % OF FULL SCALE	STEP SIZE IN dB AT CHORD ENDPOINTS	STEP SIZE AS A % OF READING AT CHORD ENDPOINTS	RESOLUTION & ACCURACY OF EQUIVALENT BINARY DAC
0	2	0.5	0.025%	0.60	6.67%	SIGN + 12 BITS
1	4	1.0	0.05%	0.38	4.30%	SIGN + 11 BITS
2	8	2.0	0.1%	0.32	3.65%	SIGN + 10 BITS
3	16	4.0	0.2%	0.31	3.40%	SIGN + 9 BITS
4	32	8.0	0.4%	0.29	3.28%	SIGN + 8 BITS
5	64	16	0.8%	0.28	3.23%	SIGN + 7 BITS
6	128	32	1.6%	0.28	3.20%	SIGN + 6 BITS
7	256	64	3.2%	0.28	3.19%	SIGN + 5 BITS

(C) Transfer characteristics.

Dynamic Range – 12 bits + Sign (72 dB)
Settling Time – 500 ns
Voltage Compliance – −5 to +18 V
8-Bit Control (7 Bits + Sign)
Multiplying Operation
Sign Control
Exponential Coding – Bell System $\mu - 255$ Law
Programmable Logic Threshold
Encode or Decode Operation
2 mA I_O ($I_O \cong 4\ I_R$)
Supplies ±15 V

(D) Key specifications and features.

Courtesy Precision Monolithics, Inc.

d/a converter.

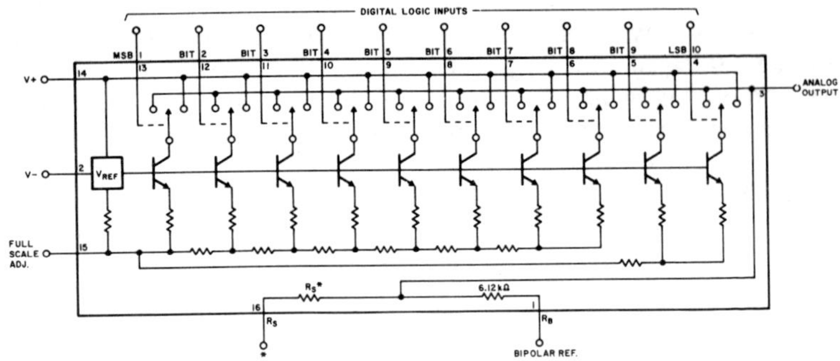

*For 10V or ±5V Operation R_S = 4.88kΩ (Package Q1, Q3, Q5, Q7)
For 5V or ±2.5V Operation, R_S = 2.44kΩ (Package Q2, Q4 or Q6, Q8)

(A) Functional diagram.

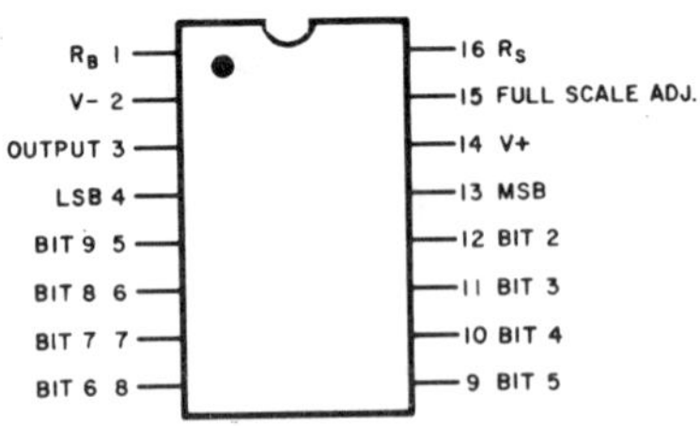

(B) Pin configuration.

Nonlinearity - 0.05 to 0.3%
TC - 15 to 120 ppm/°C (Gain)
Settling Time - 375 ns
Internal Reference
Various Accuracy/TC Options
Various Output Range Options (Includes Application Resistors)
Complementary Binary Coding
TTL, CMOS (5 V) Logic Compatibility
Supply Range ±6 to ±18 V

(C) Key specifications and features.

Courtesy Precision Monolithics, Inc.

Fig. 4-17. DAC-100 10-bit internal reference d/a converter.

The AD561

The Analog Devices AD561 10-bit internal reference d/a converter, is illustrated in Fig. 4-18. This device features pretrimmed accuracy, low TCs for gain and linearity error, and a fast settling time.

The AD561 is constructed on a single chip using the elements shown in Fig. 4-18A. These include the internal reference source of +2.5 volts, a trimmed R-2R ladder, ten-bit switches and output, plus offset scaling resistors. The logic input is binary coded, and thresholds are adjustable by choosing the positive supply voltage. The thresholds are TTL comptaible for a +5-volt supply. It uses a positive 5 to 15-volt supply, and a negative supply of 15 volts.

The AD561 full scale output current is 2 mA, and output may be developed either across a load resistor or with a current-to-voltage op amp. Output voltage compliance is −2 to +10 volts. Output current settling time for the device is 250 ns.

The output voltage range for the AD561 (using an external op amp) is 0 to +10 volts or +5 volts using the internal 5K and 2.5K scaling and offset resistors. The AD561 has a pretrimmed gain and offset accuracy error of 0.1% (typical).

The AD561 is available in several temperature ranges and accuracy grades. The AD561J and AD561K are 0 to 70°C temperature range devices; the J version with ½ LSB relative accuracy and 80 ppm/°C gain TC, the K with ¼ LSB accuracy and 30 ppm/°C gain TC. The AD561S and AD561T are −55 to 125°C devices; S version with ½ LSB relative accuracy and 60 ppm/°C gain TC, the T with ¼ LSB accuracy and 30 ppm/°C gain TC. Monotonicity is guaranteed for all devices over their respective temperature ranges.

A summary of performance highlights is given in Fig. 4-18C. The AD561 series is supplied in a 16 pin DIP. A device data sheet is included in the appendix.

The AD563

The Analog Devices AD563 12-bit internal reference d/a converter is illustrated in Fig. 4-19. This device retains the desirable features of the previously described AD562, and adds the convenience of an internal reference source of +2.5 volts.

As can be noted in Fig. 4-19A, the +2.5-volt source is made externally available at pin 6 and is jumpered to pin 4 to operate the AD563 as an integral reference d/a converter. Bipolar offsetting is accomplished by jumpering pin 8 to pin 6.

The general features and specifications of the AD563 are similar to the AD562, but there are some additional specifications added, related to the internal reference. Calibration accuracy is within 0.1% for gain and offset (both are externally adjustable to zero error) for the AD563 converter.

The AD563 is available in several temperature ranges and accuracy grades. The AD563J and AD563K are 0 to 70°C range devices, with ½ and ¼ LSB relative accuracy error, 30 and 20 ppm/°C gain error TCs, respectively. The AD563S and AD563T are −55 to +125°C range devices, with ¼ LSB relative accuracy, 30 and 10 ppm/°C gain error TCs respectively. Monotonicity is guaranteed for all devices over their respective temperature ranges.

A summary of performance highlights is given in Fig. 4-19C. The AD563 series is supplied in a 24 pin DIP.

Integral D/A Converters

An integral d/a converter is one which is functionally complete for all standard applications. It includes an internal reference, the d/a switches and associated ladder, and the output op amp required for voltage output applications.

Integral d/a converters require a minimal period of design time for their application, as they include all components necessary for the d/a

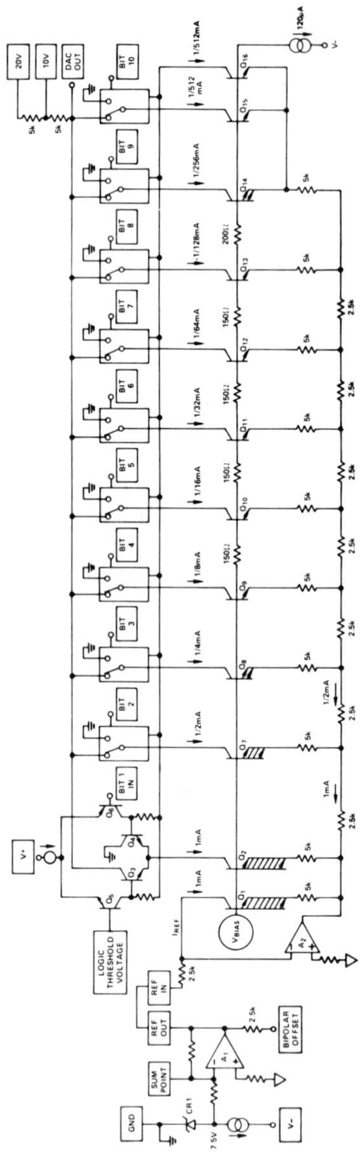

(A) Functional diagram.

Fig. 4-18. AD561 10-bit internal

function within a single package, which is specified as a system. For this reason they are attractive for general purpose use where the d/a application requirement is straightforward.

Devices are available with either binary or bcd coding, and all with standard output voltage ranges. In general, integral d/a converters are supplied to a given "as delivered" calibration accuracy, such as ±0.1% for example. These accuracy limits are related to gain and offset errors, and they are usually externally trimmable to zero error, at the user's option.

Inasmuch as integral d/a converters are complete in themselves, their application considerations are relatively limited. This makes such a converter highly attractive to the system designer. A major inherent feature of an integral d/a converter is the fact that it is specified completely, whereas with other types the user must account for each individual error source.

The DAC-02

The Precision Monolithic DAC-02 10-bit sign/magnitude d/a converter is illustrated in Fig. 4-20. This device is a complete integral d/a converter with 10 bits of resolution, plus a sign control bit. It is available in a variety of different accuracy grades.

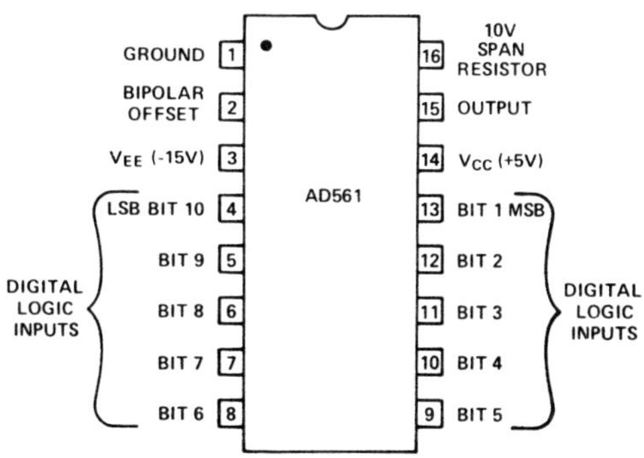

(B) Pin configuration (top view).

Nonlinearity - 0.012 to 0.05%	Internal Reference
TC Linearity - 2.5 ppm/°C (Typ)	Pretrimmed Accuracy
Gain - 30 to 80 ppm/°C (Max)	Binary Coding
Settling Time — 250 ns	Programmable Logic Threshold
Voltage Compliance - −2 to +10 V	2 mA I_o
Calibration Accuracy - 0.1° (Gain, Offset)	Supplies +5 to +15 V, −15 V

(C) Key specifications and features.

Courtesy Analog Devices

reference d/a converter.

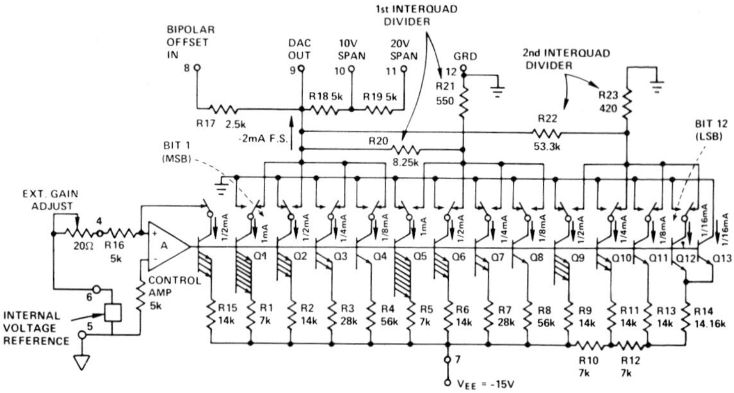

(A) Functional diagram.

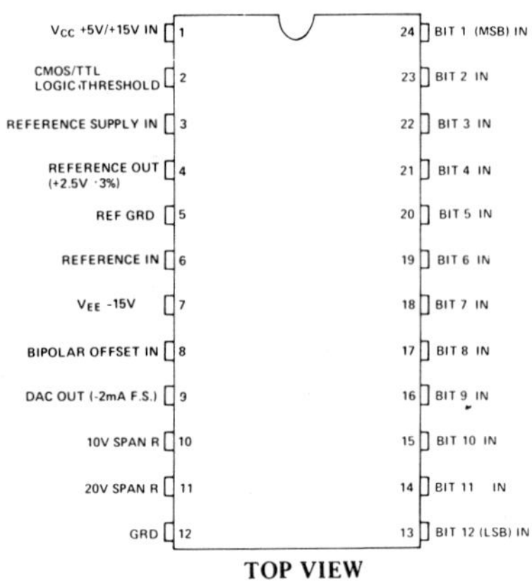

(B) Pin configuration.

Nonlinearity - 0.006 to 0.012% (¼ to ½ LSB)
TC Linearity - 2 ppm/°C (Typ)
Gain - 10 to 30 ppm/°C (Max)
Settling Time - 1.2 ns
Voltage Compliance - −1.5 to +10 V
Calibration Accuracy - 0.1% (Gain, Offset)

Internal Reference
Pretrimmed Accuracy
Programmable Ranges (Includes Application Resistors)
Binary or BCD Coding
Programmable Logic Threshold
2 mA I_o
Supplies +5 to +15 V, −15 V

(C) Key specifications and features.

Courtesy Analog Devices

Fig. 4-19. AD563 12-bit internal reference d/a converter.

As noted in Fig. 4-20A, the DAC-02 is composed of an internal reference and buffer amplifier, a 10-bit R-2R ladder, ten-bit switches, a sign bit switch, and the output amplifiers which provide the sign control/bipolar voltage capability.

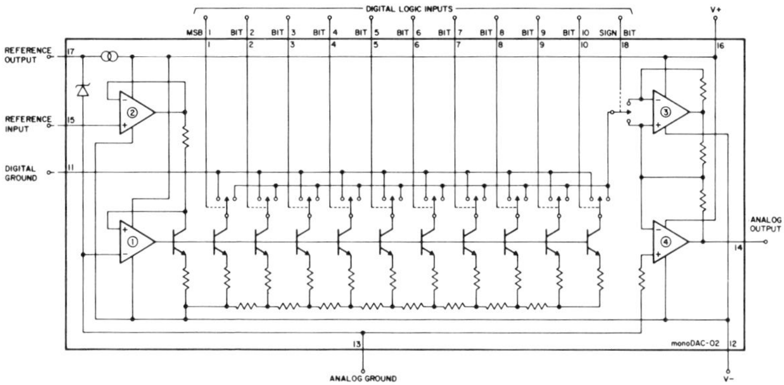

(A) Functional diagram and pin configuration.

Nonlinearity - 0.1 to 0.4%	Complete, Integral D/A
Monotonicity - 10, 9, 8 or 7 bits	Sign/Magnitude Coding
Settling Time - 1.5 μs	TTL, CMOS (5 V) Logic Compatibility
TC (Gain) - 60 to 150 ppm/°C	Multiplying Capability
Bipolar Symmetry - 0.08%	Supplies ±15 V

(B) Key specifications and features.

Courtesy Precision Monolithics, Inc.

Fig. 4-20. DAC-02 10-bit sign/magnitude d/a converter.

Logic inputs of the DAC-02 are TTL and 5-volt CMOS compatible, and binary sign/magnitude coded. Positive output voltage is obtained with a true (high) sign bit, negative with a false (low) sign bit input. Output voltage scaling is either ±10 volts or ±5 volts, and is specified by a part number suffix. Although the device includes an internal reference, it may also use an external reference (if desired), and it may be used in a multiplying mode. Voltage output settling time is 1.5 μs. Bipolar output symmetry, which determines sign/magnitude polarity accuracy is 0.08% (max). The device uses supplies of ±15 volts.

The DAC-02 series is available in a number of accuracy grades with various linearity and TCs as specified as a part number suffix. The DAC-02AC, DAC-02BC, and DAC-02CC have 10-, 9-, and 8-bit monotonicity, respectively, and all have a 60 ppm/°C max gain TC. The DAC-02DD has 7 bit monotonicity, and 150 ppm/°C max gain TC. All devices are specified for a 0 to 70°C temperature range of operation.

A summary of performance highlights is given in Fig. 4-20B. The DAC-02 series is supplied in a 18 pin DIP.

Two closely related devices are also available from Precision Monolithics. The DAC-03 is a low cost 10-bit resolution d/a converter, binary coded smiliar to the DAC-02, but without sign-magnitude operation. The DAC-04 is a 10-bit resolution twos complement d/a converter, with bipolar offset capability. These two devices are available in various accuracy grades, similar to the DAC-02.

The DAC80, DAC85

The Burr-Brown DAC80 and DAC85 series of 12-bit d/a converters is illustrated in Fig. 4-21. This device series includes complete integral

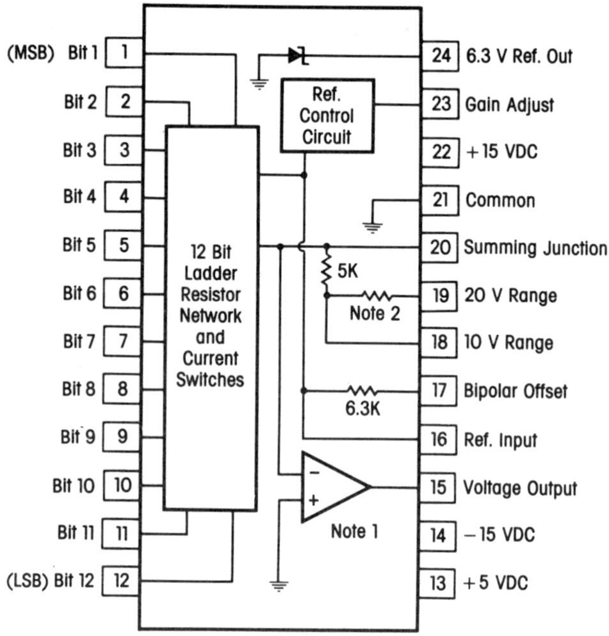

Note 1: Amplifier not included in current output models.
Note 2: 3K for CCD models, 5K for CBI models.

(A) Functional diagram and pin configuration.

Nonlinearity - ½ LSB
Settling time
  Current - 1 μs (1K Load)
  Voltage - 5 μs
TC (Gain)
  DAC80 - 30 ppm/°C (Max)
  DAC85 - 20 ppm/°C (Max)
  DAC85LD - 10 ppm/°C (Max)
Voltage Compliance - ±2.5 V
Calibration Accuracy
  Gain - 0.1%
  Offset - 0.05%

Complete, Integral D/A
Pretrimmed Accuracy
Programmable Ranges (Includes Resistors)
Complementary Binary or BCD Coding
2 mA I_o (Current Models)
Supplies ±5 V, ±15 V

(B) Key specifications and features.

Courtesy Burr-Brown Research Corp.

Fig. 4-21. DAC80, DAC85 12-bit d/a converter.

d/a converters with complementary logic coding, either binary or bcd, current or voltage outputs, and various output range options. The series is one of the most versatile and widely applicable of those available.

Fig. 4-21A shows the functional diagram and pin configuration for DAC80 and DAC85 devices which operate in the voltage output mode. Current output devices delete the op amp. The device series consists of a 12-bit set of quad switches and accompanying ladder network, an internal reference and output scaling and offset resistors. Resistor scaling can be for either binary or bcd coding, but in either case the input logic is complementary in operation. The DAC80 and DAC85 use supplies of +5 volts, and ±15 volts. Relatively high pretrimmed accuracy is achieved, 0.1% for gain and 0.05% for offset. Both are externally reducible to zero, if desired.

The DAC80 and DAC85 are very similar in operation but are packaged differently (although pin equivalents), and the DAC85 has some improved specifications. Both devices use 24 pin DIP packages, the DAC80 in ceramic, the DAC85 in an all-metal version.

Nonlinearity of all models is specified as ½ LSB and monotonicity is maintained over the respective temperature ranges. These are 0 to 70°C for the DAC80 and DAC85C, and −25 to +85°C for the DAC85 and DAC85LD. Settling time is 1 μs for current output models (into 1K), and 5 μs for voltage models. Voltage compliance of current models is ±2.5 V.

Temperature coefficients are the main accuracy differences among the models; the DAC80 has a 30 ppm/°C gain TC, the DAC85 20 ppm/°C and the DAC85LD is 10 ppm/°C.

Voltage output models are pin programmable for ranges of 0 to +5 volts, 0 to +10 volts, ±10 volts, ±5 volts, and ±2.5 volts. Current models are 0 to −2 mA, or ±1 mA.

A given device is specified by the basic part number plus two sets of suffixes—CBI for binary, CCD for bcd; V for voltage, I for current. For example a DAC80-CBI-V is a voltage output, complementary-binary coded DAC80.

Both the DAC80 and DAC85 are popular devices, and are second sourced by a number of suppliers. A DAC80 data sheet is included in the appendix.

A/D DEVICES

In this section of the chapter a wide variety of a/d converter devices are discussed, in a format similar to that of d/a devices previously covered in this chapter. This section of the chapter is further divided into three main a/d converter types represented in IC form—V/F (voltage-to-frequency) converters, integrating converters, and successive-approximation converters.

V/F Converters

There are a number of IC V/F converters available to the designer, and five representative units are illustrated in this section. Most of them operate on the charge-balancing principle.

Among the various types there are a number of different selection criteria which may govern one choice over another for a given application. Typical of such considerations are input-voltage (or current) ranges, input-signal polarity, output waveform and logic level, supply voltages, single- or dual-supply operation, frequency range, and overall accuracy. With each device these points are discussed, and key device specifications are summarized.

For all (or most) devices, there are the common V/F characteristics of integrating and inherently monotonic operation. In general, they all may

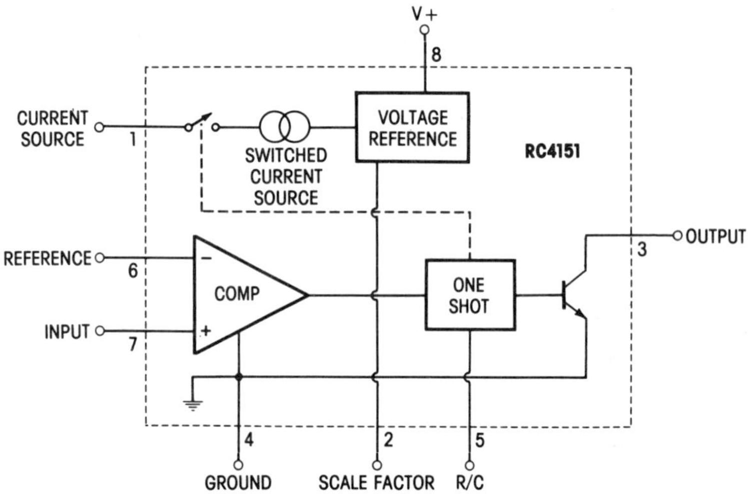

(A) Functional diagram.

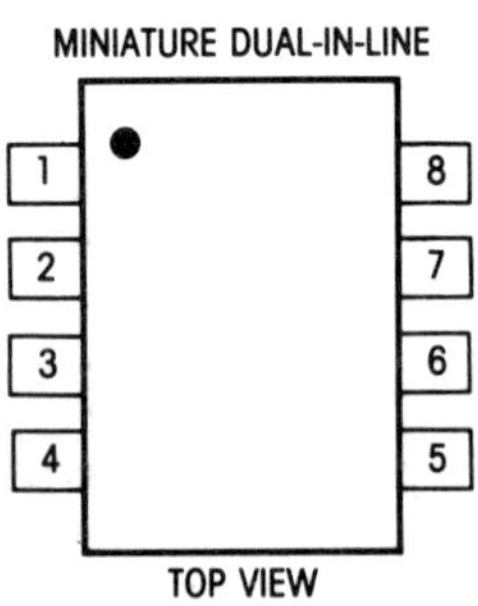

(B) Pin configuration.

TC (Gain) - 100 ppm/°C
Ground Referenced Input Capability
Programmable Scale Factor
Programmable Output Level
Single Supply, +8 to +22 V (+15 V Typical)
Internal Reference

(C) Key specifications and features.

Courtesy Raytheon Company, Semiconductor Division

Fig. 4-22. RC4151 V/F converter.

be interfaced with virtually any logic family, and operate in the range of 0 to 100 kHz. Most of the devices also operate in a F/V mode as well, but the context of discussion here is as a V/F converter.

The RC4151

The Raytheon RC4151 V/F converter is illustrated in Fig. 4-22. This device was one of the first IC V/F devices introduced, and features low cost, simplicity, and flexibility.

The RC4151 functional diagram is illustrated in Fig. 4-22A. The circuit consists of a voltage comparator, a one-shot, a switched current source, and an npn open collector output stage. It is set up for operation by use of external passive components, which may be configured for various input and frequency ranges. Output is from pin 3, in the form of negative going pulses.

The RC4151 in its basic V/F mode has a nonlinearity of better than 1%, and a typical TC of 100 ppm/°C. Linearity (and other forms of performance) can be enhanced greatly if desired, with the use of additional active components. The device operates from a single positive power supply, and accepts a ground-referenced input voltage.

A summary of performance characteristics is given in Fig. 4-22C. The RC4151 is a 0 to 70°C range device; an RM4151 is also available for −55 to +125°C. The RC4151 is packaged in an 8 pin MINIDIP package. The device is second sourced by other manufacturers. A device data sheet is included in the appendix.

The LM331

The National Semiconductor LM331 is a recently introduced V/F device which is functionally similar to the RC4151, but with improved specifications in terms of linearity, TC, and power supply range.

The 9400

The Teledyne Semiconductor 9400 V/F converter is illustrated in Fig. 4-23. This device is a complete V/F converter of the charge-balancing type, using a combination of CMOS/bipolar circuit technology for its operation.

The functional diagram of Fig. 4-23A shows the internal elements of the 9400, consisting of an integrating op amp, a voltage comparator, a 3 μs delay one-shot, and an open collector pulse-output buffer. Also included is a feature unique to this device, a divide-by-two counter stage, which drives an open collector output square wave, at the $f_o/2$ terminal. The 9400 is scaled and configured for operation with user selected external components. These may be selected for frequency ranges up to 100 kHz, and positive input currents of +10 μA full scale. The input voltage range may be scaled to any positive level desired through the use of an appropriate input range resistor.

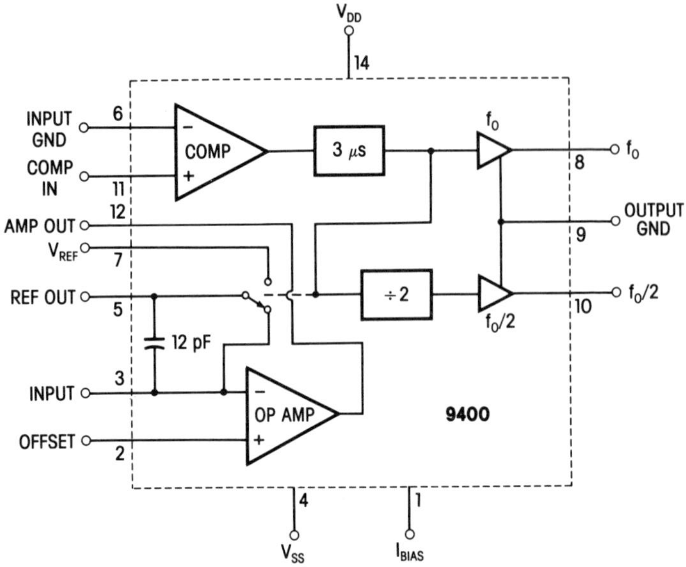

(A) Functional diagram.

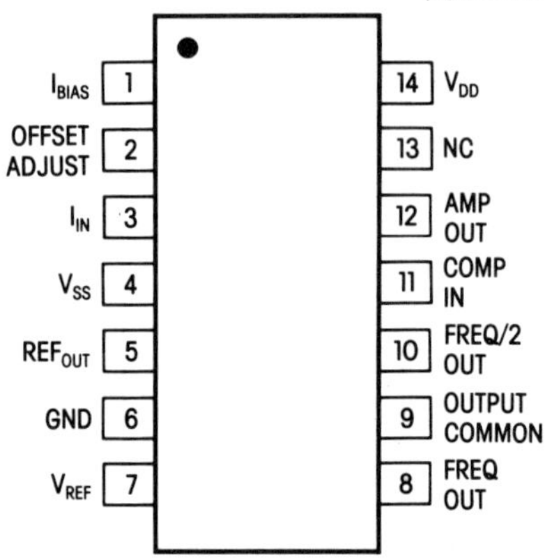

(B) Pin configuration.

Nonlinearity
10 kHz - 0.05%
100 kHz - 0.25%
Offset Voltage - 50 mV
TC (gain) - 40 ppm/°C
0 to +10 μA Input Range (Voltage User Adjusted)
Programmable Scale Factor
Programmable Output Level
Pulse (f_0) and Square ($f_0/2$) Outputs
Dual or Single Supplies, ±4 to ±7.5 V/+6 to +15 V

(C) Key specifications and features.

Courtesy Teledyne Semiconductor

Fig. 4-23. The 9400 V/F converter.

The 9400 uses either dual or single supplies in the range of ±4 volts to ±7.5 volts, with ±5 volts being rated levels. The negative supply level serves as a reference voltage for the conversion process.

The device has maximum nonlinearities of 0.05% and 0.25%, for 10 kHz and 100 kHz ranges, respectively. Offset voltage is 50 mV, but is externally adjustable to zero. Gain TC is 40 ppm/°C, max. The 9400 operates over a temperature range of 0 to 70°C.

A summary of performance specifications is given in Fig. 4-23C. The 9400 is supplied in a 14 pin DIP.

The AD537

The Analog Devices AD537 V/F converter is illustrated in Fig. 4-24. This device series is unique among IC V/F converters in that it does not operate on a charge-balancing concept, but actually employs an entirely different principle of operation.

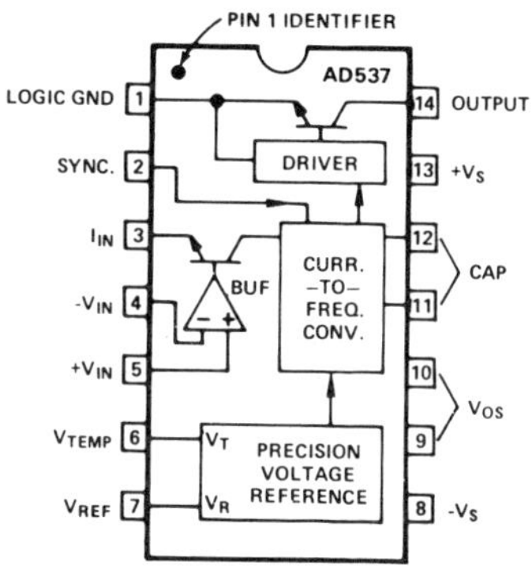

(A) Functional diagram and pin configuration.

Nonlinearity
10 kHz - 0.07 to 0.15%
100 kHz - 0.1 to 0.25%
Offset Voltage - 2 to 5 mV
TC (gain) - 50 to 150 ppm/°C
Calibration Accuracy - 5 to 7%
Ground Referenced Input
Positive or Negative Inputs
Programmable Scale Factor
Programmable Output Level
Square-Wave Output
Reference Voltage Output
Temperature Voltage Output
Single Supply, +5 to +36 V

(B) Key specifications and features.

Courtesy Analog Devices

Fig. 4-24. AD537 V/F converter.

The functional diagram of Fig. 4-24A shows the circuit blocks employed in the AD537, which operates as a linear current-controlled emitter-coupled relaxation oscillator. The current applied externally to the I_{IN} terminal drives this oscillator to a given frequency, in conjunction with an external timing capacitor over a range of four decades. An input op amp is available to afford precise current control from either voltage or current source inputs, for either positive or negative signal polarity. The offset voltage of this op amp is (optionally) adjustable to zero to optimize device dynamic range. The emitter-coupled oscillator used produces a natural output square wave, which is available from a high current (20 mA) open collector npn output stage.

Also included in the AD537 is a precision 1-volt output reference source V_{REF} at pin 7, and a temperature proportional voltage output V_{TEMP} at pin 6. These features greatly enhance the utility of the device. The AD537 operates from single power supplies of from +5 to +36 volts, or dual supplies of up to ±18 volts.

The AD537J is specified for a 0 to 70°C temperature range with a max nonlinearity of 0.15 and 0.25%, at 10 kHz and 100 kHz, respectively, and a 150 ppm/°C gain TC. The AD537K is for a similar temperature range, but with nonlinearities of 0.07% and 0.1% respectively, and a TC of 50 ppm/°C for similar conditions. The AD537S operates from −55 to +125°C, with specifications which are similar to the AD537K.

A summary of performance specifications is given in Fig. 4-24B. The AD537 is supplied in a 14 pin DIP.

The A-8400

The Intech/Function Modules A-8400 V/F converter is illustrated in Fig. 4-25. This operation of this device is a classic example of the charge-balancing V/F principle, with some useful additional features.

The functional diagram of Fig. 4-25A shows the internal circuit components of the device. The circuit accepts positive input signals, in the range of 0 to +10 volts applied to pin 4. As can be noted from Fig. 4-25, both reference and input resistors are supplied on the chip, a feature which will minimize scaling errors and drift from these components, since their drifts will tend to cancel. Alternately, external components can be used for different scale factors, such as higher or lower full scale input voltages. The output of the device is a series of positive pulses at pin 11, with width determined by R/C components attached to pin 12. This output is an open collector npn stage with a maximum high level of +6 volts. The A-8400 operates from supplies in the range of ±12 to ±18 volts, and ±15 volts for rated specifications.

Device nonlinearity is 0.025% max at 10 kHz, and 0.05% max at 100 kHz. Gain TC is 100 ppm/°C maximum, and offset voltage is 10 mV maximum. Offset is externally adjustable to zero.

A summary of performance specifications is given in Fig. 4-25C. The A-8400 is supplied in a 14 pin DIP.

The VFC32

The Burr-Brown VFC32 V/F converter is illustrated in Fig. 4-26. This device uses the charge-balancing principle to high advantage, featuring very high overall accuracy.

The functional diagram of Fig. 4-26A shows the internal configuration of the unit. The VFC32 can accept either positive or negative range input signals, applied to either the (+) IN or (−) IN terminals. The output signal is a series of negative going pulses at the output terminal. This output is an npn open collector stage.

The VFC32 maximum nonlinearity is 0.01% at 10 kHz, and 0.05% at 100 kHz. It is also rated for operation to 500 kHz, with a maximum nonlinearity of 0.2%. Offset voltage is 4 mV maximum, and gain TC 100 ppm/°C (maximum). Offset is externally adjustable to zero. The VFC32 operates from ±9 to ±20 volt supplies, with ±15 volts for rating.

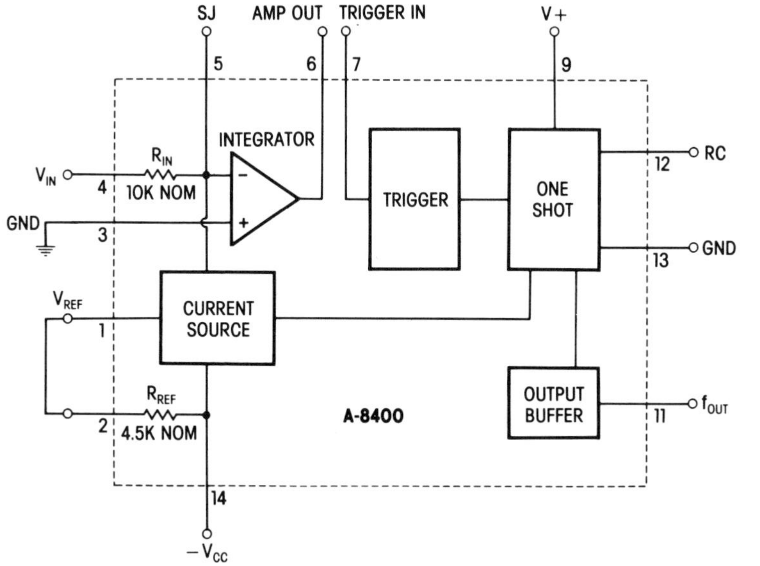

(A) Functional diagram.

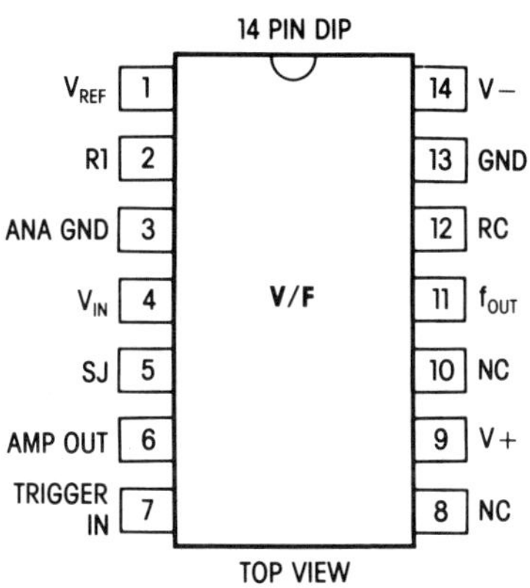

(B) Pin configuration.

Nonlinearity
- 10 kHz - 0.025%
- 100 kHz - 0.05%

Offset Voltage - 10 mV
TC (Gain) - 100 ppm/°C
0 to +10 V Input Range (Also Adjustable)
Fixed or Adjustable Scale Factor
Programmable Output Level (to +6 V)
Reference Voltage Output
Dual Supply, ±12 to ±18 V (±15 V Typical)

(C) Key specifications and features.

Courtesy Intech-Function Modules

Fig. 4-25. The A-8400 V/F converter.

The VFC32 is available in three temperature ranges; the VFC32K for 0 to 70°C, the VFC32B for −25 to +85°C, and the VFC32S for −55 to +125°C. The VFC32K is in a 14 pin DIP, the other devices are in a 10 pin hermetic T0-100 package.

A summary of performance specifications is given in Fig. 4-26C. Table 4-1 is useful as a general aid to comparison of the various de-

Table 4-1. V/F Converter Performance

Device	Input Range Volts or μA	Offset Voltage mV (max)	Linearity % (max) 10 kHz	Linearity % (max) 100 kHz	Gain Drift ppm/°C (max)	Minimum Number of Ext. Components Total (Precision)	Use as F/V?	Supplies
RC4151	0 to +10 V[1]	5 (typ)	0.02 to	1[1]	100 (typ)	7(3)	Yes	+8 to +22 V
9400[4]	0 to 10 μA	50	0.05	0.25	40	5(3)	Yes	+8 to +15 V (±4 to ±7.5 V)
AD537K[2 3]	±10 V[6]	2	0.07	0.1	50	3(2)	Yes[5]	+5 to +36 V
A-8400[2]	0 to + 10 V[6]	10	0.025	0.05	100	3(2)	Yes	±12 to ±18 V
VFC32KP	±10 V[6]	4	0.01	0.05	100	3(2)	Yes	±9 to ±20 V

1 Variable, dependent upon external components used.
2 Reference voltage externally available.
3 Temperature output.
4 $f_o/2$ output.
5 With external active components.
6 As typically used—can be extended with scaling components.

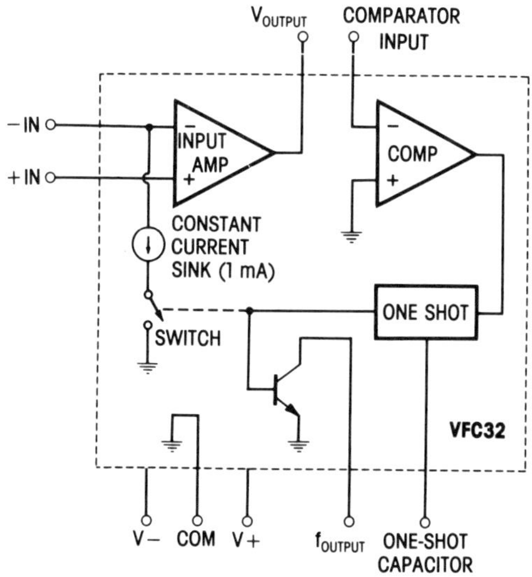

(A) Functional diagram.

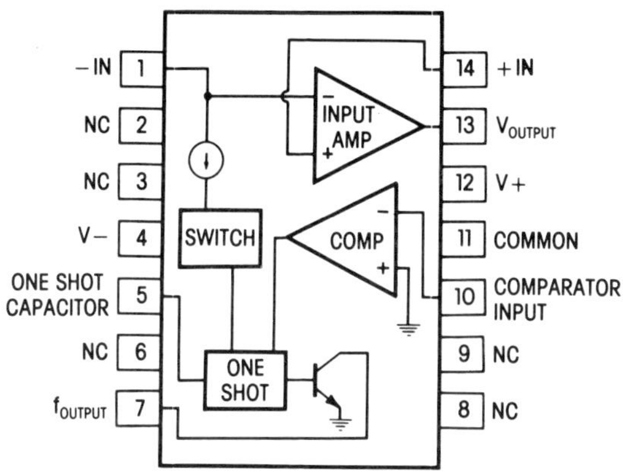

(B) Pin configuration.

Nonlinearity
  10 kHz - 0.01%
  100 kHz - 0.05%
  500 kHz - 0.2%
Offset Voltage - 4 mV
TC(Gain) - 100 ppm/°C
0 to ±10 V Input Range
  (Also Adjustable)
Adjustable Scale Factor
Programmable Output Level
Dual Supply, ±9 to ±20 V
  (±15 V Typ)
Internal Reference

(C) Key specifications and features.

Courtesy Burr-Brown Research Corp.

Fig. 4-26. VFC32 V/F converter.

vice features and performance. This table summarizes the key performance specifications of the five V/F types under discussion.

In general, the data of this table lists specifications in terms of their worst case limits, where specified as such. Also listed is the number of external components required to make a device function in the V/F mode. This is both in terms of overall count (the first number) and the number of precision components required (the bracketed number).

Integrating A/D Converter Devices

In the general area of IC integrating A/D devices there are currently two general design approaches being followed by manufacturers. These are the "chip set" or converter pair, where a number of ICs (two or more) are used to realize the function of an integrating a/d converter, and the single chip approach, where a single IC package accomplishes the a/d function, with little or no support circuitry. Generally the former types afford the most design flexibility, while the latter are suitable for more dedicated, specific applications. This section is subdivided into these two categories of IC integrating a/d converters, illustrating a number of devices in each.

The LD110/LD111(A)

The Siliconix LD110/LD111 and LD111(A) a/d converter pairs are illustrated in Fig. 4-27. The LD110/LD111 chip set was one of the first IC a/d converters introduced, and is suitable for 3½ digit a/d applications. The functional diagram is shown in Fig. 4-27A, and pin diagrams are shown in Fig. 4-27B. A more recently introduced improvement to this device is the LD111A, functionally similar to the LD111, but with increased accuracy. As can be noted from the figure, the LD111 is an analog processor chip, consisting of amplifiers and switches which accomplish the a/d analog functions, while the LD110 accomplishes the digital and timing functions.

Together, this pair of devices forms a 3½ digit a/d converter with input ranges of ±199.9 millivolts or ±1.999 volts. Auto zero and auto polarity are features of operation, and the LD111 has a nonlinearity of 0.05%. This figure is improved to 0.02% by the use of the LD111A. The devices use an external clock, reference voltage, and operate from ±12-volt supplies.

Output data is in the form of 4 digit bcd, which is multiplexed, and TTL compatible.

A summary of performance is given in Fig. 4-27C. The LD110/LD111 types operate over a 0 to 70°C range, and are packaged in 16 pin DIPs.

The LD111/LD114

The Siliconix LD111/LD114 a/d converter pair is illustrated in Fig. 4-28. As may be noted, the combination retains the functions of the LD111 analog processor, but offers increased flexibility through additional features provided by the more complex LD114 digital processor device.

The basic analog range, functioning and accuracy of this combination is similar to the LD110/LD111; this is 3½ digit operation for 0.2 or 2 volt nominal ranges. However, the LD114 has a number of additional

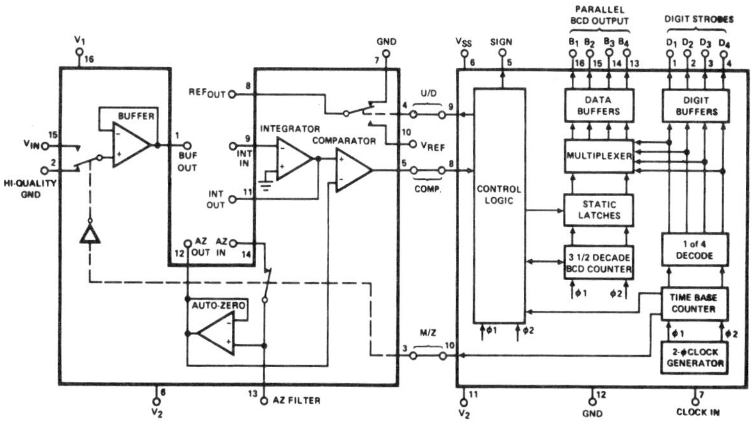

SWITCH STATES ARE FOR A LOGIC "0" AT U/D AND M/Z INPUTS.

(A) Functional diagram.

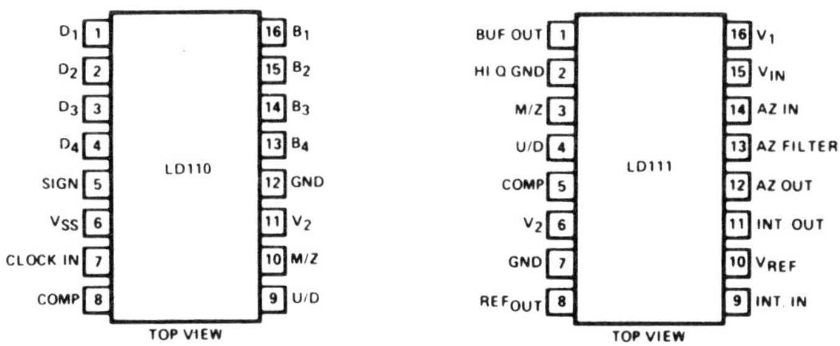

(B) Pin configurations.

Resolution - 3½ digits
Input Range - ±199.9 mV, ±1.999 V
Input Current - 4 pA
Nonlinearity - 0.05% (LD111), 0.02% (LD111A)
TC Gain - 5 ppm/°C (LD111A)
Conversion Rate - 5/sec (30.7 kHz clock)

Reference (Ext) - +2 to +11 V; +6.8 V typ.
Auto Zero
Auto Polarity
Ratiometric
Multiplexed BCD Outputs
TTL Compatible
Supplies - ±12 V

(C) Key specifications and features.

Courtesy Siliconix, Inc.

Fig. 4-27. LD110/LD111(A) 3½ digit integrating a/d converter pair(s).

features such as a clock divide by 512 output for phase-lock uses; sign-output inversion options, and inhibit and serial output options. Like the LD110/LD111, the LD111/LD114 uses ±12-volt supplies and requires external clock and reference sources. A summary of performance is given in Fig. 4-28C.

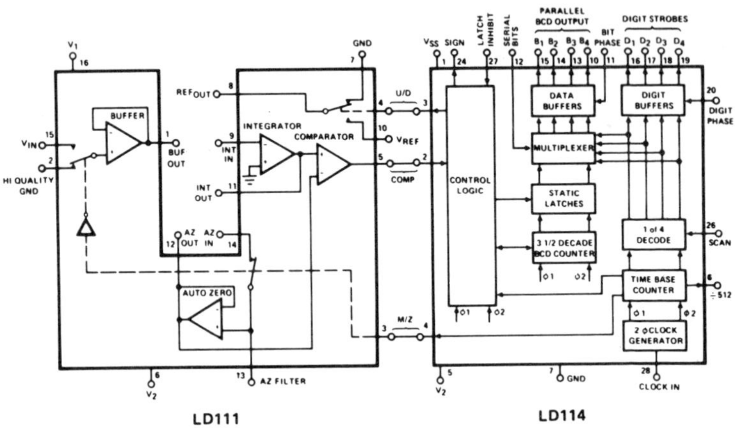

SWITCH STATES ARE FOR A LOGIC "0" AT U/D AND M/Z INPUTS.

(A) Functional diagram.

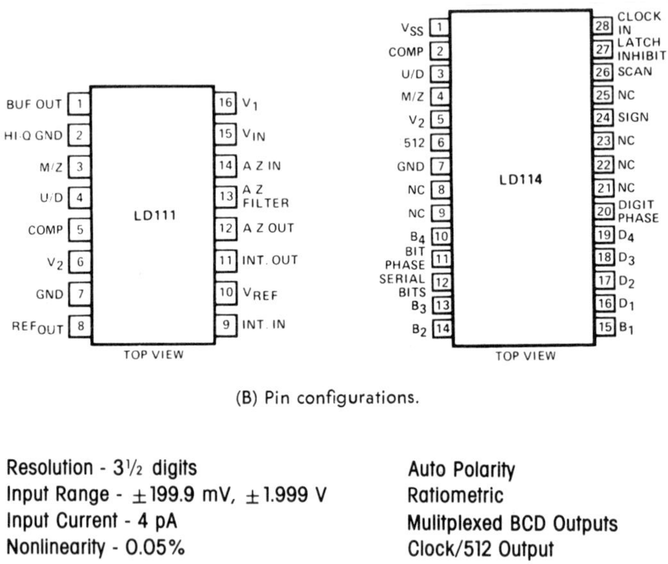

(B) Pin configurations.

Resolution - 3½ digits
Input Range - ±199.9 mV, ±1.999 V
Input Current - 4 pA
Nonlinearity - 0.05%
Conversion Rate - 5/sec. (30.7 kHz clock)
Reference (Ext) - +2 to +11 V; +6.8 V typ
Auto Zero

Auto Polarity
Ratiometric
Mulitplexed BCD Outputs
Clock/512 Output
Complemented Logic Options
Serial Output
Inhibit Function
Supplies - ±12 V

(C) Key specifications and features.

Courtesy Siliconix, Inc.

Fig. 4-28. LD111/LD114 integrating a/d converter pair.

The LD114 is supplied in a 24 pin DIP, and the devices operate over a 0 to 70°C range. The device is second sourced by other manufacturers.

The 8052/8053, 7101/7103 Series

Perhaps the widest range of a/d converter pair flexibility is contained in the 8052/8053 and 7101/7103 series of devices by Intersil. These devices, illustrated in Fig. 4-29, can be used in various combinations for differing requirements of range, resolution and data format.

The 8052 and 8052A, along with the 8053 and 8053A are analog processors. The 8052/8052A contains amplifiers and a reference voltage source, while the 8053/8053A contains analog switches. These devices can be used in combination to form a variety of a/d configurations, with resolution of up to 16 bits, or 4¾ digits.

The 8052/8052A can also be used in combination with the 7101 or 7103 for 3½ digit uses. Use of the 8052A and 7103A yields a 4½ digit converter. The various ranges of performance combinations are summarized in Fig. 4-29D.

The 8052/8052A includes an on-chip reference. An external clock is required. The 8052/8052A requires ±15-volt supplies, while the 8053/8053A and the 7101/7103 devices require +5, −15-volt supplies. The devices operate over 0 to 70°C ranges, and are supplied in DIP packages.

The LD130 and LD131

The Siliconix LD130 and LD131 integrating a/d converters are illustrated in Fig. 4-30. These devices are suitable for 3 (LD130) or 3⅓ (LD131) digit applications, with a minimum of external components.

The functional diagram of the LD130 and LD131 is illustrated in Fig. 4-30A. As can be noted from this figure, the devices contain both the analog and digital components of an a/d system which are similar in general operation to the LD110/LD111 chip set.

For operation, the LD130 or LD131 is made complete with some passive components and an external reference voltage. The LD130 is a 3 digit device with a range of ±0.999 V. The LD131 is a 3⅓ digit unit, with a ±1.499 V range. Both devices feature auto-zero and auto-polarity operation, and have a basic 0.1% nonlinearity.

Outputs are multiplexed bcd, and are TTL compatible. The devices operate from ±5-volt power supplies, and use an external clock and reference voltage. Performance is summarized in Fig. 4-30C.

The LD130 and LD131 operate over a 0 to +70°C range, and are supplied in 18 pin DIPs.

The MC14433

The Motorola Semiconductor MC14433 a/d converter is illustrated in Fig. 4-31. This device is suitable for 3½ digit operation for analog inputs of ±199.9 millivolts or ±1.999 volts. It features a low number of required external components for operation.

The functional diagram of the MC14433 is illustrated in Fig. 4-31A. As can be noted from this figure, all analog and digital circuitry is on the chip with the exception of the reference source, and a few passive components to make it complete.

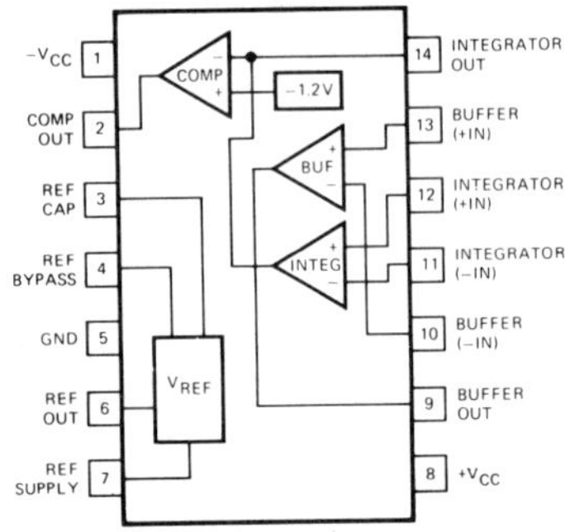

(A) 8052(A) analog signal conditioner.

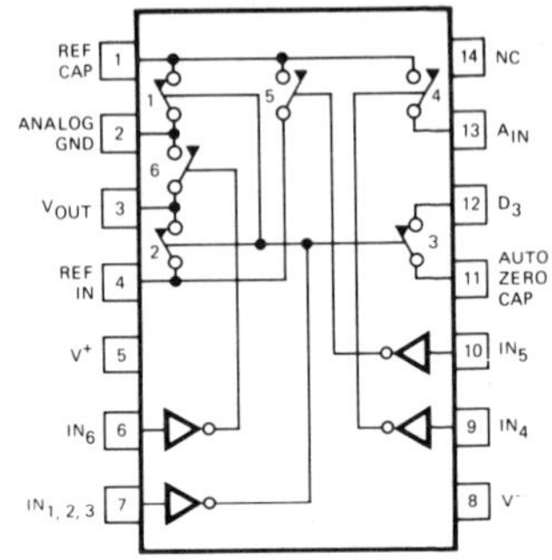

(B) 8053(A) auto zero switch network.

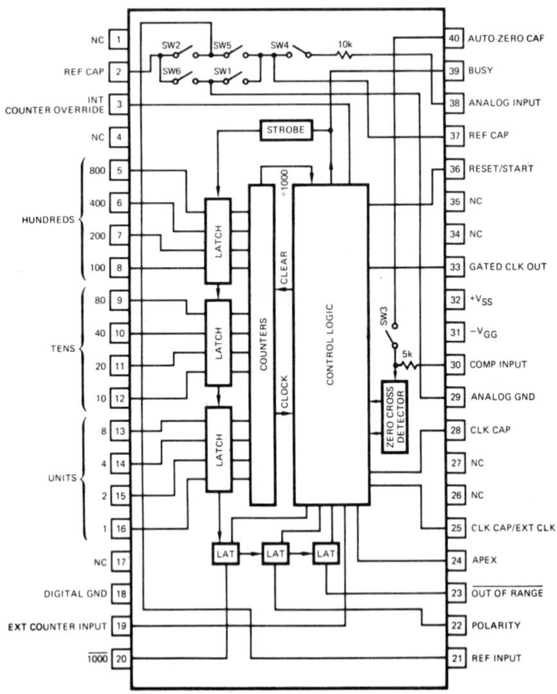

(C) 7101 digital processor.

Fig. 4-29. 8052(A), 8053(A), 7101, 7103(A)

The voltage range of the MC14333 may be selected by the reference voltage range. Operation is auto zero and auto polarity, and outputs are 4 digit data, which is multiplexed bcd, and CMOS compatible. An over-range signal is provided as well. The device operates from ±5-volt power supplies, with relatively low consumption. A summary of performance is given in Fig. 4-31C. The MC14433 operates over a −40 to +85°C range, and is supplied in a 24 pin DIP.

The 7106 and 7107

The Intersil 7106 and 7107 a/d converters are illustrated in Fig. 4-32. These devices are complete 3½ digit a/d converters on a single chip, including both reference and clock source, and direct drive for either liquid crystal (7106) or LED (7107) displays.

The functional diagram of the 7106/7107 analog section is shown in

Resolution - Up to 16 Bits, or 4¾ Digits
Input Range - to ±4 V
Input Current - 5 pA (8052)
  2 pA (8052A)
Nonlinearity - 0.01% (8052/8053/Logic, 8052/7103)
  0.005% (8052/7101)
  0.0025% (8052A/7103A)
TC Gain* - 3 ppm/°C (8052/7101, 8052/7103)
  2 ppm/°C (8052A/8053A/Logic, 8052A/7103)
Offset - 1 μV/°C (8052/8053/Logic, 8052/7101, 8052/7103)
  0.5 μV/°C (8052A/8053A/Logic, 8052A/7103A)
Conversion Rate - 3/sec (Clock = 20 kHz, 3½ Digits;
  120 kHz, 4½ Digits)
Reference (Int) - 1.75 V, TC = 50 ppm/°C (8052), 40 ppm/°C (8052A)
Auto Zero
Auto Polarity
Flexibility
Ratiometric
BCD or Binary Operation
Latched Parallel BCD Outputs - 7101 (3½ Digits)
Multiplexed BCD Outputs - 7103 (4½ Digits)
TTL Compatible - 7103
Internal Reference - 8052(A)
Internal Clock - 7101
Overrange Outputs - 7101, 7103
Underrange Output - 7103
Supplies ± 15 V 8052(A)
  +5, −15 V 8053(A), 7101, 7103

* Excluding TC of Internal Reference

(D) Key specifications and features.

Courtesy Intersil, Inc.

integrating a/d converter pair devices.

Fig. 4-32A, the pinout is shown in Fig. 4-32B. Operation of the devices is both auto zero and auto polarity, and the input ranges of ±199.9 millivolt or ±1.999 volts is selected by scaling of the reference voltage. Nonlinearity is quite low, only 0.05%. The 7106 operates from a single 9-volt supply (such as a battery), while the 7107 is suited for ±5V supplies.

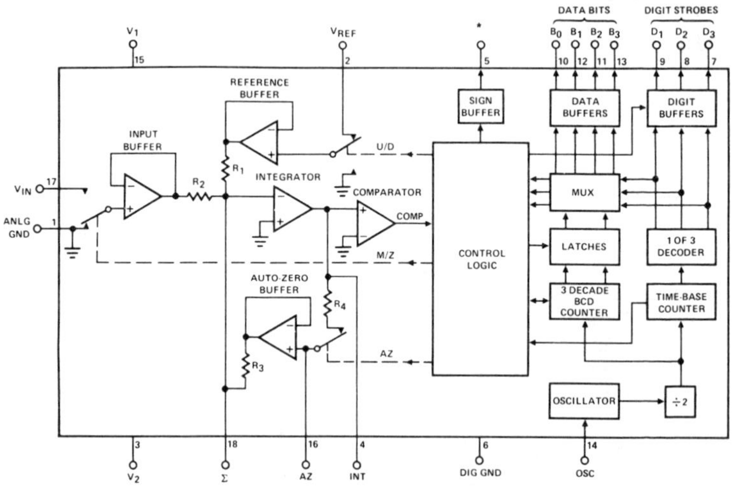

(A) Functional diagram.

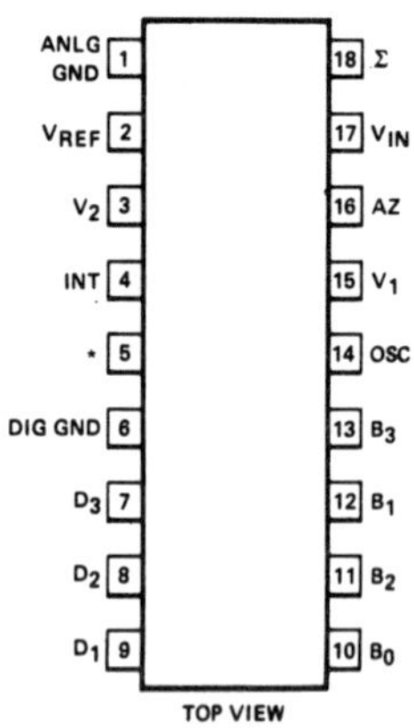

*LD130 SIGN/UR/OR
LD131 SIGN/UR/OF

(B) Pin configuration.

Resolution - 3 Digits (LD130), 3⅓ Digits (LD131)
Input Range - ±0.999 V (LD130, ±1.499 V(LD131)
Input Bias Current - 7 pA
Accuracy - 0.1%
TC Gain - 15 ppm/°C
Offset - 10 μV/°C
Conversion Rate - 5/sec (Typ)
Reference Voltage (Ext) - +2 V
Auto Zero
Auto Polarity
Ratiometric
Internal Clock
Multiplexed BCD Outputs
Over/Under Range Outputs
TTL Compatible
Supplies - ±5 V

(C) Key specifications and features.

Courtesy Siliconix, Inc.

Fig. 4-30. LD130 (LD131) 3 digit (3⅓ digit) integrating a/d converters.

The 7106/7107 operates over a 0 to +50°C range, and is supplied in 40 pin DIPs. A device data sheet is included in the appendix.

The 8700 Series

The Teledyne Semiconductor 8700 series of a/d converters is illustrated in Fig. 4-33. This series includes devices which feature 8-, 10-, or 12-bit binary operation, with both conventional and 3 state binary outputs, and 3½ digit bcd operation. The 8700, 8701, and 8702 are 8-, 10-, and 12-bit binary units respectively, with latched parallel binary outputs. The 8703, 8704 and 8705 are their counterparts with 3-state parallel binary outputs. The 8750 is a 3½ digit bcd device, with parallel outputs. All devices feature nonlinearities consistent wtih their (respective) resolution.

Analog range of the 8700 series devices is user determined for voltage, as the design is based on a 10 μA full-scale current input. Voltage range is determined by an external resistor. An external reference is required, in the form of a 20 μA current, which can be determined by a reference voltage and resistor.

The 8700 series has an internal clock, and can be used in either a free running or triggered mode. Outputs are provided in the form of busy and data valid signals. The device series requires power supplies of ±5V. A summary of performance is given in Fig. 4-33C.

The 8700 series devices are available for either −40 to +85 or 0 to +70°C operation, and they are supplied in 24 pin DIPs.

The AD7550

The Analog Devices AD7550 a/d converter is illustrated in Fig. 4-34. This device is a 13-bit binary operating unit, with twos complement scaling. The AD7550 has very low error coefficients for gain and offset, due to a "quad-slope" principle of operation. This scheme provides an automatic gain calibration phase, in addition to an auto zero phase.

Input ranges are up to 10 volts, and the device uses an external reference voltage. An internal clock is provided, and the device may be used in either a triggered or free-running mode. Outputs are 3-state binary, and can be TTL or CMOS compatible. It uses ±5 volt and +12 volt supplies. Performance is summarized in Fig. 4-34C. The AD7550 operates over a temperature range of −25 to +85°C, and is supplied in a 40 pin DIP.

Successive Approximation A/D Converter Devices

In successive approximation (SA) a/d converters there are a number of different approaches which can be undertaken by the designer. A SA a/d converter can be implemented by design from an IC d/a converter, and appropriate logic. This approach is discussed in detail in Chapter 7.

IC SA a/d devices can also be used, either directly as integral devices

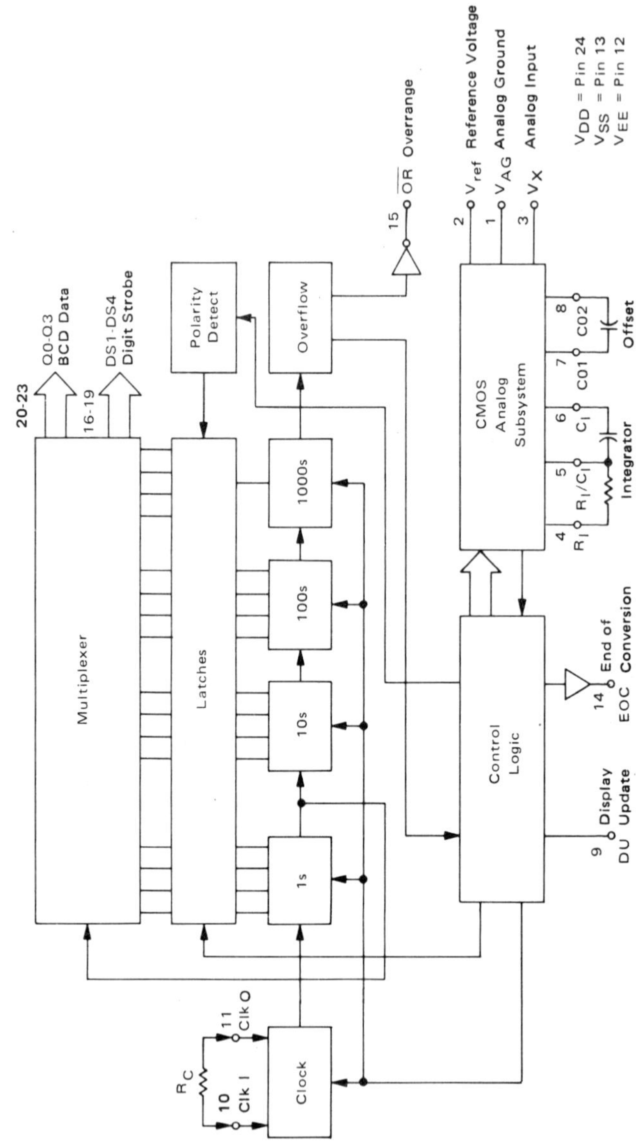

(A) Functional diagram.

Fig. 4-31. MC14433 3½ digit

which are entirely self-contained, or semicomplete, which requires the addition of an external active component, such as a comparator. Examples of both types of IC SA a/d converters are discussed in this section.

The AD7570

The Analog Devices AD7570 10-bit SA a/d converter is illustrated in Fig. 4-35. This device is a "semicomplete" SA a/d converter, requiring the addition of an external comparator and reference voltage to implement the SA a/d function.

As can be noted from the functional diagram of Fig. 4-35A, the AD7520 is composed of a 10-bit d/a converter (similar to the AD-7520), an SA register, and three-state output logic buffers. Analog input range is ±10 volts, and the unit has an internal clock. The AD 7570 operates from a single supply polarity, and can be either TTL or CMOS logic compatible at input and outputs.

The AD7520 is available in two basic accuracy grades, the AD7570J which has 8-bit linearity; and the AD7570L which has 10-bit linearity. Conversion time for 8 and 10 bit linearities is 40 and 120 μs, respectively.

A summary of performance is given in Fig. 4-35C. The AD7570 is packaged in a 28 pin DIP, and operates over a −25 to +85°C temperature range. The device is second sourced by other manufacturers.

The ADC82

The Burr-Brown ADC82 8-bit SA a/d converter is illustrated in Fig. 4-36. This device is an integral SA type high speed 8-bit a/d converter, with programmable analog input ranges.

A combined functional diagram and pinout of the ADC82 is shown in Fig. 4-36A. The ADC82 has complementary-coded logic outputs,

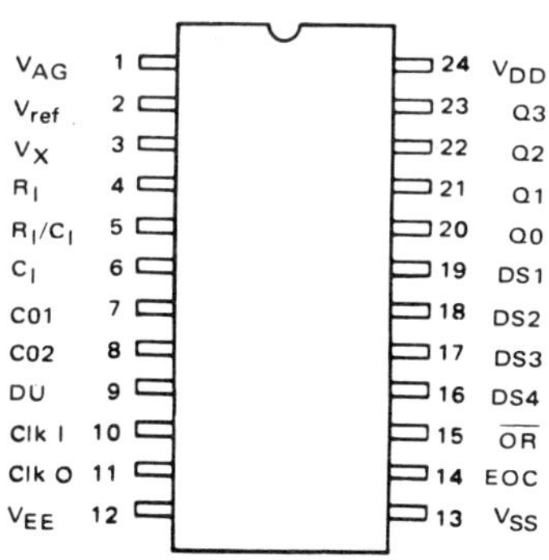

(B) Pin configuration.

Resolution - 3½ Digits
Input Range - ±199.9 mV, or ±1.999 V
Input Bias Current - 20 pA
Nonlinearity - 0.05%
Conversion Rate - 4/sec (Typ)
Reference Voltage - +0.2 to +2 V
Auto Zero
Auto Polarity
Ratiometric
Internal Clock
Over/Under Range Outputs
Multiplexed BCD
Supplies - ±5 V

(C) Key specifications and features.

Courtesy Motorola Semiconductor Products, Inc.

integrating a/d converter.

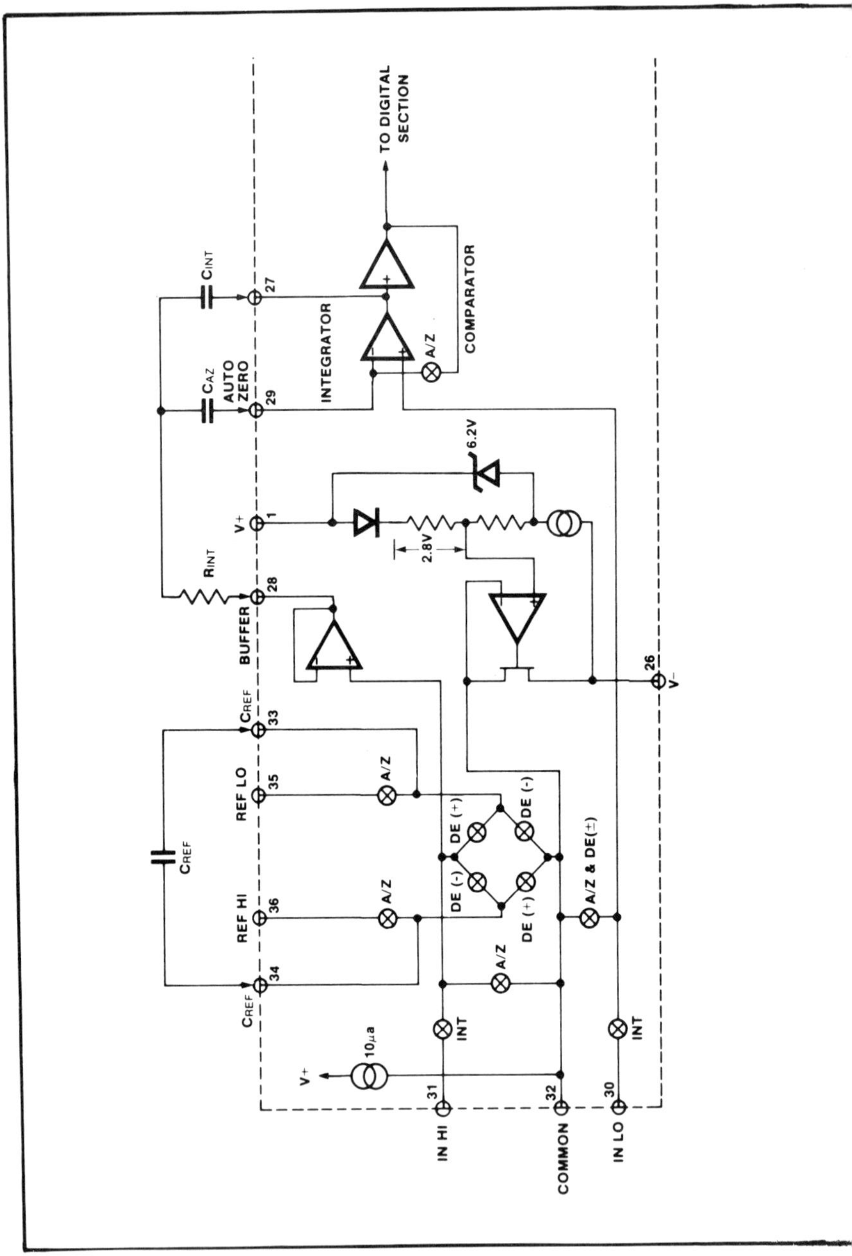

(A) Functional diagram (analog section).

Fig. 4-32. 7106/7107 3½ digit

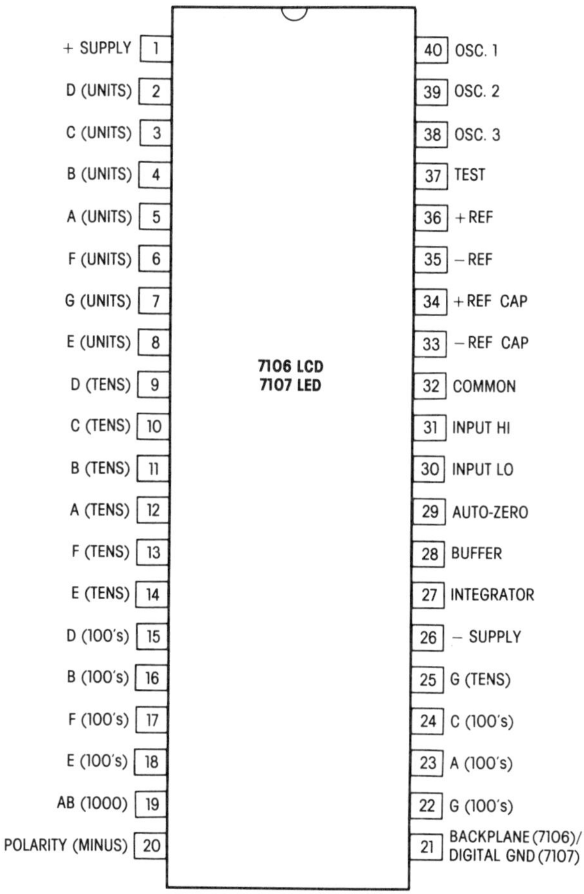

(B) Pin configuration.

Resolution - 3½ Digits
Input Range - ±199.9 mV, or ±1.999 V
Input Current - 2 pA
Nonlinearity - 0.025%
Conversion Rate - 3/sec (48 kHz clock)
Reference (Int) - 2.8 V, 100 ppm/°C
Auto Zero
Auto Polarity
Ratiometric
Internal Clock
Internal Reference
Direct Display Drive - LCD 7106, LED 7107
Supplies - Single 9 V (7106), Dual ±5 V (7107)

(C) Key specifications and features.

Courtesy Intersil, Inc.

integrating a/d converters.

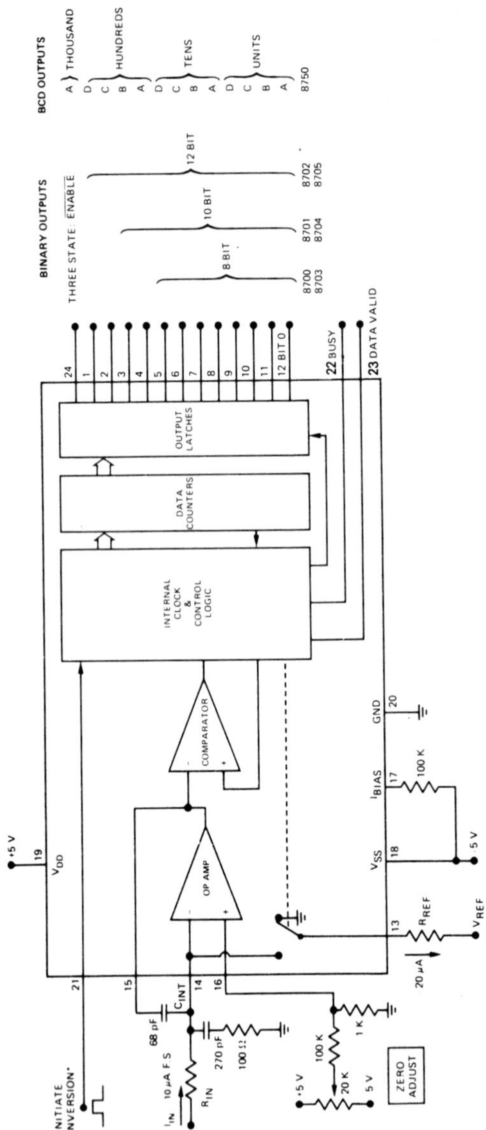

(A) Functional diagram (composite).

Fig. 4-33. The 8700 series 8-, 10-, 12-bit

pin-programmable analog input ranges of up to ±10 volts bipolar and up to 0 to +20 volts unipolar.

Without user adjustment, the ADC82 achieves an as delivered accuracy of ±1 LSB, 8 bits (±0.4%). This can be lowered at the option of

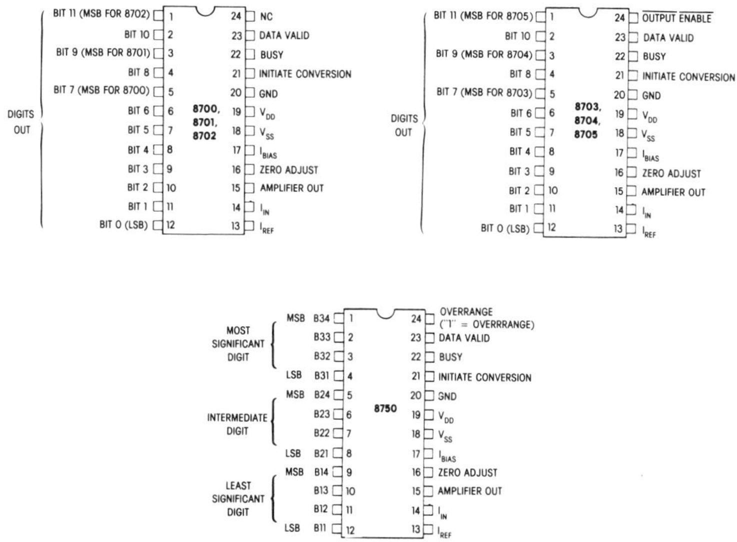

(B) Pin configuration.

Input Range - 10 μA; Voltage User Determined
Nonlinearity - ½ LSB, 8 bits; 8700, 8703
½ LSB, 10 bits; 8701, 8704
½ LSB, 12 bits; 8702, 8705
½ LSB, 3 digits; 8750
TC Gain - 25 ppm/°C
Offset - 30 μV/°C
Conversion Time - 1.25 ms; 8700, 8703
5 ms; 8701, 8704
20 ms; 8702, 8705
10 ms; 8750
Reference (Ext) - 20 μA ; Voltage User Determined
Free Running or Triggered Operation
Binary or BCD Scaling - 8700-8705, 8750
Latched Parallel Outputs - All Devices
3 State Outputs - 8703/8704/8705
LPTTL, 5 V CMOS Compatible
Supplies - ±5 V

(C) Key specifications and features.

Courtesy Teledyne Semiconductors

binary, 3½ digit integrating a/d converters.

the user by trimming. An 8-bit conversion is completed in 2.8 μs using the internal clock. Outputs are in serial and parallel form. All inputs and outputs are TTL compatible.

The ADC82 uses +5 and ±15 volt power supplies. A summary of performance is given in Fig. 4-36B. The ADC82AG operates over a

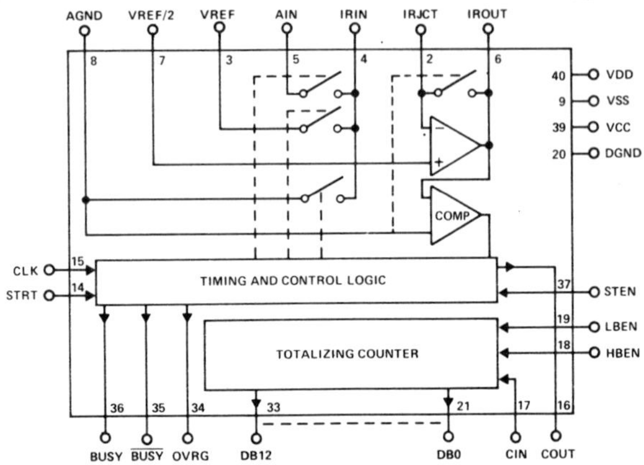

(A) Functional diagram.

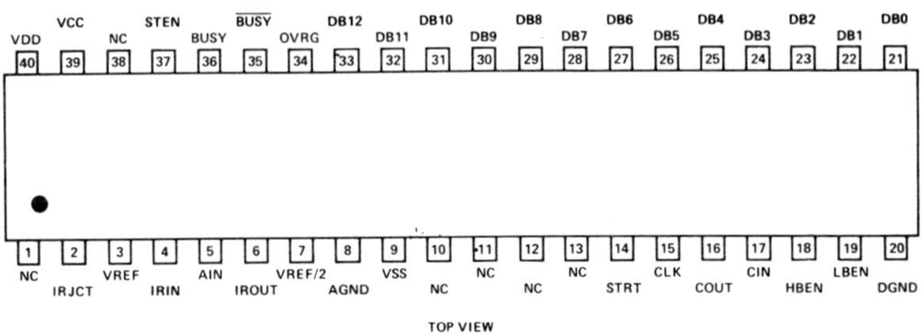

(B) Pin configuration.

Resolution - 13 Bits
Nonlinearity - ½ LSB, 13 Bits
TC Gain - 1 ppm/°C
Offset - 1 ppm/°C
Conversion Time - 40 ms (1 MHz Clock)
Free-Running or Triggered Operation
Auto Zero

Auto Gain Correction
2's Complement Coding
Internal Clock
3 State Outputs
TTC, CMOS Compatible
Supplies - ±5 V, +12 V

(C) Key specifications and features.

Courtesy Analog Devices

Fig. 4-34. AD7550 13-bit binary integrating a/d converter.

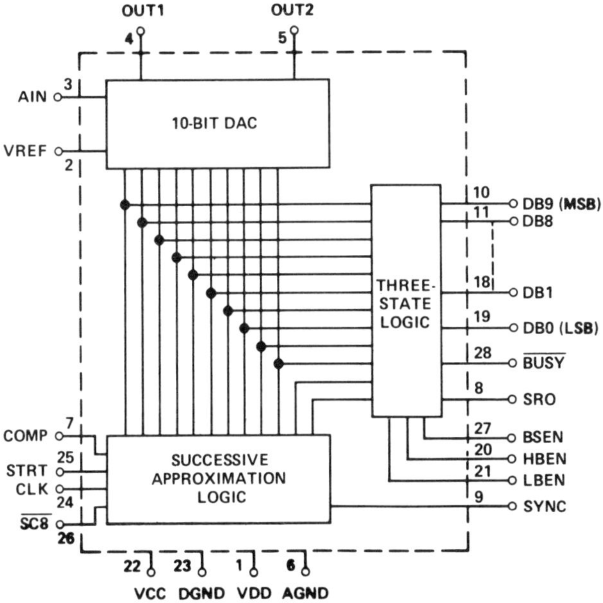

(A) Functional diagram.

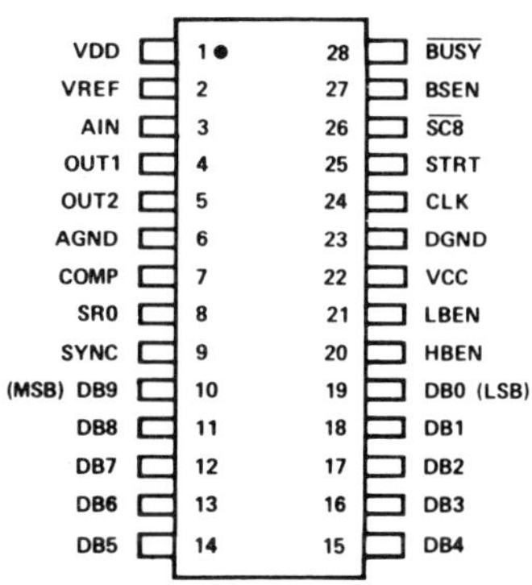

(B) Pin configuration.

Nonlinearity - ½ LSB, 8 or 10 Bits
Conversion Time - 40 μs (8 Bits, Max)
120 μs (10 Bits, Max)
TC (Gain) - 10 ppm/°C (Max)
8 or 10 Bit Resolution/Accuracy
±10 V Analog Range
Ratiometric Operation
Internal Clock (External Optional)
Short Cycle Capability
Three-State Outputs
Serial Output
TTL or CMOS Input/Output Logic (Programmable)

(C) Key specifications and features.

Courtesy Analog Devices

Fig. 4-35. AD7570 10-bit successive approximation a/d converter.

−25 to 85°C temperature range and is ceramic packaged; the ADC-82AM operates over the same range, in an all metal package. Both units are in 24 pin DIPs.

The ADC80

The Burr-Brown ADC80 12-bit SA a/d converter is illustrated in Fig. 4-37. This device is an integral 12-bit medium speed SA a/d converter, with programmable input ranges.

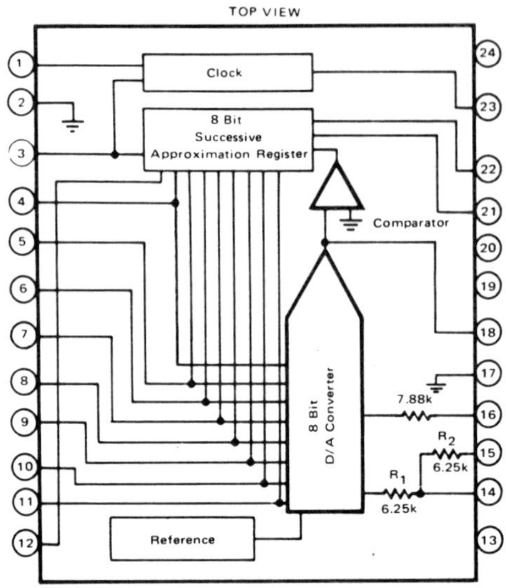

1. Clock Out	24. +5V
2. Digital Common *	23. Convert Command
3. Status	22. Clock In
4. Bit 8 (LSB)	21. Serial Out
5. Bit 7	20. −15V
6. Bit 6	19. +15V
7. Bit 5	18. Comparator Input
8. Bit 4	17. Analog Common
9. Bit 3	16. Bipolar Offset
10. Bit 2	15. R_2 (20V Range)
11. Bit 1 (MSB)	14. R_1 (10V Range)
12. $\overline{\text{Bit 1}}$ ($\overline{\text{MSB}}$)	13. Gain Adjust

* Internally connected to case on ADC82AM

(A) Functional diagram and pinout.

Nonlinearity - ½ LSB (8 Bits)
Monotonicity - Guaranteed Over Temp.
Conversion Time - 2.8 μs
TC (Gain) - 40 ppm/°C (Max)
Programmable Input Ranges
Internal Clock (External Option)
Serial and Parallel Outputs
Complementary Binary (TTL) Logic
+5, ±15 V Power Supplies

(B) Key specifications and features.

Courtesy Burr-Brown Research Corp.

Fig. 4-36. ADC82 8-bit successive approximation a/d converter.

A combined functional diagram and pinout of the ADC80 is shown in Fig. 4-37A. The ADC80 has complementary-binary coded logic outputs and pin-programmable analog input ranges of up to ±10 volts bipolar, and up to 0 to +10 volts unipolar.

Without user adjustment the ADC80 has gain and offset errors of ±0.1% and 0.05% respectively. These errors can be lowered (optionally) by user trim. Gain TC is 30 ppm/°C (maximum) and the ADC80 is monotonic over its temperature range of −25 to 85°C.

Conversion time of the ADC80 is 25 μs for 12-bit operation, and 21 μs for 10-bit operation. The unit can be short-cycled for 10-bit opertion, and comes in 12- and 10-bit accuracy grades for the ADC80-12, and ADC80-10, respectively.

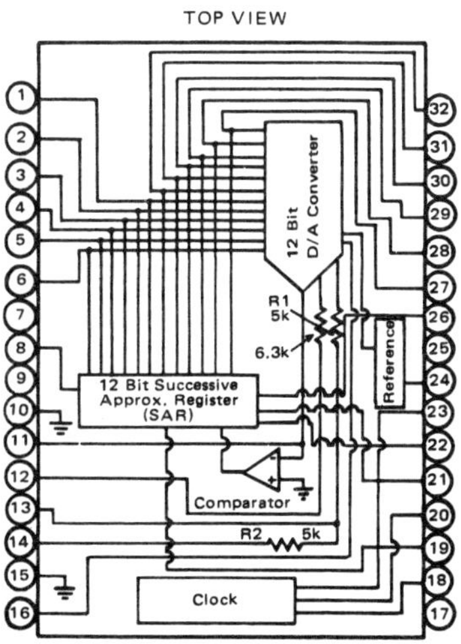

1. Bit 6	32. Bit 7
2. Bit 5	31. Bit 8
3. Bit 4	30. Bit 9
4. Bit 3	29. Bit 10 (LSB-10 Bits)
5. Bit 2	28. Bit 11
6. Bit 1 (MSB)	27. Bit 12 (LSB-12 Bits)
7. +5V Analog Supply	26. Serial Out
8. Bit 1 (MSB)	25. -15V
9. +5V Digital Supply	24. Ref. Out (+6.3V)
10. Digital Common	23. Clock Out
11. Comparator IN	22. Status
12. Bipolar Offset	21. Short Cycle
13. R1 10V Range	20. Clock Inhibit
14. R2 20V Range	19. External Clock
15. Analog Common	18. Convert Command
16. Gain Adjust	17. +15V

(A) Functional diagram and pinout.

Nonlinearity - ½ LSB (12 or 10 Bits)
Monotonicity - Guaranteed Over Temp.
Conversion Time - 25 μs (12 Bits), 21 μs (10 Bits)
TC (Gain) - 30 ppm/°C (Max)
Calibration Accuracy - 0.1% (Gain),
0.05% (Offset)
Programmable Input Ranges
Short Cycle Capability
Parallel and Serial Outputs

(B) Key specifications and features.

Courtesy Burr-Brown Research Corp.

Fig. 4-37. ADC80 12-bit successive approximation a/d converter.

Outputs are available in both serial and parallel form, and all logic input/outputs are TTL compatible. An external clock can be optionally used.

The ADC80 uses +5, ±15 volts power supplies. A summary of performance is given in Fig. 4-37B.

The ADC80AG is supplied in a 32 pin ceramic package. The device is second sourced by other manufacturers. A device data sheet is included in the appendix.

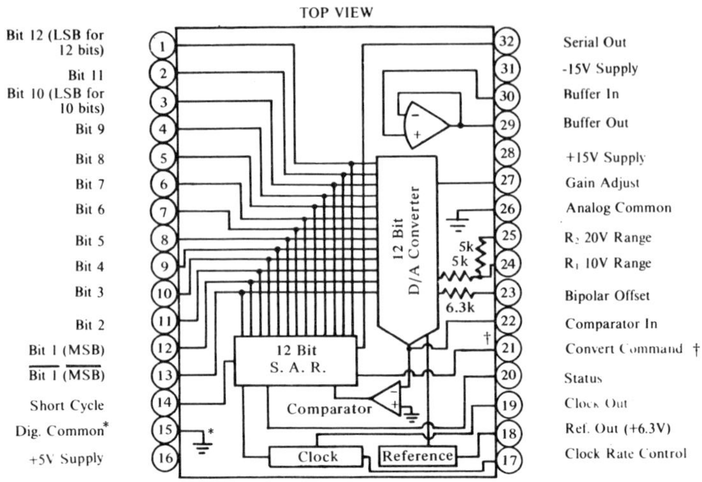

* Digital Common is internally connected to case. (ADC85 and ADC85C only)
† If an external clock is used, connect the clock to Pin 21 (conv. command).

(A) Functional diagram and pinout.

Nonlinearity - LSB (10 or 12 Bits)
Monotonicity - Guaranteed Over Temp.
Conversion Time - 10 μs (12 Bits), 6 μs (10 Bits)
Calibration Accuracy - 0.1% (Gain), 0.05% (Offset)
Programmable Input Ranges
Input Buffer Amplifier Option
Short Cycle Capability
Clock Rate Control
Parallel and Serial Outputs

(B) Key specifications and features.

Courtesy Burr-Brown Research Corp.

Fig. 4-38. ADC84/85 12-bit successive approximation a/d converters.

The ADC84/85

The Burr-Brown ADC84 and ADC85 SA a/d converters are illustrated in Fig. 4-38. These devices are integral SA type high speed 12- or 10-bit a/d converters with programmable analog input ranges.

A combined functional diagram and pinout of the ADC84/85 is shown in Fig. 4-38A. The ADC84/85 have complementary coded logic outputs, and pin-programmable analog input ranges of up to ±10 volts bipolar and up to 0 to +10V unipolar. An optional input amplifier can be inverted in the analog path to obtain a high input impedance.

The ADC84 and ADC85 come in two basic accuracies of 10 and 12 bits respectively, represented as -10 or -12 suffixes. The ADC85 grade is for −25 to 85°C use (10 bits) or 0 to 50°C (12 bits). The ADC85

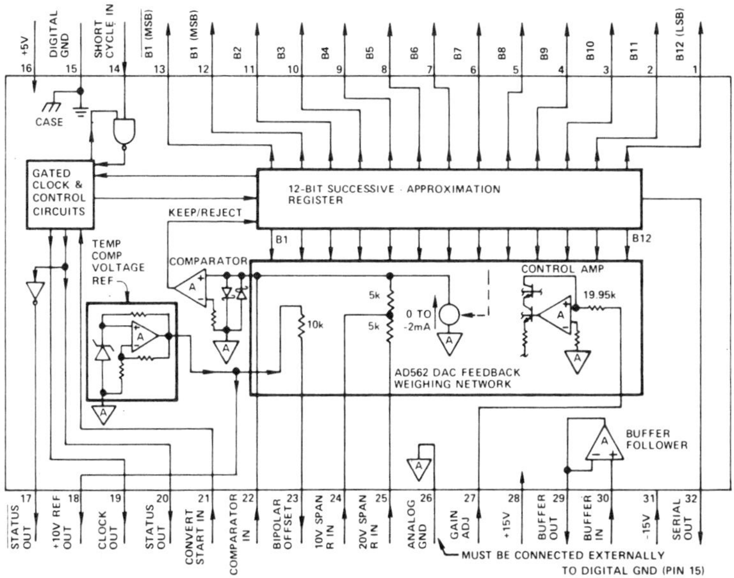

(A) Functional diagram and pinout.

Nonlinearity - ½ LSB, 12 Bits
No Missing Codes
  AD572AD - 0 to 70°C
  AD572BD - −25 to 85°C
  AD572SD - −55 to 125°C
Conversion Time - 25 μs

Calibration Accuracy - 0.05%
Programmable Input Ranges
Input Buffer Amplifier Option
Buffered Reference Output (+10 V)
Short Cycle Capability
Parallel and Serial Outputs

(B) Key specifications and features.

Courtesy Analog Devices

Fig. 4-39. AD572 12-bit successive approximation a/d converter.

is for 0 to 70°C or 0 to 50°C use with 10 and 12 bits, respectively. The ADC85 is housed in a metal package. The ADC84KG is similar, but housed in a ceramic package.

Without user adjustment the ADC84/85 achieves accuracies of 0.1% and 0.05% for gain and offset. With trim, these errors are reducible. A 12-bit conversion in the ADC84/85 is completed in 10 μs; 10-bit conversion can be done in 6 μs. The devices can be short-cycled, and have a clock rate control input.

Outputs are available in both serial and parallel form, and all logic input/outputs are TTL compatible. An external clock can be used. The ADC 84/85 uses +5, ±15 volt power supplies. A summary of performance is given in Fig. 4-38B. The ADC84/85 are supplied in 32 pin packages.

The AD572

The Analog Devices AD572 12-bit SA a/d converter is illustrated in Fig. 4-39. This device is an integral 12-bit medium-speed high-accuracy SA a/d converter, with programmable input ranges.

A combined functional diagram and pinout of the AD572 is shown in Fig. 4-39A. The AD572 has binary-coded logic outputs, and pin-programmable analog input ranges up to ±10 volts bipolar, and 0 to +10 volts unipolar. An optional input amplifier can be inserted in the analog path for high input impedance.

The AD572 is available in three basic temperature ranges; the AD-572AD for 0 to 70°C; the AD572BD for −25 to 85°C; and the AD-572SD for −55 to 125°C. All devices feature 12-bit, no-missing code performance over their respective ranges.

Without user adjustment the AD572 achieves 0.05% accuracy for gain and offset; this can be reduced further by user trimming. A 12-bit conversion in the AD572 is completed in 25 μs, and the device can be short-cycled for shorter conversion times with less resolution. Outputs are available in serial and parallel form, and all logic input outputs are TTL compatible. A summary of performance is given in Fig. 4-39B.

The AD572 is supplied in a 32 pin metal package. A device data sheet is included in the appendix.

REFERENCES

1. Bowers, J. O. "Monolithic Modules for D/A Systems Applications." *Electronic Instrument Digest,* Vol. No. 9, September 1970.
2. Brown, C. H.; Marshall, W. "Sixteen-Bit Conversion Gets a Lift From IC Technology." *Electronics,* September 25, 1972.
3. ———. "12-Bit DAC Design Can be Routine—16 Bits Is More Difficult, But Achievable." *Analog Dialogue,* Vol. 5, No. 3, June 1971.
4. Cecil, J. B. "A CMOS 10-Bit D/A Converter." *ISSCC Digest of Technical Papers,* February 1974.
5. ———; Howard, W. G. "A Fully Monolithic 6-Bit Multiplying D/A Converter." *ISSCC Digest of Technical Papers,* February 1971.
6. ———; Whitmore, J. "A 10-Bit Monolithic CMOS D/A Converter." *Analog Dialogue,* Vol. 8, No. 1, 1974.
7. Craven, R. B. "An Integrated Circuit 12 Bit D/A Converter." *ISSCC Digest of Technical Papers,* February, 1975.
8. ———; Harris, S. "12-Bit IC D/A Converter." *Analog Dialogue,* Vol. 8, No. 2, 1974.
9. Dooley, D. J. "A Complete Monolithic 10-Bit D/A Converter." *ISSCC Digest,* February 1973.
10. Dufort, M., Fullagar, D. *Low Cost Digital Panel Meter Designs.* Application Bulletin A023, May 1977. Intersil, Inc., Cupertino, California.

11. Evans, L.; ———. "A Low Cost 4½ Digit A/D Converter." *Proceedings of EE Times IC Applications Conference,* 1976.

12. Gilbert, B. "A Versatile Monolithic Voltage-to-Frequency Converter." *IEEE JSSC,* Vol. SC-11, No. 6, December, 1976, P. 852.

13. Grandbois, G. *Applying the LD130 +3 Digit A/D Converter.* Application Note AN76-5, January 1977. Siliconix, Incorporated.

14. ———. *Function and Application of 3½ Digit A/D Converter Set.* Application Note AN74-1, January 1977. Siliconix, Incorporated.

15. Kress, D. "Highest Performance 10-Bit IC DAC," *Analog Dialogue,* Vol. 11, No. 1, 1977.

16. Maddox, E. "Current-Steering Chip Upgrades Performance of D/A Converter." *Electronics,* April 4, 1974.

17. O'Neil, W. *Digital-to-Analog Converter Circuits Using the 8018.* Intersil Applications Bulletin A010, June 1972. Intersil, Inc., Cupertino, California.

18. Osgood, S. "Using the 8700 Series CMOS A/D Converters." *Proceedings of EE Times IC Applications Conferences,* 1976.

19. Pastoriza, J. J.; Spofford, W. R. Jr. *Designing With the μDAC Monolithic IC D/A Converter Components.* Application Note, September 1970. Analog Devices, Inc., Norwood, Mass.

20. Ritmanich, W. "13-Bit Monolithic CMOS A/D Converter." *Analog Dialogue,* Vol. 10, No. 1, 1976.

21. Schmoock, J. C. "The Monolithic Voltage to Frequency Converter." *Proceedings of EE Times IC Applications Conference,* 1976.

22. Schoeff, J. A. "A Monolithic Companding D/A Converter." *ISSCC Digest of Technical Papers,* February, 1977.

23. Tatro, R. D. "Which Hybrid Converter: Single Switch or Quad?" *Electronics,* July 15, 1974.

5

Data Conversion Peripheral Components

In this chapter a number of a/d, d/a system components which perform functions peripheral to converters themselves are studied. All of these components are necessary to the implementation of a successful conversion system, but are not necessarily integral to any specific a/d or d/a type. They may be used with many different types.

Discussed here are such components as the instrumentation amplifier, the analog switch, and the sample hold amplifier. Each of the component categories is discussed first in a conceptual sense, then in the light of representative IC devices, and finally within representative circuitry applying the components. At the end of the chapter are listed bibliographic references for further information.

INSTRUMENTATION AMPLIFIERS

One of the most useful circuit applications in data acquisition signal processing is the differential-input dc amplifier configuration, more commonly known as the "instrumentation amplifier." Due to the inherent rejection of common-mode noise and the linear amplification of desired signals that such an amplifier provides, this form of circuit and variations of it find widespread use in conversion systems. Dc and low frequency signals are commonly received from a transducer, scaled and/or amplified, and then transmitted in a single-ended mode to an a/d converter. Since the desired signal may have a full-scale amplitude of typically only a few millivolts, while common-mode noise may be several volts, such factors as amplifier input offset, drift, and noise are critical to overall accuracy. Equally important is the common-mode rejection of the amplifier configuration. Several instrumentation amplifier circuits are discussed in this section, with the unique and distinguishing features of each highlighted.

A general functional symbol for an instrumentation amplifier (IA) is shown in Fig. 5-1. This circuit performs the function of amplifying the differential input signal V_1-V_2 by a gain factor, G. In so doing the balanced input signal is transformed into single-ended form at the amplifier output. Amplifier specifications which are highly relevant to overall accuracy are those that cause departure from the idealized gain expression shown.

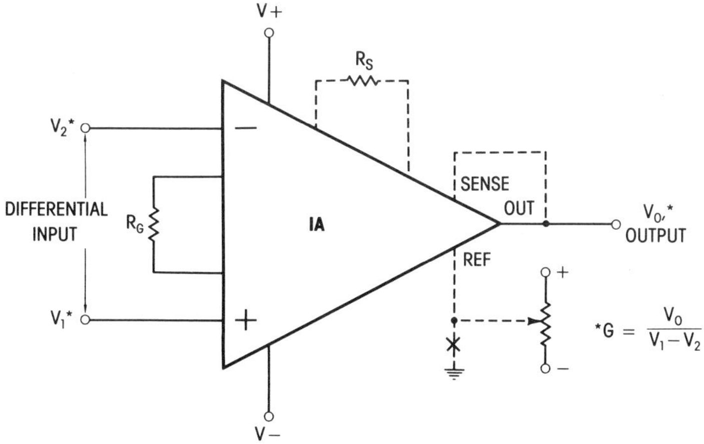

Fig. 5-1. Functional diagram of an instrumentation amplifier.

An instrumentation amp, unlike an op amp, is a *committed* device; that is, it is specifically configured for the optimum amplification of differential signals, with (normally) the only user variable parameter being its gain factor, G.

IAs usually include two operating options, which are made available by the SENSE and REFERENCE terminals. The SENSE terminal is the feedback-path input to the IA output stage, and may be used to correct for wiring drops between the output terminal and the load, if remote. In the simplest form of application it is connected as shown. This terminal may also be used to scale an IA for additional gain by using a feedback attenuator, or to allow use of a current booster within the feedback loop.

The REFERENCE terminal is grounded as shown for normal ground referred output signals, but may also be used to *offset* the output signal by a fixed level. For such uses the desired offset voltage is applied to the reference terminal and the output level then appears referred to this voltage.

Gain is manipulated in an IA by one or two gain scaling resistors. As shown here in single resistor form, R_G adjusts the operating gain to the desired level. The specific gain equation will vary somewhat with individual IA devices, but gain is usually inversely proportional to R_G. A second gain set resistor (R_s) is also often used. Gain is usually proportional to this resistor. Gain related error specifications of IA devices are exclusive of these resistors, therefore the user must select accurate and low drift components for their functions.

IA Specifications

Key IA performance specifications are: gain error, gain nonlinearity, gain stability, common mode rejection, input bias current, offset voltage (and its drift), and noise. In a high performance IA these error factors may be held to as low as 0.01% (or less). However, some parameters, such as gain and offset drift with temperature, may accumulate to worst case magnitudes as high as 0.1% or more. This drift is largely correctable by an automated calibration cycle in computer based systems, as long as the drift does not introduce an accompanying nonlinearity drift. Therefore, cmrr and nonlinearity usually emerge as the largest irreducible error sources in an IA design.

Gain Error

This error factor indicates the deviation from the ideal gain equation, and represents a simple scaling error. It may be calibrated out by an adjustment of the scaling resistor (R_G or R_s).

Nonlinearity

Gain nonlinearity is the error produced by changes in the basic differential input-to-output transfer function, or more specifically, its departure from a constant slope. Ideally, the incremental changes in an input voltage will always produce a corresponding output increment scaled by a constant gain factor.

Typically nonlinearity is specified over the full differential input voltage range, usually ±10 volts max. This will of course decrease for higher gains, as output voltage is normally rated at ±10 volts. Nonlinearity will vary as a function of gain.

Gain TC

This is the temperature coefficient of gain. It will vary somewhat with gain changes, in a given IA. Typically gain TC is on the order of a few ppm/°C.

Common Mode Rejection Ratio (CMRR)

Cmrr is highly important because of the fact that the desired differential mode signal is often seen at the IA input atop a larger common mode signal. Therefore if the differential signal is on the order of a few millivolts, a common mode signal of several volts can cause serious error, if the cmrr is not sufficiently high. Typical common mode input voltage ranges are ±10 volts for rated specifications, with cmrr itself on the order of 70 dB or more. Common mode rejection ratio varies quite strongly with gain, and increases at highest gains.

Offset Voltage

The offset voltage of an IA is one of the more confusing parameters, because in some instrumentation amplifiers it must be separated into two components. This is because some IA units utilize a two-stage design technique, each stage of which has an associated offset voltage. In these types offset voltage is given in terms of "input offset" and "output offset." This simply means that the output offset term is constant and independent of gain; but, the input offset, when referred to the output will yield an offset which is variable according to the operating gain. The total output offset is the sum of these two components, and is the "offset" voltage which should be used to calculate offset related error as a percent of output voltage.

Typical input offset voltage figures are a few millivolts, while output offset may be as high as several hundred millivolts. Offset voltage is in most cases adjustable to zero, and often separate trims are provided for each component. The exact procedure varies from one device to another.

Offset Voltage Drift

Offset voltage drift is the temperature-induced change in offset voltage. It is also separated into input and output terms, like the offset voltage itself. Also like offset voltage, drift must be summed appropriately and referred to the output to arrive at a composite drift figure which is a percentage of the output voltage.

Input Bias Current

Input bias current, or the dc bias current into the (+) and (−) inputs, is important in an IA to minimize error voltages produced across unequal source impedances in the input path. For a given source unbalance (1K is a typically specified figure), the two bias currents will produce an equivalent differential error voltage. Therefore, lower bias currents allow proportionally higher source unbalance for a given error.

The importance of a minimum bias current in practice comes about because there is often little the designer can do about source impedances —he must live with what exists. Typical input bias currents are in the range of 1 to 100 nA for bipolar input IAs. FET input devices lower this to below 100 pA (at room temperature). Note that in all cases a dc path *must* be provided at both IA input terminals.

Noise

Noise is the random variation in output which will exist at a given gain level. It is usually specified in rms μV within a given bandwidth, but may be further broken into input and output referred components. Input noise is multiplied by the gain for output referral, then rms summed with the output component for the total output noise. Alternately, input referred noise for various gains may be given.

Instrumentation Amplifier Designs

There are many different circuit configurations which can be utilized to build an instrumentation amplifier. Many of these are based on op amps and matched resistor networks, and a discussion of the various op-amp schemes may be found in the author's *IC Op Amp Cookbook,* published by Howard W. Sams & Co., Inc.

One of the highest performing op-amp circuits is shown in Fig. 5-2. This circuit uses a pair of matched input op amps in a differential follower-with-gain configuration. This first stage provides gain for differential mode signals, but unity gain for common mode signals. The

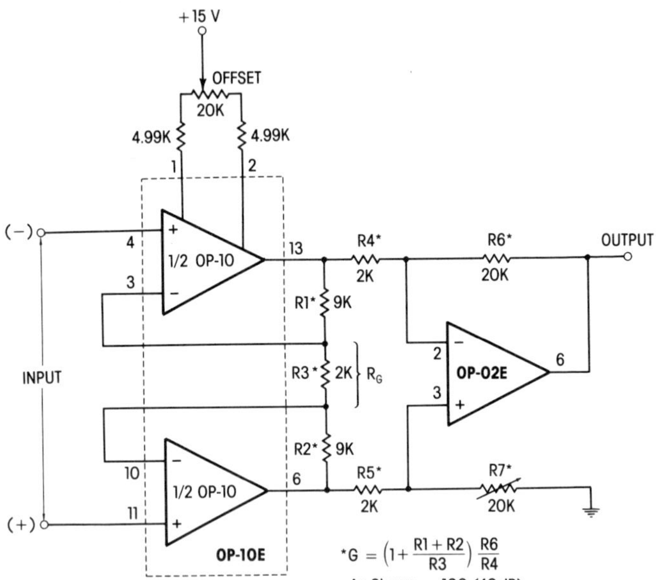

$$*G = \left(1 + \frac{R1 + R2}{R3}\right) \frac{R6}{R4}$$

As Shown = 100 (40 dB)
Adjust R3 For Gain
Trim R7 For Best CMRR
Input Offset ≤ 100 μV
Input Offset Drift ≤ 1 μV/°C
Input Bias Current: 4.5 nA (Max)
Gain Nonlinearity: 0.008% (Max)

Fig. 5-2. Op-amp instrumentation amplifier.

second stage provides additional differential gain, and rejects the common mode signal.

The overall performance of this circuit is dependent upon two major factors. One of these is the match of the second stage bridge resistors, which critically affects cmrr. In this stage the ratio of R6/R4 and R7/R5 must be carefully matched, both initially and with temperature. This resistor bridge should be either an ultralow TC close tolerance wirewound combination, or a laser-trimmed low TC thin film network.

The second major factor is the cmrr and drift of the input stage op-amp pair, which should match. The op-amp type shown is specified for cmrr and TC drift match, which satisfies this requirement. Offset is trimmed in one side only, which matches the offset and drift of the two "halves." The result is a low net offset and drift.

A summary of performance is given in the figure. For different operating gains, R3 is adjusted to suit, and as shown gain is 40 dB. This resistor is analogous to the "R_G" gain-adjust resistor shown in Fig. 5-2.

IC Instrumentation Amplifiers

IC instrumentation amplifiers assume both the form of the op-amp based design described previously, and specialized ICs designed just for the IA function. Both types of designs are described below.

In Table 5-1 are listed the key performance specifications of three IC instrumentation amplifiers; the LF352 (National Semiconductor) and the AD521K and AD522C (Analog Devices). These performance specs are listed in a worst case fashion, minimum or maximum as appropriate, except where noted. All three devices feature comparable supply voltages of ±15 volts, and input-output ranges of ±10 volts. All have user accessible SENSE and REFERENCE pins.

This table is essentially self-explanatory, but some comments are in order. Total output-referred offset voltage must be computed for the LF352 and AD521 types by adding the gain multiplied by the (listed) input offset to the (listed) output offset. A similar comment applies to offset drift. The 522, being a complete pretrimmed package, specifies offset as a single figure referred to the input.

Functional Diagrams

In Fig. 5-3 the functional diagrams of the respective three units are shown. The LF352, shown in Fig. 5-3A is packaged in a 16 pin DIP, and uses two gain set resistors, R_G and R_s. Both REFERENCE and SENSE terminals are provided for flexibility. Two external compensation components, R_c and C_c, are required for this device.

There are input and output voltage offset adjustment networks usable with the LF352, but their implementation is fairly complex. For clarity they have been omitted here; the reader is referred to the device data sheet for exact details.

Table 5-1. Instrumentation Amplifier Performance Data

	LF352	AD521K	AD522C
Gain Range	1-1000	1-1000	1-1000
Gain Error	0.2%	0.25-0.004G(typ)	G = 1 : 0.05% G = 1000 : 0.2%
Nonlinearity	0.1%	0.1%	G = 1 : 0.001% G = 100 : 0.005%
Gain TC	3 ppm/°C (typ)	(3 ± 0.05 G) ppm/°C (typ)	G = 1 : 2 ppm/°C G = 1000 : 50 ppm/°C
CMRR ($\Delta R_s = 1$ K) Gain = 1 Gain = 1000	 65 dB 105 dB	 70 dB 100 dB	 80 dB 110 dB
Offset Voltage Input[1] Output[1]	 30 mV 400 mV	 1.5 mV[1] 200 mV[1]	100 μV[1]
Offset Voltage Drift Input Output	 10 μV/°C (typ) 600 μV/°C (typ)	 5 μV/°C 150 μV/°C	 G = 1 : 10 μV/°C (RTI) G = 1000 : 0.5 μV/°C (RTI)
Input Bias Current	40 pA	40 nA	10 nA
Sense, Reference Pins?	Yes	Yes	Yes
Comments	Requires external compensation 2 gain resistors	2 gain resistors	Guard drive available shielded care 1 gain resistor

1 Adjustable to zero.

The AD521K is shown in Fig. 5-3B and is packaged in a 14 pin DIP. It also uses two gain set resistors. REFERENCE and SENSE terminals are provided. No other external components are required for operation for this unit, with the exception of the offset adjustment pot, which is optional. If desired or necessary, additional compensation may be added (optional).

The AD522 shown in Fig. 5-3C is a high accuracy instrumentation amplifier design conceptually similar to Fig. 5-2. Its high performance as delivered is achieved by on-chip laser trimming to the rated accuracy, as the device uses both laser trimmed op amps and resistor networks. This unit is packaged in a shielded 14 pin DIP, and REFERENCE and SENSE terminals are provided, in addition to a guard drive line used to bootstrap input shielded lines.

An optional offset trim is shown, but is not recommended except for the most stringent applications, as the device has only a (typical) 50 μV offset to begin with.

Instrumentation Amplifier Applications

In this section a few circuit applications which highlight the methods of effectively applying IAs are discussed. Some of these are the more classic converter oriented IA uses; others show specific points of usefulness unique to IAs.

Resistance Bridge Amplifier

The amplification of resistive bridge transducer outputs is a very common instrumentation problem. An IA circuit configuration which solves this problem is illustrated in Fig. 5-4.

Here the resistance bridge is composed of the four resistors of nominal value R, with one of these being sensitive to the physical quantity being measured. Examples would be strain-gauge bridges, thermistor bridges, etc.

The output signal from the bridge is amplified by the high input impedance IA, here an AD522C. The amplified output voltage is of the form shown where the term α is the deviation of the sensitive arm of the bridge. In general, the left term of this expression is the bridge output, the right term the IA gain.

Application points important here are the use of the AD522C guard terminal to drive the shield of the input line. In general shielded cables will usually be found necessary to minimize pickup; however, the capacitance present between the conductors and shield can seriously affect ac cmrr response. This problem is alleviated by driving the shield with a voltage level equal to the common-mode voltage, or more simply put, "bootstrapping" the shield.

To a lesser extent, the connections to R_G may desirably be bootstrapped. If possible R_G should be close to the AD522, and the PC lay-

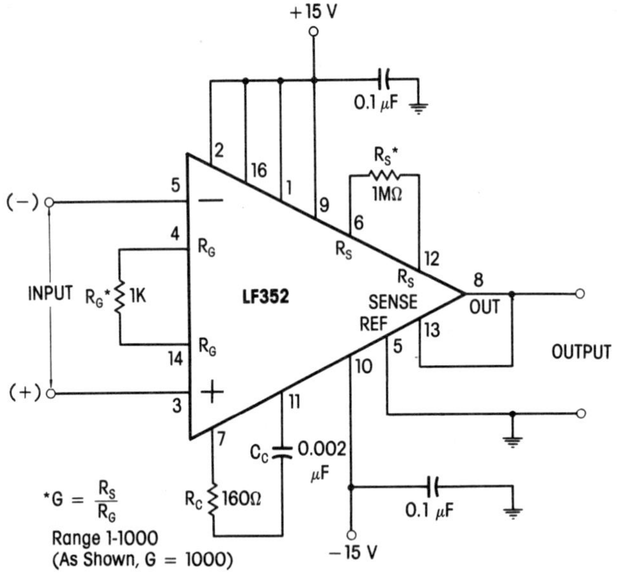

(A) LF352 (16 pin DIP)

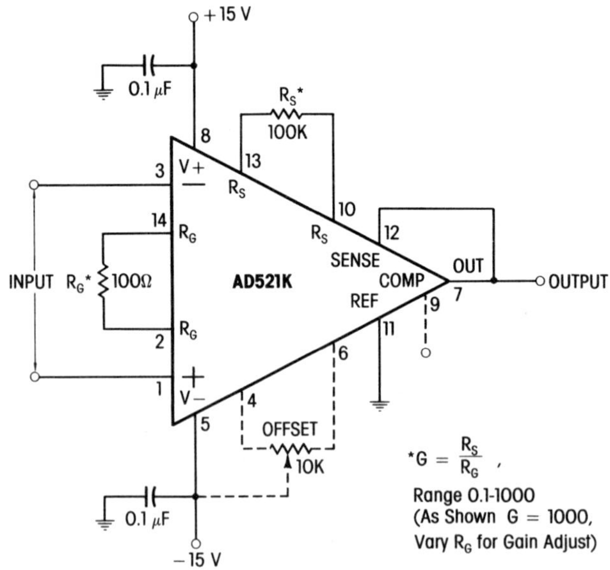

(B) AD521 (14 pin DIP).

Fig. 5-3. Examples of

out can use PC track guard conductors on either side of the R_G connections, with the guard tracks tied to pin 13. The pinout of the module is designed to accommodate this practice.

The excitation voltage for a resistive bridge should be stable, as it is actually a coefficient in the expression for output voltage. Here it is 10, corresponding to the AD581s 10-volt output. A locally derived reference source such as A2 can be quite attractive from a system viewpoint, as it minimizes noise and pickup. Further, since the voltage excitation for the bridge determines the system scale factor, the system "gain" is easily trimmed for calibration with the AD581.

Active Guard Drive

For IA devices that do not furnish a guard output terminal, the scheme of Fig. 5-5 can be used to drive a shield. In this circuit the gain set resistor R_G is split into two equal series-connected resistors, which causes the common mode voltage to appear at the midpoint. This point is buffered by a FET input op amp A2, which then drives the shield from its low impedance output. The op amp used for A2 need not be highly accurate in terms of drift or cmrr, but it should have high input impedance and good drive for capacitive loads. The LF355 (or LF356) are good choices for this use.

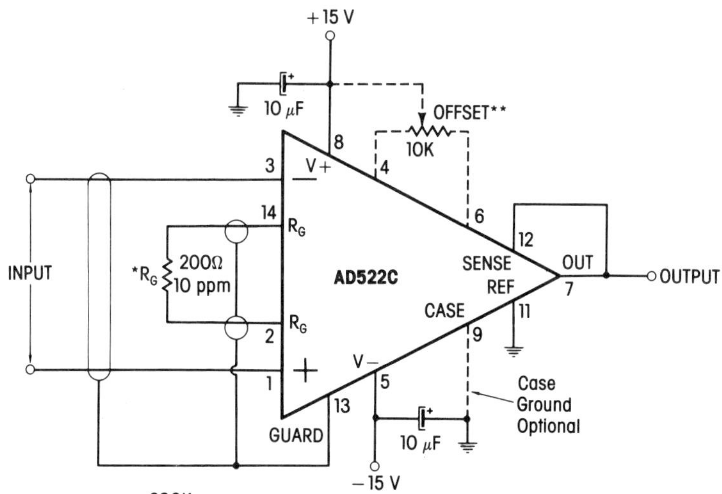

$$^*G = 1 + \frac{200K}{R_G},$$

Range 1-1000 (As Shown, G = 1000)

**Use Offset Trim Only When Necessary, TCR pot < 25 ppm/°C

(C) AD522 (14 pin DIP).

instrumentation amplifiers.

For best results R_G should be physically close to the IA device, and if desired it too can also be bootstrapped from A2's output. Similarly, A2 should be close to R_G.

Output Current Boost

Fig. 5-6 illustrates the use of the SENSE terminal of an IA to allow use of a current booster amplifier within the feedback loop. This technique allows a current swing up to the rated level of the buffer device used. Here A2 is an LH0002C, which has a 100 mA output rating. This device in general should possess unity voltage gain; otherwise, it may compromise IA stability.

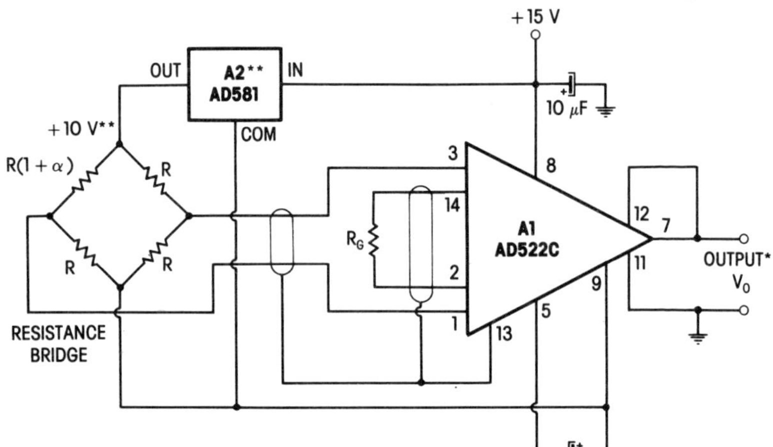

$$*V_O = \left(\frac{10\,\alpha}{4\left(1+\frac{\alpha}{2}\right)}\right)\left(1+\frac{200K}{R_G}\right)$$

**Trim AD581 For Desired Bridge Excitation And Calibration (See CH3)

Courtesy Analog Devices

Fig. 5-4. Resistance bridge amplifier.

The SENSE terminal (13) of the IA is connected directly to the output terminal of the circuit where it feeds the load. This will also correct for any wiring drop, if the path from A2 is lengthy. Similarly, the REFERENCE terminal (5) should connect to the end of R_L where it reaches circuit common.

High Gain IA Configuration

Some IAs can be used to achieve overall gains in excess of 1000 by using feedback dividers to the SENSE terminal. This can be an effective technique, as it may eliminate the necessity for an extra amplifier in some cases.

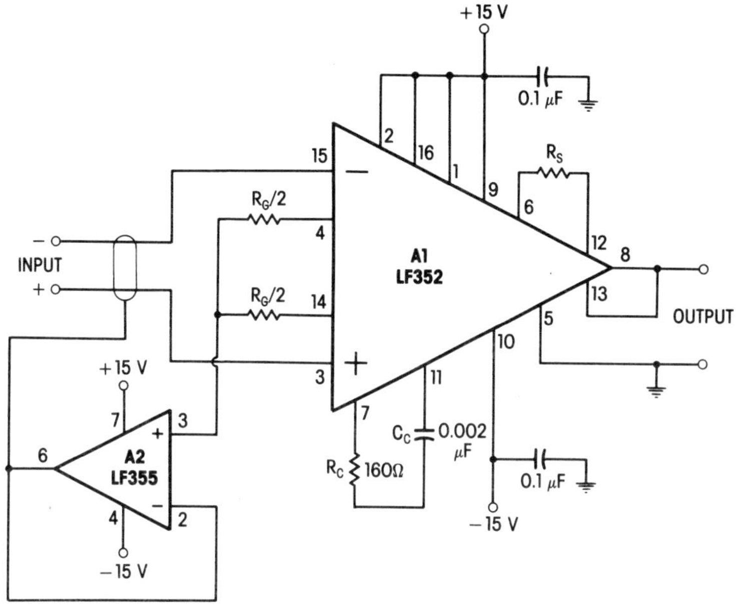

Courtesy National Semiconductor Corp.

Fig. 5-5. Active guard drive.

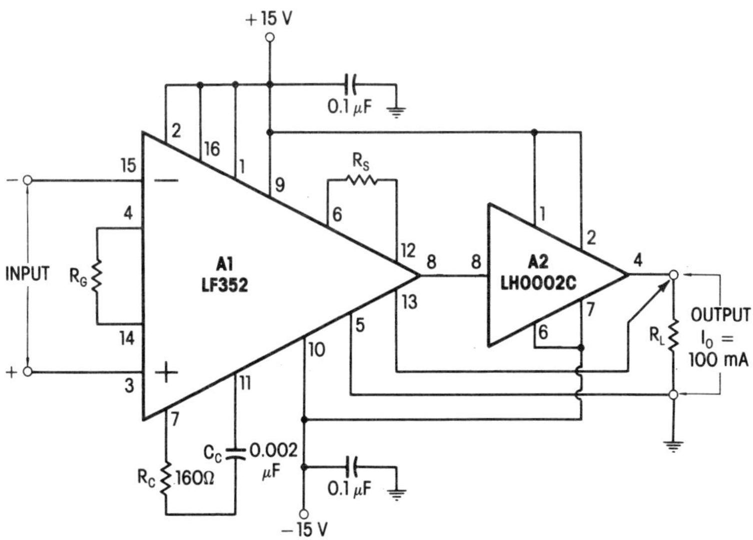

Courtesy National Semiconductor Corp.

Fig. 5-6. Output current boost.

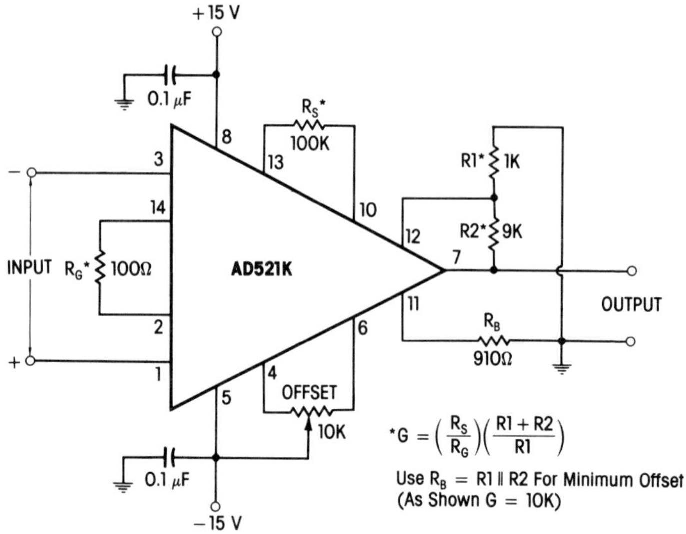

Courtesy Analog Devices

Fig. 5-7. High gain instrumentation amplifier.

The method is shown in Fig. 5-7 using the AD521K instrumentation amplifier. Here R1 and R2 are the feedback divider which causes the output stage gain to be (R1 + R2)/R1, rather than unity as in the normal connection. Note the new overall gain expression for this hookup in the figure. R1 and R2 should be stable resistors, similar to R_s and R_G, consistent with the overall accuracy requirements.

With the addition of R1 and R2, an additional offset voltage will be generated, due to REFERENCE and SENSE input bias currents. This can be minimized by using R_B, a bias compensation resistor which is equal to the parallel equivalent resistance of R1 and R2.

The reference terminal can still be used for offsetting purposes with this circuit, if desired. However the gain from REFERENCE to output terminal in this case is now raised by the same factor as signal gain, so smaller offsets are allowable at the REFERENCE input.

Due to the very high overall gain, an offset null is advisable for this circuit.

Current Monitor

Often it is required to monitor a current flowing in a conductor, where the common mode voltage may be high with respect to ground. The circuit of Fig. 5-8 shows how this can be implemented with an instrumentation amplifier.

Here a current sampling resistor R_x is inserted in the current path to be monitored. Its value is chosen to drop a relatively small voltage, so as not to disturb the circuit being fed. Generally this will be below 1 volt; here, a 0.1-volt level is chosen.

With values shown, R_x will drop 0.1 volt with a 10 A current flowing. The IA is set up for a gain of 100; therefore, the differential voltage across R_x will produce a 10-volt output with 10 A flowing in R_x.

Due to the high common mode rejection of the IA, the common voltage appearing at R_x can range ±10 volts with respect to circuit common, without loss of accuracy.

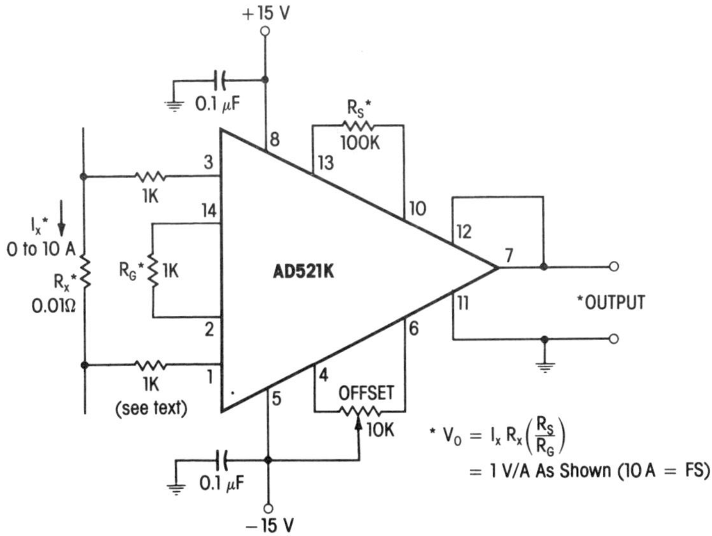

Fig. 5-8. Current monitor.

For other values of current, R_x should be adjusted accordingly. The circuit as shown will work with good accuracy down to about a 10 μA current level. If it is desired to go appreciably lower, the LF352 can be used, as its 3 pA (typical) input current is very low.

If voltages can exist on the line monitored, without power applied to the IA, it would be prudent to include IK series input resistances (shown dotted), for safety against input faults. This is not necessary in the AD521K, as it has internal protection. However it may be in other IA types.

Floating Input Voltmeter

Fig. 5-9 shows how an IA can be used to implement a voltmeter with a "floating" input. This circuit makes use of the natural differential voltage input operation of the IA, in conjunction with the operation of the SENSE terminal.

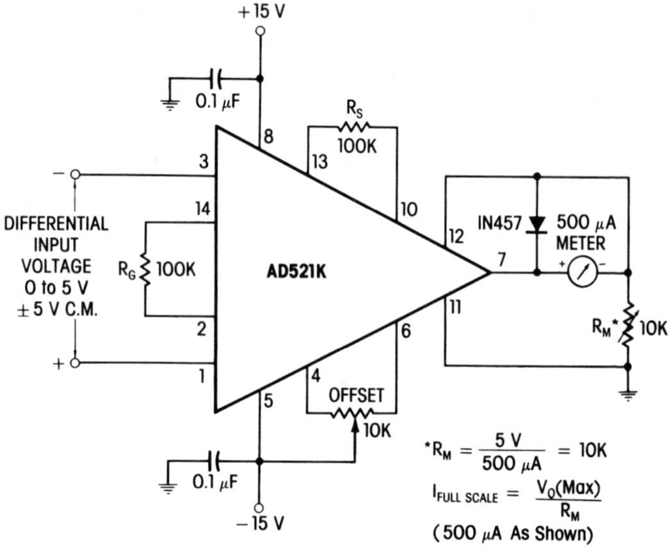

Fig. 5-9. Floating input voltmeter.

The input and gain scaling of this circuit is more or less conventional, and serves to translate a floating 5-volt input into a 5-volt potential appearing across R_M.

With the current meter connected as shown, and the SENSE terminal monitoring the voltage across R_M, the current in R_M and the meter will be linearly proportional to the input voltage. Thus R_M can be adjusted to calibrate for the desired full scale current. As shown here a 500 μA meter is used, so R_M is nominally 10K.

Many options are possible with this circuit. One would be to switch select R_s or R_G for various input sensitivities. The offset should be adjusted for zero meter defection, at the highest gain used.

Another option would be to use the IA offset adjust capability to provide a half-scale meter offset, which would allow voltages to be measured without polarity reversal of leads.

Of course, the circuit can also be used as the "front end" of a digital voltmeter circuit as well, by connecting the output in a conventional fashion to feed the digital voltmeter.

Floating Voltage-to-Current Converter

The circuit of Fig. 5-10 is interesting in that it can translate a floating (up to ±10 volts) differential voltage into a proportional current. This current can be referred to ground, or to voltages above or below ground up to ±10 volts. Similar circuits are possible with op amps (for instance the Howland current source), but they require five precision resistors.

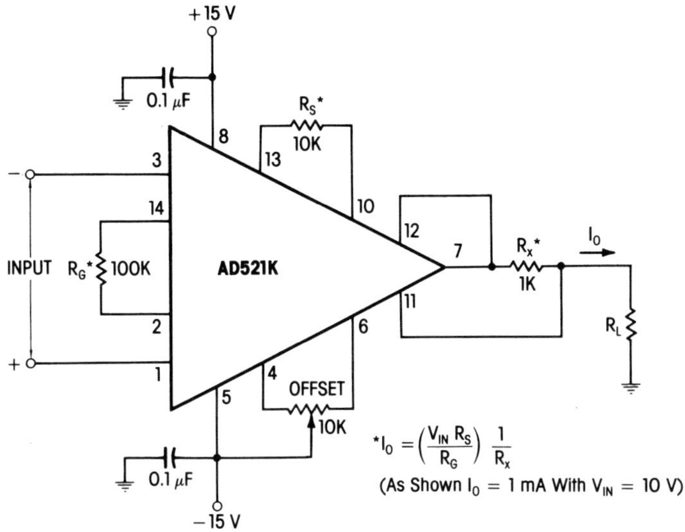

Fig. 5-10. Floating input voltage-to-current converter.

The circuit here uses only three precision resistors, and they need not be matched.

In operation the instrumentation amplifier is set up for an input to output voltage transfer which will result in 1 volt across R_X. This is to maximize the voltage compliance of the circuit. Here a 10-volt input results in 1 volt across R_X; thus, with R_X equal to 1 mA, I_o is equal to 1 mA.

It should be noted that since the reference and sense terminals can swing ±10 volts, this circuit can deliver up to ±9 volts to R_L. R_L need not be grounded; it can be referred to any voltage within this range. However, the absolute level as seen at either the SENSE or REFERENCE pin must be less than ±10 volts in all cases.

Because of the relatively low output voltage of 1 volt, an offset adjustment is recommended for best results. Output offset drift is the major limitation of this circuit, being about 50 ppm/°C due to the instrumentation amplifier.

ANALOG SWITCHES

An analog switch is a linear circuit element which is used for a wide variety of switching and control functions in data conversion systems. Its circuit function is generally just as its name implies, the switching of analog signals, with the on/off state usually being accomplished by control logic.

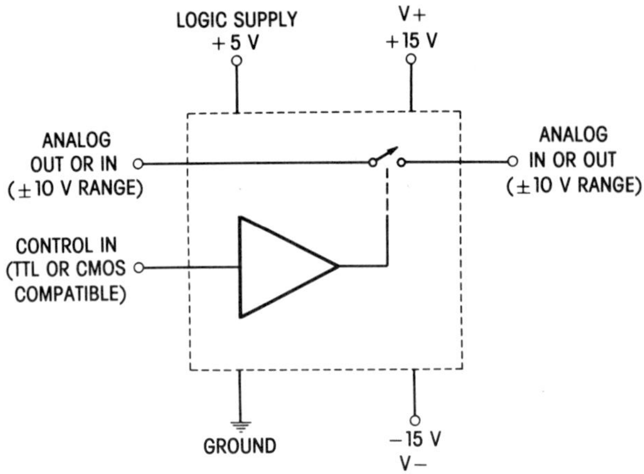

Fig. 5-11. Analog switch functional diagram (general).

The hardware typically used for analog switches are ICs which use various forms of semiconductor switching elements in the analog path. Integral to these ICs are also driver circuits which interface the logic level control input to the switch itself. Typically, analog switches can control signal levels up to ±10 volts, with switching times of a few microseconds. They are usually *bilateral* devices, passing signals in either direction, with low distortion and low signal degradation. On resistances are typically on the order of 100 ohms or less, while off state resistance is on the order of 10^{10} ohms, or leakages of 1 nA or less. Analog switches have inherently zero voltage offset, as they are basically voltage controlled resistors.

"Contact" configurations come in a wide variety of options, from the most basic spst arrangements, up to dual 1 of 8 selector (multiplexer) configurations. A general functional diagram for an analog switch is shown in Fig. 5-11. The switching element itself is represented by the switch symbol, and its terminals may be considered to be isolated (floating) for signal voltages up to ±10 volts. The driver, which controls the state of this switch, accepts a TTL or CMOS logic input and translates it into the on/off voltage levels required by the actual switch(es) in use. It is this driver circuit which requires the power supplies, usually ±15 volts, but these voltage levels can vary for some types of switches. A +5 volt logic supply may also be required, as may a ground terminal, but is not necessary in all cases. Some specialized types of switches require no supply voltages at all.

Switch Types

Fig. 5-12 shows in much simplified form the various semiconductor devices which are commonly used as analog switches.

JFET Switches

In Fig. 5-12A are shown JFET devices, which can be of either n- or p-channel designs. From an application standpoint, the main attribute of a JFET switch is its low "on" resistance, plus the fact that this resistance remains constant with analog voltage level.

PMOS Switches

In Fig. 5-12B is shown a PMOS switch. This type of switch operates in an enhancement mode, requiring the presence of a bias voltage to be on. Its disadvantage is an appreciable variation in "on" resistance with voltage, and (usually) a requirement for odd power supply voltages, +10 volts and −20 volts being common.

CMOS Switches

In Fig. 5-12C is shown a CMOS switching architecture. This is simply a pair of n and p MOS transistors paralleled, with complemented control drive, that is with both units being either on or off together.

CMOS swtiches eliminate the drastic variation in "on" resistance with signal which is characteristic of PMOS, due to the complementing NMOS and PMOS device characteristics. This type of switch's advantage is an analog signal swing capability which extends to the supply levels used, or ±15 V (usually).

A strong disadvantage of early CMOS switches (and some types still available) was a sensitivity to "latch-up," a usually destructive failure mechanism which can occur when the analog signal level exceeds either supply level. This phenomenon has been largely conquered in more recent types of CMOS switches. But, the user should be aware of it in general, and ensure that the device contemplated for use is "latchup free," or "overvoltage protected." Designs which feature floating body or dielectrically isolated construction are protected against latchup. Regard specifications carefully, as there are application features which distinguish even the protected CMOS switches.

Virtual Ground Switches

The virtual ground switch is illustrated in Fig. 5-12D. This is actually a specialized form of p-channel JFET switch. It is designed for simplicity and low cost, and requires no driver circuit or power supply voltages. The JFET element used is designed for a cutoff voltage of 5 volts or less, which is a voltage swing compatible with TTL logic directly.

By placing the JFET at the virtual ground of an op amp (not part of the switch itself but necessary) the FET sees very small signal levels. Thus a 5-volt logic level can fully control its conduction state. This is accomplished for 5-volt swings by using a conventional TTL gate, and adding a pull-up resistor, such as R_L. Parts available in this switch family type are either 5-volt or 15-volt logic-drive compatible; thus, 15-volt driven parts would use an open collector 15-volt gate, with R_L referred to a +15-volt supply. For either 5-volt or 15-volt sensitivity units the low or ground logic state turns the switch ON, the high logic state is OFF.

It can be noted from the use of the op amp and resistors R_F and R_{IN} that this switching configuration must be *user defined* for input impedance and gain. This is accomplished similarly to any other op-amp inverting circuit, with allowances for the FETs on resistance of about 100 ohms. Also note that this form of switch is *not* bilateral in operation.

Positive Signal Switches

A positive signal switch is shown in Fig. 5-12E. As may be noted, this type of switch also uses a p-channel JFET, but with the analog signal input range restricted to the 0 to +10-volt range. With a logic control level of +15 volts, a 5-volt cutoff voltage JFET can control levels up to +10 volts, thus the term positive signal switch. This type of switch also requires no power supply.

ANALOG SWITCHING DEVICE CHARACTERISTICS

Table 5-2 summarizes the various types of analog switching elements, and discusses typical specifications of representative units. This table is for the most part self-explanatory, and summarizes also the key pros and

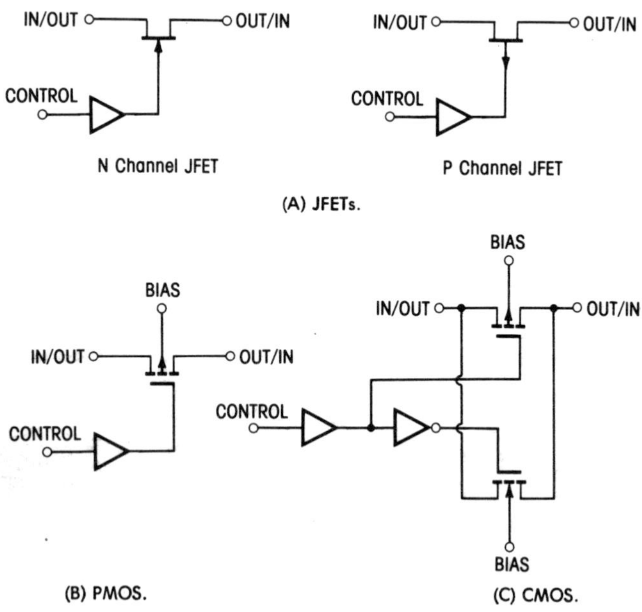

Fig. 5-12. Analog

cons of each type. The specifications listed in the table are those of the specific device example listed.

Analog Switch Specifications

Analog switches have several operation-defining specifications which are the most important keys to their successful selection and use. They are briefly described below.

Analog Voltage Range

This parameter defines the range of analog voltages which can be applied to the terminals of the switching element, with completely specified on/off control by the logic section. This voltage range is a strong function of the switch technology in use, and indirectly dictates the supply voltages which must be used. CMOS types for instance usually can control analog levels up to the supply lines, or typically ±15 volts. JFET types require 5 to 8 volts of negative bias below the most negative analog swing, or a −15 to −18-volt supply to handle a −10-volt analog level. A bias of about 10 volts more negative is required in PMOS switches, which typically require a −20-volt supply.

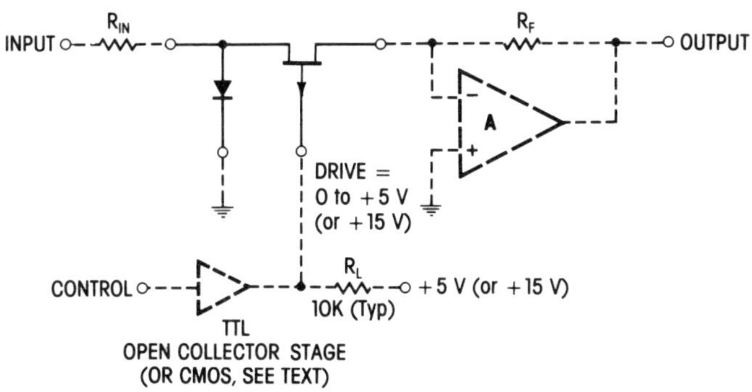

(D) Virtual ground switch.

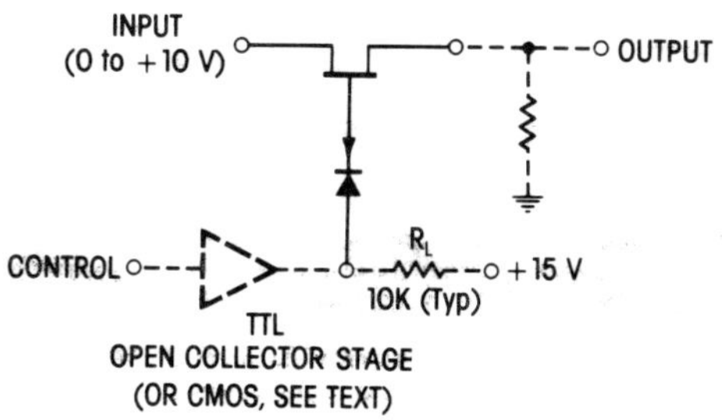

(E) Positive signal switch.

switching elements.

Table 5-2. Analog Switch Characteristics

Switch Type	JFET	PMOS	CMOS	Virtual Ground	Positive Signal
Representative PNs	DG120, 130, 140 series (old) DG180, 190 series	DG170 series MM554, 554 3705	DG200/201, DG300 series IH5040 series AD7510 DI series HI1800 series CD4016/CD4066	IH5009 series AM97C09 series	IH5025 series
Typical characteristics	DG126 (old) DG181 (new)	DG 172	DG200	IH5009 series	IH5025 series
Signal handling:	±10 V ±7.5 V	±10 V	±15 V	±2 mA[1]	0 to +10 V (+20 V max)
Ron	10 to 100 Ω	150 Ω	< 100 Ω	100 Ω	90 Ω
Ron match	30%	20%	30%	50%	50%
Switching times					
ON	1 μs 180 ns	0.5 μs	1 μs	0.150 μs	0.1 μs
OFF	2 μs 150 ns	1 μs	0.5 μs	0.3 μs	0.1 μs
Break-before-make	No Yes	No	Yes	No	No
Drive required	TTL TTL[3]	TTL	TTL, CMOS (3)	+5 or +15 V	+15 V
Supplies[2]	+12 V, −18 V (old) ±15 V (new)	+10 V, −20 V	±15 V	None	None
Advantages:	Low Ron Constant Ron vs. signal Fast switching (180/190)	Fail safe	High signal levels Fail safe Low power	No supplies Inexpensive	No supplies Inexpensive
Disadvantages:	Odd supplies or limited analog range	Δ Ron vs. signal Odd supplies	Latchup prone (older units)	Low impedance Accuracy resistor dependent Unilateral switching Requires op amp	Unipolar +15 V drive

1 Current based operation, voltage set by series resistor.
2 For a ±10 V analog signal control capability.
3 Programmable logic threshold.

Full range levels of ± 10 volts will therefore require consideration of the supply levels to be used, to ensure compatibility of a given switching technology or specific device.

On Resistance

This important analog switch parameter can affect a large number of overall performance parameters. Among these are acquisition and/or settling time, offset voltage error, gain accuracy (and its TC), and nonlinearity. Generally, a lower on resistance is better, and JFET types currently yield a lowest on resistance.

All analog switches have a temperature coefficient of resistance; this should be considered when the switch is used in combination with gain-scaling resistors. Also, in some types, the on resistance varies with analog voltage; this should also be considered for high accuracy applications. Examples of this are PMOS types, and to a much smaller extent, some CMOS types.

Leakage Currents

Leakage currents are those currents which effectively appear between the switch terminals, and either the supplies or ground. This leakage is usually given as a single lumped parameter, typically on the order of 1 nA at 25°C, although it varies from type to type.

Leakage usually doubles for every 10°C rise in temperature; thus, it will be worst at elevated temperatures. This high temperature leakage will result in an offset voltage error, when multiplied by the switch on resistance (or on resistance plus source resistance, if significant). Leakage tends to be higher in multiplexer analog switches, as they have many channels paralleled at their output.

Switching Times

Analog switches exhibit delay between logic input and actual switched analog output, for both on-to-off and off-to-on transitions. This delay may range from 100 ns or less, to 1 to 2 μs, dependent upon the driver technology used in a given switch. Usually the switches themselves are faster than the drivers. This is evident in driverless types of switches such as the virtual ground and positive signal types, which switch is about 100 ns.

The relative time delays for switching off and on can be important if a break-before-make operation is desired. In such case the off switching time must be faster than the on time, otherwise there will be an overlap interval where two switches can be on simultaneously.

Higher speed switches are useful for sample/hold applications, where aperture time (and its uncertainty) is usually a critical parameter. They are also useful in minimizing spurious outputs, such as "glitches," or other time-related output voltage errors.

Switch Capacitance

Analog switch output capacitance can influence a number of performance parameters. At the output for instance, the shunt capacitance forms a low-pass filter in conjunction with the on resistance (plus any source resistance). This filter will have an associated settling time which will limit the overall data rate, and/or accuracy.

Since it forms a high pass filter in conjunction with the load resistance, capacitance from input to output of the switch will determine the leakage or feedthrough in the off state. Thus for high frequencies there will be a minimum off-state attenuation.

In sample-hold applications (see sample hold section), capacitance from the switch control input gate to output will result in a charge transfer. This error can be compensated by inducing an equal, but opposite, charge through the use of a small capacitance and reverse-phase signal. An interesting note relevant to this effect is that CMOS switches tend to minimize charge transfer (to a degree), because they use equal and opposite voltage swings to turn off the n- and p-MOS switching devices.

Logic Thresholds

Currently available analog switches tend to be either (or both) TTL or CMOS logic compatible, but to be entirely safe in a given circumstance data sheets should be carefully consulted on drive requirements. This is simply because there are so many interpretations of what constitutes TTL or CMOS "compatibility."

True TTL compatibility implies a fully defined switch state for a logic input "1" level of +2 to 5 volts, a "0" level of 0 to +0.8 volt. High, or 1-state required input levels *higher* than 2 volts are *not* worst case TTL compatible, and to ensure safety should use a pullup resistor to the +5 volt logic level.

CMOS input levels can be confusing also, as CMOS logic can operate from +5 to +15 volts. In general, when driven from CMOS the voltage input threshold of the driven device (the analog switch) should be desirably centered in the range of source logic swing. This is not absolutely necessary, however, as long as the CMOS driving stage can guarantee worst case "0" and "1" levels at the current load level of the driven device.

Some analog switches offer logic voltage programmability through the use of jumpers and package pins. Consult specific data sheets in such circumstances to ensure complete compatibility.

Representative Diagrams

Fig. 5-13 illustrates by way of specific device examples each type of analog switch in a function diagram. Also shown in this figure are the control states for each switch, with appropriate comments as necessary.

A popular early JFET switch series is the DG120, 130, 150 series (Siliconix). In this series there are many different contact options. Four of these types shown in Fig. 5-13A are the DG126, DG134, DG-152, and DG154.

The DG134 and DG126 are dual spst and dual dpst types respectively. They are both contained in a 14 pin DIP, with the spst unit deleting use of four pins. These devices use a +12-, −18-volt power supply to switch signals of ±10 volts. Logic control is described by the control table.

The DG154 and DG152 are similarly arranged units, which operate from ±15-volt supplies and can switch signals up to ±7.5 volts.

More modern JFET type switches are the DG181, DG182 types (Siliconix) also shown in Fig. 5-13A. The DG180-190 series is much faster in speed, and features break-before-make operation, and offers various on resistance levels and contact arrangements.

The DG172 PMOS switch (Siliconix) is illustrated in Fig. 5-13B. This switch is arranged in a 4 in, 1 out configuration, suitable for multiplexer applications. It uses +10-, −20-volt power supplies to switch signals of ±10 volts. The logic control definitions are shown in the table, and the device requires a +4- to +5-volt logic level at the V_L terminal, which may also be used as an ENABLE control if desired. This device is packaged in a 14 pin DIP.

The DG200 (Siliconix) is a representative CMOS switch, arranged in a dual spst setup (Fig. 5-13C). This device uses supplies of ±15 volts, and can switch analog levels of this magnitude. It is overvoltage protected, and packaged in a 14 pin DIP.

The DG200 logic control is as indicated in the table. This device has a V_{REF} pin, which when left open establishes a TTL or 5-volt CMOS logic input compatability. The voltage at this pin may be externally adjusted via a resistive divider for other logic levels, if desired.

The IH5021/IH5022 (Intersil) are single channel virtual ground switches from the IH5009 series, and are illustrated in Fig. 5-13D. A device of this type is shown in a complete circuit as it would typically be used. The 10K input and feedback resistors set the on state gain to unity, when the control input is low.

It can be noted that this particular switch uses a second JFET, with its gate-source shorted. This is a "compensating" FET, which causes the feedback path resistance ($R2 + R_{on}$) to be equal to the input path ($R1 + R_{on}$). This is done in applications requiring highest accuracy; for modest requirements, this FET need not necessarily be used. Parts in the series have the option of including or excluding the compensating FET. Both devices come in an 8 pin MINIDIP package. The IH5022 requires a +5-volt drive; the IH5021 a +15-volt drive.

The IH5009 series was originally introduced by Intersil, but is additionally supplied by National Semiconductor under the AH5009 series designation. National Semiconductor also supplies their own proprietary version of virtual ground switches, the AM97C09 series. These devices are functionally similar to the 5009 series, but designed to be compatible with CMOS drivers, eliminating the need for pull-up resistors.

Positive signal switches are illustrated in Fig. 5-13E. Here is shown the IH5033/IH5034 (Intersil), a two input, common output switch. These devices are best (although not necessarily so) used with a buffer op amp, such as is shown.

The logic control is as shown, but it should be noted that only a single switch is allowed to be on at a given time.

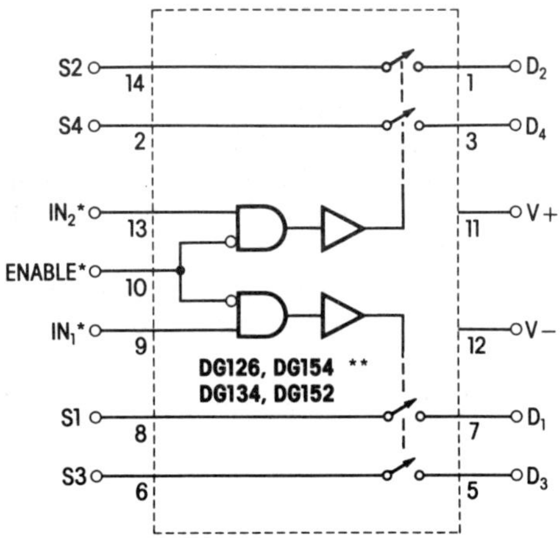

*CONTROL TABLE

CONTROL			SWITCH STATES			
IN_1	IN_2	EN	CH1	CH3	CH2	CH4
0		0	OFF	OFF		
1		0	ON	ON		
X		1	OFF	OFF		
	0	0			OFF	OFF
	1	0			ON	ON
	X	1			OFF	OFF

**SELECTION GUIDE

DUAL SPST	DUAL DPST	SUPPLIES
DG134	DG126	+12, −18 V
DG152	DG154	±15 V

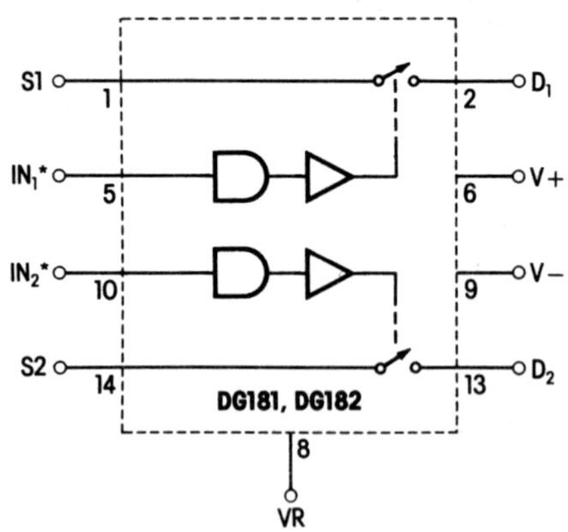

*CONTROL TABLE

CONTROL		SWITCH STATES	
IN_1	IN_2	CH_1	CH_2
1		OFF	
0		ON	
	1		OFF
	0		ON

V_R = 0V (GND)
1 = $\geq$ 2 V
0 = $\leq$ 0.8 V

(A) JFET switches—DG126, DG154, DG134, DG152, DG181, DG182 (14 pin package)

Fig. 5-13. Functional diagrams

Analog Switch Pinout/Functional Diagrams

The general characteristics of each analog switch type have been described in the context of representative devices from each group. To supplement this information, the information contained in Fig. 5-14 shows the pinouts and functional diagrams of major popular analog switch family types. This information may be used as a form of "mini-catalog" of available types, as a guide to switches available. For more complete specific information on each switch type consult the individual device data sheets.

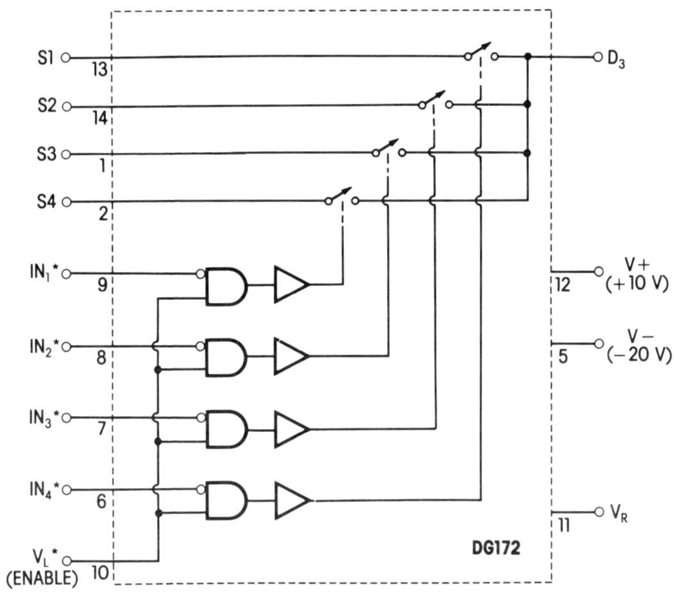

* CONTROL TABLE

CONTROL					SWITCH STATES			
IN_1	IN_2	IN_3	IN_4	V_L	CH_1	CH_2	CH_3	CH_4
0	1	1	1	1	ON	OFF	OFF	OFF
1	0	1	1	1	OFF	ON	OFF	OFF
1	1	0	1	1	OFF	OFF	ON	OFF
1	1	1	0	1	OFF	OFF	OFF	ON
X	X	X	X	0	OFF	OFF	OFF	OFF

V_L = ENABLE $\geq$ 4 V
V_L = INHIBIT $\leq$ 0.4 V
CONTROL AS SHOWN WITH V_R = 0 V

(B) PMOS switch, DG172 (14 pin DIP).

of representative switches.

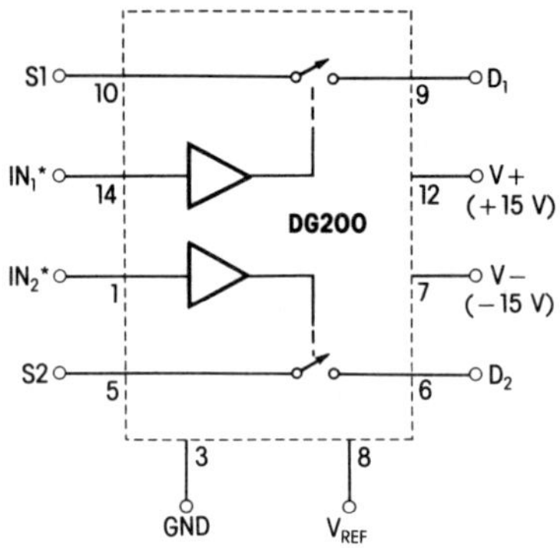

*CONTROL TABLE

CONTROL IN_1	CONTROL IN_2	SWITCH STATES CH_1	SWITCH STATES CH_2
1		OFF	
0		ON	
	1		OFF
	0		ON

CONTROL AS SHOWN WITH V_{REF} OPEN
1 = +2.4 to +15 V
0 = 0 to +0.8 V

(C) CMOS switch, DG200 (14 pin DIP).

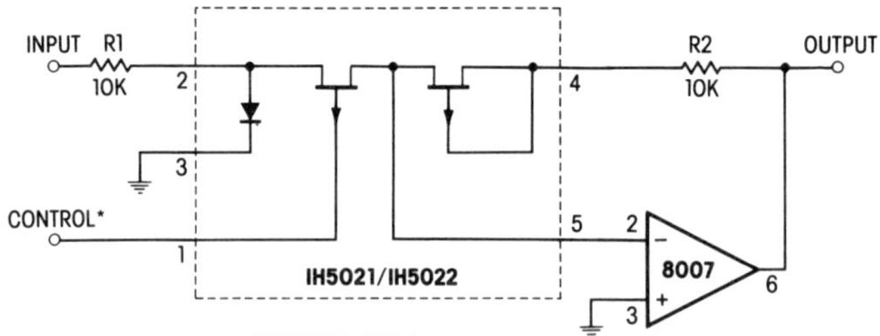

*CONTROL TABLE

CONTROL	OUTPUT
1	OFF
0	ON

1 = +5 V IH5022
1 = +15 V IH5021

(D) Virtual ground switch, IH5021/5022 (8 pin DIP).

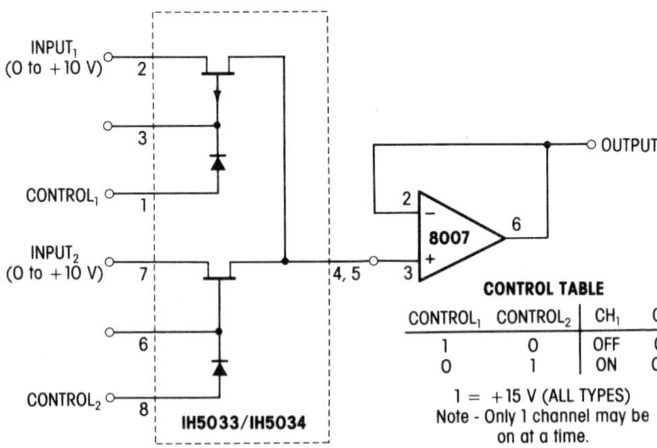

CONTROL TABLE

$CONTROL_1$	$CONTROL_2$	CH_1	CH_2
1	0	OFF	ON
0	1	ON	OFF

1 = +15 V (ALL TYPES)
Note - Only 1 channel may be on at a time.

(E) Positive signal switch, IH5033/IH5034 (8 pin DIP).

Fig. 5-13. Functional diagrams of representative switches. (cont)

LF112XX/LF123XX/LF133XX Series

In Fig. 5-14A are shown the National Semiconductor LF112XX/LF123XX/LF133XX series of quad JFET switches. These units are of a p-channel JFET construction, and feature a full ±10-volt analog signal handling range, using standard ±15-volt supply voltages. Logic control is TTL compatible, and they feature break-before-make switching action. A disable input is provided on most types. The LF11201 is a pin compatible replacement for the DG201 type, the LF11202 is similar but with an opposite logic control state.

IH5040 Series

Fig. 5-14B illustrates the Intersil IH5040 series of CMOS switches. This line of devices is the broadest available family of CMOS switches, featuring a wide variety of functions. All operate from ±15-volt supplies and feature analog control ranges to greater than ±10 volts. They feature break-before-make switching, and a logic control which is TTL and CMOS compatible. They are overvoltage protected, and have an antilatchup design through the use of a "floating body" process. Specific types are offered as pin for pin replacements of earlier JFET types, such as the DG180/190 series.

The IH5040 series is second-sourced by Harris Semiconductor as their HI-5040 series. This series uses a dielectric isolation process which is inherently antilatchup, and offers lower leakages. Analog signal ranges extend to the level of the supplies (normally ±15 volts).

DG300 Series

Fig. 5-14C illustrates the Siliconix DG300 series of CMOS switches. This line of devices features fast, break-before-make switching (150 ns typical), with low power dissipation. Analog signal range is ±15 volts with ±15-volt supplies. The famliy is divided into two classes, devices which have TTL control characteristics and those with CMOS control characteristics. Switching functions are the same in both classes.

The HI-1800A

Fig. 5-14D illustrates the Harris Semiconductor HI-1800A, a general purpose 4 channel CMOS switch. This device has a variety of control options. It features the antilatch and low leakage properties of the dielectric isolation process, and uses TTL control logic.

AD7510DI Series

Fig. 5-14E illustrates the Analog Devices AD7510DI series of CMOS switches. This series features low leakage, antilatchup and overvoltage protection, due to the use of a dielectric isolation process. Input control characteristics are TTL and CMOS compatible.

CD4016, CD4066

Fig. 5-14F illustrates the RCA CD4016 and CD4066 CMOS 4 channel analog switches. Although these devices are restricted to a 15 volt total analog range (0 to +15 volt or ±7.5 volt) and a 15-volt total supply, they are very popular due to their wide availability and low cost. The CD4016 has a typical on resistance of 300 ohms at +15 volts, while the CD4066 has an 80 ohm on resistance. Otherwise the types are similar. Signal control is "on" for a logic 1 control state.

The DG508

The Siliconix DG508 8 channel CMOS multiplexer analog switch is illustrated in Fig. 5-14G. This device features a 1 of 8 analog signal selection function, and has an on-chip logic decoder. Control inputs are TTL compatible, using three bits for address, plus a master enable bit. Switching is break-before-make, and signal range is ±15 volts, with ±15-volt supply. The device uses a latchproof process.

The DG508 is second sourced. One noteworthy unit is the Harris HI-508A, which uses the dielectric isolation process, and guarantees no signal channel interaction in the event of power loss in the presence of signal inputs.

The DG506

The Siliconix DG506 16 channel CMOS multiplexer is illustrated in Fig. 5-14H. This device features a 1 of 16 analog signal select function with an on chip decoder. Control inputs are TTL compatible, four bits for address plus an enable bit. Signal handling characteristics are generally similar to the DG508.

The DG506 is also second sourced by Harris as the HI-506A, which features the analog characteristics described above for the HI-508A.

The DG509

The Siliconix DG509 4 channel differential multiplexer is illustrated in Fig. 5-14I. This device features a differential (or dual) 1 of 4 analog signal select function, with a decoder. Control inputs are TTL compatible, two bits for address plus an enable bit. Signal handling characteristics are generally similar to the DG508, with the exception that a *pair* of analog signal channels are active for each control state.

The DG509 is second sourced by Harris as the HI-509A.

The DG507

The Intersil IH5009 series of virtual ground switches is illustrated in Fig. 5-14J. This device features a differential 1 of 8 analog signal select function, with a decoder. Control inputs are TTL compatible, three bits for control, plus an enable bit. Signal handling characteristics are

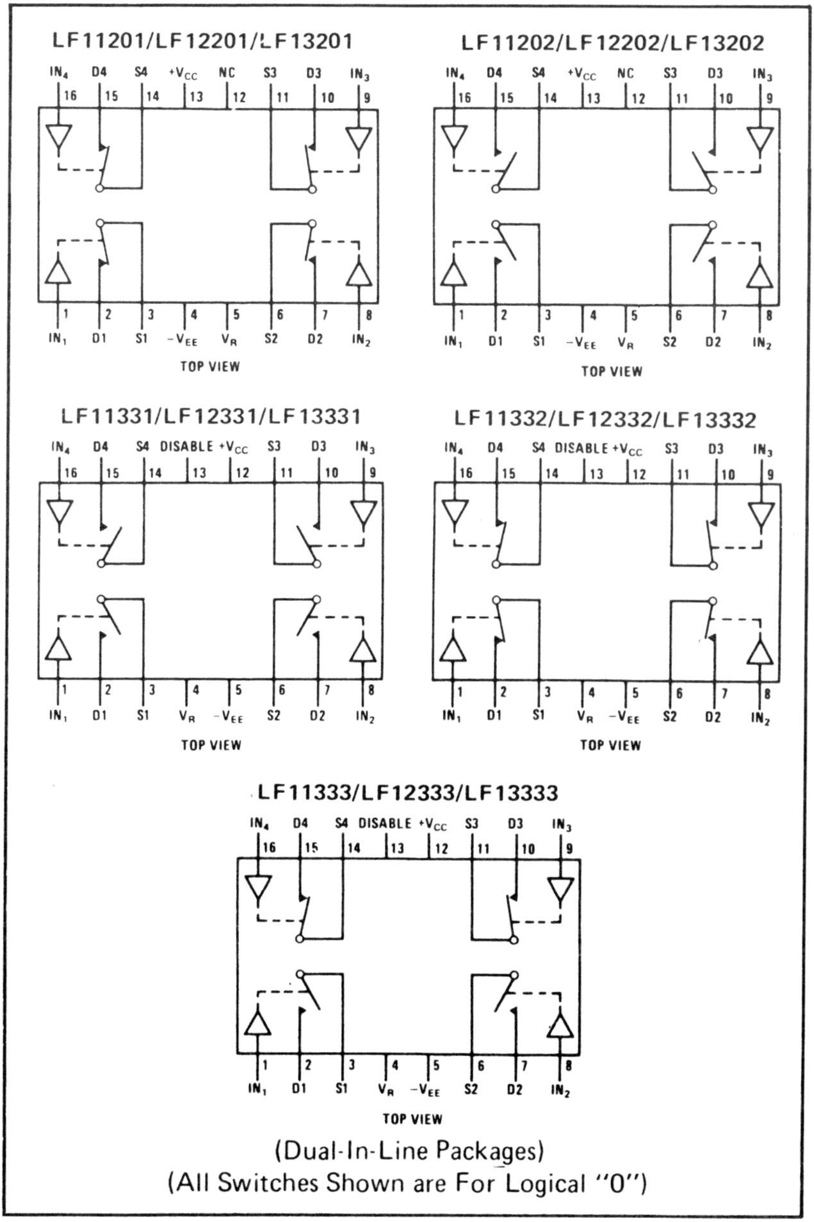

Courtesy National Semiconductor Corp.

(A) LF112XX, LF113XX series.

Fig. 5-14. Analog switch family pinouts/functional diagrams.

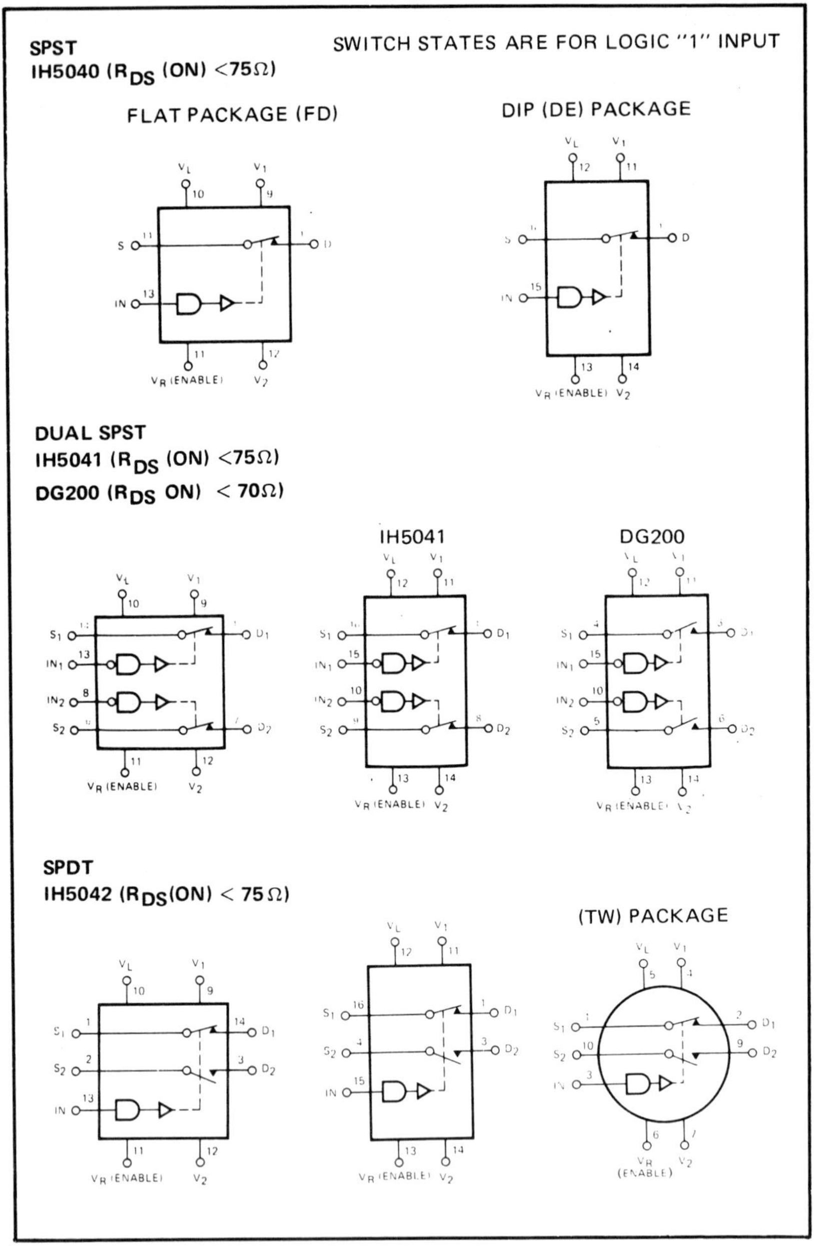

(B) IH5040 series.

Courtesy Intersil, Inc.

Fig. 5-14 (cont). Analog switch family pinouts/functional diagrams.

DUAL SPDT
IH5043 (R_{DS} (ON) <75Ω)

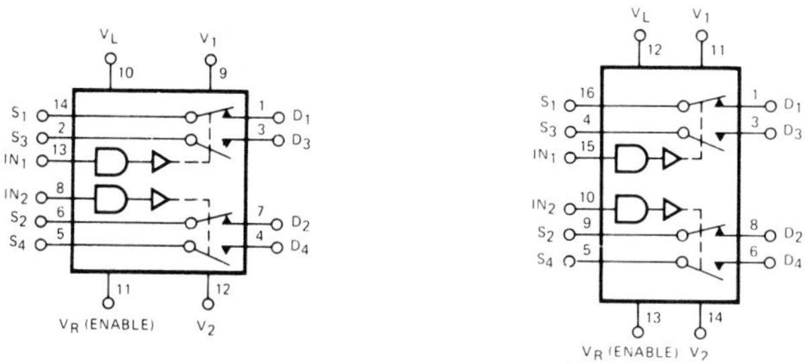

DPST
IH5044 (R_{DS} (ON) <75Ω)

(TW) PACKAGE

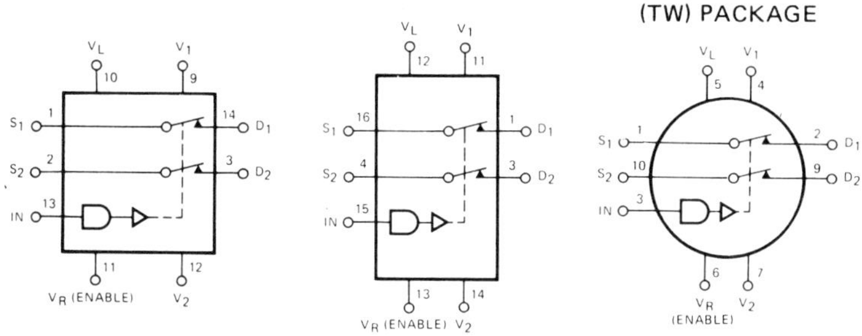

DUAL DPST
IH5045 (R_{DS} (ON) <75Ω)

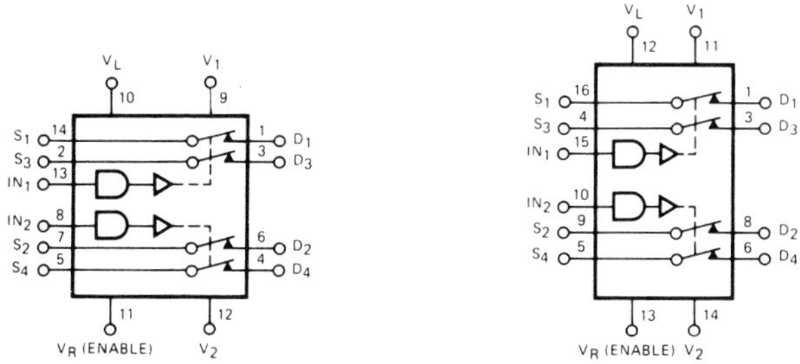

(B) IH5040 series. (cont)

Courtesy Intersil, Inc.

Fig. 5-14 (cont). Analog switch family pinouts/functional diagrams.

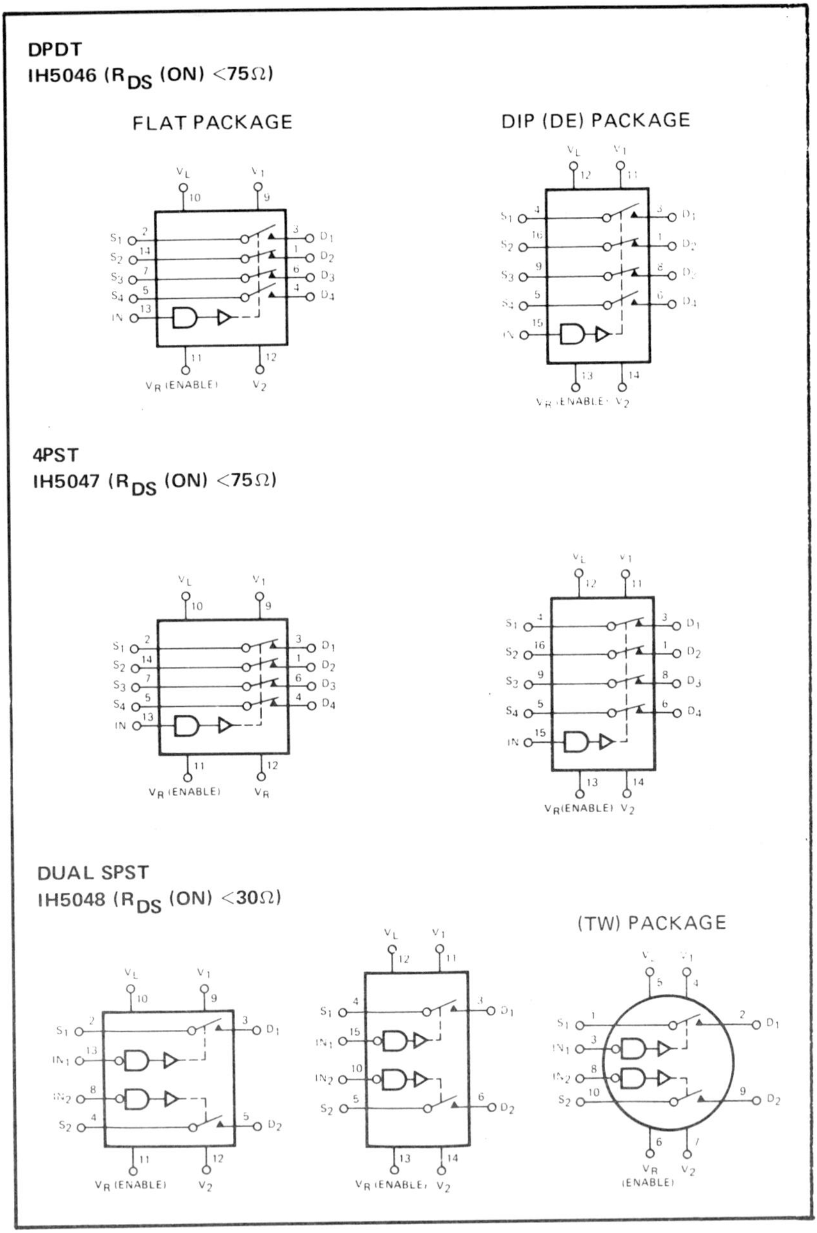

Courtesy Intersil, Inc.

(B) IH5040 series. (cont)

Fig. 5-14 (cont). Analog switch family pinouts/functional diagrams.

DUAL DPST
IH5049 (R_{DS}(ON) < 30 Ω)

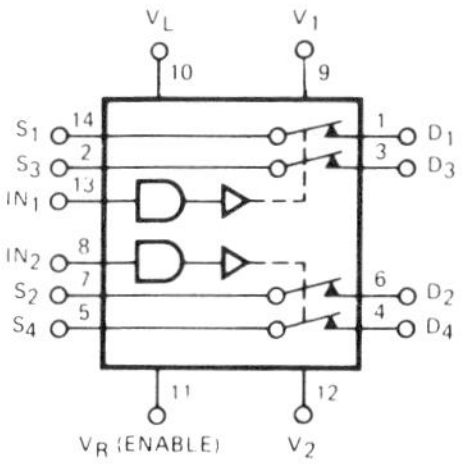

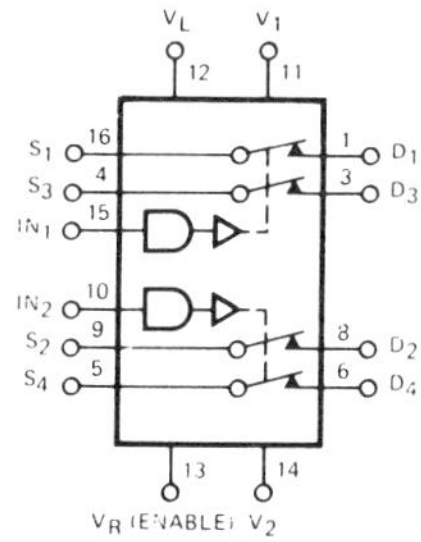

SPDT
IH5050 (R_{DS}(ON) < 30 Ω)

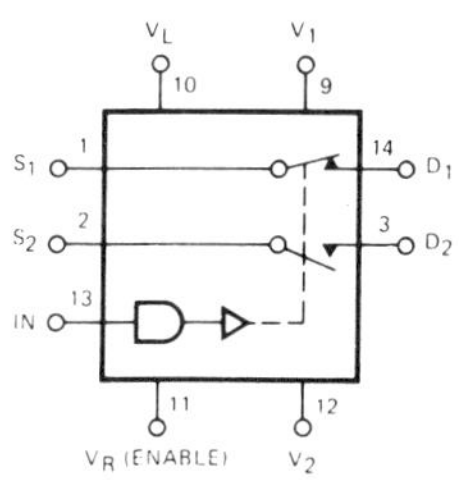

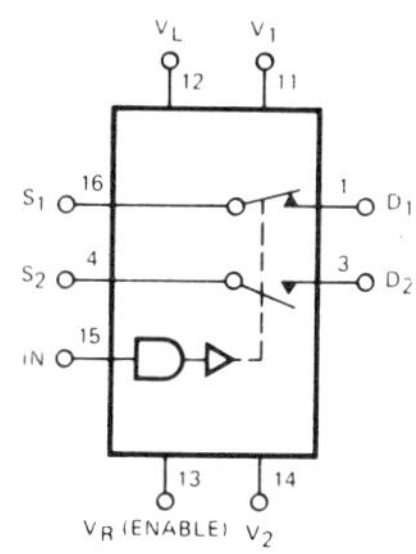

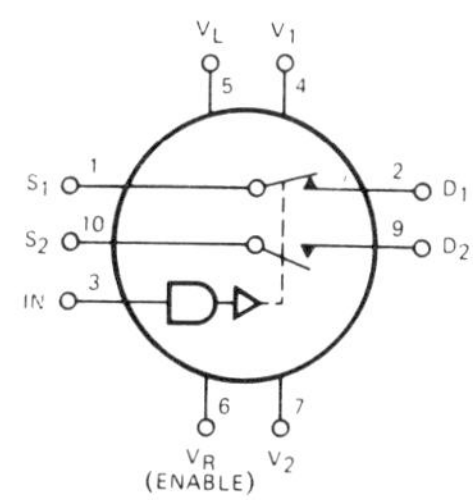

DUAL SPDT
IH5051 (R_{DS} (ON) <30Ω)

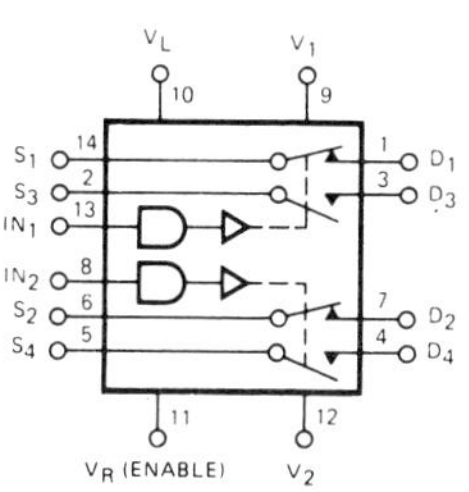

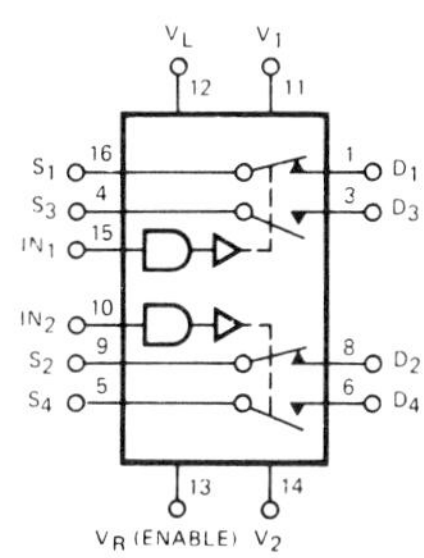

Courtesy Intersil, Inc.

(B) IH5040 series. (cont)

Fig. 5-14 (cont). Analog switch family pinouts/functional diagrams.

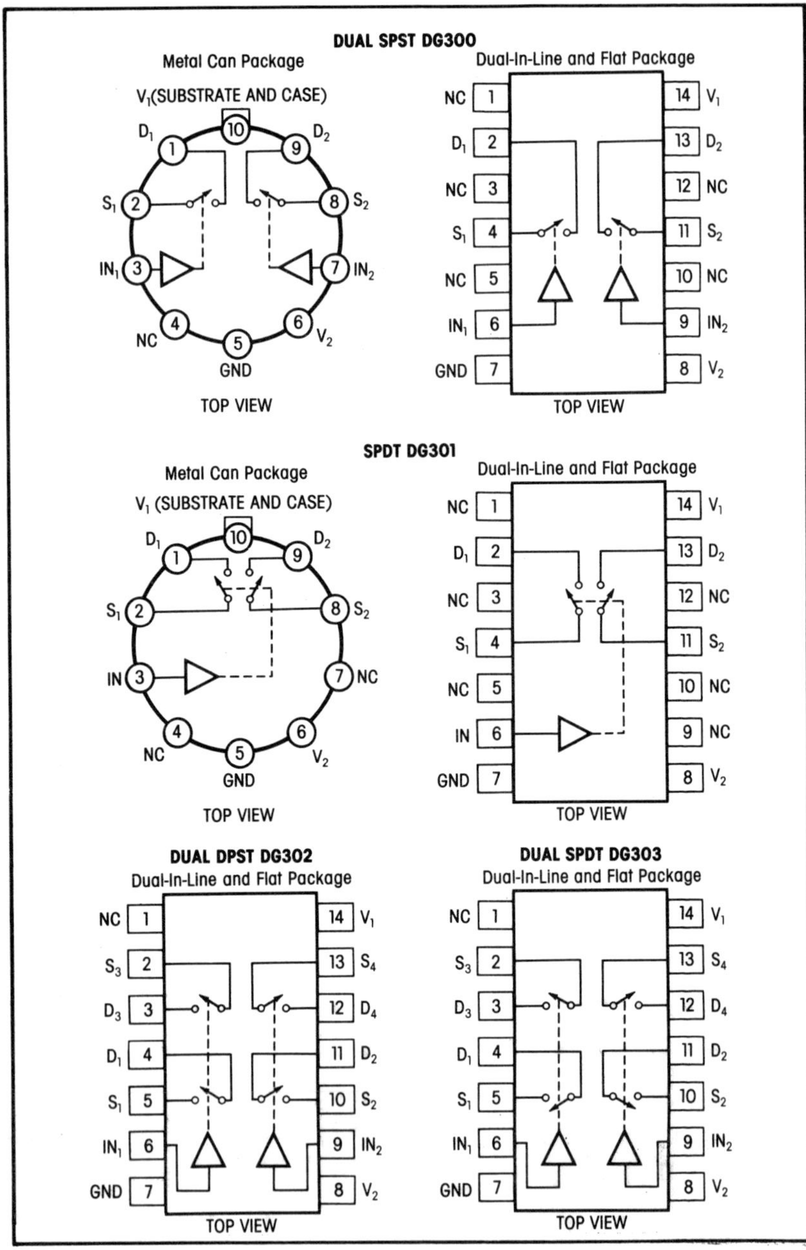

Courtesy Siliconix, Inc.

(C) DG300 series.

Fig. 5-14 (cont). Analog switch family pinouts/functional diagrams.

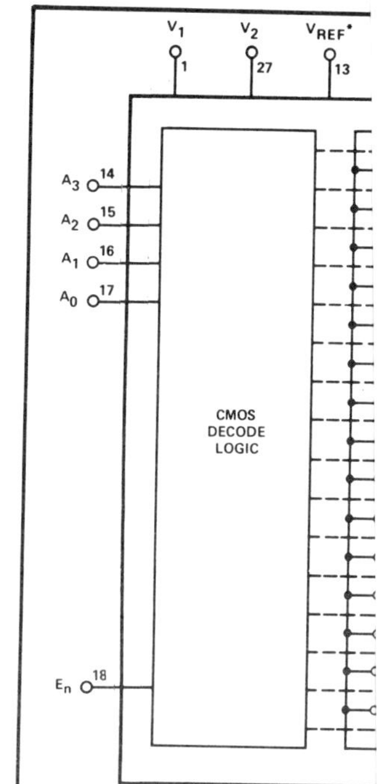

FUNCTIONAL DIAGRA

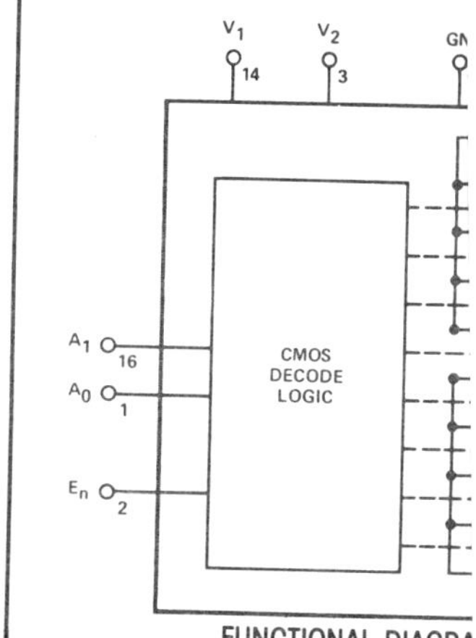

FUNCTIONAL DIAGRA

Fig. 5-14 (cont). Analog s

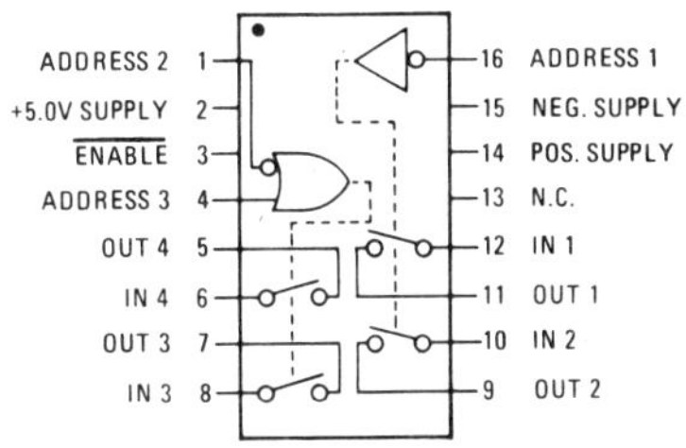

TRUTH TABLE

INPUT ADDRESS				SWITCH CHANNELS			
A1	A2	A3	$\overline{EN}$	1	2	3	4
L	X	X	L	ON	ON		
H	X	X	L	OFF	OFF		
X	L	X	L			ON	ON
X	X	H	L			ON	ON
X	H	L	L			OFF	OFF
X	X	X	H	OFF	OFF	OFF	OFF

H≥+4.0V L≤+0.4V

Courtesy Harris Semiconductor

(D) HI-1800A.

AD7510DI
AD7511DI

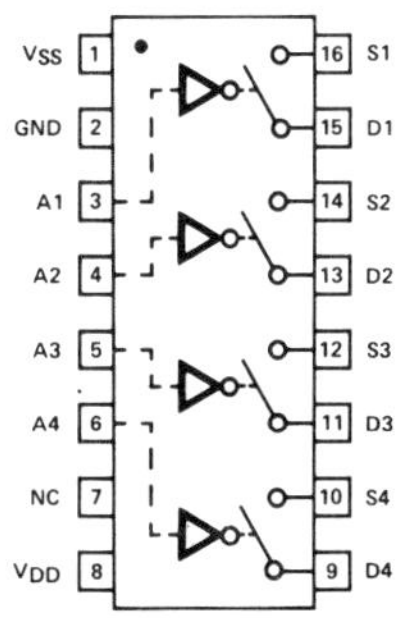

AD7512DI

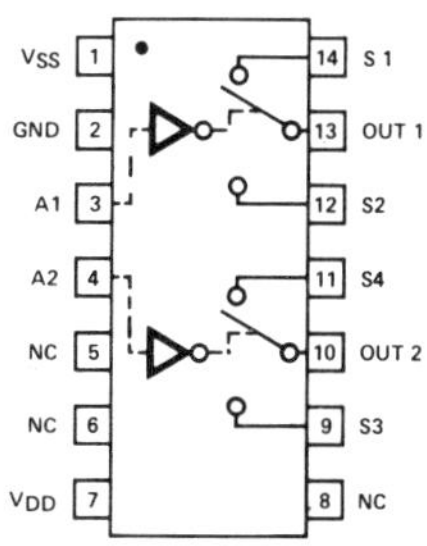

CONTROL LOGIC

AD7510DI: Switch "ON" for Address "HIGH"

AD7511DI: Switch "ON" for Address "LOW"

AD7512DI: Address "HIGH" makes S1 to Out-1 and S3 to Out-2

Courtesy Analog Devices, Inc.

(E) AD7510DI series.

Fig. 5-14 (cont). Analog switch family pinouts/functional diagrams.

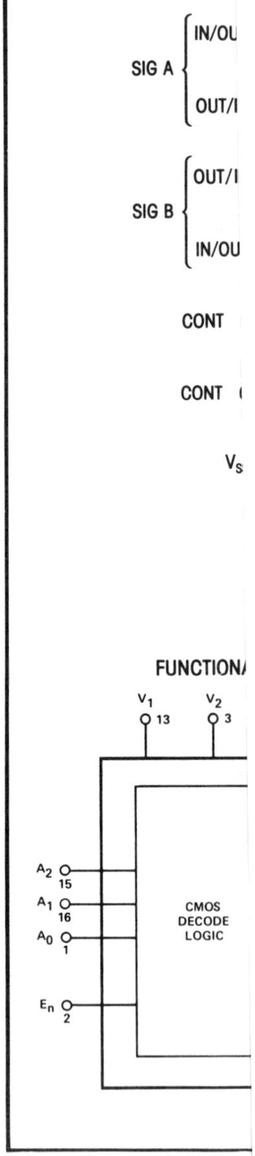

Fig. 5-14 (cont).

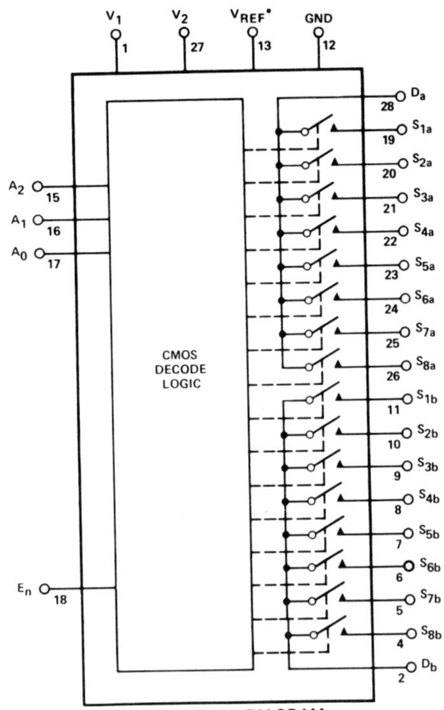

FUNCTIONAL DIAGRAM

TRUTH TABLE

A_2	A_1	A_0	En	ON SWITCH PAIR
X	X	X	0	NONE
0	0	0	1	1
0	0	1	1	2
0	1	0	1	3
0	1	1	1	4
1	0	0	1	5
1	0	1	1	6
1	1	0	1	7
1	1	1	1	8

Logic "1" = $V_{AH} \geq 2.4$ V
Logic "0" = $V_{AL} \leq 0.8$ V

*OPTIONAL (NORMALLY LEFT OPEN)

Courtesy Siliconix, Inc.

Fig. 5-14 (cont). Analog switch family pinouts/functional diagrams.

HI-1818A
PIN OUT

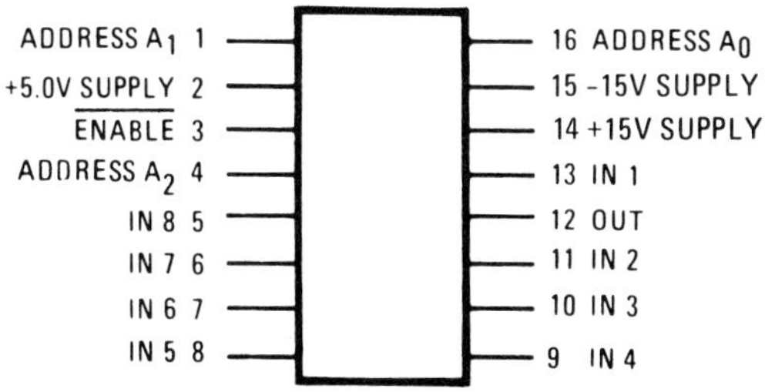

TRUTH TABLE

ADDRESS A_2	A_1	A_0	$\overline{EN}$	"ON" CHANNEL
L	L	L	L	1
L	L	H	L	2
L	H	L	L	3
L	H	H	L	4
H	L	L	L	5
H	L	H	L	6
H	H	L	L	7
H	H	H	L	8
X	X	X	H	NONE

HI-1828A
PIN OUT

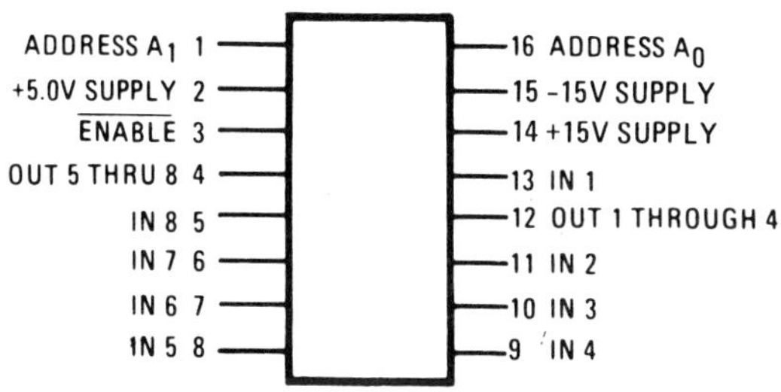

TRUTH TABLE

ADDRESS A_1	A_0	$\overline{EN}$	"ON" CHANNELS
L	L	L	1 and 5
L	H	L	2 and 6
H	L	L	3 and 7
H	H	L	4 and 8
X	X	H	NONE

(K) HI-1818A/1828A.

Courtesy Harris Semiconductor

Fig. 5-14 (cont). Analog switch family pinouts/functional diagrams.

FOUR CHANNEL

IH5009 ($R_{DS(ON)} \leq 100\Omega$)
IH5010 ($R_{DS(ON)} \leq 150\Omega$)
14 PIN DIP

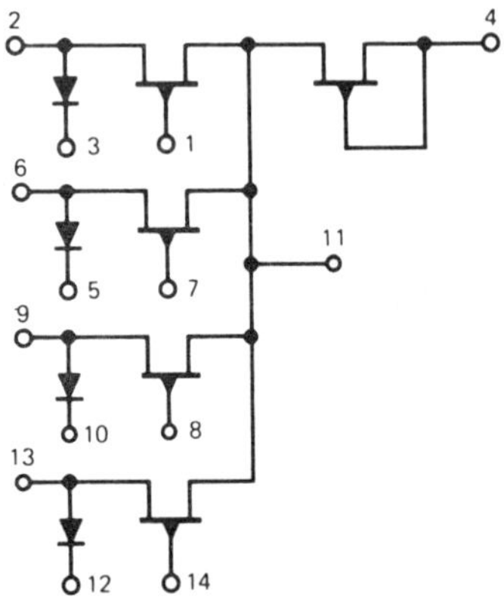

IH5011 ($R_{DS(ON)} \leq 100\Omega$)
IH5012 ($R_{DS(ON)} \leq 150\Omega$)
16 PIN DIP

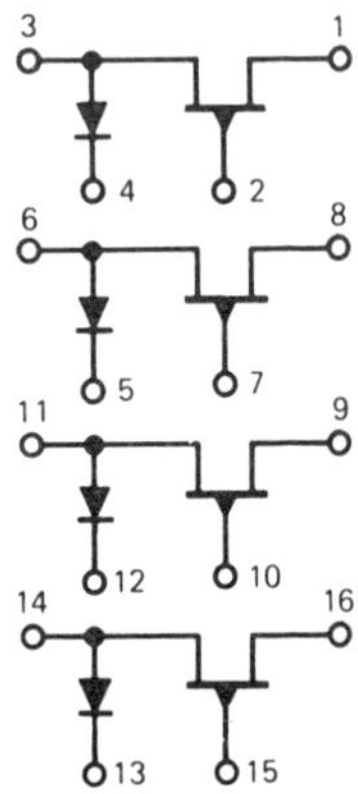

THREE CHANNEL

IH5013 ($R_{DS(ON)} \leq 100\Omega$)
IH5014 ($R_{DS(ON)} \leq 150\Omega$)
14 PIN DIP

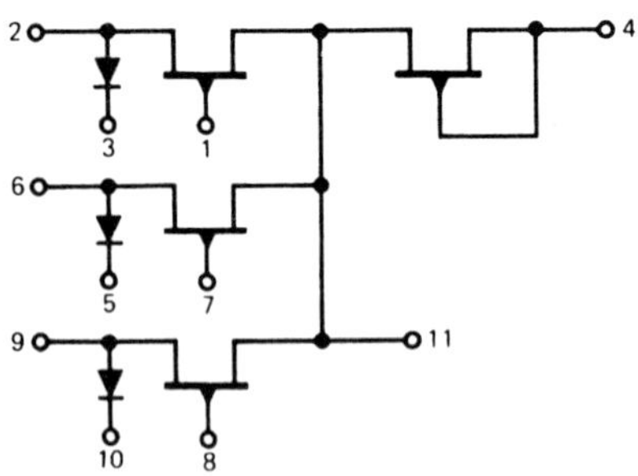

IH5015 ($R_{DS(ON)} \leq 100\Omega$)
IH5016 ($R_{DS(ON)} \leq 150\Omega$)
16 PIN DIP

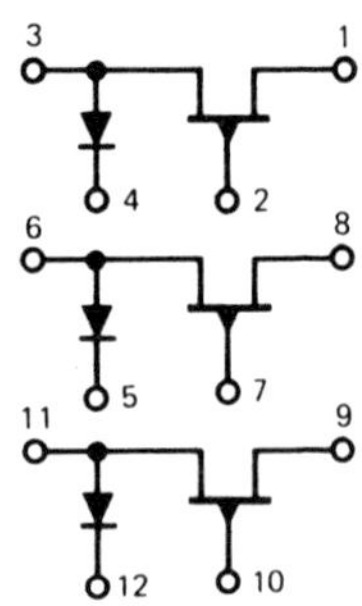

Courtesy Intersil, Inc.

(L) IH5009 series.

Fig. 5-14 (cont). Analog switch family pinouts/functional diagrams.

TWO CHANNEL

IH5017 ($R_{DS(ON)} \leq 100\Omega$)
IH5018 ($R_{DS(ON)} \leq 150\Omega$)
8 PIN DIP

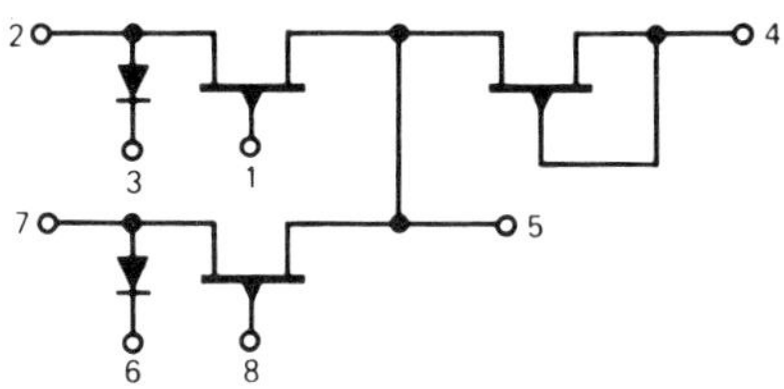

IH5019 ($R_{DS(ON)} \leq 100\Omega$)
IH5020 ($R_{DS(ON)} \leq 150\Omega$)
8 PIN DIP

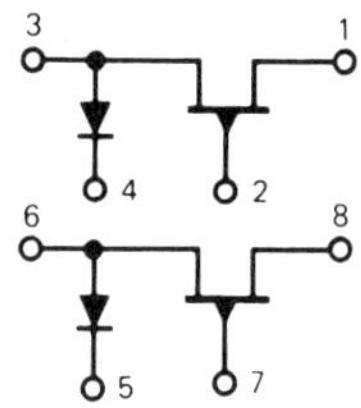

SINGLE CHANNEL

IH5021 ($R_{DS(ON)} \leq 100\Omega$)
IH5022 ($R_{DS(ON)} \leq 150\Omega$)
8 PIN DIP

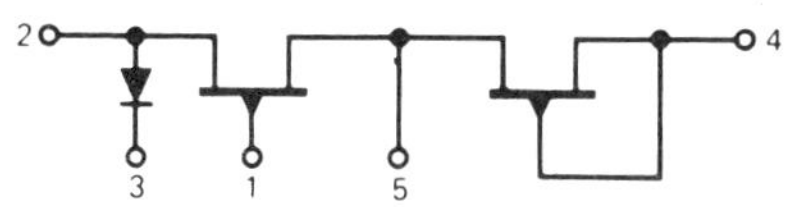

IH5023 ($R_{DS(ON)} \leq 100\Omega$)
IH5024 ($R_{DS(ON)} \leq 150\Omega$)
8 PIN DIP

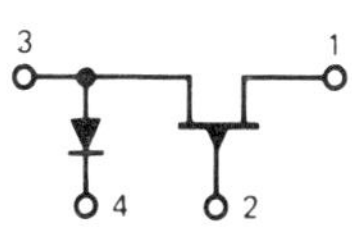

Courtesy Intersil, Inc.

(L) IH5009 series. (cont)

Fig. 5-14 (cont). Analog switch family pinouts/functional diagrams.

Four Channel

AM97C09CN ($R_{DS(ON)} \leq 100\Omega$, 10–15V CMOS)
AM97C10CN ($R_{DS(ON)} \leq 150\Omega$, 5–10V CMOS)
AM9709CN, AH5009CN ($R_{DS(ON)} \leq 100\Omega$, 15V TTL)
AM9710CN, AH5010CN ($R_{DS(ON)} \leq 150\Omega$, 5V TTL)

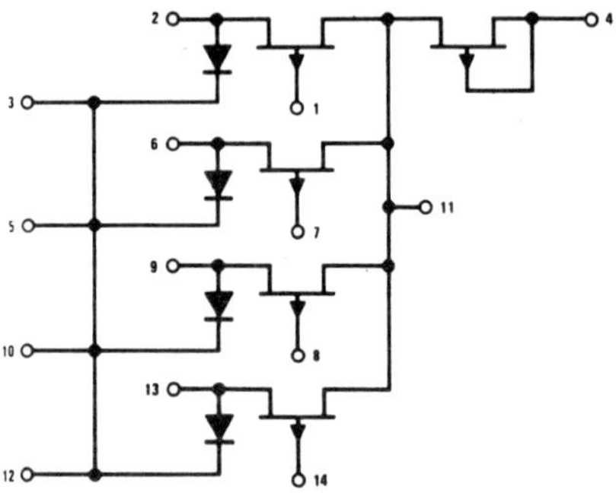

14-Pin DIP

AM97C11CN ($R_{DS(ON)} \leq 100\Omega$, 10–15V CMOS)
AM97C12CN ($R_{DS(ON)} \leq 150\Omega$, 5–10V CMOS)
AM9711CN, AH5011CN ($R_{DS(ON)} \leq 100\Omega$, 15V TTL)
AM9712CN, AH5012CN ($R_{DS(ON)} \leq 150\Omega$, 5V TTL)

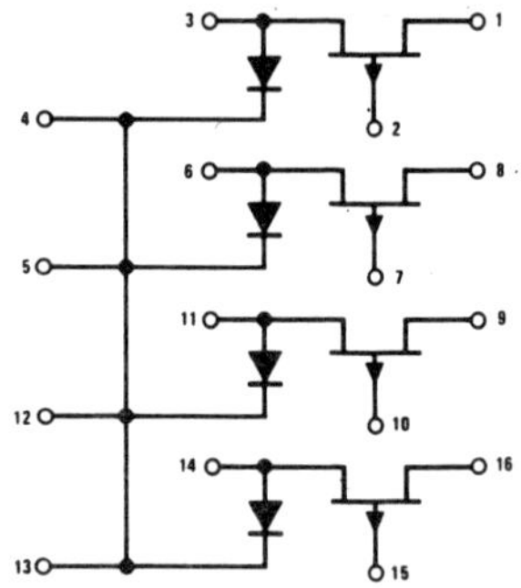

16-Pin DIP

Three-Channel

AH5013CN ($R_{DS(ON)} \leq 100\Omega$, 15V TTL)
AH5014CN ($R_{DS(ON)} \leq 150\Omega$, 5V TTL)

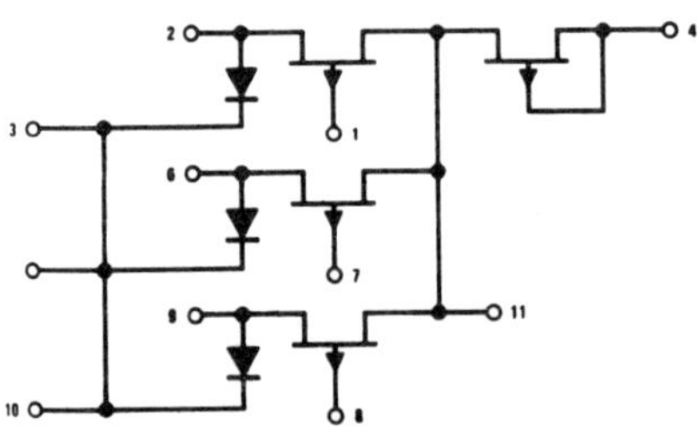

14-Pin DIP

AH5015CN ($R_{DS(ON)} \leq 100\Omega$, 15V TTL)
AH5016CN ($R_{DS(ON)} \leq 150\Omega$, 5V TTL)

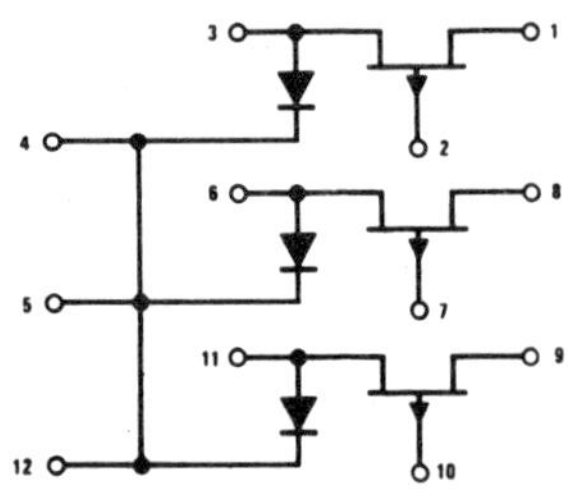

16-Pin DIP

Courtesy National Semiconductor Corp.

(M) AM97C09, AH5009 series.

Fig. 5-14 (cont). Analog switch family pinouts/functional diagrams.

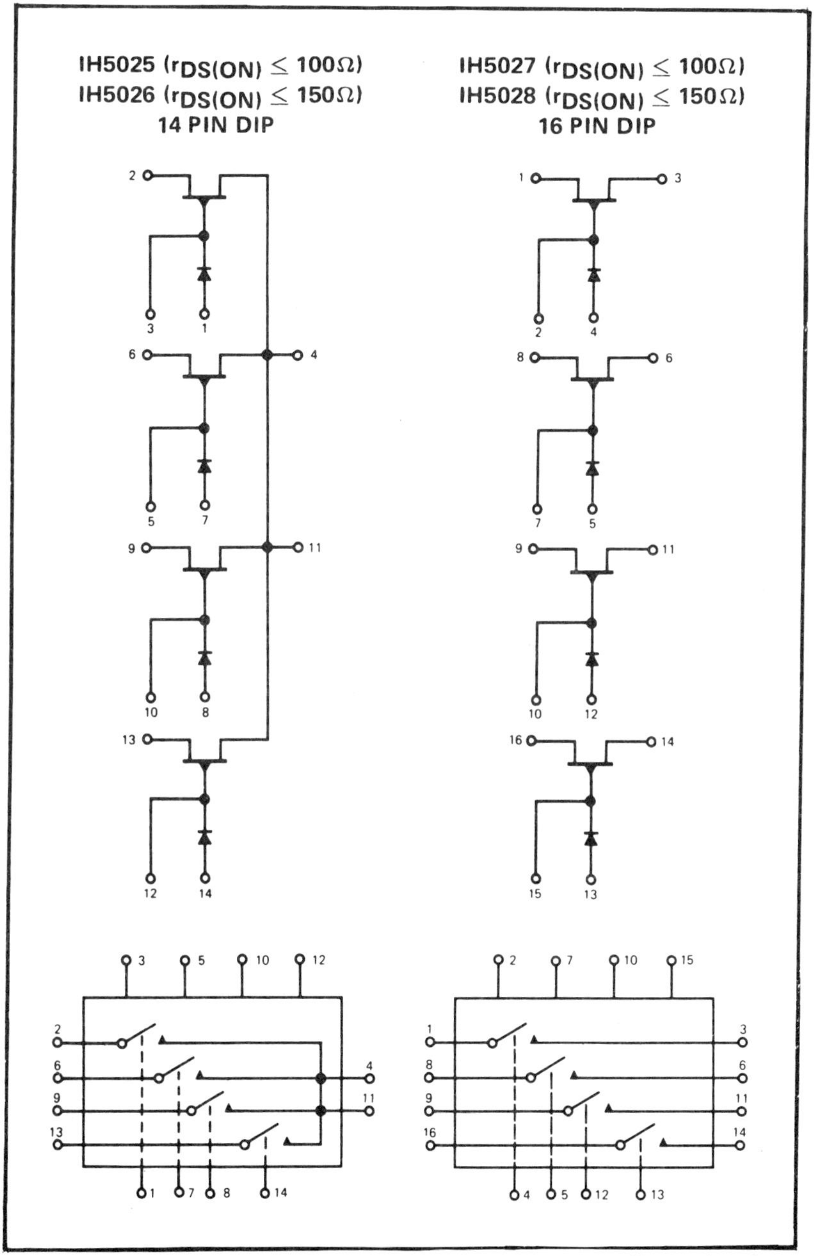

Courtesy Intersil, Inc.

(N) IH5025 series.

Fig. 5-14 (cont). Analog switch family pinouts/functional diagrams.

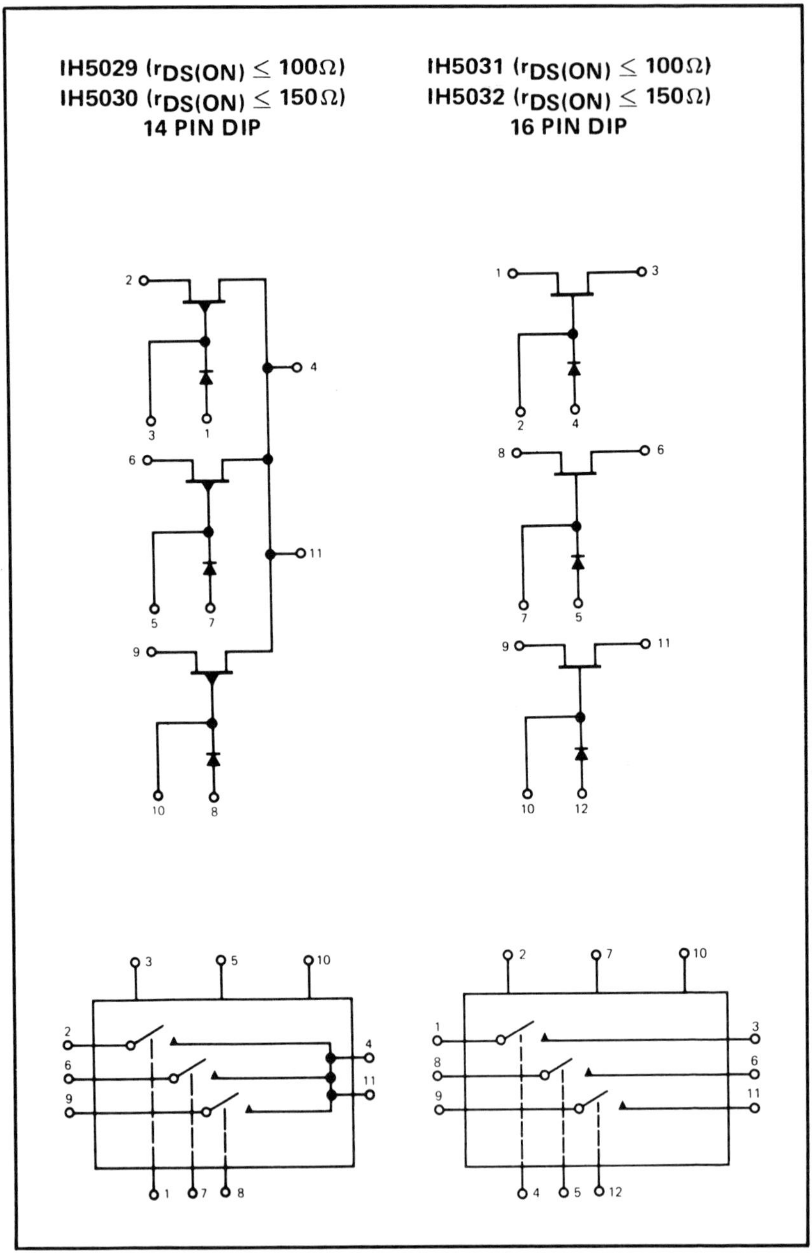

(N) IH5025 series. (cont)

Courtesy Intersil, Inc.

Fig. 5-14 (cont). Analog switch family pinouts/functional diagrams.

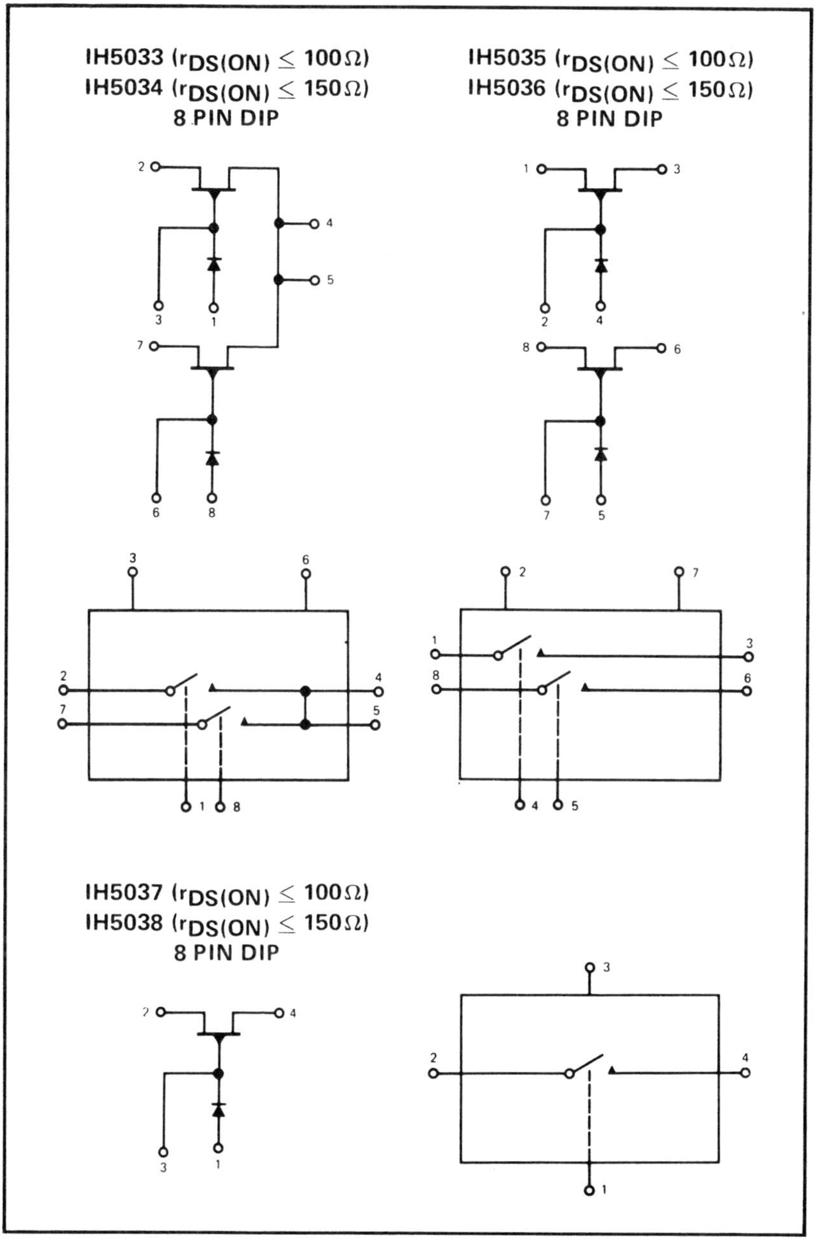

(N) IH5025 series. (cont)

Courtesy Intersil, Inc.

Fig. 5-14 (cont). Analog switch family pinouts/functional diagrams.

similar to the DG509, except for the numbers of differential channels.

The DG507 is second sourced by Harris as the HI-507A.

The HI-1818A/HI-1828A

The Harris HI-1818A and HI-1828A are illustrated in Fig. 5-14K. These devices are 8-channel CMOS multiplexers configured in a single 1 of 8, and differential 1 of 4 patterns, respectively. Both have on chip decoders and use 3- (or 2-) bit digital addresses, plus an enable bit. Logic inputs are TTL, and the devices handle a ±15 volt analog signal with ±15-volt supplies. They are processed using dielectric isolation, and thus feature inherent overvoltage protection and latchproof operation.

IH5009 Series

The Intersil IH5009 series of virtual ground switches is illustrated in Fig. 5-14L. This diagram essentially illustrates the various options available in the family. Their salient electrical characteristics were discussed previously.

AM97C09 Series

The National Semiconductor AM97C09 (and AH5009) series of virtual ground JFET switches are illustrated in Fig. 5-14M. Analog signal characteristics of these devices are similar to the IH5009 series. The AM97C09 (and other odd numbered series devices) are designed for 10- to 15-volt CMOS control, while the AM9710 (and even numbered units) are designed for 5-volt CMOS control.

IH5025 Series

The Intersil IH5025 series of positive signal JFET switches is illustrated in Fig. 5-14N. This diagram shows the various options; the electrical characteristics have been discussed previously.

Applications of Analog Switches

Selected applications examples will be used to illustrate typical methods of using analog switches in data conversion systems. This will not be a detailed treatment of the various uses of analog switches, as the extremely broad range of application possibilities places such a tabulation beyond the scope of this book. For further details on analog switches, the reader is referred to the references at the end of the chapter.

Multiplex Systems

A very common requirement in data acquisition and conversion systems is the multiplexer (MUX). A multiplexer is nothing more than an analog switch array designed to sequentially scan a number of input channels; typically 1 of 4, 1 of 8, 1 of 16, etc. Common devices also can be operated in reverse; that is they can transmit one signal sequentially to 4, 8, or 16 points. A commonly seen option is a differential type multiplexer, which is one with dual sets of switches; dual 1 of 4, 1 of 8, etc.

In a system, multiplexers are typically operated as shown in the simplified block diagrams of Fig. 5-15. In Fig. 5-15A, a 1 of 2 MUX selects either one of two single ended analog signals and transmits it to the amplifier. The amplifier then amplifies the selected signal to the desired full scale level. The output of this amplifier represents the signal in a scaled format, compatible with the range of S/H amplifiers and a/d converters in use.

Both the channel address and gain are logic controllable; thus, a computer or microprocessor can readily handle a large number of different channels with varying signal levels, etc. Practical conversion systems may use as many as 16 or 32 channels with gains arranged in either binary or decade scaled increments.

In Fig. 5-15B a differential MUX/amplifier system is shown. A differential MUX operates similar to a single-ended one, except two transmission channels are selected for each signal input. The signal is carried differentially through to the amplifier, usually an instrumentation amplifier (IA). The IA amplifies the differential signal, rejecting common mode noise in the process, and transforms it into an equivalent single-ended format scaled to the required level. Channel address and gain logic is similar to the single-ended case just described. Examples of MUX and IAs operating as described follow below.

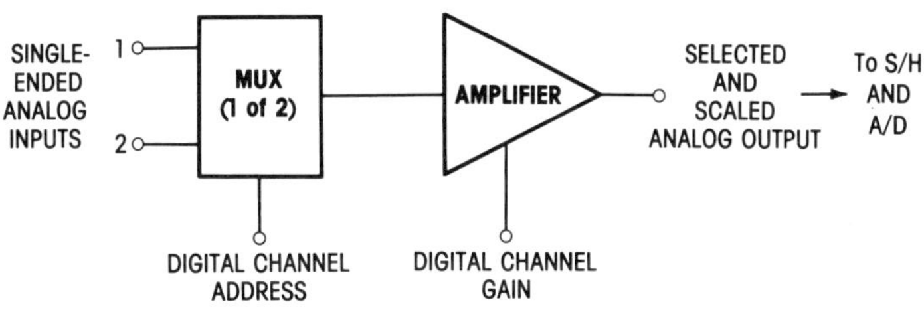

(A) Single ended.

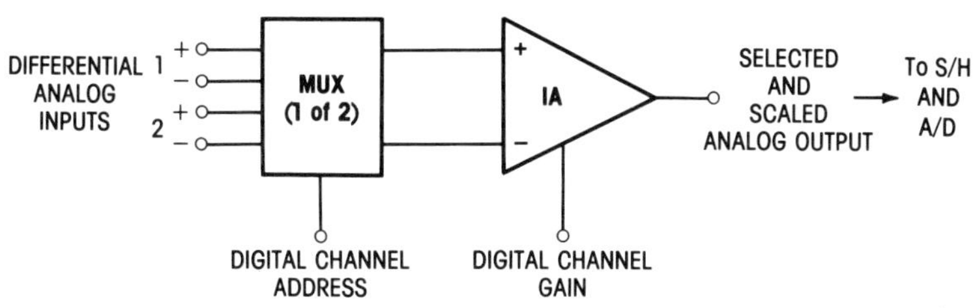

(B) Differential.

Fig. 5-15. Block diagrams of multiplex systems.

Multiplexer hardware, as noted above, can take on many different forms, such as 1 of 4, 1 of 8, etc. Some examples of these are shown in Fig. 5-16, which illustrates use of the Siliconix DG series of analog switches and multiplexers. Of course, the general concept shown here can be extended to many other such devices.

Fig. 5-16A illustrates a 1 of 4 sequencing MUX as built up from individual logic ICs and a general purpose 4 channel analog switch, the DG201. The MUX is sequenced at a rate determined by the clock input to the 2-bit binary counter. This counter is decoded by a series of three input gates, and the active input of the DG201 will close a designated channel when all 1s appear at the decoder inputs.

While this is a workable operating system, large numbers of switching points are more efficiently handled by MUX devices with higher levels of integration. This is what analog switch MUXs do for the designer—they place the hardware of Fig. 5-16A into a single package.

Fig. 5-16B illustrates an 8-channel sequencing MUX using a DG508 which is driven by a 74C93 counter, with three bits only used. This hookup sequences similar to the 4 channel just described except there are now 8 channels and 8 data slots. It is more simple in that a single logic package is used, and the decoder is integral to the DG508 MUX.

A 16-channel sequencing MUX is shown in Fig. 5-16C using a DG506 and a 74C93 counter. In this case the counter is connected for 4-bit operation by cascading its A and B sections. Although CMOS counters are shown here, they could also be the like numbered device TTL counterpart, for 5-volt logic systems.

A differential 1 of 4 sequencing MUX using a DG509 is shown in Fig. 5-16D. The logic sequencing drive in this case is handled by a dual flip-flop connected as a 2-bit ripple counter, similar to Fig. 5-16A.

Finally, an 8-channel sequencing MUX using the DG507 is shown in Fig. 5-16E. Again a 74C93 counter is used, connected for 3-bit operation. This circuit, like the circuit of Fig. 5-16D, can also use a 7493 counter drive for 5-volt logic systems.

For all of the preceding sequencing MUX systems the analog data on time slot simply will be equal to the period of one clock pulse, i.e., 10 μs for a 100-kHz clock. They all may also be used in reverse for DEMUX applications.

Programmable Gain Amplifier

In Fig. 5-17 is shown a circuit useful as both a programmable gain amplifier, and as a two-input signal selector. This circuit is a practical realization of the conceptual diagram of Fig. 5-15A.

The DG200 performs as the input signal selector, selecting either V_{in1} or V_{in2} to be amplified by the op amp. This particular use of an analog switch, where it looks directly into a high input impedance (the op amp) minimizes errors due to switch-on resistance; thus, it is suitable for high accuracy use. The DG200 is logic compatible with either TTL or CMOS inputs, and the analog input can range up to ±10 volts (or the full common mode range of the op amp used).

Amplifier gain is selected by choosing one of the four feedback paths through the DG201 analog switch. Ranging downward on the diagram, these paths correspond to gains of 1, 10, 100, and 1000, respectively. In general, gain is set by the equation shown; thus, R_F and R_G can be varied as required for other gains. These resistors should, of course, be stable precision types for best accuracy. This gain switching method also uses the analog switch to best advantage, as there is no signal current flowing in it to cause errors.

Gain input logic can be either TTL or CMOS, but should allow only one channel to be active at a given time to yield the gains shown. For both the DG200 and 201, the switches are closed for a low logic input.

Overall errors of this circuit are determined mostly by the op amp used. This device should be optimized in terms of drift, offset, speed, and so forth, to suit the particular application.

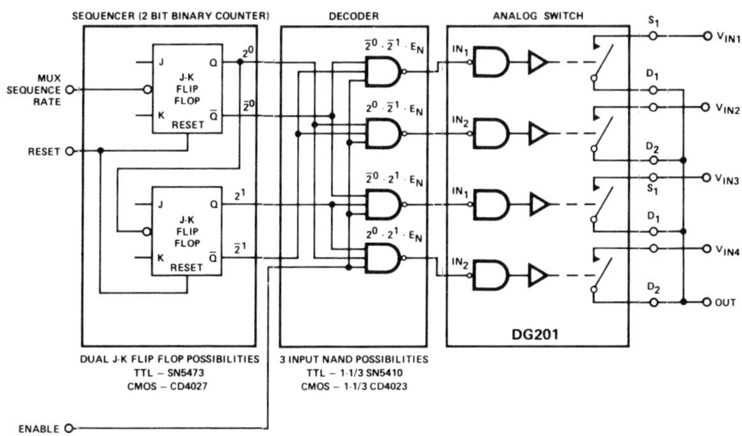

TRUTH TABLE

ENABLE	MUX SEQUENCE RATE	SEQUENCER OUTPUT 2^0	2^1	SWITCH STATES (– DENOTES OFF) SW1	SW2	SW3	SW4
0	0	0	0	–	–	–	–
1	0	0	0	ON	–	–	–
1	1 pulse	1	0	–	ON	–	–
1	2 pulses	0	1	–	–	ON	–
1	3 pulses	1	1	–	–	–	ON
1	4 pulses	0	0	ON	–	–	–

(A) A 4 channel sequencing MUX.

Courtesy Siliconix, Inc.

Fig. 5-16. Types of multiplexer systems.

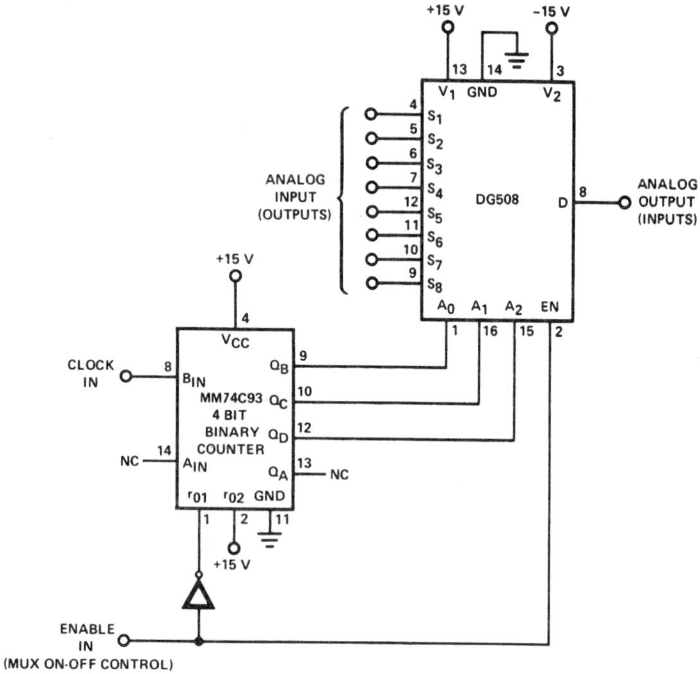

TRUTH TABLE

ENABLE	MUX SEQUENCE RATE	MUX INPUTS A_0	A_1	A_2	DG508 SWITCH PAIR STATES (– DENOTES OFF) S1	S2	S3	S4	S5	S6	S7	S8
0	0	X	X	X	–	–	–	–	–	–	–	–
1	0	0	0	0	ON	–	–	–	–	–	–	–
1	1 pulse	1	0	0	–	ON	–	–	–	–	–	–
1	2 pulses	0	1	0	–	–	ON	–	–	–	–	–
1	3 pulses	1	1	0	–	–	–	ON	–	–	–	–
1	4 pulses	0	0	1	–	–	–	–	ON	–	–	–
1	5 pulses	1	0	1	–	–	–	–	–	ON	–	–
1	6 pulses	0	1	1	–	–	–	–	–	–	ON	–
1	7 pulses	1	1	1	–	–	–	–	–	–	–	ON
1	8 pulses	0	0	0	ON	–	–	–	–	–	–	–

(B) An 8 channel sequential MUX (DEMUX).

Fig. 5-16 (cont). Types of

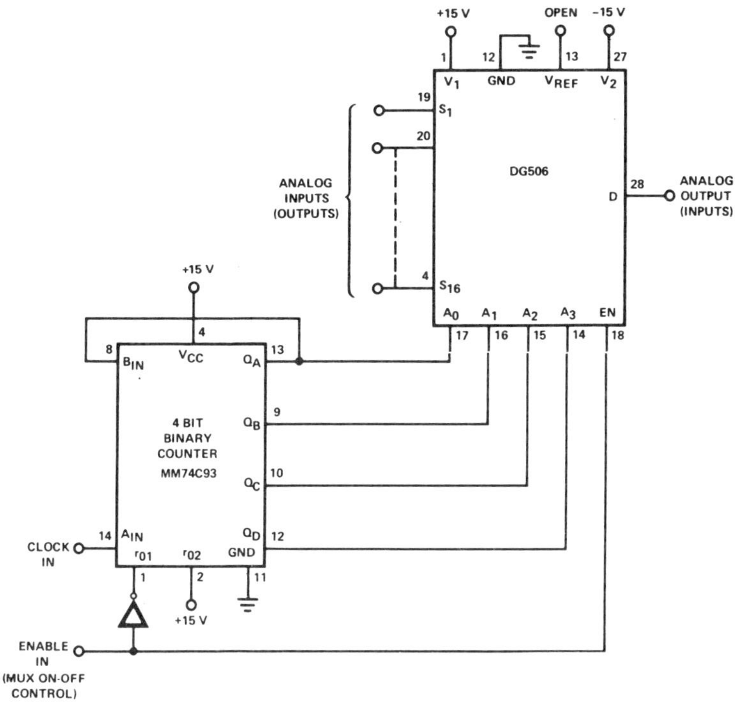

TRUTH TABLE

ENABLE	MUX SEQUENCE RATE	MUX INPUTS A_0	A_1	A_2	A_3	DG506 SWITCH STATES (– DENOTES OFF) S1	S2	S3	S4	S5	S6	S7	S8	S9	S10	S11	S12	S13	S14	S15	S16
0	0	X	X	X	X	–	–	–	–	–	–	–	–	–	–	–	–	–	–	–	–
1	0	0	0	0	0	ON	–	–	–	–	–	–	–	–	–	–	–	–	–	–	–
1	1 pulse	1	0	0	0	–	ON	–	–	–	–	–	–	–	–	–	–	–	–	–	–
1	2 pulses	0	1	0	0	–	–	ON	–	–	–	–	–	–	–	–	–	–	–	–	–
1	3 pulses	1	1	0	0	–	–	–	ON	–	–	–	–	–	–	–	–	–	–	–	–
1	4 pulses	0	0	1	0	–	–	–	–	ON	–	–	–	–	–	–	–	–	–	–	–
1	5 pulses	1	0	1	0	–	–	–	–	–	ON	–	–	–	–	–	–	–	–	–	–
1	6 pulses	0	1	1	0	–	–	–	–	–	–	ON	–	–	–	–	–	–	–	–	–
1	7 pulses	1	1	1	0	–	–	–	–	–	–	–	ON	–	–	–	–	–	–	–	–
1	8 pulses	0	0	0	1	–	–	–	–	–	–	–	–	ON	–	–	–	–	–	–	–
1	9 pulses	1	0	0	1	–	–	–	–	–	–	–	–	–	ON	–	–	–	–	–	–
1	10 pulses	0	1	0	1	–	–	–	–	–	–	–	–	–	–	ON	–	–	–	–	–
1	11 pulses	1	1	0	1	–	–	–	–	–	–	–	–	–	–	–	ON	–	–	–	–
1	12 pulses	0	0	1	1	–	–	–	–	–	–	–	–	–	–	–	–	ON	–	–	–
1	13 pulses	1	0	1	1	–	–	–	–	–	–	–	–	–	–	–	–	–	ON	–	–
1	14 pulses	0	1	1	1	–	–	–	–	–	–	–	–	–	–	–	–	–	–	ON	–
1	15 pulses	1	1	1	1	–	–	–	–	–	–	–	–	–	–	–	–	–	–	–	ON
1	16 pulses	0	0	0	0	ON	–	–	–	–	–	–	–	–	–	–	–	–	–	–	–

Courtesy Siliconix, Inc.

(C) A 16 channel sequential MUX (DEMUX).

multiplexer systems.

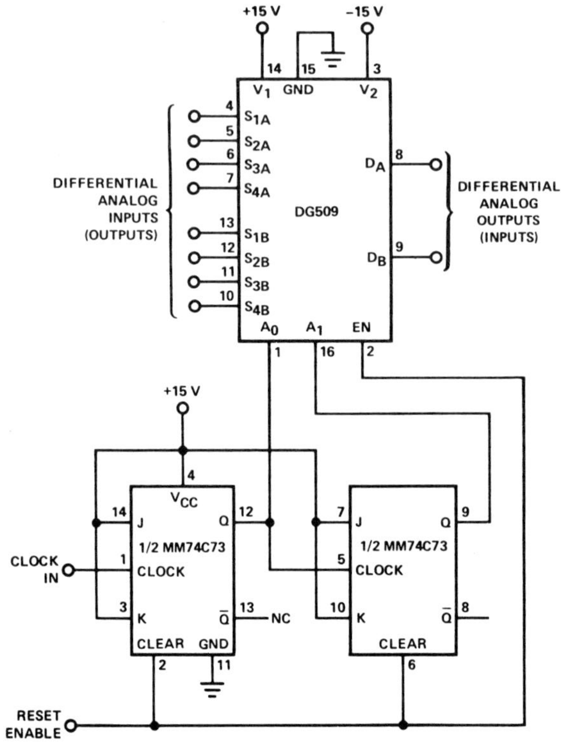

TRUTH TABLE

ENABLE	MUX SEQUENCE RATE	MUX INPUTS A_0	A_1	DG509 SWITCH PAIR STATES (– DENOTES OFF) S1	S2	S3	S4
0	0	0	0	–	–	–	–
1	0	0	0	ON	–	–	–
1	1 pulse	1	0	–	ON	–	–
1	2 pulses	0	1	–	–	ON	–
1	3 pulses	1	1	–	–	–	ON
1	4 pulses	0	0	ON	–	–	–

(D) Differential 4 channel sequential MUX/DEMUX.

Courtesy Siliconix, Inc.

Fig. 5-16 (cont). Types of

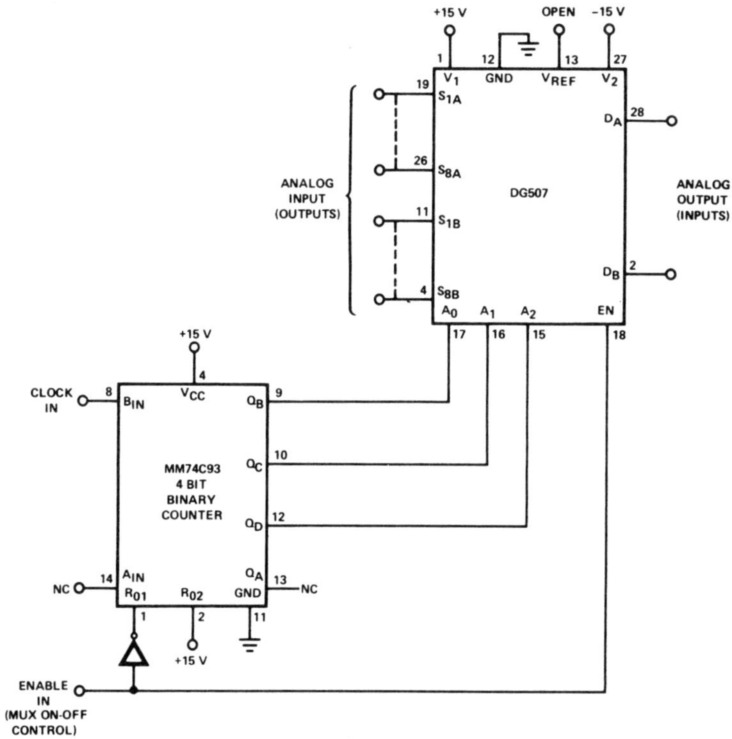

TRUTH TABLE

ENABLE	MUX SEQUENCE RATE	MUX INPUTS A_0	A_1	A_2	DG507 SWITCH PAIR STATES (– DENOTES OFF) S1	S2	S3	S4	S5	S6	S7	S8
0	0	X	X	X	–	–	–	–	–	–	–	–
1	0	0	0	0	ON	–	–	–	–	–	–	–
1	1 pulse	1	0	0	–	ON	–	–	–	–	–	–
1	2 pulses	0	1	0	–	–	ON	–	–	–	–	–
1	3 pulses	1	1	0	–	–	–	ON	–	–	–	–
1	4 pulses	0	0	1	–	–	–	–	ON	–	–	–
1	5 pulses	1	0	1	–	–	–	–	–	ON	–	–
1	6 pulses	0	1	1	–	–	–	–	–	–	ON	–
1	7 pulses	1	1	1	–	–	–	–	–	–	–	ON
1	8 pulses	0	0	0	ON	–	–	–	–	–	–	–

(E) Differential 8 channel sequential MUX (DEMUX).

Courtesy Siliconix, Inc.

multiplexer systems.

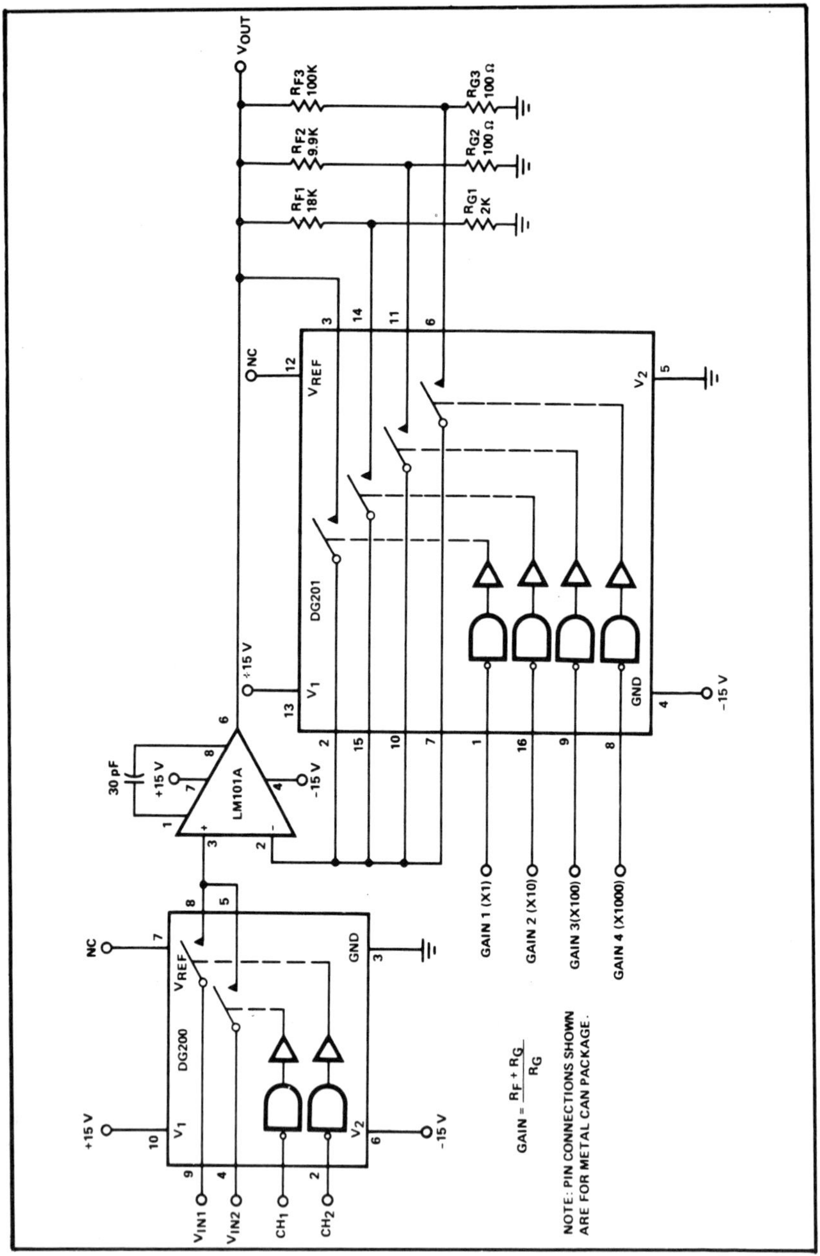

Courtesy Siliconix, Inc.

Fig. 5-17. Precision amplifier with digitally programmable inputs and gains.

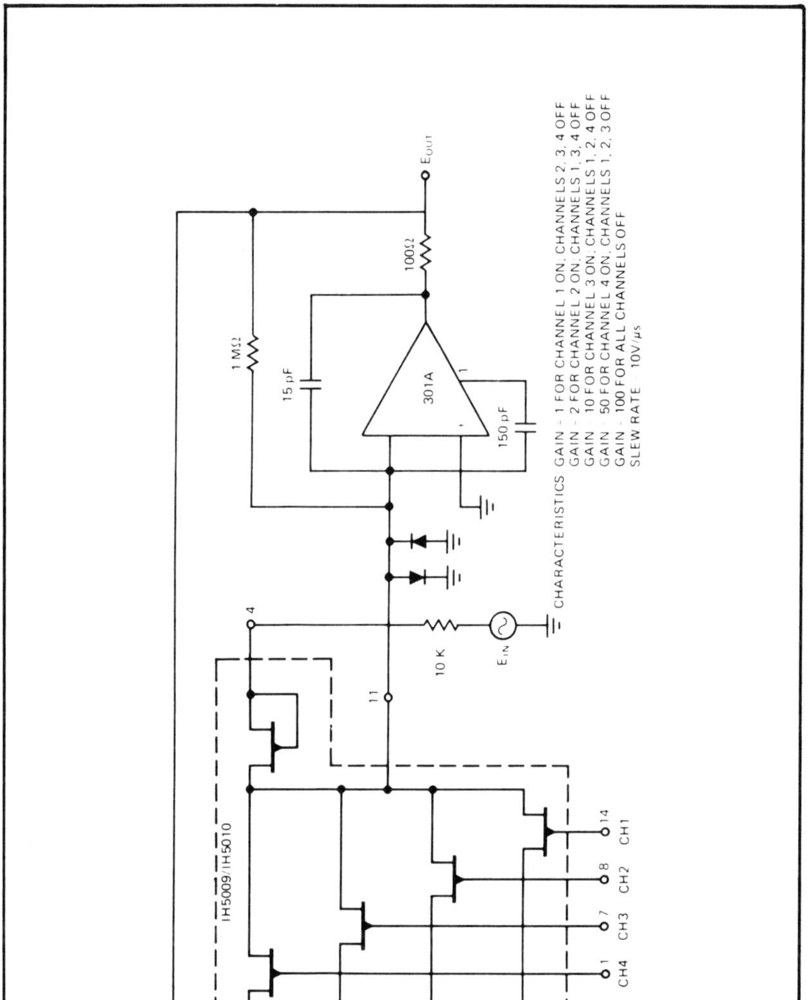

Courtesy Intersil, Inc.

Fig. 5-18. Gain ranging circuit.

Gain Ranging Circuit

Fig. 5-18 shows another type of programmable gain circuit using the IH5009 virtual ground switch. In this circuit the input is applied to pin 4 through a 10K resistor. This uses the compensating FET in the input path, and the remaining four FETs are switched to select the feedback resistor around the op amp. Note that there is a fixed 1 megohm feedback path which appears in parallel with the feedback path switch selected. The gain equation in general is R_F/R_{in}. Resulting gains are as shown in the table, and range from 1 to 100.

This circuit is attractive for audio or other ac signal use, as the 301A op amp is capable of high slew rate and bandwidth when operated in its feedforward mode (as shown). These factors in combination with the circuit's economy make it very useful.

Instrumentation Amplifier With Digitally Selectable Inputs and Gain

The circuit of Fig. 5-19 illustrates a method of differential signal selection and gain control which follows the conceptual diagram of Fig. 5-15B. This circuit uses DG300 series CMOS analog switches and is CMOS logic programmable for signal selection and gain.

The DG306 is used as a differential channel signal selector, selecting either differential input A or input B. The CMOS inverter driving pin 9 provides a complemented signal; the remaining devices serve just as buffers. It should be noted that there is no loading on the switch in this circuit as both outputs feed high Z inputs; in this case the (+) and (−) inputs to the 3 op amp IA. This minimizes input errors.

An op-amp based instrumentation amplifier such as is formed here by the Siliconix LI44 triple op amp is easily gain programmed by changing R1, without affecting CM response. Although there is significant current flowing in this path, the gain errors due to the analog switch-on resistance can be minimized by using an on resistance which is low with regard to the value of R1. For the DG305 on resistance of 75 ohms, the minimum R1 value used is 20 times higher. In general, it is desirable to keep R1 as high as practical with regard to the switch resistance, as the switch errors will then be reduced proportionately.

In this configuration the op amp used is a programmable characteristic type; thus, input bias current, power dissipation, slew rate, etc., can be adjusted to suit specific requirements. With the values as noted, the circuit features low power operation, but parameters can be tailored to optimize other characteristics if desired.

Low Cost 4 Channel Multiplexer

Although typically instrumentation applications require high accuracy, there are uses for analog switches where the lowest cost, consistent with reasonable accuracy, is of high importance. For these types of uses

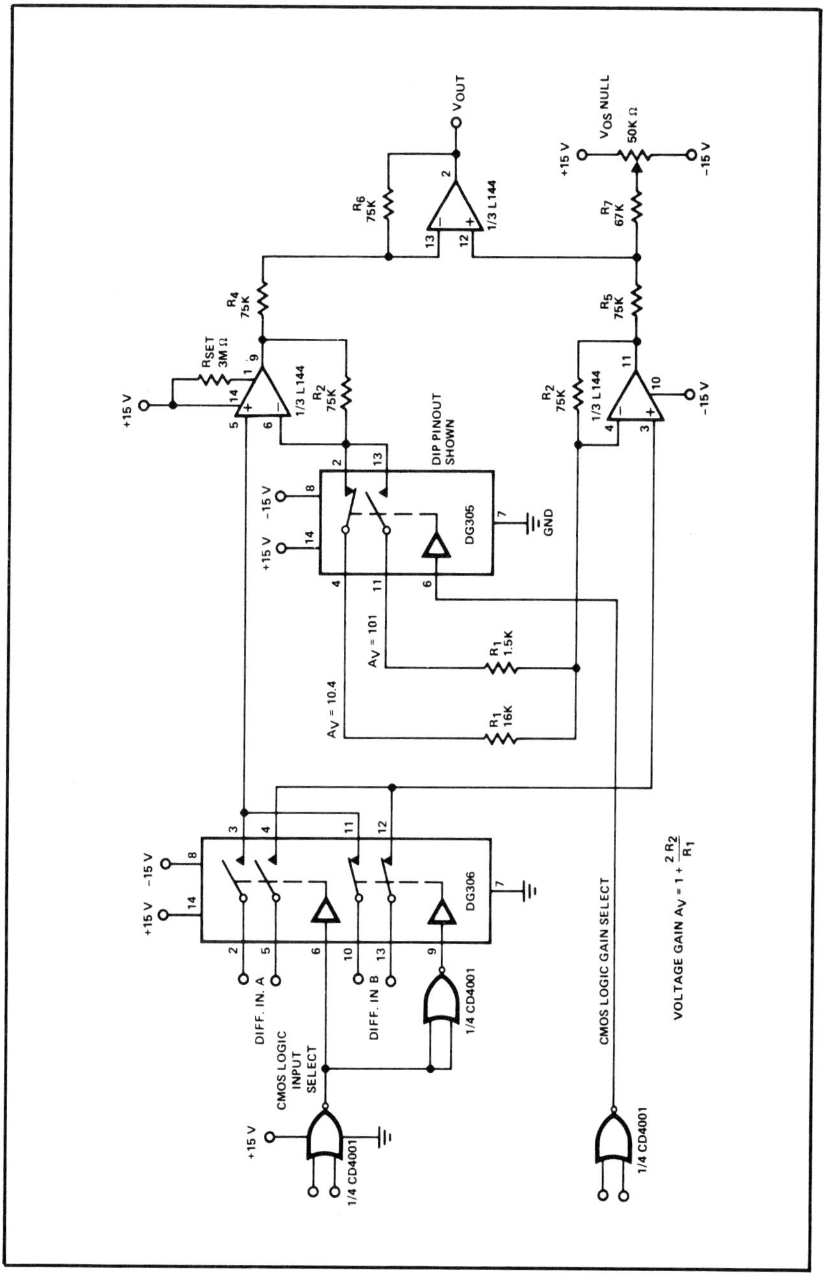

Courtesy Siliconix, Inc.

Fig. 5-19. Low-power instrumentation amplifier with digitally selected inputs and gain.

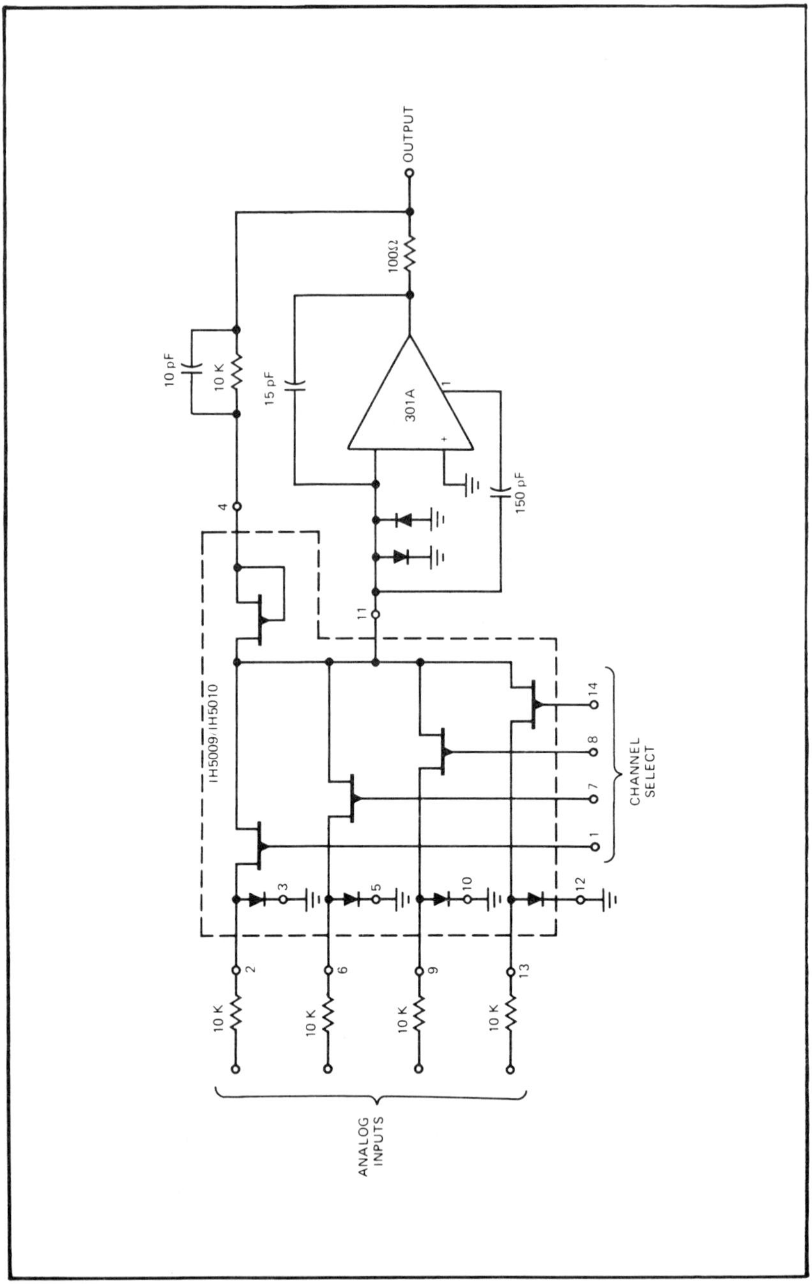

Courtesy Intersil, Inc.

Fig. 5-20. Low-cost 4 channel multiplexer.

the virtual ground JFET switch is useful. A virtual ground switch based multiplexer is shown in Fig. 5-20.

This four input multiplexer uses the IH5009 switch as an input channel selector, with gain fixed at unity in each channel. Input impedance is 10K, and input and feedback resistors should be matched for best accuracy. Gain error (exclusive of resistor match) is 0.5% with the IH5009; this can be improved through use of switches with better match, available as selections with accuracy down to 0.05%. Resistor errors can be most economically minimized by using readily available matched arrays.

This circuit can also be used as a signal mixer if desired, by allowing more than one channel to be on at a given time. Like the previously described programmable gain amplifier using virtual ground switches, this circuit is well suited to ac signals due to its excellent bandwidth and slew rate.

Polarity Reversing Amplifier

The circuit of Fig. 5-21 illustrates how the polarity (or sign) of an analog signal can be controlled by a logic level, through use of an analog switch. This function is often required in data normalization, or in sign + magnitude converters.

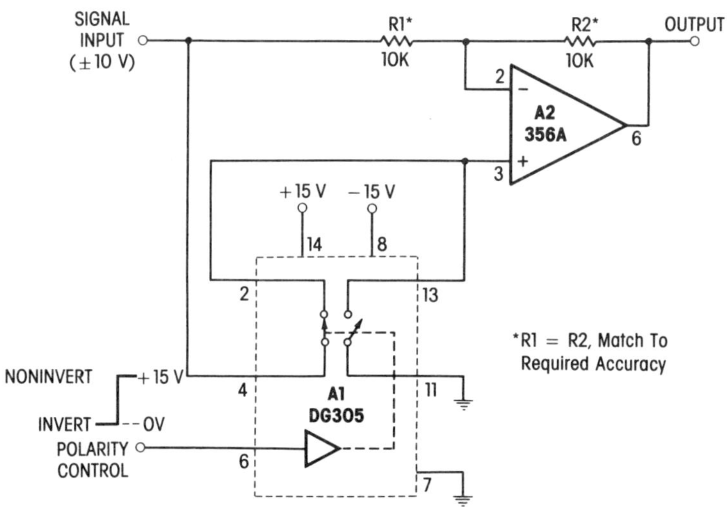

Fig. 5-21. Polarity reversing amplifier.

In this circuit A1 is an analog switch which programs the op-amp stage A2 to either an inverting or noninverting state of unity gain. A1, an spdt unit, connects the (+) input of A2 to either analog ground or to the analog signal input line. When connected to the signal input (switch state shown) the circuit is inherently a unity gain follower. When grounded, A2 becomes a unity gain inverter with gain set by R2/R1.

The circuit has low errors using the devices shown. Leakage of the analog switch has negligible effect, and the DG300 series devices are high speed units. Overall speed is limited by the device selected for A2, with the settling time of the 356A being typically 1.5 μs to 0.01%, for a 10-volt step. The dc bias-current errors of A2 are negligible, as the 356 device is a FET input unit. Offset voltage is the main dc error, being 0.04% maximum for a 10-volt scale signal. This may be reduced to zero by offset nulling, if desired.

For any op amp used, the match of resistors R1 and R2 will determine the accuracy of the inverting gain mode. These units should be matched (or trimmed) to the desired accuracy, and preferably should have low tracking TCs as well. As noted, matched arrays of thin film or other precision types are attractive for this type of application.

The DG305 is a CMOS drive compatible unit; a DG301 may be substituted if desired for TTL drive sources. Other spdt analog switches may be used for this function also, as it is relatively insensitive to on resistance or leakage effects.

SAMPLE/HOLD AMPLIFIERS

A sample-and-hold (S/H) circuit is one which has two basic operational states; one where an input signal is sampled and simultaneously transmitted to the output, and a second in which the last value sampled is held until the input is sampled again.

A sample/hold is necessary in data conversion systems when the a/d converter being used must be presented with a constant value analog input during its conversion process. An example of such use is with a successive approximation converter. In an ideal sense, the S/H "freezes" the last instantaneous voltage prior to its hold command, and presents this voltage unaltered to the a/d converter, which converts it to the corresponding digital word. In practice of course there are many error factors which come into play in building (or choosing) a S/H, thus this section will discuss these basic considerations, as well as representative devices.

Basic Operation

The basic circuit operation is diagrammed in Fig. 5-22. Here an input signal is applied to electronic switch S1 and depending on the state of S1, will be either transmitted to C_H or be blocked. The state of switch S1 is controlled by the sample/hold control line. When S1 is closed, the input signal appearing across C_H is buffered by A1, and also appears at the output. If S1 is connected for a period of time while the input is vary-

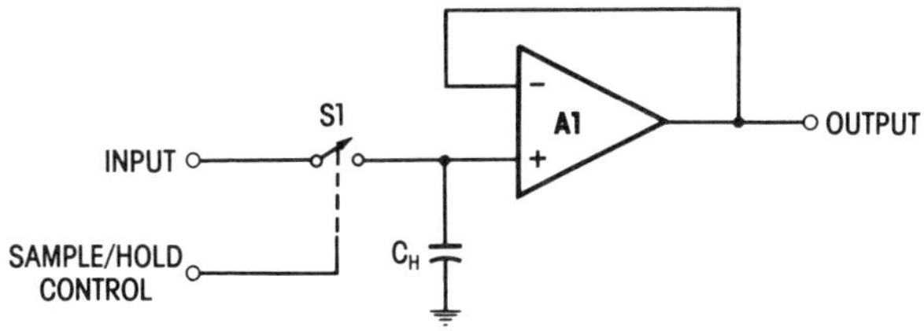

Fig. 5-22. Basic sample/hold circuit.

ing, as with an ac waveform, the operation may also be said to be *tracking;* any input changes are transmitted to the output. When S1 is opened, the latest input voltage value is retained on C_H as a charge, and A1 continues to read this voltage until the next sample period. This action is illustrated by the waveforms in Fig. 5-23. Sample/hold circuits are used for a wide variety of signal-processing functions, such as data-handling systems, analog/digital interfaces, analog memory functions, etc.

The S/H operational waveforms of Fig. 5-23 are highly ideal and assume perfect switching, perfect tracking, and ideal holding characteristics. In practice, S/H errors are found to exist for each of the four states of the device. These states are: (1) hold-to-sample transition; (2) sample; (3) sample-to-hold transition; (4) hold. These errors are now defined and illustrated below.

Hold-to-Sample Errors

Acquisition Time—Acquisition time is the time required for the S/H to acquire and track the input signal after the sample command. It is usually specified for a full scale level change (−10 to +10V, or 0 to ±10V). The output must assume its appropriate level within a rated error band, for example 0.01% or 0.1%. This operation is shown in Fig. 5-24A.

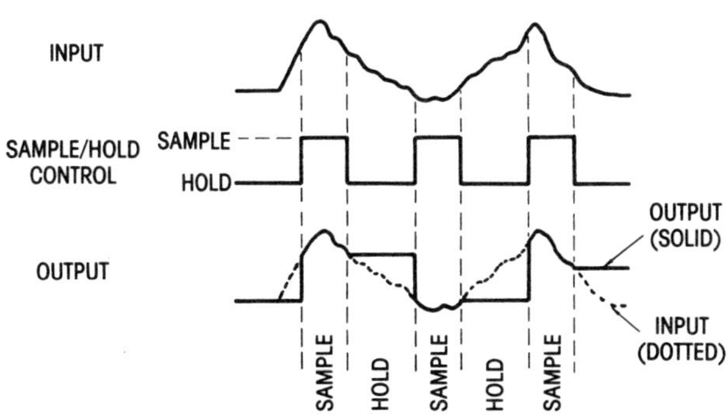

Fig. 5-23. Sample/hold operational waveforms.

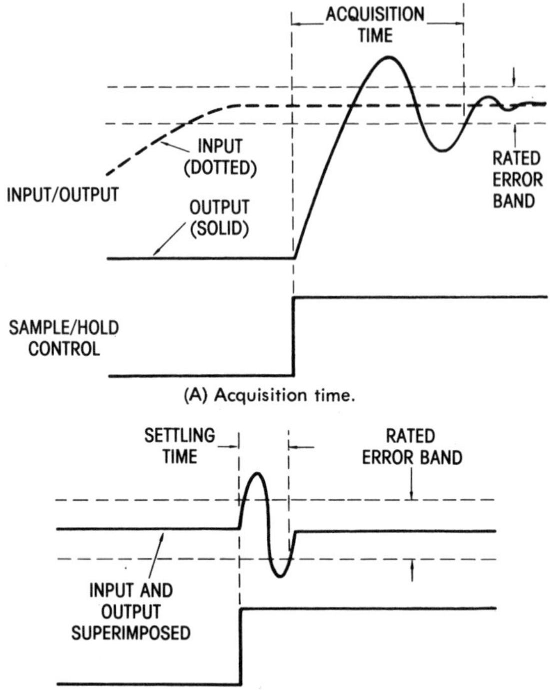

(A) Acquisition time.

(B) Hold-to-sample transients.

Fig. 5-24. Hold-to-sample errors.

A major portion of acquisition time for large step hold-to-sample changes will be taken up by a slewing interval, after which the output will then settle to within rated accuracy.

Acquisition time is the major hold-to-sample error component, and is the primary determinant of how fast a conversion system can be operated. Typical times are on the order of several microseconds to accuracies of 0.1% or better, but there are of course both faster and more accurate units. Acquisition time is strongly dependent upon the value of the holding capacitor used, as this (usually) influences slew rate.

Hold-to-Sample Transient—Illustrated in Fig. 5-24B, the hold-to-sample transient is a switching transient produced at the time of H to S transfer. Its amplitude may be well in excess of the rated accuracy; thus, it must be allowed to die out before the output voltage can be considered valid.

This will necessitate a settling time interval after the H to S command, which must be allowed for in system timing. However, in practice the settling time associated with H to S transients will typically be much less than acquisition time. Therefore, a time interval equal to the acquisition time will then take into account the H to S transient error.

Sample Errors

During the sample interval, the S/H device is tracking the input signal in a fashion which is very much like an op amp. Since most S/H amplifiers reduce to, or are equivalent to an op-amp voltage follower, their sample mode errors can be calculated in a similar fashion.

Nonlinearity—This component is the dynamic deviation from the ideal numerical gain as the unit is exercised over its rated signal output range, usually ±10 volts. Its largest component is usually input stage common-mode error. Note that a true gain (scaling) error can be adjusted out by a system scale calibration, whereas a nonlinearity is a nontrimmable error.

Typical nonlinearity figures are 0.01% or better over the ±10V signal range.

Offset—Offset is the dc shift which occurs between input and output, with the input grounded. It is usually adjustable to zero via an optional trimpot. A typical dc offset specification will be on the order of ±2 mV, or less.

Offset temperature drift is another nonreducible error component, as it will cause errors as temperature varies. Typical drifts are on the order of 10 μV/°C. Alternately, a maximum offset voltage over the temperature range can be specified.

The offset voltage will vary with supply voltage to some degree, and this should also be specified. Typically, supply rejection is on the order of 80 dB, or 100 μV/V.

Settling Time—Settling time is applicable to the sample mode, for rapid input voltage changes. When tracking an input signal a S/H is governed by dynamic limitations, similar to other op-amp configurations.

Settling time is governed by slew rate and small signal bandwidth, with slewing rate being dominant for large scale step changes. Typical slew rates are 5V/μs, and settling times in the 5-10 μs region. As noted above under acquisition time, the exact specifications are strongly dependent upon the holding capacitor.

Sample-to-Hold Errors

Aperture Time—Aperture time is the time which elapses between the hold command and the actual opening of the S/H switch. With a rapidly changing input voltage, this time will then necessarily create an error in the voltage held. The voltage error resulting will be equal to the change in input voltage over the aperture time interval. See Fig. 5-25A.

As an example of its effect, a 1V/μs input slew rate signal sampled by a S/H with a 100 ns aperture time will result in a 100 mV sampled error due to aperture time. This sort of error is significant; therefore, it is usually compensated for by an *advance* in system timing by the amount of the nominal aperture time.

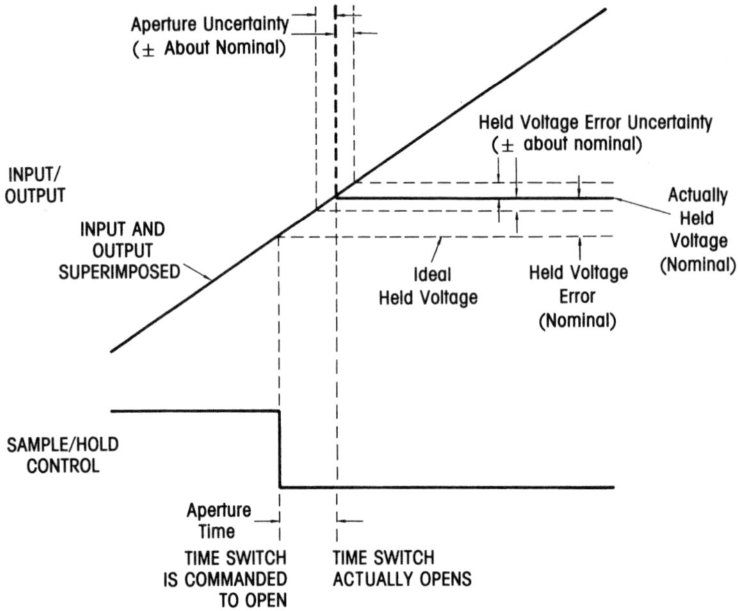

(A) Aperture time error.

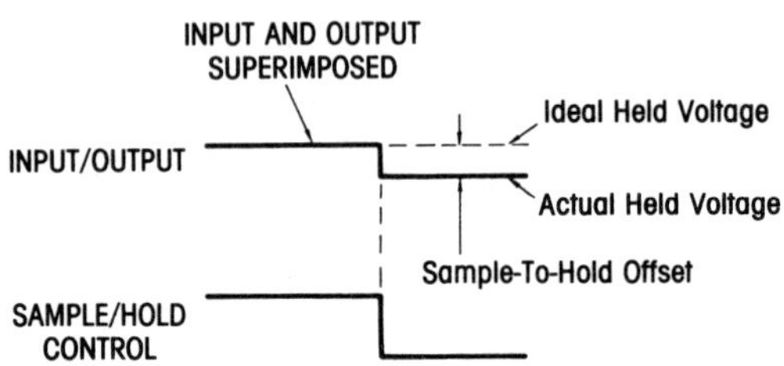

(B) Sample-to-hold offset.

Fig. 5-25. Sample-to-hold errors.

This leaves the remaining error as aperture time uncertainty (or jitter). This is simply pulse-to-pulse variation in the actual switch timing, which places the ultimate limit on aperture time related error. For the 1 V/μs slew rate example, a 10 ns aperture time uncertainty would result in a ±10 mV voltage uncertainty.

Obviously the lower these two figures, the better. The times quoted for illustration are typical for medium speed solid-state S/H circuits and IC units.

Sample-to-Hold Offset—Sample-to-hold offset (also called "jump" or "step") is a transient charge injected into the hold capacitor when the S/H switches from sample to hold. It is caused by the finite switch capacitance, which feeds through a portion of the control signal into the hold capacitor by capacitive divider action. It is reduced with larger value hold capacitors and increased with smaller ones. The effect is illustrated in Fig. 5-25B.

This effect, which may be on the order of several millivolts, can in some cases be compensated by feeding an antiphase control signal into the hold capacitance via a small coupling capacitor.

Sample-to-hold offset may not always be specified as such, particularly in IC units which use external capacitors. In such case a specification is often in charge terms, in units of pC. The sample-to-hold offset may be then calculated as

$$S/H\ offset_{(V)} = \frac{Charge\ (pC)}{C_H\ (pF)}$$

A 10 pC charge transfer with a 1000 pF C_H would for example cause an S/H offset of 10 mV.

Settling Time—This is time for the S/H output to settle to within its rated accuracy following the hold command. It includes the time required for any switching transient to die out.

Hold Errors

Droop—Droop, also called tilt, is the change in the hold voltage (ΔV) during the hold interval (Δt). It is caused by the net leakage currents into (or from) the hold capacitor. It is governed by the simple charge relationship

$$\frac{\Delta V}{\Delta t} = \frac{I_L}{C_H}$$

where

I_L is the leakage,
C_H is the hold capacitor value.

It is illustrated in Fig. 5-26A.

With a given device the current I_L is given by specification, but C_H is under user control. Thus droop can be adjusted by C_H, becoming smaller as C_H is increased.

Note that the value of I_L which should be used to calculate droop should account for any temperature dependence of bias current, such as exists for example in FET input buffers.

This is a tradeoff relationship however, for as C_H increases this will also result in longer acquisition times. It is therefore usually desirable that the leakage current I_L be held as small as practical to minimize droop. Typically, this is the input bias current of a FET buffer amplifier, 50 pA or less. Since this bias current increases with temperature, droop will be worst at high temperatures.

Feedthrough—Feedthrough is the leakage of signal across the S/H switch in the hold (off) state caused by the switch incapacitance. Since

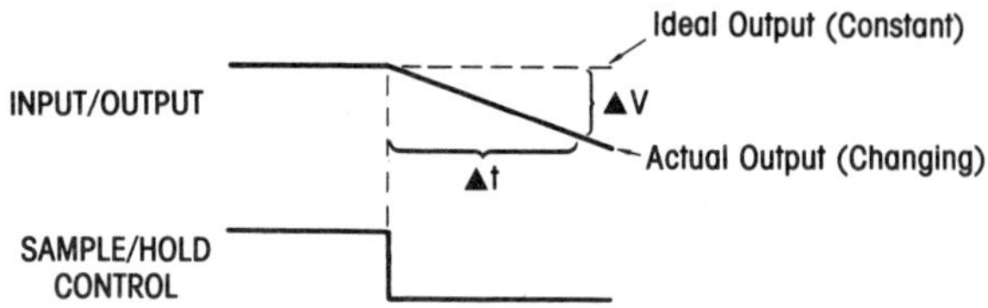

(A) Droop.

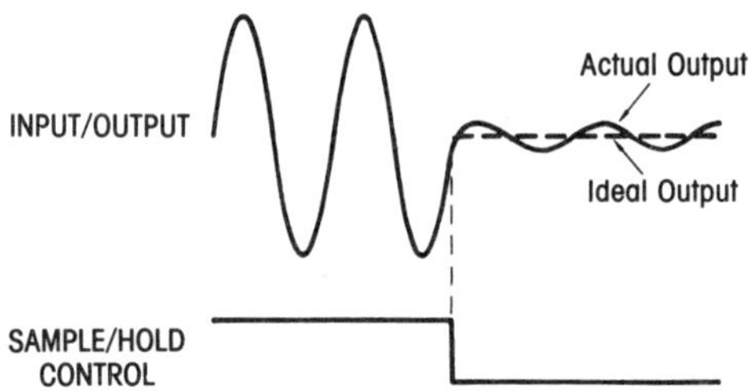

(B) Feedthrough.

Fig. 5-26. Hold errors.

the switch capacitance and the hold capacitance form a voltage divider, feedthrough decreases for higher values of hold capacitance, like droop and S/H offset. The effect of feedthrough is shown in Fig. 5-26B.

Feedthrough is usually specified at a full scale peak-to-peak sine wave input at a high frequency, such as 10kHz. Typically it will be on the order of 80 dB or more.

Dielectric Absorption—In certain capacitor types, the dielectric does not completely release all of its energy after a charge/discharge cycle. The result of this phenomenon is an error in stored voltage after a period of time in the hold mode. This effect is minimized in certain dielectrics, such as teflon, polystyrene, and polypropelyne, and these types should be used with S/Hs for best results.

Drift—A S/H can, and often does, have drift characteristics different in the hold mode than in the sample mode. This is because in the hold the output terminal sees only the drift of the hold buffer amplifier, whereas in the sample mode it sees either the input amplifier alone, or the composite drift of two series amplifiers.

S/H Design Types

Although there are great numbers of S/H devices available, they tend to fall into only a few categories of design types, illustrated below.

Buffered Input

Only a step removed from the elementary circuit of Fig. 5-22, the circuit of Fig. 5-27A is a practical S/H amplifier type. Here A1, C_H, and

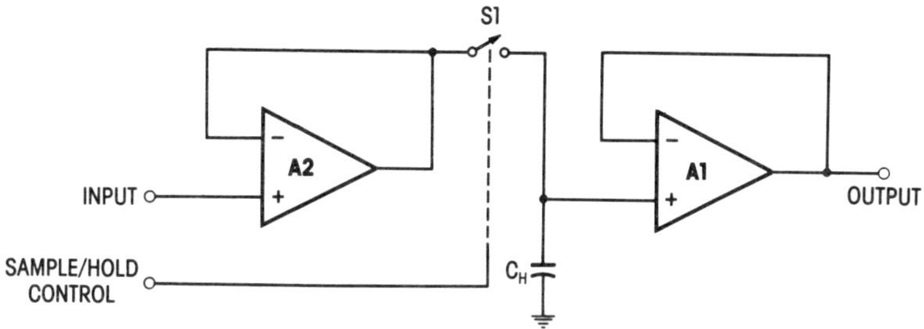

(A) Buffered input.

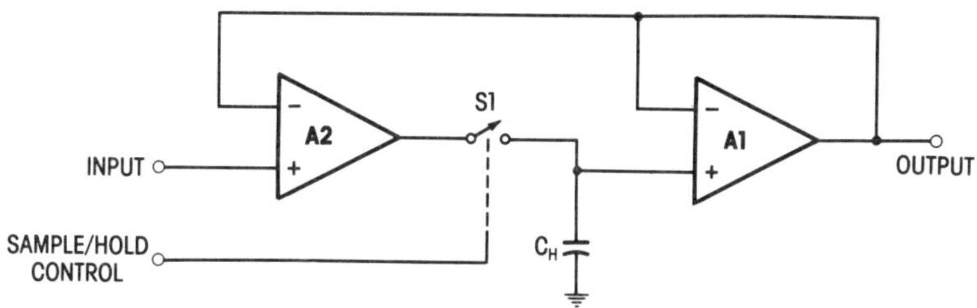

(B) Feedback type, floating switch.

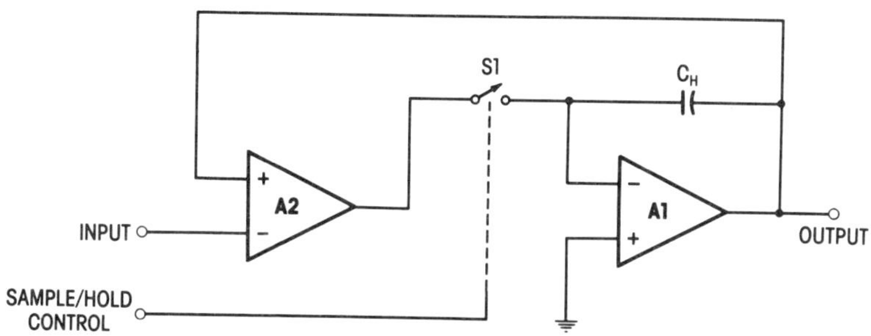

(C) Feedback type virtual ground switch.

Fig. 5-27. S/H design types.

the switch operate as described before, but an input stage buffer is also added, A2.

The addition of the buffer provides increased charging current for C_H, with accompanying faster acquisition time, without loading of the source. Both amplifiers must have high slew rates, fast settling times, and both low offset voltages and drift for good accuracy. A1 should be a FET input device to minimize loading of C_H, but this is not necessarily true for A2.

The S/H configuration of Fig. 5-27B has the advantage that feedback is returned around *both* amplifiers in the sample mode. As a result, the errors of A1 are minimized for the sample state, although they do appear in the hold mode. This circuit has potentially higher accuracy, but with some sacrifice in overall settling characteristic due to multistage loop dynamics.

The circuit of Fig. 5-27C is similar in that it connects feedback around the two op amps, and in this regard has similar advantage. The switch here operates at the virtual ground input of A1 simplifying its drive and leakage requirements.

All three of these S/H types are represented by IC sample/hold devices.

IC Sample/Hold Types

In Table 5-3 is listed a brief summary of the key performance specifications of five IC sample/hold types. Of these units, four are intended to be used with user selected hold capacitors; the AD582, the LF398, the HA2425, and the SMP-81. The SHC80 is a complete integral S/H unit, while some of the others feature various user option connections.

For the purposes of a common comparison, the specifications as listed here are for a C_H of 1000 pF, for all applicable units. Further details of operation are noted in the individual device commentary which follows. The specifications listed are those of the most accurate commercial (0 to 70°C) temperature range unit.

The AD582K

The AD582K is an IC S/H which is functionally similar to the diagram of Fig. 5-27C. It is connected for operation as shown in Fig. 5-28A, with a 14 pin DIP package.

Both (+) and (−) input terminals of the AD582 are user accessible for hookup options. The connection shown here is for a noninverting, unity gain S/H. Offset adjustment (optional) is accomplished with a 10K pot, and the logic control as shown is hold-high, sample-low and is TTL compatible.

This device has an acquisition time of 25 μs to an accuracy of 0.01%, for a 10 volt step signal.

The LF398

The LF398 is an IC S/H which is functionally similar to the diagram of Fig. 5-27B. This is apparent from its block diagram shown in Fig. 5-28B. The LF398 is internally connected for the noninverting unity gain configuration, and is an 8 lead TO-99 package.

The schematic diagram of the device and its connection are shown in Fig. 5-28C, which also indicates the (optional) offset control connection. The logic input as shown is TTL compatible, with a sample-high, hold-low sense.

This device has an acquisition time of 6 μs to an accuracy of 0.01%, for a 10 volt step signal.

Table 5-3. Typical S/H Performance Data

	AD582K[1]	LF398[1]	HA2425[1]	SHC80KP	SMP-81E[2]
Acquisition Time, 10 V to 0.01%	25 μs	6 μs	8 μs	10 μs (max)	3.5 μs (0.1%)
20 V to 0.01%	—	—	10 μs	12 μs (max)	—
Aperture time	150 ns	125 ns	50 ns	40 ns	50 ns
Aperture time uncertainty	15 ns	—	5 ns	1 ns	—
S/H offset	5pC, or 5 mV	10 mV	10pC or 10 mV	2 mV (max)	0.3 mV
Droop	50 μV/mS (max)	15 μV/mS	5 μV/mS	0.5 mV/ms (max)	60 μV/ms
Feedthrough	0.005%	0.009%	0.001%	0.02%	0.0015%
Nonlinearity	0.01%	0.004%	0.02% (max)	0.015 (max)	—
Offset	2 mV[3]	1 mV[3]	3 mV[3]	2 mV[3]	0.3 mV[3]

1 With external $C_H = 1000$ pF.
2 With external $CH = 5000$ pF.
3 Adjustable to zero.

The HA2425

The HA2425 is an IC S/H which is functionally similar to Fig. 5-27B. It is connected as shown in Fig. 5-28D, and is furnished in a 14 pin DIP.

Both the (+) and (−) input terminals of the device are available, allowing flexibility in hookup. As shown it is connected for the non-inverting, unity gain S/H configuration. Note that the two pins of the device package immediately adjacent to the C_H terminal are intended to be used for a guard connection. This is accomplished by tying them to the output pin, which reduces the hold state leakage current into C_H, and thus minimizes droop.

The optional offset control uses a 100K pot, as shown. Logic sense is fixed, hold-high and sample-low, and is TTL compatible.

This device has an acquisition time of 8 μs to an accuracy of 0.01%, for a 10 volt step signal.

The SMP-81

The SMP-81 is an IC S/H which is also functionally similar to Fig. 5-27B. It is connected as shown in Fig. 5-28E, and is furnished in a 14 pin DIP.

As can be noted, the SMP-81 is similar in some regards to the HA-2425, and is pinned nearly alike. Key differences include a committed feedback connection for unity gain, a differential adjustable bias S/H control input, and the addition of a "supercharger" amplifier, which enhances slew rate during acquisition, lowering acquisition time appreciably.

The device is internally trimmed for optimum operation with a 5000 pF capacitor, and will acquire a 10-volt signal to 0.1% in 3.5 μs. Offset is pretrimmed to a low value of 0.3 mV; it can also be adjusted externally, as shown.

Logic is TTL compatible with the V_{LC} pin grounded as shown, with the sense indicated.

The SHC80

The SHC80 is a hybrid IC S/H which is functionally similar to Fig. 5-27B. It is connected as shown in Fig. 5-28F, and is contained in a 14 pin DIP.

This device is complete with an integral 820 pF C_H and requires no external components for proper operation. It is internally strapped for the noninverting, unity gain configuration.

User options are offset control, which uses a 2 to 5K pot connected as shown. The logic sensitivity can be programmed by the connections to pins 9 and 10. As shown with pin 9 connected to +5 volts, it is TTL compatible. Optionally, pin 9 can be wired to +15 volts through a 27K resistor for TTL compatibility. Logic S/H control sense is sample-high,

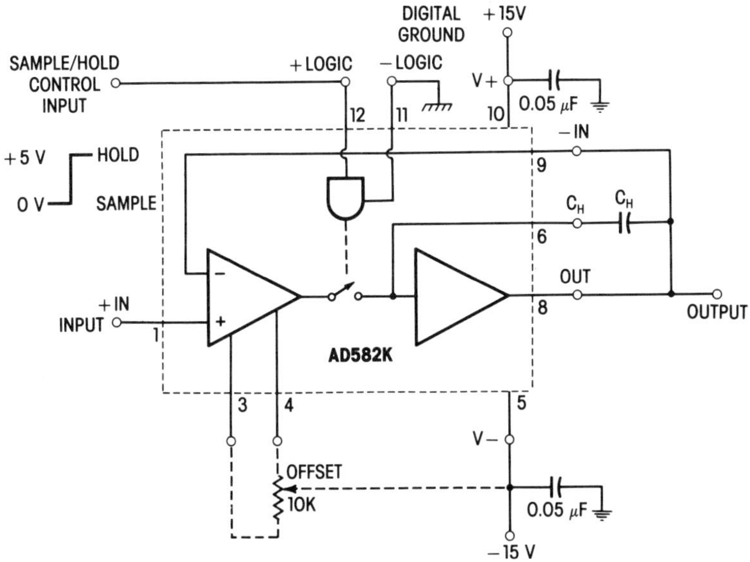

(A) Connection of AD582 (14 pin DIP).

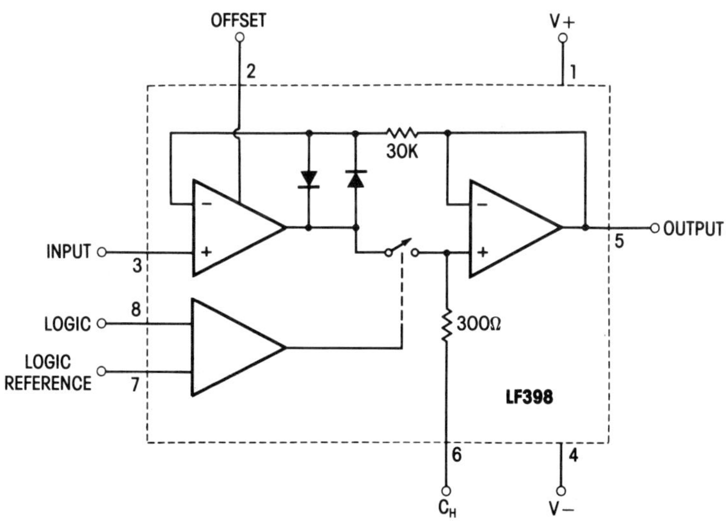

(B) Functional diagram LF398 (8 pin TO-99 package).

Fig. 5-28. IC S/H device hookups.

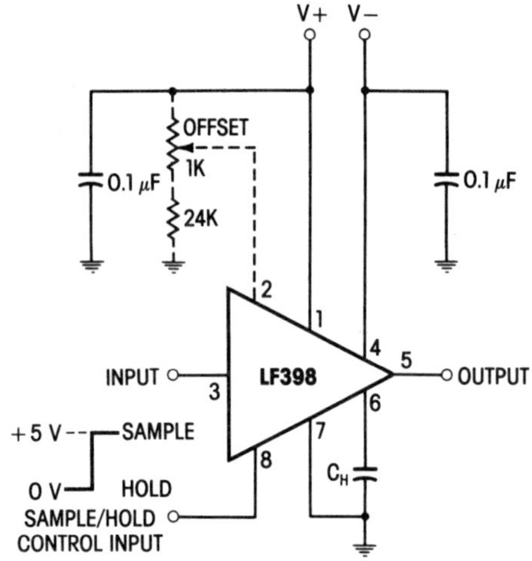

C) Schematic of LF398.

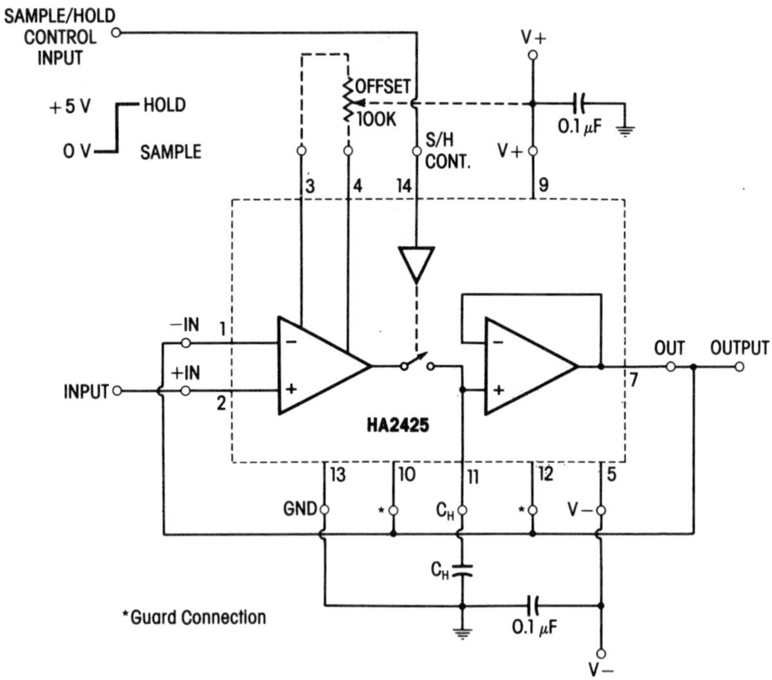

(D) Connection of HA245 (14 pin DIP).

Fig. 5-28 (cont.) IC S/H

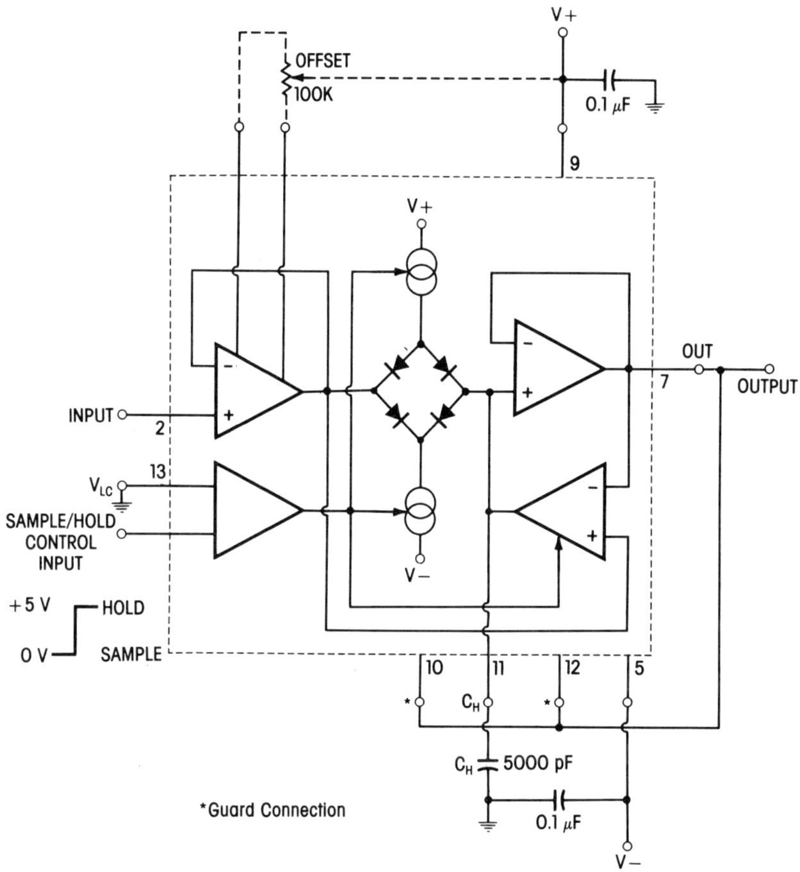

(E) Connection of SMP-81 (14 pin DIP).

device hookups.

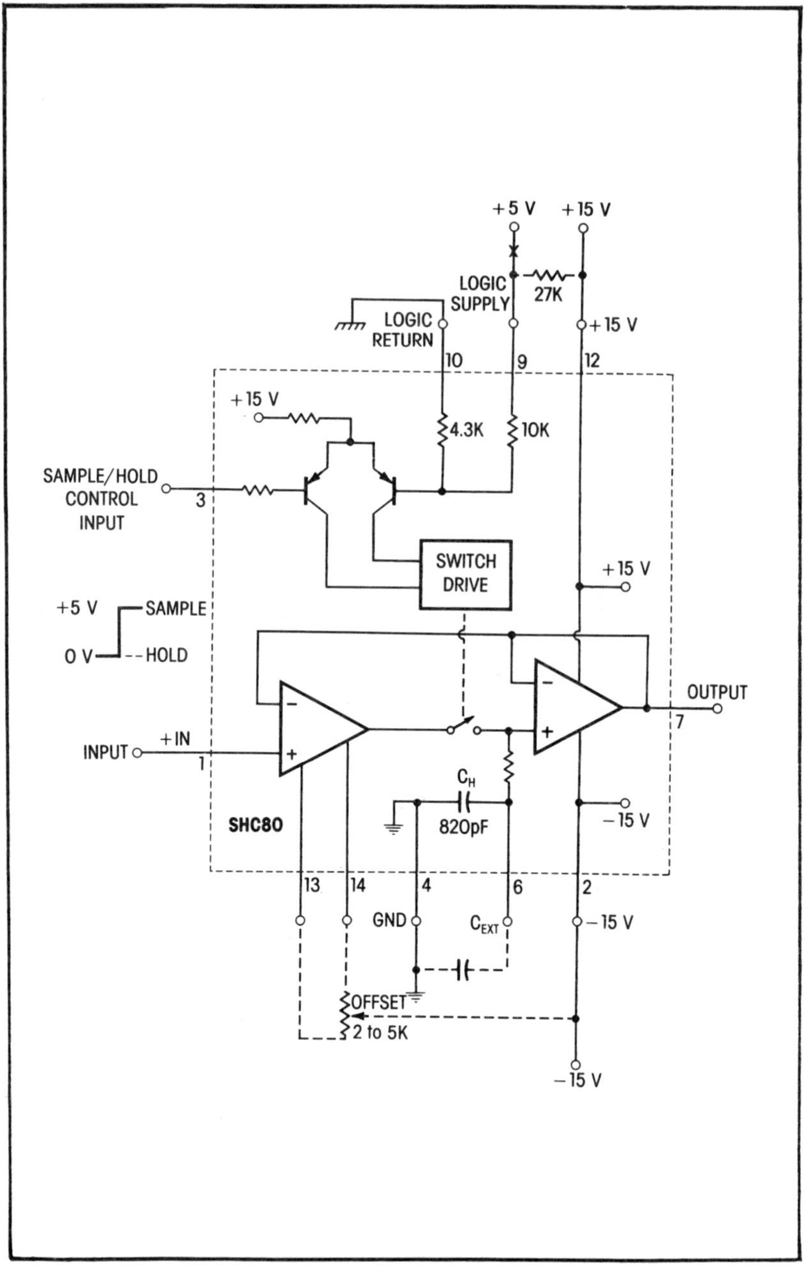

(F) Connection of SHC80 (14 pin DIP).

Fig. 5-28 (cont). IC S/H device hookups.

hold-low. When desired, additional holding capacitance can be connected to the C_H pin.

This device has an acquisition time of 10 μs to an accuracy of 0.01%, for a 10 volt step signal.

REFERENCES

1. Analog Devices Application Staff. *Seminar on Analog CMOS*. Analog Devices, 1973. Analog Devices, Inc., Norwood, Massachusetts.

2. *Applications Manual for Computing Amplifiers*. Philbrick Researches, Inc., Dedham, Massachusetts, 1966.

3. Boag, T. R. "Designing With Monolithic FET Switches." *Electronic Products,* January 15, 1973.

4. Borlase, W. *Application and Analysis of the AD520 Monolithic Data Amplifier*. Analog Devices Application Note, June 1972. Analog Devices, Inc., Norwood, Massachusetts.

5. Brokaw, A. P. "Two Instrument IC's Sum Six Inputs." *Electronics,* August 7, 1975.

6. ———. "You Can Compare Two Independent Signal Levels With Only One IC." *EDN,* April 5, 1975.

7. Bruck, D. S. *Data Conversion Handbook*. Hybrid Systems Corp., Burlington, Massachusetts, 1974.

8. Clayton, G. B. *Linear Integrated Circuit Applications*. MacMillan Press Ltd., 1975.

9. Cohen, J. "Solid State Signal Switching: It's Getting Better All the Time." *EDN,* November 15, 1972.

10. Demrow, R. *Evolution from Operational Amplifier to Data Amplifier*. Analog Devices Application Note, September 1968. Analog Devices, Inc., Norwood, Massachusetts.

11. Dixon, G. *Analog Switches in Sample and Hold Circuits*. Application Note AN74-2, May 1974. Siliconix, Incorporated.

12. Dobkin, R. C. *Instrumentation Amplifiers*. National Semiconductor Linear Brief LB-21, June 1973. National Semiconductor Corp., Santa Clara, California.

13. Evans, A. D. "The Basics of Using FETs for Analog-Signal Switching." *EDN,* May 20, 1973.

14. Flood, J.; Pujol, H. L. *Guide to Better Handling and Operation of CMOS Integrated Circuits*. RCA Application Note ICAN 6525, May 1976. RCA Solid State Div., Somerville, New Jersey.

15. Fullagar, D. "Analog Switches." *Proceedings of Electronic Products Magazine Seminar,* December 1973.

16. ———. "Analog Switches Replace Reed Relays." *Electronic Design,* June 21, 1973.

17. ———. "Analog Switches-Rugged, Reliable and Fast." *Electronic Products,* September 17, 1973.

18. ———. *A New CMOS Analog Gate Technology.* Intersil Application Bulletin. April 1973. Intersil, Inc., Cupertino, California.

19. ———. "Don't Neglect the Solid-State Switch." *EDN,* March 1, 1972.

20. ———. *The IH5009 Series of Analog Switches.* Intersil Application Bulletin A004, January 1972. Intersil, Inc., Cupertino, California.

21. ———. *Understanding and Applying the Analog Switch.* Intersil Application Bulletin A003, April 1973. Intersil, Inc., Cupertino, California.

22. ———; Henry, R. "Low-Cost, Dual FET Input Op Amp Provides New and Versatile Tool for Designers." *EDN,* September 20, 1973.

23. Gordon, B. M. "The ABC's of A/D and D/A Converter Specifications." *EDN,* August 15, 1972.

24. Graeme, J. G. *Applications of Operational Amplifiers—Third-Generation Techniques.* McGraw-Hill Book Co., New York, 1973.

25. Haynes, D.; Ott, W. E. "IC's Give New Dimensions to Instrumentation Amplifiers." *EDN,* February 20, 1976.

26. Johnston, R. "Analyzing the Dynamic Accuracy." *Electronic Design,* September 13, 1973.

27. Jones, D. "A New Generation of Linear IC's," *Proceedings of Electronic Products Magazine IC Seminar,* December 1973.

28. ———. *Application of a Monolithic Sample-and-Hold/Gated Operational Amplifier.* Harris Semiconductor Application Note AN517, March 1974. Harris Semiconductor, Melbourne, Florida.

29. ———. *CMOS Analog Multiplexers and Switches; Applications Considerations.* Harris Semiconductor Application Note AN520, June 1976. Harris Semiconductor, Melbourne, Florida.

30. Jung, W. G. *IC Op Amp Cookbook.* Howard W. Sams & Co., Inc., Indianapolis, 1974.

31. ———. "Op Amps Multiplex Analog Signals Without Need for FET Switches." *Electronic Design,* June 21, 1974.

32. Kress, D. "Monolithic S/H Amplifiers: IC Flexibility Comes to the S/H Scene." *Electronic Products,* March, 1977.

33. ———; Wawner, D. *Monolithic Sample-Hold Amplifiers.* Analog Dialogue Vol. 10, No. 1, 1976.

34. Miller, W. "For Tough Measurements, Try IAS." *Electronic Design,* August 2, 1974.

35. Mroz, I. J. *Analog Switches and Their Applications,* Chapter 4. Siliconix Incorporated, 1976.

36. ———. *DG300 Series Analog Switch Applications.* Application Note AN76-6, October 1976. Siliconix, Incorporated.

37. Ott, W. E. *Instrumentation Amplifiers.* Burr-Brown Application Note AN-75, August 1974.

38. Patstone, W.; Dunbar, C. "Choosing A Sample- and Hold Amplifier Is Not As Simple As It Used To Be." *Electronics,* August 2, 1973.

39. Pouliot, F. "Simplicity Amplifier Selection." *Electronic Design,* August 2, 1973.

40. Riskin, J. "Differential Instrumentation Amplifier." *Analog Dialogue,* Vol. 10, No. 2, 1976.

41. Ritmanich, W. "Protected Analog Switches." *Analog Dialogue,* Vol. 10, No. 2, 1976.

42. Ruggeri, L. "Multiplex Signals the Analog Way." *Electronic Design,* July 19, 1974.

43. Sanchez, A. "Understanding Sample-Hold Modules." *Analog Dialogue,* Vol. 5, No. 4.

44. Santen, J. "Evaluating the Tradeoffs: Hybrid vs Monolithic S/H Amplifiers." *Electronic Products,* March, 1977.

45. Schmid, H. *Electronic Analog/Digital Conversions.* Van Nostrand Reinhold, New York, 1970.

46. Seino, K. C. "Low Drift ICs Form Instrumentation Amplifier." *Electronics,* May 1, 1975.

47. Shaeffer, L. "Attenuating Transients in Analog FET Switches." *Electronics,* October 3, 1974.

48. Sheingold, D. *Analog Digital Conversion Handbook.* Analog Devices, Inc., Norwood, Mass., 1972.

49. Sherwin, J. S. "Cut Transients in FET Analog Switches." *Electronic Design,* April 17, 1972.

50. Siliconix Applications Staff. *Analog Switches and Their Applications.* Siliconix, Incorporated, 1976.

51. Skipper, J. D. "Gain Set Accurately With a Single Resistor In a High-Performance Differential Amplifier." *Electronic Design,* September 13, 1974.

52. Smith, J. I. *Modern Operational Circuit Design.* John Wiley & Sons, Inc., New York, 1971.

53. Thibodeaux, E. "Getting the Most Out of CMOS Devices for Analog Switching Jobs." *Electronics,* December 25, 1975.

54. Timko, M.; Maxwell, B.; Brokaw, A. P. "Monolithic Differential Instrumentation Amp." *Analog Dialogue,* Vol. 9, No. 1, 1975.

55. Tobey, G. E. "Ease Multiplexing and A/D Conversion." *Electronic Design,* April 12, 1973.

56. ———; Graeme, J. G.; Hullsman, L. P. *Operational Amplifiers—Design and Applications.* McGraw-Hill Book Co., New York, 1971.

57. Torrero, E. A. "Focus on IC Analog Switches and Multiplexers." *Electronic Design,* September 1, 1975.

58. Wait, J. V.; Huelsman, L. P.; Korn, G. A. *Introduction to Operational Theory and Applications*. McGraw-Hill Book Co., New York, 1975.

59. Wilenken, R. "Beware of CMOS-Switch Failure Modes." *Electronic Design*, March 15, 1975.

60. Zis, J. "Applications of Recently Announced High Technology Linear Integrated Circuits." *Electronic Products Magazine IC Seminar Proceedings*, December 1973.

PART II

IC Converter Applications

6

Basic D/A Converter Applications

D/A LOGIC INTERFACING

The digital inputs of IC d/a converters are reasonably straightforward, and usually present no great application difficulties in interfacing with a logic control system. There are, however, areas of interfacing that can raise questions as to the most efficient means of a practical solution. These are logic buffers, logic-level conversion, and data-storage techniques. The discussion which follows illustrates simple means to these problems, which are applicable to virtually any IC d/a device. The thoughtful application of such interfacing techniques can greatly enhance the flexibility of most d/a converters.

Logic Buffering

The most elementary form of logic interface is a buffer or inverter used to reconstitute logic levels or increase fanout in a given logic system. This usually takes the form of a bank of inverters, 1 stage per bit of resolution for the d/a converter in use. The reasons can not be only to restore degraded logic levels and/or lower input currents from the driving source, but for code conversion and so on. The last reason is a predominant one in practice, as many IC d/a converters have complementary-binary coding arrangements. Two logic buffering solutions effective for TTL or CMOS systems are shown in Fig. 6-1.

Hex inverters are attractive for inverting buffers as they minimize package count. For TTL, any of the 7404 family types can be used,

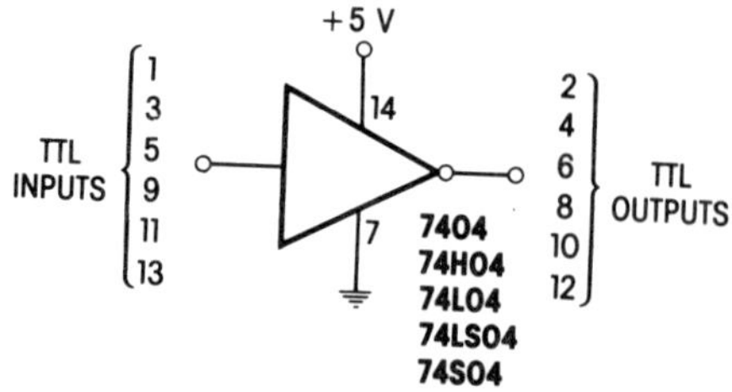

(A) TTL to TTL.

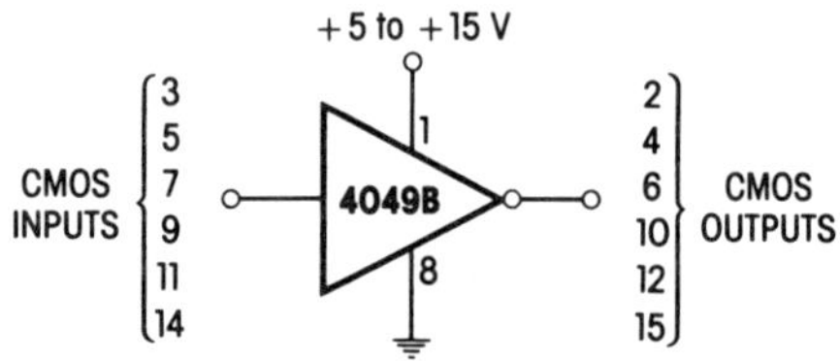

(B) CMOS to CMOS.

Fig. 6-1. Buffer/inverters.

depending upon the required speed. For CMOS systems in the +5 to +15-volt range, the 4049B hex inverters are useful, as shown in Fig. 6-1B. Another hex inverter type is the 74C04, a CMOS unit, pinned as shown in Fig. 6-1A. These two types of buffers are used where input and output logic and supply levels are similar; that is TTL to TTL, or CMOS to CMOS.

For noisy applications, Schmitt trigger input types devices can be useful, as they provide an input hysteresis. Two functional examples suitable for TTL or CMOS systems are shown in Fig. 6-2.

Hex Schmitt inverters are available for either TTL or CMOS operation, as shown in Fig. 6-2A. For the devices shown the pinouts are identical (that of the 7404). Take care to use the appropriate device and supply voltage level, as is indicated for TTL and CMOS operation.

For gated applications, a NAND Schmitt function is shown in Fig. 6-2B. This may be for either a TTL or CMOS device. Note however the pin number differences, as well as the supply ranges.

Level Shifters

A situation may often arise where it is desirable to use a d/a device which is not directly compatible with the IC logic family in the remainder of the system. This requires the use of a logic level-shift stage. Several devices for such use are illustrated in Fig. 6-3 that are capable of CMOS to TTL and TTL to CMOS conversion.

To convert from CMOS to TTL (high to low) logic levels, two device types are shown in Fig. 6-3A. For noninverting uses, the 4050B is used;

for inverting use, the 4049B is used. Either device is powered from a supply of +5 volts which yields a TTL output drive (2 loads). The logic input levels can exceed the +5-volt supply, without harm, up to +15 volts. Other devices (not illustrated) which can perform this function are the 74C901 (inverting) and the 74C902 (noninverting).

For conversion from TTL to CMOS (low to high) several devices are shown in Fig. 6-3B. One is a quad input noninverting device. This unit, the 40109B, is operated with two different supplies, which define the input and output levels, +5 volts and +15 volts respectively. For this application the ENABLE inputs would be tied to +5 volts.

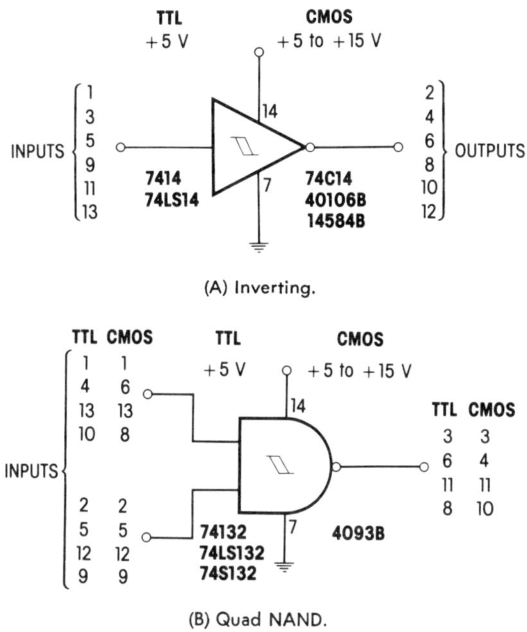

Fig. 6-2. Schmitt trigger buffers.

Also shown is a hex inverting TTL to CMOS interface using an open collector high voltage TTL inverter, the 7406. The output will in its high logic state be the referral level used (+5 to +15 volts). The noninverting stage shown in Fig. 6-3B is similar, except it uses a 7407, an noninverting device.

There are a multitude of other IC devices which can operate as level shifters, but they all cannot be presented here. General device families which will be found useful are peripheral drivers and interface devices; catalogs from manufacturers such as Fairchild, Motorola, National, and Texas Instruments can be consulted for further details.

Code Conversion

For a given number of bits, binary coded converters have greater resolution and thus greater potential accuracy than bcd coded ones. A 10-bit binary device for instance has 1024 possible codes, while comparable bcd resolution and accuracy requires 3 digits (12 bits).

Advantage can be taken of this factor to use a 10-bit binary coded d/a converter as a 3 digit bcd converter. The key to the implementation

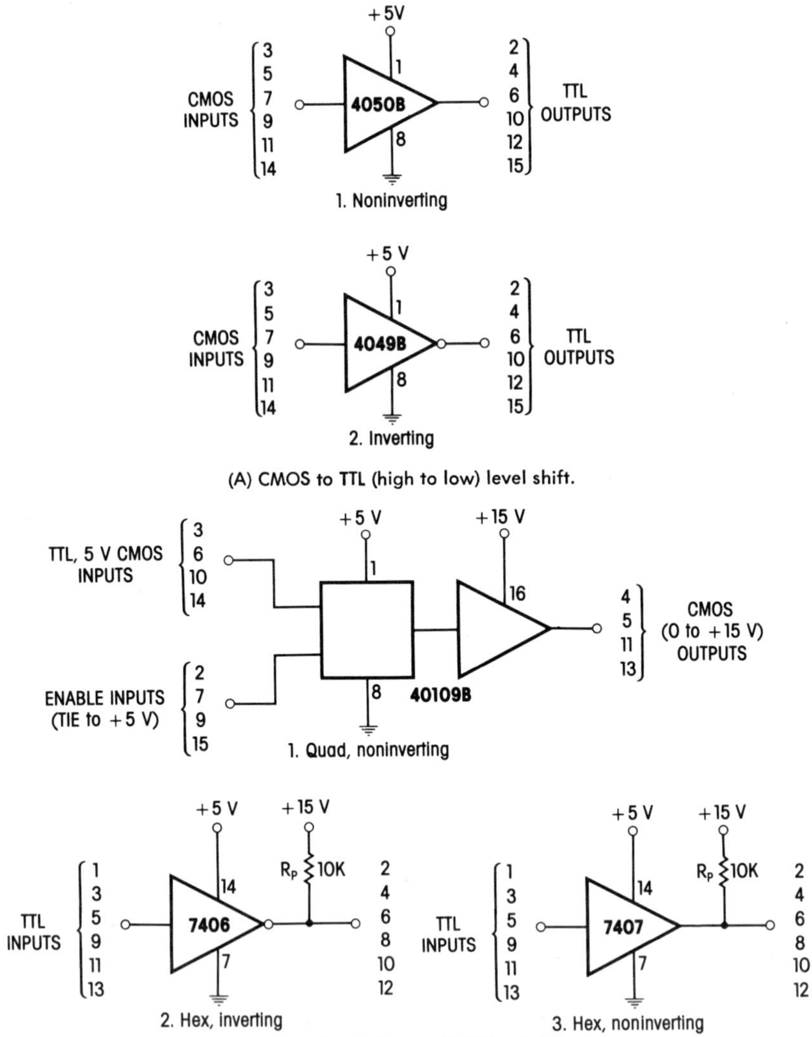

(A) CMOS to TTL (high to low) level shift.

(B) TTL to CMOS (low to high) level shift.

Fig. 6-3. Logic level

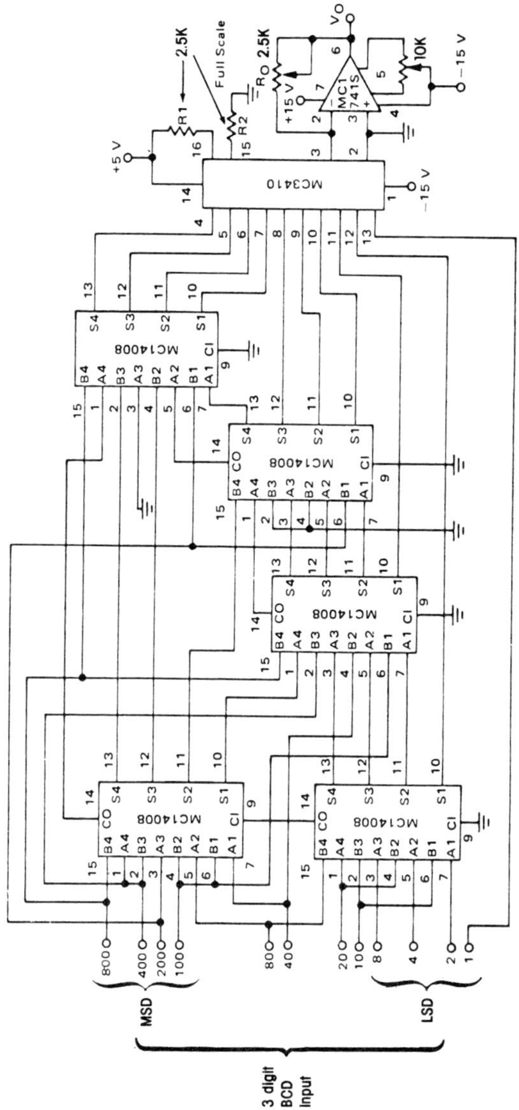

(C) Bcd-to-binary code converter.

and code converters.

lies in a bcd-to-binary code converter, which precedes the binary d/a converter. The two components together form a bcd converter system. A 3 digit bcd-to-binary converter is shown in Fig. 6-3C driving an MC-3410 10-bit d/a converter. Overall, this circuit functions as a 3 digit bcd converter.

The logic translation is performed here by five MC14008 CMOS adders, which convert the twelve bcd coded inputs into ten equivalent binary outputs, which in turn drive the MC3410.

In this example, the MC3410 is scaled for +9.99-volt full scale output, with R1, R2 and R_0 all 2.5K resistors. The output op amp is trimmed for zero via the 10K pot, with all zeros in, and R_0 is adjusted for the full scale voltage with a 1001 1001 1001 input.

Data Storage Elements

The previous discussion has been concerned with the relatively simple forms of data interfacing where only logic buffering or level shifting is the requirement. Another type of interfacing, which can more often be required, is d/a input data storage. This data storage can involve shifting or conversion of the data format, timing buffers, and other forms of short-term memory. IC logic devices which are generally useful in this area are latches, flip-flops, and registers. Application examples follow and are discussed in terms of both TTL and CMOS devices.

Fig. 6-4 illustrates in general concept how a d/a converter is used with a data storage register. This storage register can take on many forms, but its general function is to acquire input data when so directed by a timing control signal, and hold (store) this data until commanded

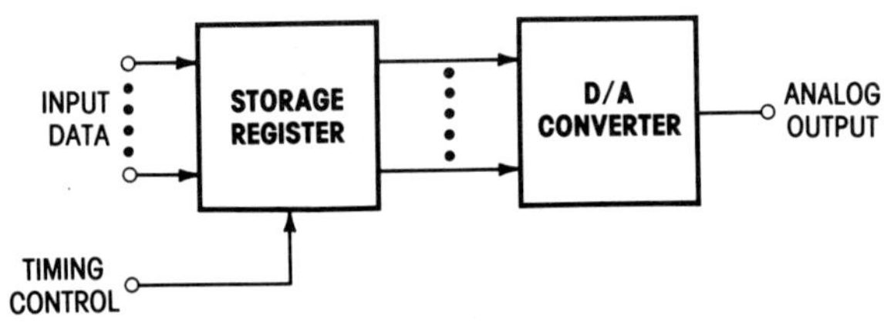

Fig. 6-4. A d/a converter with input register.

to acquire new data. When in the storage or hold mode, the register delivers a constant data output to the d/a converter, which in turn outputs a steady analog output. Input data to the register can change while it is in the hold mode, without affecting the d/a output, as it is locked out by the storage register.

The details of various means of implementing this function (all of which are useful) are illustrated in the following discussion.

Data Latches

The simplest means of storing digital data is by a set of R-S flip-flops or latches, one latch per bit of d/a input data. There are a number of TTL and CMOS MSI elements available which can accomplish this function. These devices are available in various arrangements of bits, from 4 to 10 per package, and with various control options. Examples are listed in Tables 6-1 and 6-2.

Table 6-1. Data Storage Elements

	Device	No. of Bits	Comments
Latches	7475	4	Transparent, Complementary Outputs
	7477	4	" True Outputs
	9314	4	" "
	74100	8	" "
	9308	8	" "
	74116	8	" "
	74LS363, 373	8	" "
	9334/74259	8	" " 3 Bit Addressable
	8200, 8201	10	Various Options
	8202, 8203		
"D" Flip-flops	74173		Clocked
	74175	4	"
	74174	6	"
	74273	8	"
	74LS379	4	"
	74LS378	6	
	74LS377	8	Clocked
Registers	74164	8	Serial In, Parallel Out
	74198, 199	8	Paralllel Or Serial In, Parallel Out
	8273	10	Serial In, Parallel Out

A typical TTL device is the 7475, a quad latch with complementary outputs, shown in Fig. 6-5. This device has two ENABLE inputs, which each control two latches, as shown. In this circuit (and all other latches mentioned in this section), the latch operation is said to be "transparent." This simply means that the output data will follow the input, whenever the ENABLE input is activated. When the ENABLE input is deactivated, the latch holds the data last present at the data input.

Because of this transparent mode of operation, latches such as the 7475 and others operate within a d/a system in an alternate hold/follow mode. For the 7475, the outputs follow the inputs when the ENABLE inputs are high. The data is held when the ENABLE input(s) are taken low.

The 7475 is a TTL device; a similarly operating (but not identical) CMOS part is the 4042. Tables 6-1 and 6-2 summarize some of the available TTL and CMOS data storage elements, and a number of other latches are listed.

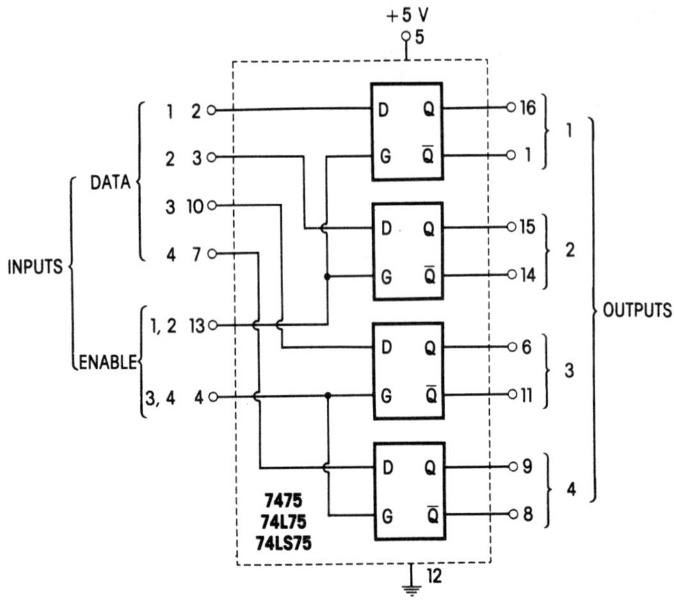

Fig. 6-5. The 7475 quad latch.

Table 6-2. CMOS Data Storage Elements

	Device	No. of Bits	Comments
Latches	4042 4099 14508	4 8 8	Transparent Transparent Transparent
"D" Flip-flops	74C175 74C173/4076 74C174	4 4 6	Clocked Clocked Clocked
Registers	4015 4034 4094 74C164	8 8 8 8	Serial In/Parallel Out Serial In/Parallel Out Serial In/Parallel Out Serial In/Parallel Out

An application example of a hold/follow type input register using a pair of 7475 devices is shown in Fig. 6-6. This circuit as shown has an 8-bit capability, and the option of either true or complemented output data operation. It can be expanded to a greater number of bits by adding more 7475 devices, if desired.

More efficient system operation is usually achieved for fewer overall packages however, and Tables 6-1 and 6-2 should be consulted for the best device combination for a given application.

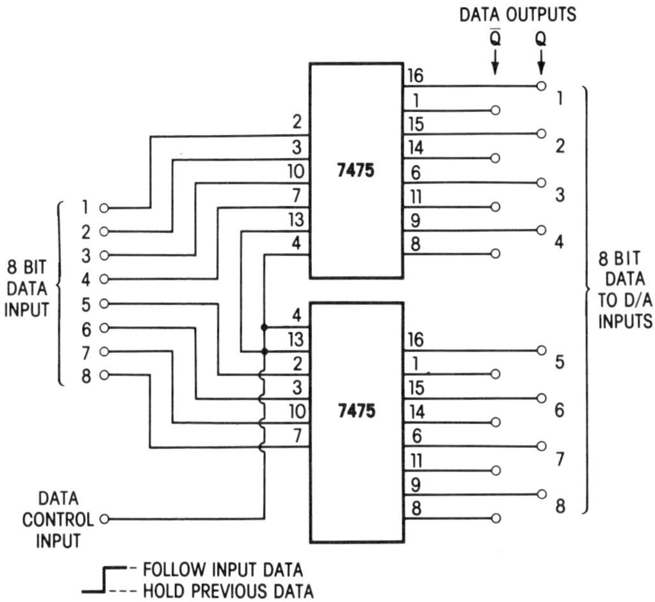

Fig. 6-6. Hold/follow input register (true or complement output).

"D" Flip-Flops

A form of data storage which is generally more desirable than a latch is the type-D flip-flop. A D flip-flop output will only change states when the clock input changes states, and will hold the input data present at the last clock pulse at all other times. This "clocked" data transfer operation makes a D flip-flop storage register much more useful for d/a use, as it decreases sensitivity to spurious operation, to a great degree.

Fig. 6-7 illustrates an example of a clocked flip-flop register using 8551/74173 (for TTL) or 74C173/4076 (for CMOS) quad D flip-flop devices. All four devices listed are pin and functional equivalents, but the appropriate supply and interfacing considerations should be applied for TTL or CMOS, as applicable.

These latches have 3-state outputs, the status of which is controlled by pins 1 and 2. Here, a (wired) low state yields continuous data from this register. The devices also have two DISABLE inputs; data can be loaded with both low, it is disabled when either is high. The data appearing at the inputs is entered into the register on the positive clock transition, and is locked out at other times. The register may be reset to zero by a positive pulse at the CLEAR input.

Although suitable for 8-bit d/a converters as shown, this register can be extended, using additional D flip-flop devices. Tables 6-1 and 6-2 list a number of other suitable devices which operate similarly.

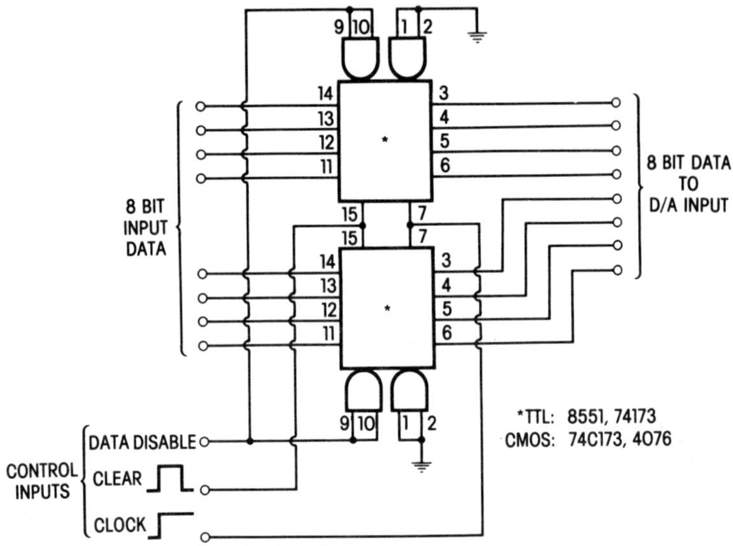

Fig. 6-7. Clocked input register.

Serial Input Data Registers

In some instances it may be more advantageous to transmit and receive the digital input data in serial form. However, since virtually all d/a converters have parallel logic inputs, this necessitates a serial-to-parallel conversion. Again, good advantage can be taken for this application of TTL or CMOS MSI logic elements.

Fig. 6-8 illustrates a scheme using double buffered registers. The first stage register is the 74164 (TTL) or 74C164 (CMOS), which receives the incoming data in serial form, and loads it into the first 8-bit register. The 74164 parallel outputs feed a second register, consisting here of a pair of 74175 (TTL) or 74C175 (CMOS) units. This register operates as a clocked, parallel loaded register.

The general sequence of operation is to load the serial data into the 74164, then at some later time parallel load it into the 74175s. Note that because of the double buffering, serial input data can be updated independent of output data, and at very high speeds. This data can then be shifted into the second register when appropriate (and/or convenient).

There are many other devices which can perform similar functions when used in suitable combinations (see tables). An interesting option is the CMOS 4094, an 8-bit serial-to-parallel converter, with holding register and 3-state output, illustrated in Fig. 6-9. This single device accomplishes the entire serial-to-parallel conversion and double buffer task for 8 bits. It also has a 3-state output, and may be bus operated with other similar units. And, it may be extended, by cascading, for more than 8 bits.

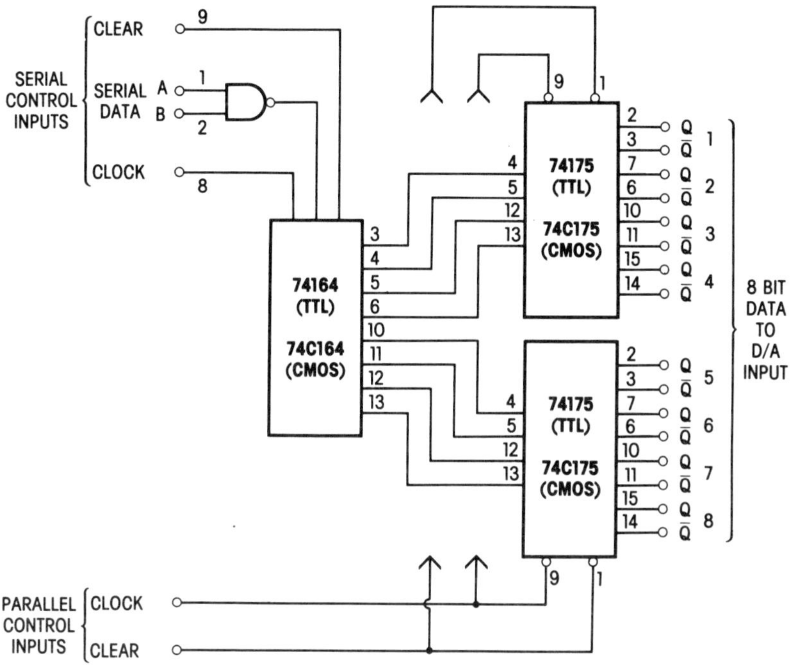

Fig. 6-8. Serial input with holding register.

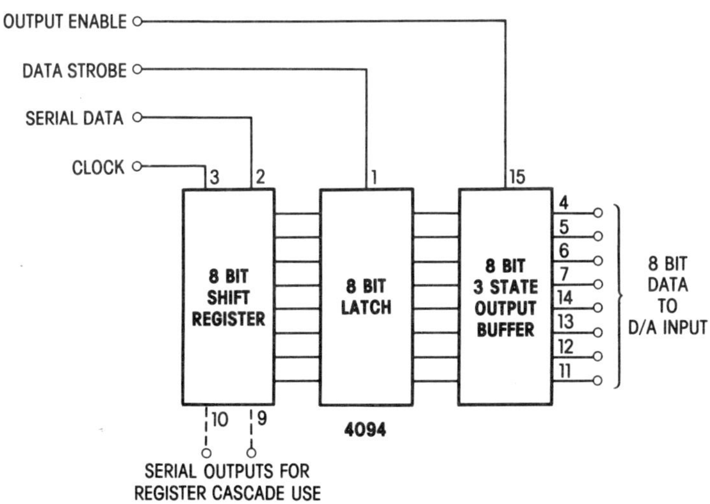

Fig. 6-9. Serial input holding register with 3-state output.

Data Bus Receivers

A popular transmission technique is the bus scheme, where a number of data transmitters and receivers time-share a common bus (or buses). At appropriate times, data is placed onto the bus by a transmitter; at the same time designated receivers simultaneously are enabled and receive the data. During periods of data transmission to other receivers, the local receivers are disabled, so as to ignore undesired data.

Two basic schemes illustrating data bus receivers that are appropriate for d/a input data uses are illustrated in Fig. 6-10.

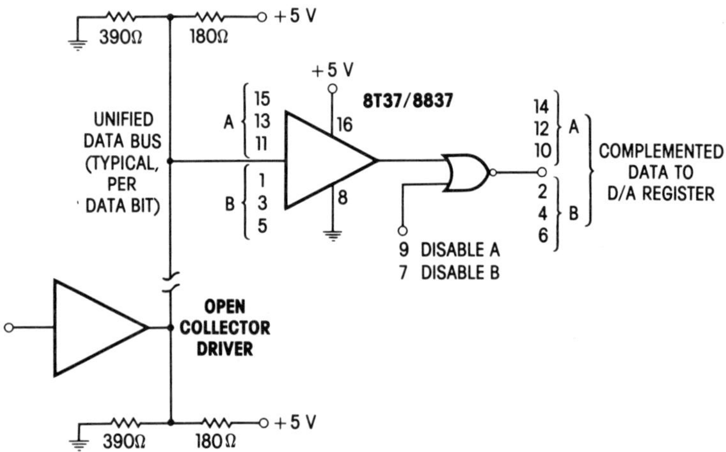

(A) Unified data bus.

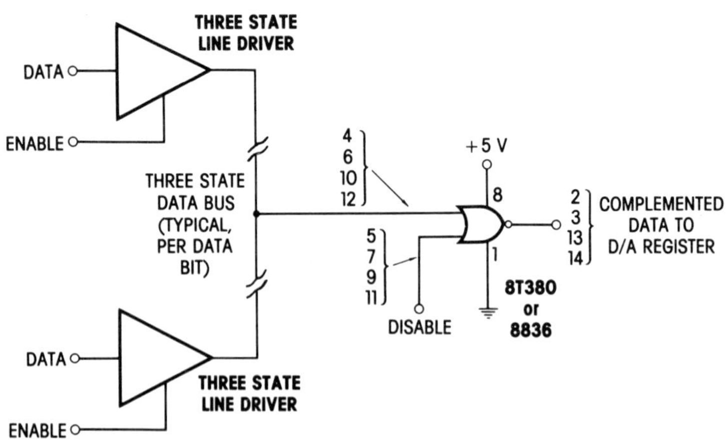

(B) 3-state data bus.

Fig. 6-10. Data bus receivers.

Fig. 6-10A shows a unified data bus receiver. In this arrangement the data bus is a single-ended 120-ohm nominal impedance line. It is terminated by a 120 Thevenin equivalent impedance at both ends, referenced to a voltage of +3.4 volts. Drive onto the bus is from open collector TTL transmitter stages. The line is a TTL "1" for a transmitter off state; a transmitter being on pulls the line down to a "0" state. Typically, flat multiconductor cabling is usually used, at distances up to about 50 feet.

To minimize loading, receivers to be used with this bus should be high input impedance. Two suitable devices are the 8T37 and 8837, which are hex receivers with 3 bit common disable inputs. They provide input hysteresis for increased noise immunity, and feature TTL output levels. The data output is complemented from the input, and the receivers are enabled by a low disable input.

As a d/a input receiver, this scheme would most effectively be interfaced at its output with one of the already described parallel input registers that would hold the received data between transmissions.

Another very popular data busing scheme is the "3-state" bus. This technique uses drivers which have a third output state in addition to the familiar high or low; a state which is off, or high impedance. Disabled drivers thus do not load the line, and the single high output driver which is enabled drives the common bus. The 3-state bus is time shared, and appropriate transmitter/receiver addressing must be used. The 3-state scheme is advantageous in that it offers good noise immunity. A large number of 3-state logic drivers and devices are available, and this data system may be considered almost a standard one.

Receiving devices for 3-state uses are many in number, but a simple example is shown in Fig. 6-10B. This arrangement is like the unified bus, in that similar input loading considerations apply. Here a quad single-ended receiver is used, the 8T380 or 8836. These units are like the hex unit shown in Fig. 6-10A, except for the quad arrangement and separate disable inputs. For d/a converter use, this receiver would also be best used with a holding register. Either hex or quad receiver devices can be used with either system, whichever is most appropriate for the number of bits being received, or most efficient in package count.

Differential Data Recivers

For transmission of data over appreciably long distances and isolation of common mode noise, differential data transmission is the preferred method. It is briefly described here, and is shown in Fig. 6-11.

This circuit uses a dual line receiver from the 8820 series of devices. These units are designed for 180-ohm differential lines and can reject several volts of common mode noise while outputting valid digital data. Two receivers per device are available, and they operate from a +5 volt supply with TTL outputs. The devices have a strobe input which dis-

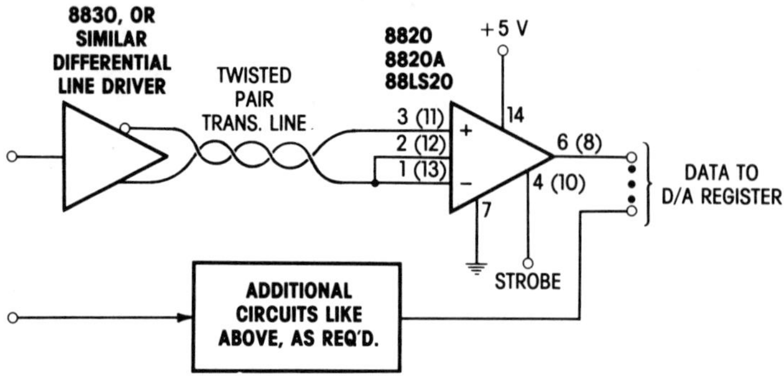

Fig. 6-11. Differential data receiver.

ables the output when low. A companion differential transmitter chip is the 8830 series.

Like the data bus receivers previously discussed, this receiver can be used with a register to hold received data. However, it need not necessarily be so operated, as each differential line is a hard committed path with continuous real time control from the source.

D/A ANALOG INTERFACING

Just as some d/a converters may require interface considerations for their logic inputs, many more may, and often do, require interfacing at their analog outputs. This is simply because of the fact that a great many IC d/a devices have outputs in the form of a current, which requires interfacing to produce a voltage output. This section of the chapter examines two types of d/a analog output interfaces, and discusses various application points pertinent to optimizing performance for differing uses. This information, which is largely general, can be applied to virtually any IC d/a device, once the basic principles are understood.

Output Buffer Amplifiers

By far the most common application task is providing a satisfactory voltage output buffer amplifier configuration for marriage to a current output d/a device. The two basic means by which this may be accomplished are shown in Fig. 6-12.

Inverting Amplifiers

In Fig. 6-12A is shown an inverting output buffer amplifier, so named because the op-amp buffer amplifier operates in an inverting, or virtual ground mode. This configuration is also known as a transresistance mode amplifier, or as more often seen, a current-to-voltage converter. It is so named because it produces an output voltage in buffered form which is

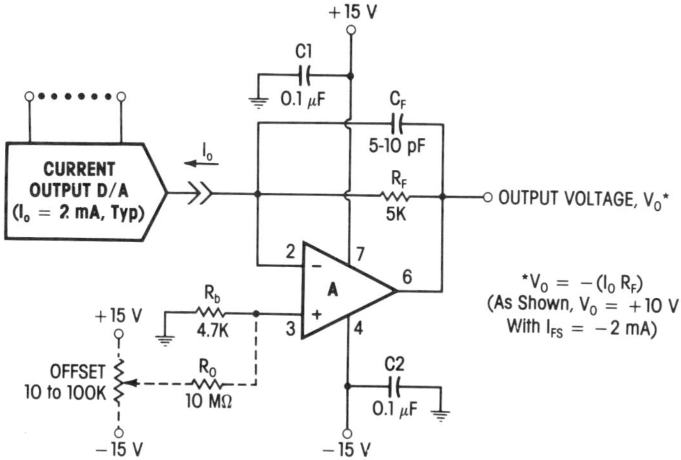

(A) Inverting output buffer (current-to-voltage converter).

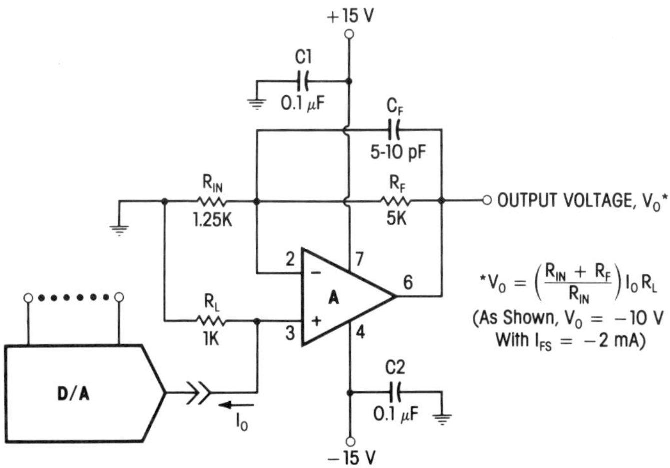

(B) Noninverting output buffer.

Fig. 6-12. Output buffer amplifiers for current output d/a converters.

the product of the d/a output current, I_o, and the feedback resistance, R_F. The equation which describes its operation is simply

$$V_o = -I_oR_F$$

With most common op-amp types there will be very small error factors in this basic equation due to op-amp gain. The major error factors are op-amp offset voltage (and its drift) and speed related errors, due to device slew rate and settling time.

This particular configuration is the most popular for use with IC d/a converters. First of all it is inherently the fastest, as it holds the d/a output at a low-impedance virtual ground, and it also eliminates voltage swing at the d/a output. Given a sufficiently fast op amp, this hookup will come closest to realizing the inherent speed and settling time of the d/a converter in use in a buffered, voltage output form. Aside from the amplifier choice (to be covered), speed may be generally optimized by use of relatively low impedance feedback path resistance values, such as R_F values in the range of 1K to 5K. C_F, a small value feedback capacitor across R_F, compensates for the net effect of d/a output, amplifier input, and stray capacitance. The value range indicated is only intended as a general guideline; this capacitor is best optimized within a given final circuit using representative components, where it can be adjusted for minimum settling time. The entire circuit is of course constructed using good high frequency practices, such as short direct signal path wiring, high quality low inductance grounds, minimal stray capacitances, and rf quality bypasses for the d/a converter in use and the op amp.

In conjunction with this line of thought, any reactive component in R_F should be avoided. Although wirewound resistances are capable of extremely fine dc precision, they are generally difficult to control for fast settling amplifier uses. Accordingly, thin film array types or metal film units are best for R_F for high speed, fast settling uses.

The circuit also shows a general form of offset voltage null adjustment which is applicable to virtually any op amp. This method is shown dotted, indicating its optional use as may be required. An alternate form of offset control is to use the trim method appropriate to the particular op amp employed. R_b, if used, or the op amp (+) input should be returned to a high quality, noise-free reference point (analog ground).

The bias compensation resistance R_b is appropriate for use if relatively high precision (> 8 bits) is the goal, and a bipolar input stage device is used for the op amp. This resistance, which is of a value nominally equal to R_F, compensates for amplifier bias current induced offset voltage, and is appropriate if the bias current of the device in use can exceed 100 nA and high accuracy is important. It can be deleted (shorted) for less than 10 bit uses, and also whenever a FET input amplifier is used.

In practice, virtually all IC current output d/a devices are current sinking, which causes I_o to flow in the direction indicated. Also, most units tend to be 2-mA full scale devices (although some available types are scaled for greater or lesser currents). Accordingly, the 5K R_F value shown will result in a basic 0 to +10-volt range, voltage output d/a converter. Smaller voltage-range limits are usually best accomplished by reducing R_F so as to maintain I_o at the level specified. This will also help reduce settling time. As can be noted, this circuit inherently produces positive range output voltages.

In choosing an amplifier for this configuration, one of the most important performance parameters is its settling time specification. Accordingly, Table 6-3 summarizes settling time performance for a wide range of IC op amps suitable for use in this type of circuit. This table is organized in terms of inverting amplifiers, as used here, and also noninverting amplifiers.

Table 6-3. D/A Output Amplifier Settling Time (Typical)

	Inverting Mode (10 V Step)		Noninverting Mode (10 V Step)	
Device	0.1%	0.01%	0.1%	0.01%
MC1741S	3 μs	N.S.	3 μs	N.S.
NE535	3 μs	N.S.	3 μs	N.S.
NE538[2]	1.2 μs	N.S.	1.2 μs	N.S.
NE531[1]	—	1.5 μs	—	1.5 μs
LM318[1]	0.8 μs	1.2 μs	N.S.	N.S.
CA3140[3]	1.4 μs	4.5 μs	1.4 μs	4.5 μs
OP-01[1]	0.7 μs (1 MAX, 5 V)		NOT APPLICABLE	
OP-16[3]	0.7 μs	1.7 μs	N.S.	N.S.
OP-17[2,3]	0.4 μs	1.5 μs	N.S.	N.S.
LF356[3]	—	1.5 μs	N.S.	N.S.
LF357[2,3]	—	1.5 μs	N.S.	N.S.
AD507[1,4]	0.9 μs	2 μs	0.9 μs	2 μs
AD509[1,4]	0.2 μs	1 μs	0.2 μs	1 μs
HA2625[1,4]	1.5 μs	N.S.	1.5 μs	N.S.
HA2525[1,4]	0.33 μs	N.S.	0.33 μs	N.S.

1 With recommended compensation for unity gain.
2 Decompensated device (not usable at unity gain).
3 FET input.
4 External compensation required.

The settling times given are as typically specified for a 10-volt output step, which represents the worst case (full range) output change for the circuit for Fig. 6-12A. The times are given in terms of both 0.1% (½ LSB for 9 bits) and 0.01% (½ LSB, 12 bits) error bands, a range which spans most applications. A 10-bit accuracy band would be 0.05%, and the corresponding settling time would be somewhat less than 0.01% (or greater than 0.1%).

In general, this data should best be used as a guide to relative performance, as appreciable settling time variation can exist due to layout variations, d/a capacitance, and many other factors. The best guide to settling time is that measured for a specific amplifier, under specific conditions, with a specific d/a device. Only when the measurement is done in this manner can results be predictably reproduced by the user with a high confidence factor. Information of this nature from a number of manufacturers is contained in a section which follows, which gives specific device examples, along with measured results.

Noninverting Output Buffer

A noninverting form of buffer for a current output d/a converter is shown in Fig. 6-12B. This circuit uses a greater number of gain determining resistances, and is not inherently as fast as the inverting one, yet it has some distinguishing features which could be useful to certain applications.

In this circuit the d/a current output develops a voltage across the load resistor, R_L. The full scale voltage will be simply I_0R_L, as the high impedance (+) input of the op amp represents a negligible loading factor. R_L is a precision component, with appropriate attendant stability. For the 1K value shown, the required d/a output voltage compliance will be relatively small, or 2 volts for a 2 mA full scale d/a converter. This magnitude of voltage will satisfactorily accommodate most d/a devices. Those with smaller compliance ranges might perhaps best be used with an inverting op amp to eliminate the compliance voltage requirement.

The 1K R_L value also has the benefit of minimizing the d/a output settling time, which for most devices, will be dominated by the RC time constant determined by R_L and the d/a converter plus shunt capacity. In practice, this capacitance will be on the order of 20 pF; therefore, the 1K-20 pF time constant will be in the range of 20 ns. Settling time of this RC node to ½ LSB accuracy must be calculated by allowing a sufficient number of time constants to achieve the desired accuracy. For example, 8-bit (0.2%) accuracy requires 6.2 time constants, 12 bits (0.01%) requires 9.2. The settling time would then be on the order of 125 ns for 8 bits (0.2%), or 180 ns for 12 bits (0.01%). This explanation is just for the purpose of general illustration; in practice, the settling time to be expected should be as specified for the particular d/a device, as it is rated for a 1K load. Typically, settling times to 0.1% accuracy for 1K load conditions will be an the order of 150 to 200 ns.

The settling time discussed thus far for this circuit is only a part of the entire story, however, as the amplifier must also be considered for its settling time, which will lengthen the overall settling time as seen at the output. In general, the net settling time of a configuration such as this with two individual times will be the root sum of the squares (RSS) of the individual settling times. Thus, if the d/a output settles in 200 ns, and the op amp also requires 200 ns, the net settling time will be 282 ns (as one example). It may not always add in such a straightforward fashion in practice, but the point should be appreciated that the net time will always be greater than either time viewed individually. It is this factor which makes this noninverting configuration inherently slower, at least theoretically.

Practical results can change the perspective of this viewpoint, however, when the gain of the op amp is considered. It can be noted that this configuration operates with a gain of 5, thus raising the d/a converter's 2-volt output to 10 volts. When operating at a gain of 5, the op amp need not be compensated for unity gain; in fact, to do so slows its response unnecessarily. Therefore it is advantageous from a speed standpoint to use a decompensated device for the amplifier. The basic requirement is that the amplifier be stable at a minimum gain of 5 (for this gain of 5 configuration). For many devices, compensation at a gain of 5 results in appreciably increased slew rates and bandwidth, and thus reduced settling time. Examples of this type of device may be noted from Table 6-3, such as the NE538, AD509, HA2625 and HA2525, LF 357, and OP-17.

However the operation of the amplifier at a gain of 5 also increases its offset voltage by an equal factor, and this may make some devices less desirable for high dc accuracy uses. A 2-mV input offset device becomes a 10-mV offset unit when referred to the output. This effect can, of course, be minimized by use of offset voltage nulling, but the drift of the device (×5 referred to the output) will still be a factor for wide temperature ranges.

The gain determining resistors used with the op amp should follow the general guidelines already discussed with regard to accuracy, stability, etc. R_F and R_{IN} need not necessarily match or track R_L, but they should track *one another* to stabilize the op amp gain factor. Although it may appear that the collection of values indicated is an impractical one, it should be noted that such a range (and ratio) are easily realized in a thin-film network array, such as, for example, an array of 5K units. This approach can be both simple and advantageous, as it gives inherent close tracking in a single package.

From an overall system standpoint this circuit yields a noninverted output voltage. As mentioned previously, most IC d/a converters are current sinking, therefore this results in a negative output voltage d/a converter, as opposed to the inverting output buffer which results in positive output voltages.

Performance Data for Specific D/A and Amplifier Combinations

As examples applying the general principles set forth previously, this section discusses the performance of combinations of various manufacturers' devices, as tested within a specific circuit, with known conditions. The details of individual operation are as outlined in each figure, and it may be generally assumed (unless noted otherwise) that operation is for ±15 volt supplies, and good design layout practices are followed. If these stipulations are upheld, the performance quoted should be readily repeated.

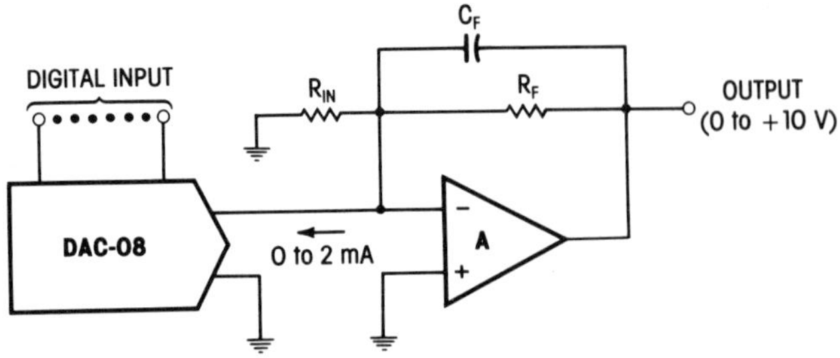

VALUE/PERFORMANCE TABLE

A	R_{IN}	R_F	C_F	SETTLING TIME*
OP-16	10K	5K	25 pF	750 ns
OP-17	1.25K	5K	8 pF	380 ns

*To 1/2 LSB (8 Bits), 10 V Step

Courtesy Precision Monolithics, Inc.

(A) DAC-08 + OP-16, OP-17.

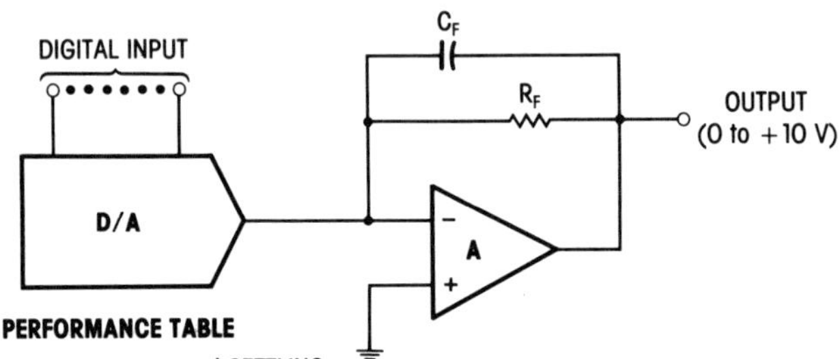

PERFORMANCE TABLE

D/A	A	SETTLING TIME*
AD1408	AD301A**	840 ns
(8 bits)	AD518	560 ns
	AD509	300 ns
AD561	AD301A**	1.5 μs
(10 bits)	AD518	1 μs
	AD509	470 ns
AD7520	AD301A**	1.6 μs
(10 bits)	AD518	1.1 μs
	AD509	640 ns
AD562/3	AD301A**	5.1 μs
(12 bits)	AD518	3.2 μs
	AD509	1.6 μs

*To 1/2 LSB, 10 V Step (For Respective Accuracies)
**Feed Forward Operation, Minimal C_L

Courtesy Analog Devices, Inc.

(B) AD1408-AD62/3, + AD301A-AD509.

Fig. 6-13. Typical performance data of specific device combinations.

Fig. 6-13A shows a connection of a DAC-08, a 0 to 2 mA current output d/a converter, to a current-to-voltage converter op amp, using either the OP-16 or OP-17 high speed op amp. This circuit is generally similar to the one in Fig. 6-13B except for the use of R_{IN}, a resistive pad from the d/a output/summing point to common.

This resistor accomplishes two purposes. The first of these is to compensate for the d/a converter to ground capacitance, C1. This is accomplished when the product of R_{IN}-C1 is equal to R_F-C_F; in essence, the feedback divider for these conditions is nonreactive. This allows the overall dynamic response to be determined by the op amp alone, and yields lowest settling times. If desired, C_F can be "tweaked" about the nominal value to optimize settling time for a given layout condition.

As can be noted, the settling times to ½ LSB of 8 bits are quite low, only 380 ns for the OP-17, and 750 ns for the OP-16. As described earlier, the penalty which accommodates a high gain connection such as this is increased output offset voltage. For 8-bit use, such as shown here, this factor has little practical significance, as both op-amp devices have offset sufficiently low as to render this contribution negligible, or < ½ LSB worst case.

Fig. 6-13B illustrates various performance combinations of analog devices, converters and op amps, for 8- through 12-bit combinations.

High Current Buffers

Most IC op amps are specified for load currents of 5 mA, which with a feedback (d/a) current of 2 mA as typically encountered, leaves only 2 to 3 mA of current available for load drive. To supply output currents appreciably higher than a few milliamperes, a unity voltage gain current buffer stage can be used, as shown in Fig. 6-14.

To maintain the overall accuracy of which the d/a converter is inherently capable, a buffer such as this must be used within the feedback loop, with feedback returned from its output to R_F. The device used should be capable of very high slew rates and low phase shift, as well as appreciable power output, so as not to compromise the speed of the d/a converter in use. Both of the indicated A2 devices have these characteristics, and can supply over 100 mA to a load.

The two units, the Harris HA2635 and National Semiconductor LH-0002CN are pinned differently, but both perform similar functions. For protection of the buffer, current limit resistors can be used to limit maximum load current to a safe level. Heavier than normal supply line bypassing is appropriate to minimize supply transients resulting from high load currents.

Deglitchers

In some d/a applications, output glitches can cause serious problems. While glitches can be held down by careful d/a circuit design and logic timing alignment, there are circumstances where even the best (lowest) residual glitches must be removed by other means. An often used approach to this problem is to follow a glitching d/a converter with a sample/hold "deglitcher" circuit. A block diagram of such a configuration is shown in Fig. 6-15.

The basic function of the S/H in this application is to switch temporarily into a "hold" mode when d/a glitches are present, then to reacquire (sample mode), when the glitch has died out. This effectively

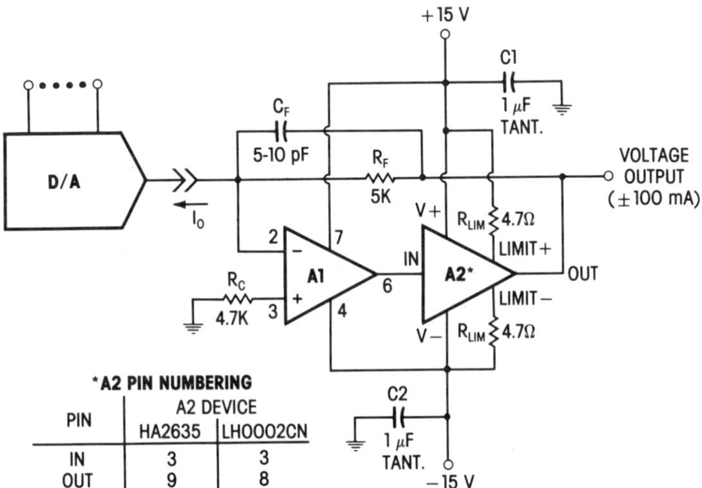

***A2 PIN NUMBERING**

PIN	A2 DEVICE HA2635	A2 DEVICE LH0002CN
IN	3	3
OUT	9	8
V+	7	2
V−	2	4
LIMIT +	8	1
LIMIT −	1	5

Fig. 6-14. High current output buffer.

opens the d/a-to-output transmission path for just the duration of the glitch time, so it cannot appear at the output. This may be better understood by referring to the timing diagram.

For a steady logic input to the d/a converter, the S/H is held in a sample (tracking) mode. When the d/a converter is to be updated to a new output level, the logic signal fires a delay one shot, which generates a short output pulse, which in turn switches the S/H to the hold state. During this time period there is no S/H output change, as it will continue to output the previous d/a output level, voltage V1. The d/a converter, however, will generate a glitch in going from voltage V1 to V2.

When the d/a output has settled to the new correct level of V2, the S/H comes out of hold and reacquires the d/a output, and also settles to voltage level V2. The output is now valid at the V2 level, and no

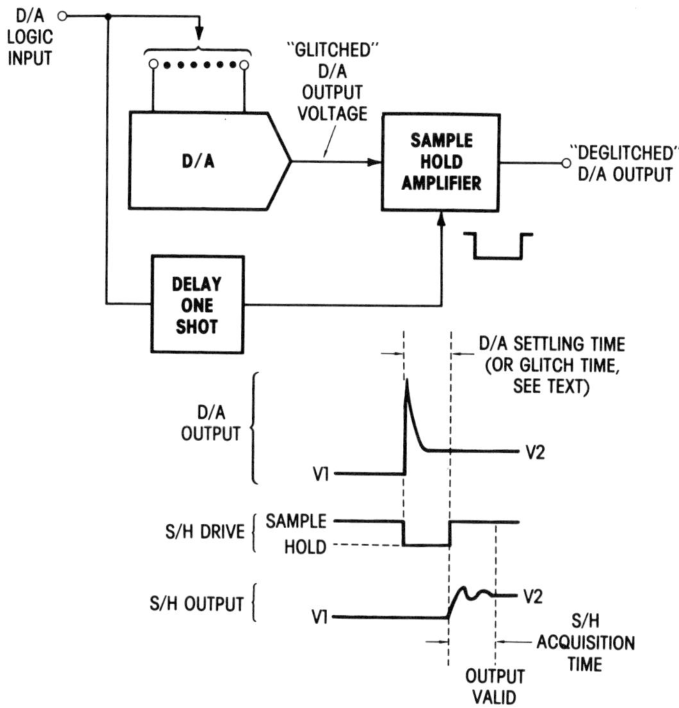

Fig. 6-15. Block diagram of a S/H deglitcher.

glitch has been transmitted (other than whatever switching transients may be inherent to the S/H in use).

A key point of operation of this scheme is that the ultimate d/a output is delayed in time, by the amount necessary to blank the glitch, plus the S/H settling time. A worst case timing alignment would be a new total settling time, which is the sum of the d/a settling time plus that of the S/H.

In practice, however, this time can be shortened appreciably by adjusting the S/H hold interval to be just as long as that of the worst expected glitch. This will usually be at a major carry point (011 - - 1 to 100 - - , or vice versa) and can be verified by exercising the d/a through this transition dynamically. The S/H timing can then be optimized for the shortest time to blank the output glitch. Settling time of the S/H to the final accuracy level must still be considered, however, and will be timed as shown. The minimum overall delay (net settling time) of this deglitcher will then be the time of the longest d/a glitch, plus the S/H settling time.

As shown in this simplified block diagram, the S/H and d/a converter are separate closed loop systems which can comprise overall accuracy because their errors will add. In many cases, however, the S/H errors can be minimized by including it *within* the d/a loop, as the output buffer amplifier. Many IC S/H devices with uncommitted feedback terminals can operate in this manner; for example, the AD582 and HA 2425.

D/A Sign Changer Circuits

A common d/a analog interface is the sign changer, a circuit which converts unipolar, magnitude only scaled d/a outputs to a bipolar, sign plus magnitude scaled operation. Two circuits which accomplish this are shown in Fig. 6-16.

In Fig. 6-16A is shown a circuit which accepts voltage inputs in the range of 0 to +10 volts, and converts them either to 0 to −10 volts, or the same 0 to +10-volt range, according to a sign bit input. Here sign control is accomplished by A1, a high speed op amp, which is driven by A2, a CMOS array configured as an analog switch.

When the sign bit input is low, the analog switch connects the (+) input of A1 to the 0 to +10 volt input d/a output at R1, and A1 functions as a noninverting amplifier passing the input signal to the output with a precise gain of unity. When the sign bit is a 1, the analog switch grounds the input of A1 converting it to an inverter with a gain set by

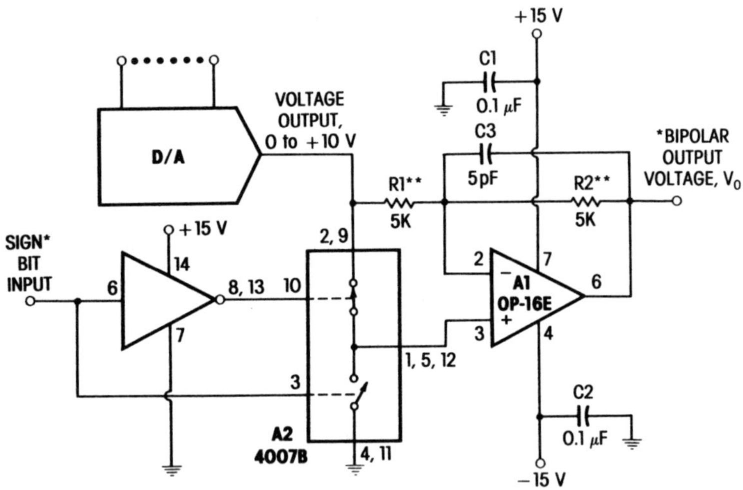

$V_O+ = V_{IN}$

$V_O- = -\frac{R2}{R1} V_{IN}$

***CONTROL TABLE**

SIGN BIT	OUTPUT VOLTAGE
0	0 to +10 V
1	0 to −10 V

(A) Voltage input.

Fig. 6-16. Unipolar to bipolar

R2/R1 as unity. Thus, the output signal is a replica of the input, with a sign of (+) or (−).

Error of the A1 stage is minimized by using an OP-16E, a low input offset device. For even 12-bit applications, both offset or CM error of this device is less than ½ LSB. The main error of the stage appears in the inverting mode, and will be due to R1-R2 mismatch. For 12-bit use, these resistors should be matched to 0.005% or better, with low tracking errors. Settling time of the stage is 2.5 μs for 0.01% accuracy, and 750 ns for 0.1% accuracy. If only 8-bit performance is necessary, a 356 type unit will be satisfactory for A1. The circuit can be used with any unipolar positive 10-volt range output d/a converter.

A circuit which is a bit more specific in its range of use is the current-driven sign changer of Fig. 6-16B. This circuit accepts a current of 0 to −2 mA full scale, as controlled by a d/a converter, and adds sign control, producing an overall 0 to ±10-volt buffered output.

In operation, the current output d/a converter A1 produces an output current which is relatively independent of the load resistance. This

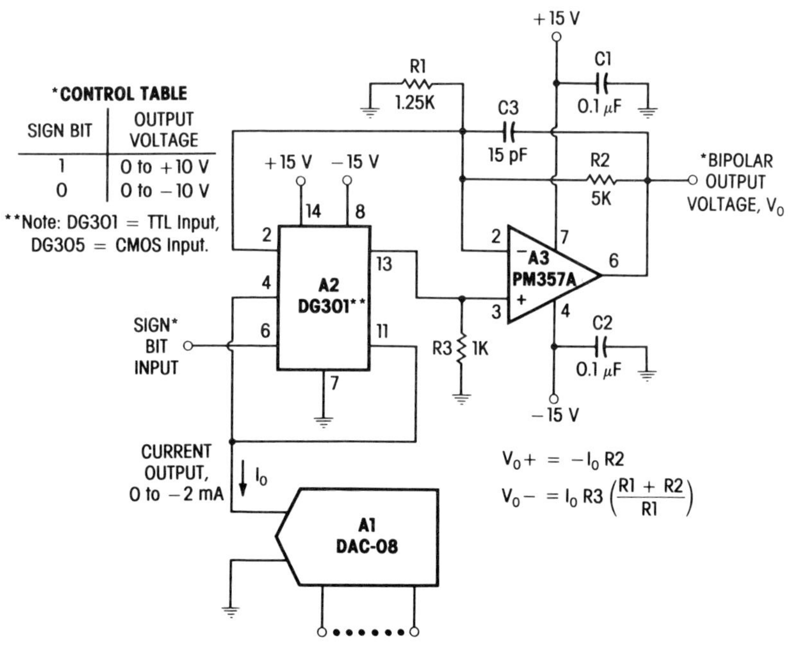

(B) Current input.

range sign changer circuits.

fact is the key to operation of the circuit, as analog switch A2 steers the d/a output current to either the (−) or (+) input of A3. The on resistance of the switch, although variable and temperature dependent, will *not* affect calibration as long as the d/a converter has a high output impedance, 20 megohms or more. This feature is a characteristic of such devices as the DAC-08, DAC-20 and others.

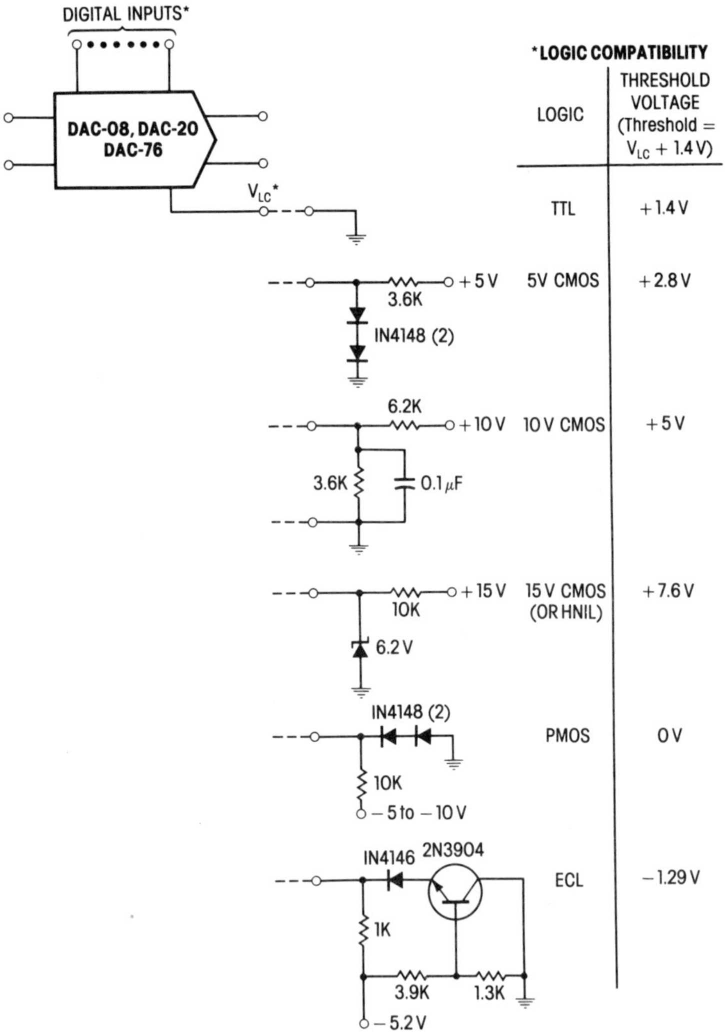

(A) DAC-08, DAC-20, DAC-76.

Fig. 6-17. Logic

The output op amp is configured as a gain-of-5 stage, allowing a fast decompensated device to be used for A3. Settling time of this circuit is on the order of 600 ns to 0.1%, for the worst case full scale (−10 to +10 volts) change. It is appreciably faster for smaller changes particularly when in the inverting mode, or for positive output voltages.

The sign control input follows the table shown in the figure. Either a DG301 or DG305 may be used, dependent upon the logic drive source. TTL or CMOS.

MULTIPLYING D/A CONVERTERS

This section of the d/a applications section is devoted to the detailed points of consideration necessary for optimum use of multiplying d/a converter types. Devices are discussed in terms of reference input connections, inputs and outputs, power supplies and their conditioning, calibration, coding and scaling, and other pertinent design information.

Housekeeping Details

With any converter device, there are certain "housekeeping" circuit details which must be attended before the device can function as intended. With d/a converters, these details include power supply conditioning, reference connections, compensation capacitors (in some cases), logic level programming, and other such necessary facts of life. They are discussed briefly in light of the various device types.

Logic Programming

Logic input compatibility can be a concern in many d/a applications and some d/a devices offer the convenience and flexibility feature of logic threshold programmability. In essence, this is the ability to adapt the logic input sensitivity to match various logic families as drive sources. Two general methods are shown in Fig. 6-17.

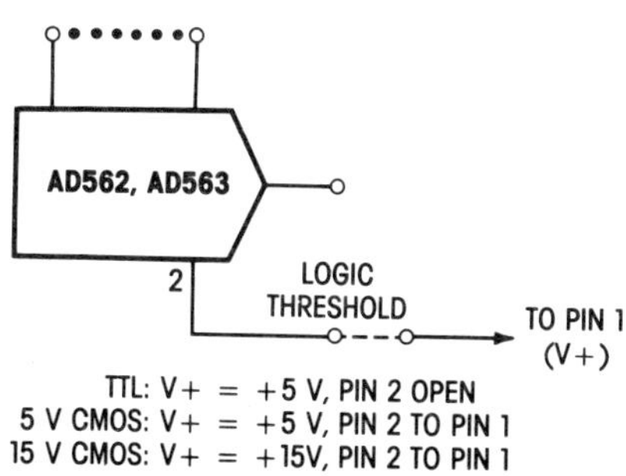

(B) AD562, AD563.

input programming.

In Fig. 6-17A is shown the logic programming scheme for the DAC-08, DAC-20, and DAC-76 type d/a converters. These units have a logic threshold adjustment pin, designated V_{LC}, that determines the logic input threshold of all bit inputs. The logic threshold is by design 1.4 volts greater than the potential of the V_{LC} pin.

For a typical case of TTL input drive, the threshold is set to +1.4 volts, simply by grounding the V_{LC} pin. Other logic levels can be accommodated by biasing this pin up or down in dc level, as appropriate. Specific connections are shown in the figure, and the table indicates the corresponding logic family and resulting threshold voltage(s). The digital inputs of these devices can swing from −10 to +18 volts, without damage.

In Fig. 6-17B is shown the logic programming for the AD562 and AD563 units. These units are programmed for either TTL/5-volt CMOS, or 15-volt CMOS by jumpering (or leaving open) pins 1 and 2, and selecting the appropriate positive supply level. The table indicates the connection details. The AD561 (not shown) has a similar logic adjustment capability, except adjustment in its case is simplified to only the supply voltage, without jumper options. Logic is 15-volts CMOS compatible for 15-volt supplies; TTL and 5-volt CMOS compatible for 5-volt supplies.

Compensation and Power Connections

Some IC d/a converters require external compensation components, in a fashion similar to some IC op amps. These details, along with the general rules and connections for power supply conditioning are shown in Fig. 6-18.

The compensation capacitor for 1408 and DAC-08 types is similar, a 75-pF capacitor from pin 16 to V− (pin 3). The optimum capacitance can be adjusted in some cases to a capacity other than that shown. For general purpose uses, where the reference channel speed is not criti-

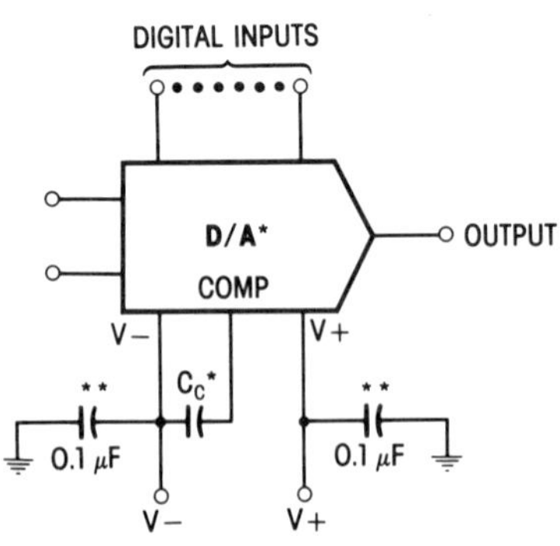

*COMPENSATION TABLE

D/A DEVICE	C_C
1406	20 pF Typical, 14 to 13
1408	75 pF Typical, 16 to 3
DAC-08	75 pF Fast / 0.01 µF Medium } 16 to 3

** Power Supply Bypassing Recommended For All D/A Devices - See Text.

Fig. 6-18. Compensation/power connections.

cal, a 0.01-μF value will suffice. The 1406 is compensated with a 20-pF capacitor from pin 14 to pin 13. As a good design practice, power supply pin bypassing should be a routine procedure for *all* converter devices. This should take the form of locally installed 0.1-μF rf quality units, such as ceramic disc, as a minimum. For some devices, such as the more complex hybrid IC units, additional capacitance such as 1-μF tantalums are recommended. Consult data sheets for specific guidelines in these cases. In the circuits which follow, supply lines and their bypassing may not always be shown (for clarification or simplicity), but they should be understood to be implicitly necessary.

The power supply DC levels for d/a devices are generally flexible, and can accommodate typical system levels of +5 volts and/or ±15 volts. Specific levels and/or ranges are mentioned with the discussions of the individual devices.

Reference Inputs

Most IC multiplying d/a converters use a reference input circuit such as is shown in Fig. 6-19. This circuit employs an input op amp A, which forces equal value currents to flow in transistors Q1 and Q2. This basic scheme was described earlier in the "d/a devices" section.

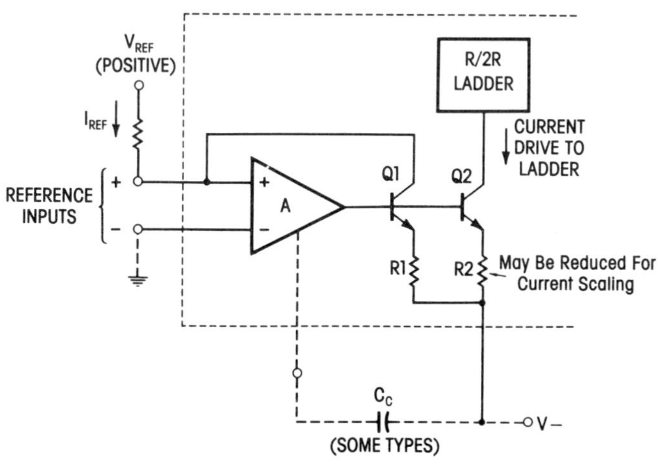

Fig. 6-19. Reference input configuration.

Key application points of this scheme include the fact that the current drive to the d/a ladder will be equal to I_{REF}, the input reference current. This is most generally the case; in some instances, the ladder drive current is scaled up by reducing R2. This results in higher ladder and output current, usually by a factor of 2 or 4, when applied. A typical value of reference current is 2 mA, and this is also the most typical full

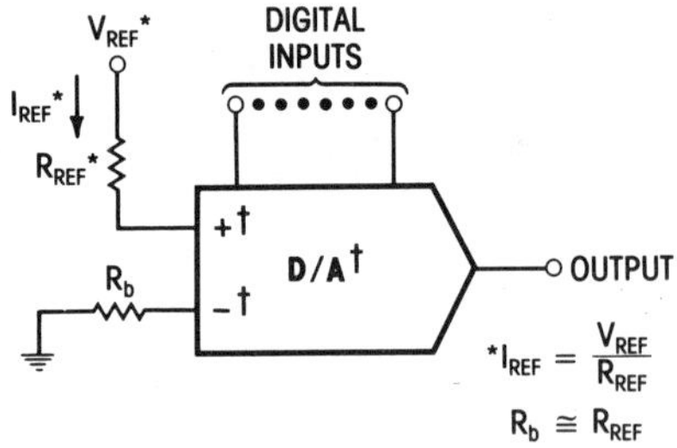

(A) Positive reference voltage.

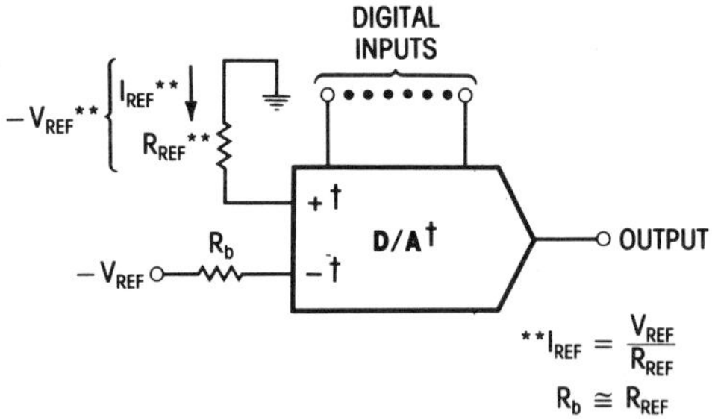

† REFERENCE INPUTS

D/A DEVICE	PINS (+)	PINS (−)	I_{REF} (mA)
1406	12	13	2
1408/1508/ 3408	14	15	2
3410/3510	16	15	2
DAC-08	14	15	2
DAC-20	14	15	2
DAC-76	11	12	0.5
AD562††	4 or 5	3	0.5

††Includes Reference Resistor

(B) Negative reference voltage.

Fig. 6-20. Reference current setup.

scale output current. In some types the reference amplifier requires external compensation, as generally indicated here by C_c.

Fig. 6-20 shows specific details of how the reference current can be established for this type of d/a converter, for both positive and negative reference voltages. Pin numbers for representative devices are indicated in the table.

For use with positive references, the amplifier (+) input is tied to the reference voltage through a resistor R_{REF}. The (−) input is grounded through an equal value resistance, R_b. The reference resistor R_{REF} must be a stable type, for good overall stability. The stability of compensation resistor R_b is of secondary importance, and it need only nominally match

R_{REF} (in some cases it may even be eliminated by grounding the (−) input directly). Full scale calibration of the d/a converter can be accomplished by variation of either R_{REF}, or the voltage V_{REF}. The reference voltage is derived from a stable source, such as previously described in Chapter 3. Supply voltages are not generally recommended for reference use.

Negative reference voltages can also be accommodated by this scheme, as shown in Fig. 6-20B. Here a voltage of V_{REF} (negative) is applied to the (−) input through resistance R_b, while the (+) input is grounded through R_{REF}. Although polarity of the applied voltage is different, this hookup also causes a current of I_{REF} to flow in R_{REF}, as was true in Fig. 6-20A. Similar selection considerations apply to R_{REF}, R_b, and the reference voltage.

In all d/a converters of this type, the reference current must always flow *into* the (+) input. The reference current may vary over a fairly wide range for multiplying uses, up to 40/1 in some types. Levels of reference current for specified operation are given in the table. Dynamic variation of reference current is useful for ac or multiplying applications. These cases will be covered later.

Output Structures

Multiplying d/a converters of the type under discussion all have current outputs, which sink a current I_o into the designated output pin. This current (I_o) will in general be the reference current multiplied by the fraction represented by the digital input. It may also be scaled by K, the current scaling factor. This factor K is normally unity, but may also be 2 (or 4) in some types. Output structures which illustrate this factor are shown in Fig. 6-21.

In Fig. 6-21A is shown the most simple form of multiplying d/a output, which is a current source represented as I_o. I_o is shown in equation form, as it would be for a full scale digital input and a reference current of I_{REF}. The current scaling factors for various devices are shown.

Devices with complementary differential outputs such as the DAC-08 and DAC-20 have outputs which appear as in Fig. 6-21B. The current I_{FS} for the DAC-08 is as shown for binary operation. The two outputs are complementary in operation; their sum is constantly equal to I_{FS}. The DAC-08 and DAC-20 have scaling factors of 1; the DAC-76 has a scaling factor of 4 (nominally).

This group of devices has outputs which are true current sources; that is they feature high output impedance and a wide range of voltage compliance. This fact is true to some extent with the others, but is not necessarily a rule.

The output structure of the AD7520 R/2R ladder d/a converter is shown in Fig. 6-21C. This device has an output I_{out1} which is designed to be used solely with an op-amp current to voltage converter. As such

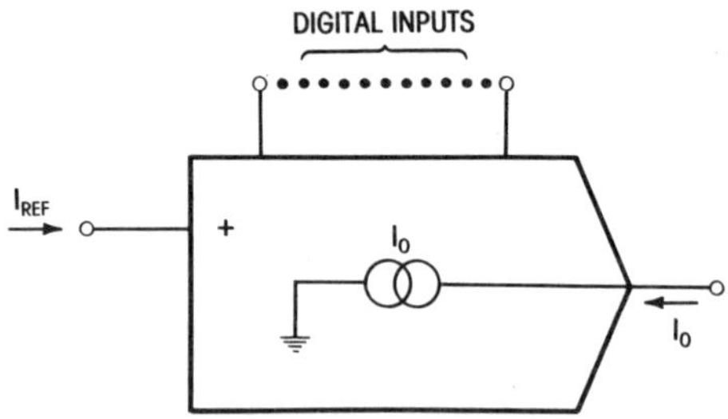

$$I_O = \frac{2^{n-1}}{2^n} K\, I_{REF}$$

Where K is current scaling factor
K = 1 for 1406, 1408, 3408
= 2 for 3410, 3510
= 4 for AD562

(A) Single ended.

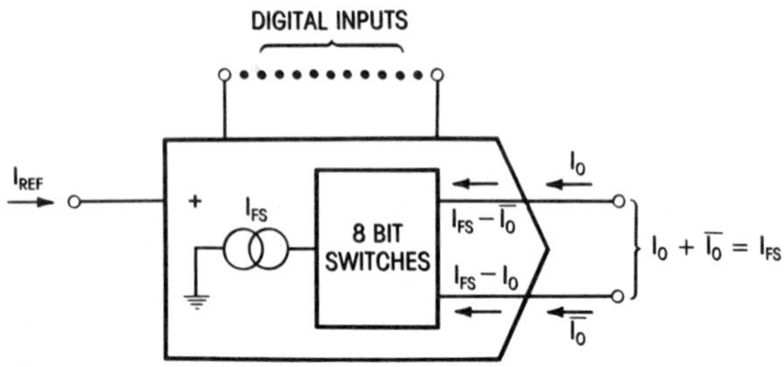

$$I_{FS} = \frac{2^{n-1}}{2^n} K\, I_{REF}$$

K = 1 for DAC-08, DAC-20 (bcd)

(B) Differential

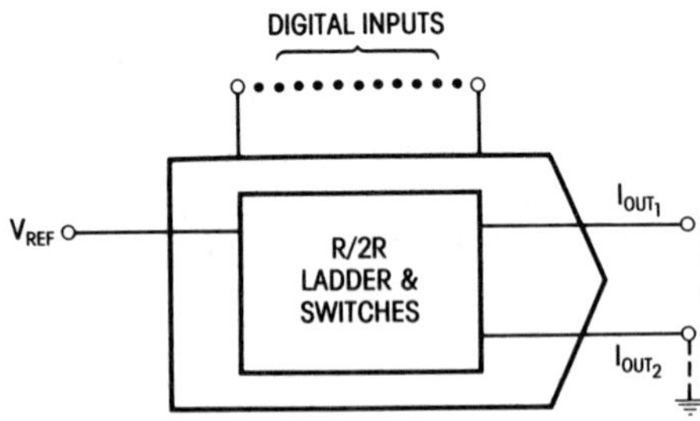

$$I_{OUT_1} + I_{OUT_2} = I_{REF}$$

$$I_{REF} = \frac{V_{REF}}{10K}$$

$$I_{OUT_1} = \frac{2^{n-1}}{2^n} K \frac{V_{REF}}{10K}$$

K = 1 for AD7520 Series

(C) R/2R ladder.

Fig. 6-21. Output structures.

it has essentially zero compliance. The I_{out2} terminal is normally grounded, but the current here is actually the complement of that appearing at I_{out1}.

Output Interfaces

Various means exist to convert the output of current output d/a converters into a useful voltage. Which particular method is most appropriate depends upon the nature of the application, and the specific device being considered for use. Several simple methods of interfacing are illustrated in Fig. 6-22.

In Fig. 6-22A is shown the simplest means of using a current output d/a converter, a resistive load to ground. The output will simply be

$$E_o = I_o R_L$$

where,

R_L is the load resistance.

Obviously this method is simple, but it is as sensitive to loading as the output impedance is to the value of R_L. For low load resistances such as 100 to 1K, this may not present a problem. The output can be buffered, if desired, by a voltage follower (dotted) which removes this restriction.

The d/a converter used must have an output voltage compliance *greater* than the desired signal swing, otherwise the circuit will not function properly. This is important in view of the fact that many d/a converters have only a volt or two of compliance.

For best accuracy and stability, it is desirable that R_L be a stable resistor type. If this resistor can be made to track the reference resistor used with the d/a converter, optimum stability will result as their individual drifts will tend to cancel. Indeed, this consideration of relative stability is most important.

Fig. 6-22B shows a means of extending the voltage compliance range of a current output d/a converter by using a common base level-shift transistor. This transistor reduces the voltage seen at the d/a converter output to less than a volt, yet its output can swing voltages up to the breakdown level of the transistor type used.

The circuit has essentially the same output voltage, but slightly reduced due to the finite beta of Q1. This can readily be compensated for by an overall scaling adjustment.

Output impedance is again the value used for R_L. A disadvantage is the fact that the voltage appears now referred to V+, rather than ground. The stage can be buffered if desired, similar to Fig. 6-22A. For further details, see the MC1406, MC1408, and MC3410 data sheets.

In Fig. 6-22C is shown the classic current-to-voltage configuration, which is the most popular all around circuit for use with d/a converters. This stage reduces the required d/a compliance range to zero, thus it

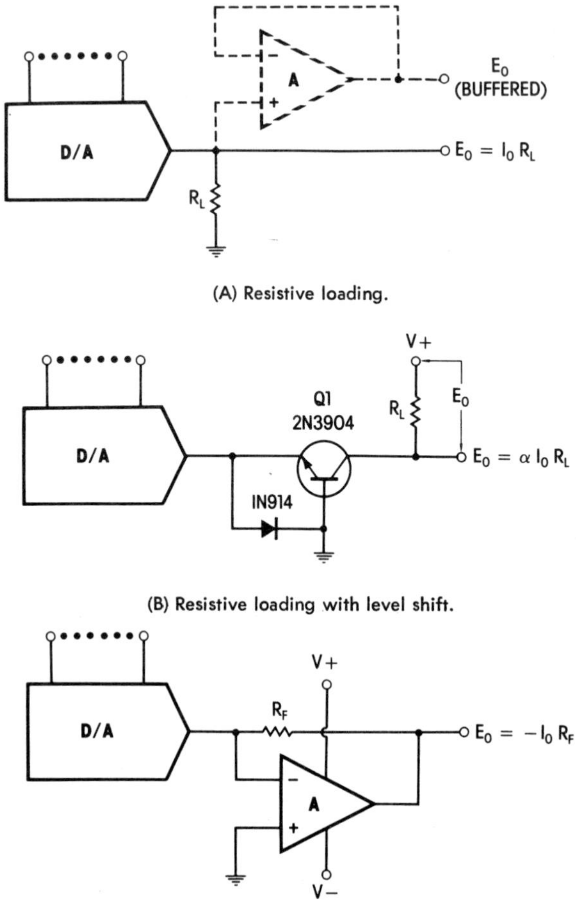

(A) Resistive loading.

(B) Resistive loading with level shift.

(C) Current-to-voltage converter.

Fig. 6-22. General output interfaces.

can be used with virtually any current output d/a converter. Further, it is highly accurate, and yields a precise buffered voltage. Its main limitation is a practical one—the speed rating of the op amp used.

Output voltage range is determined by the swing of the op amp, and its supplies. With the typically used ±15-volt supplies, 0- to 10-volt ranges are readily accommodated.

Calibration and Trim Methods

Virtually all d/a circuits require calibration means of some sort to achieve rated accuracy. For 6- or 8-bit applications where 1 LSB is on the order of 10 to 40 mV, this may be a relatively simple task, but it still must be considered. The two major calibration trims which must be dealt with are offset and gain, and they are shown in Fig. 6-23.

Fig. 6-23A shows a multiplying type current output d/a converter used with an external op amp, operating in a current-to-voltage mode. The configuration is (overall) a unipolar, voltage output, binary scaled d/a converter. For this example, scaling is shown for a 0- to +10-volt full scale output.

Of the two types of calibration adjustments, offset tends to be the smaller source of error, as typical op-amp offsets are on the order of a few millivolts, and the zero scale current output of the d/a converter is negligible. For up to about 10-bit systems, these few millivolts may often be ignored, depending upon the output voltage range. This is simply due to the fact that 5 millivolts (as an example) while being only a small percent of 10 volts, is a greater percent of 2.5 volts.

Where offset voltage is to be minimized, a bias compensation resistor such as R_b should be used with the op amp, if the input currents of the device can be much in excess of 100 nA. The actual input offset can be

Table 6-4. Calibration Voltages for Popular D/A Converter Scalings

Unipolar					Bipolar			
Binary				BCD	Binary			
Bits	Voltage Range	1 LSB	Full Scale (11...11)	Full Scale (1001....1001)	Bits	Neg. Scale (00...00)	1 LSB	Pos. Scale (11...11)
	2.5	0.039	2.461			−2.5	0.078	2.422
6	5	0.078	4.922		6	−5	0.156	4.844
	10	0.156	9.844			−10	0.313	9.688
	2.5	0.010	2.490	2.475		−2.5	0.020	2.480
8	5	0.020	4.980	4.950 } 2 digit	8	−5	0.039	4.961
	10	0.039	9.961	9.900		−10	0.078	9.922
	2.5	0.0024	2.49			−2.5	0.0049	2.495
10	5	0.0049	4.995		10	−5	0.0098	4.990
	10	0.0098	9.990			−10	0.0195	9.980
	2.5	0.0006	2.4994	2.4975		−2.5	0.0012	2.4988
12	5	0.0012	4.9988	4.9950 } 3 digit	12	−5	0.0024	4.9976
	10	0.0024	9.9976	9.9900		−10	0.0049	9.9951
For all unipolar converters, zero scale = . . . 00 = 0.000 V					For all bipolar converters, midscale = 100 . . 00 = 0.000 V			

adjusted to zero by the recommended trim for the op amp, such as the 10K pot shown with the wiper to V−. This is the most standard technique, but it is not universal. When possible, use the specific recommended method for the amplifier to be used. Alternately, the more general technique shown can be used, which will operate with any op amp (even if it has no internal offset null method).

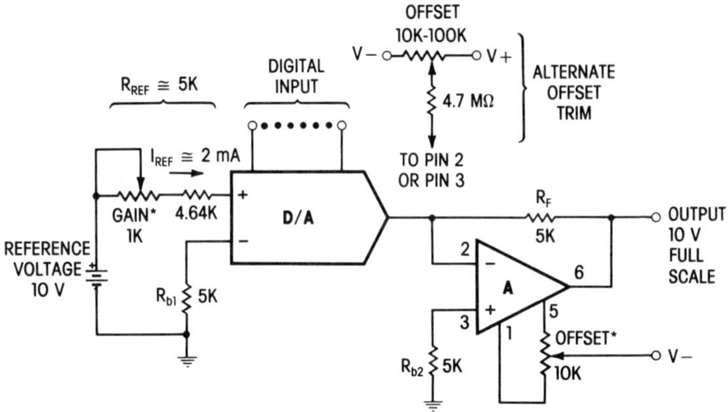

***TRIM PROCEDURE (Binary, 10 V full scale out)**

1. Offset: Adjust OFFSET control for 0.000 V out, with 00• • •00 digital input.
2. Gain: Adjust GAIN control for $\left(\frac{2^n - 1}{2^n}\right)$ 10 V out, with 11• • •11 digital input.

(See Table 6-4)

(A) Unipolar.

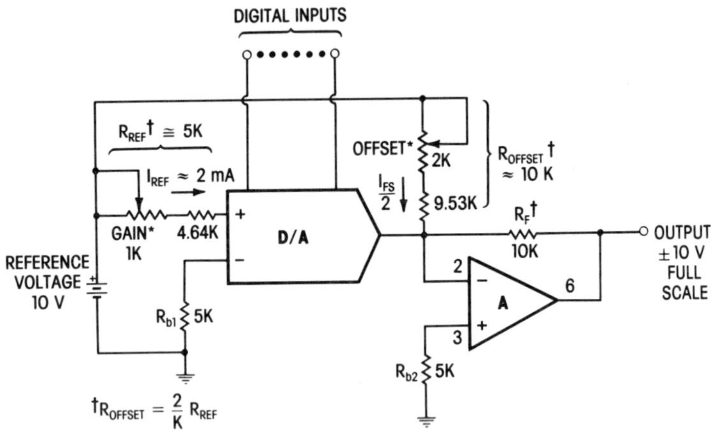

$†R_{OFFSET} = \frac{2}{K} R_{REF}$

$R_F = \left(\frac{V_{FULL\ SCALE}}{V_{REF}}\right) R_{OFFSET}$

Where K = Current scaling factor of D/A

***TRIM PROCEDURE (Offset binary, ±10 V out)**

1. Offset: Adjust OFFSET control for −10.000 V out, with 00• • •00 digital input.
2. Gain: Adjust GAIN control for $\left(\frac{2^n - 1}{2^n} \cdot 20\right) - 10$ V out, with 11• • •11 digital input.

(See Table 6-4)

(B) Bipolar.

Fig. 6-23. Calibration trim methods.

As a procedural trim, offset should be adjusted first, as an offset can appear as a gain error. It is simply adjusted for zero output voltage with a zero digital code input (or all 1s, for complementary coded devices).

Gain calibration may in theory be accomplished by variation of either V_{REF}, R_{REF}, or R_F. In practice it is normal to adjust either R_{REF} or R_F, as most reference sources are fixed (and often arbitrary) in level. With a fixed R_F of 5K, I_{REF} is adjusted about its nominal 2 mA level to yield the correct full-scale output voltage. Table 6-4 is a summary of popular d/a calibration levels; for an 8-bit 10-volt scaled d/a converter the gain calibration point would be 9.961 volts.

It should be noted that the range of calibration adjustment must accommodate the tolerance limits of the d/a converter, the resistors used for R_{REF} and R_F, and possibly V_{REF}. Some IC multiplying d/a types have calibration tolerances up to 5 percent; therefore the adjustment range should be on the order of 10 percent, to accommodate all factors. Stable, multiturn low temperature coefficient trimmers are recommended for ease and permanence of adjustment. This comment is applicable to both gain and offset trims.

Where possible, an error-minimization technique which can be employed to advantage with multiplying d/a converters is to use reference and scaling resistors which match and track with temperature. This can be accomplished with thin-film resistor arrays, which are readily available in the range of values typically used (1K to 10K).

Fig. 6-23B shows trim methods for bipolar scaled multiplying d/a converters, again using an op amp operating in the current-to-voltage mode. This figure also shows generally how the bipolar offset method is achieved using this type of device.

For offset binary or bipolar operation, the d/a converter must output the negative full scale limit for an all zero input. For this condition the d/a converter contributes no current output, and the offset is accomplished by a resistor shown here as R_{offset}. This resistor operates with R_F to generate the negative full scale voltage because for this d/a code condition, the circuit reduces to a simple inverting amplifier.

The reference, feedback, and offset resistors have a definite relationship to yield a correctly scaled bipolar code. In general, the offset resistor is twice the value of the reference resistor, divided by the current scaling factor of the d/a converter. For a unity scaling factor (see Fig. 6-21), the offset resistor is simply twice the reference resistor. The desired objective is that the offset resistor contribute exactly half of the full scale d/a output current.

The feedback resistor is selected as the value of the offset resistor multiplied by the ratio of full scale output voltage to the reference voltage. In the case here this ratio is 1, therefore R_F is the same as the reference resistor, or 10K. For arbitrary reference voltages (such as 6.4 volts for instance) R_F would change accordingly.

The trim procedure for bipolar operation is still sequenced as offset and gain, in that order. However, in this case the op-amp offset is not adjusted; the value of the offset resistor is trimmed for the desired full scale negative voltage with all zeros in. In the case here this is −10.000 volts; for other levels see Table 6-4.

Gain is adjusted for the appropriate positive scale voltage, with all ones in. For an 8-bit, ±10-volt output example, this would be +9.992 volts.

Similar comments apply to the trim components used in this circuit, as well as the bulk resistances used for R_{REF}, R_{OFFSET}, and R_F. For bipolar converters, offset, and its drift, are multiplied by the output op amp; therefore, the bias compensation resistor should be used unless the amplifier is a FET or other low current type.

Unipolar/Bipolar Multiplying D/A Application Circuits

With the details of various device housekeeping and other basic considerations having been discussed, this section now addresses multiplying d/a converters as applied to the tasks of both unipolar and bipolar d/a circuits. The following series of circuits are full working systems, illustrating each device as it is most readily applied to these two types of ap-

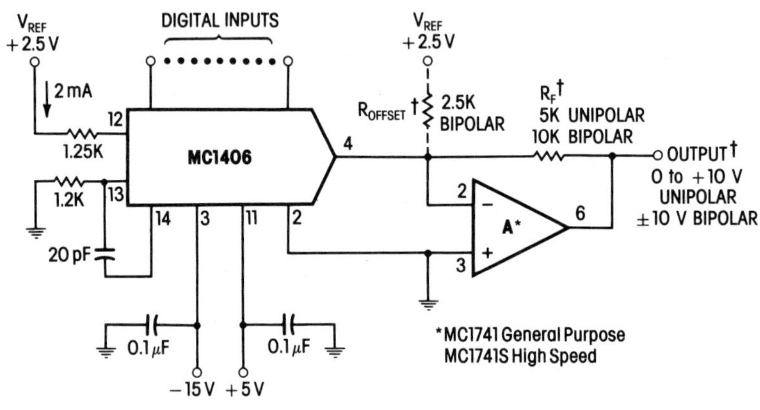

† R_F SELECTION TABLE

OUTPUT RANGE		R_F	
Unipolar	Bipolar	1406/1408	3410
0 to +10 V		5K	2.5K
	±10 V	10K	5K
0 to +5 V		2.5K	1.25K
	±5 V	5K	2.5K
0 to +2.5 V		1.25K	0.625K
	±2.5 V	2.5K	1.25K

(A) Using 6-bit MC1406 series (complementary binary).

Fig. 6-24. The MC1406, MC1408, MC3410

plications, which are, of course, the most popular forms of use. Examples are shown for each device, and the circuits range from resolutions of 6 to 12 bits. In general, each device is shown in manufacturer-recommended configurations, using suggested peripheral devices, such as reference sources and output op amps. The reader can apply these circuits either directly as shown, or adapt them further to more specific requirements.

MC1406, MC1408, MC3410 Series

In Fig. 6-24 are shown the circuits applicable to the MC1406, MC1408, and MC3410 devices suitable for 6-, 8-, and 10-bit uses, respectively.

In Fig. 6-24A the MC1406 is shown as used in a 6-bit unipolar or bipolar d/a converter. A reference voltage of +2.5 volts is used, such as

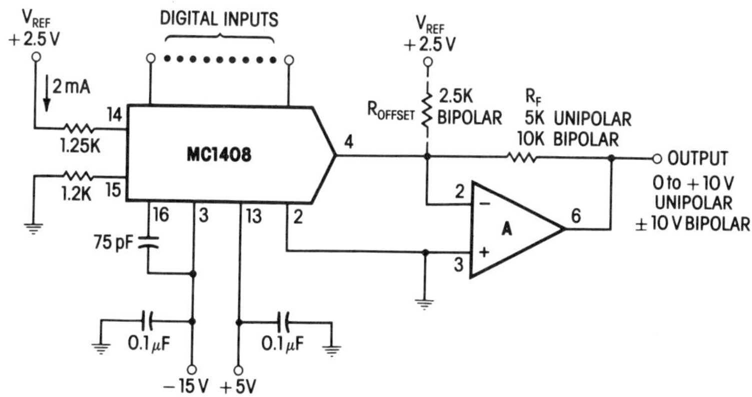

(B) Using 8-bit MC1408 series.

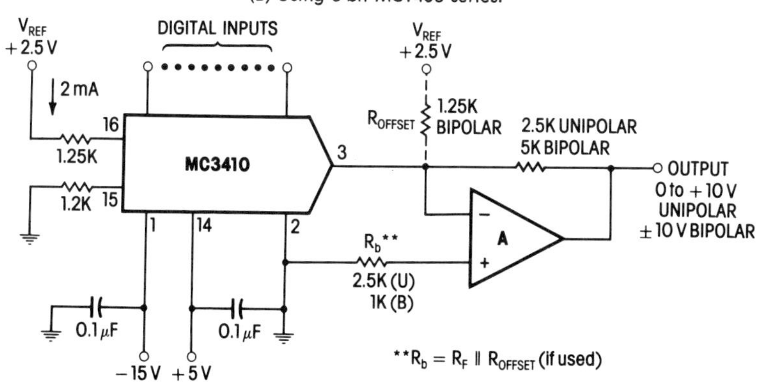

(C) Using 10-bit MC3410 series.

series voltage output d/a converters.

would be provided by the MC1403 reference. The output amplifier can be either the MC1741 for general purpose or nonhigh speed uses, or the MC1741S for high speed or fast settling applications. These reference/op-amp devices are also applicable to the MC1408 and MC3410 circuits which follow.

To apply the MC1406 (and the others) to a specific output voltage range, a table of feedback resistor values is given. These values apply for a reference voltage of 2.5 volts, a reference current of 2 mA, and an offset resistor of 2.5 K (for bipolar use). In the example as shown, the schematic values are for a 0- to +10-volt unipolar output, or ±10 volts for bipolar. As can be noted from the table, R_F is reduced for lower output ranges.

Calibration and offset trims can be applied, as generally described with Fig. 6-23. The MC1406 is coded for complementary-binary operation.

The MC1408 series, suitable for 8-bit uses, is shown in Fig. 6-24B. This configuration is virtually identical to the MC1406, with the exception of the increased resolution and accuracy, and different compensation.

The MC3410 series, for 10-bit requirements, is shown in Fig. 6-24C. This device uses a 2 mA reference current as the others, but has different offset and feedback resistors due to its current scaling factor of 2. Feedback resistor values are shown in the table of Fig. 6-24A. A bias compensation resistor is suggested for 10-bit operation, selected as shown. The MC3410 requires no compensation components. Digital inputs of all devices are TTL compatible, and the MC1408 and MC3410 are also 5 volt CMOS compatible.

DAC-08 Series

The DAC-08 series of 8-bit binary d/a converters is illustrated in Fig. 6-25. This device, being generally more flexible due to its high voltage compliance current outputs, and differential output operation, has a variety of use options which can achieve unipolar and bipolar operation. Circuits shown here are for a reference current of 2 mA, and a +10-volt reference voltage, as would be available from a REF-01 reference source.

In Fig. 6-25A is shown a hookup for a unipolar, negative output voltage d/a converter. Output E_o here is a negative going voltage with a full scale level of −9.961 volts (see table). Output E_o is the complement of $\overline{E}_o$, and will reach full scale with an all 1s input. Either (or both) outputs may be used, as desired. If only a single output is used, the opposite output should be grounded. Logic input may be programmed as per Fig. 6-17; here it is shown TTL compatible.

The output impedance of Fig. 6-25A is 5K, which may be buffered if desired by an op-amp voltage follower. This option is illustrated in

Fig. 6-29B, using an OP-02 device, for medium to slow speed uses. High speed uses can use an OP-16 or PM356 device.

As shown here, the connection is for a binary, negative output d/a converter. Complementary binary negative output is achieved by using the DAC-08 $\overline{I}_o$ terminal, and grounding I_o.

Fig. 6-25C shows how the extended voltage compliance of the DAC-08 can be used to generate a simple, unbuffered bipolar d/a converter. With all zeros in, the E_o output will be a +10 volts; with all ones it will go to −9.920 volts. From Table 6-4, it can be noted this is the basic bipolar range, with the sign inverted from conventional use. The $\overline{E}_o$ output delivers the complement of this bipolar range. As previously, either (or both) can be used, and the configuration can be buffered with a voltage follower, similar to Fig. 6-25B.

Fig. 6-25D shows a buffered output scheme which is unique to the DAC-08. This is a symmetrical offset binary d/a converter, using a differential to single-ended converter. In this application the load resistors are matched. The output has a range of ±9.960 volts, and is symmetrical about ground.

Although this particular code is unlike the standard bipolar code in that it does not have a unique code for an analog zero output, it is symmetrical with respect to the scale ends, which may be an advantage. For high speed the OP-16, or PM356 can be used as the amplifier.

For all of these DAC-08 circuits, calibration can be achieved using the general technique of Fig. 6-23. However, an alternate method of gain trim recommended by the manufacturer is shown in Fig. 6-25E, along with an alternate offset trim method useful for voltage follower op amps.

The gain trim here uses a single untrimmed resistor for R_{REF}, which is advantageous. The potentiometric operation of the GAIN control minimizes its contribution to TC.

The offset trim method (applicable to any op amp) can be used when the device in use does not have an offset trim provision. It simply sums a small variable offset in series with R_L, and is also applicable to other d/a configurations of this type.

DAC-20 Series, 2 Digit BCD D/A Circuits

The DAC-20 bcd d/a converter is very similar to the DAC-08 in basic concept, but with bcd scaling. A series of 2 digit bcd d/a circuits is shown in Fig. 6-26, using the DAC-20.

In Fig. 6-26A are shown positive output voltage 2 digit bcd d/a converters using the DAC-20. Both of these configurations operate in the classic current-to-voltage output conversion mode, and feature a buffered voltage output. They also use a +10 volt reference source, such as the REF-01. Fig. 6-26A, Part 1 is a straight bcd coded converter, while Part 2 is a complementary bcd. The op amp shown is a fast settling unit; for more general use, the OP-02 can be used.

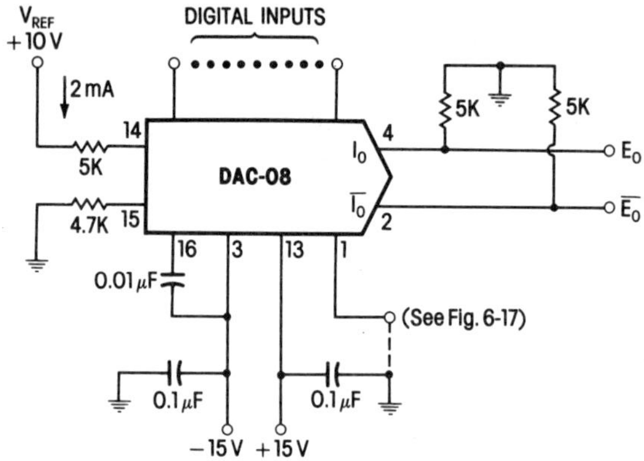

	B1	B2	B3	B4	B5	B6	B7	B8	I_O mA	$\overline{I_O}$ mA	E_O	$\overline{E_O}$
FULL SCALE	1	1	1	1	1	1	1	1	1.992	.000	-9.960	.000
FULL SCALE-LSB	1	1	1	1	1	1	1	0	1.984	.008	-9.920	-.040
HALF SCALE+LSB	1	0	0	0	0	0	0	1	1.008	.984	-5.040	-4.920
HALF SCALE	1	0	0	0	0	0	0	0	1.000	.992	-5.000	-4.960
HALF SCALE-LSB	0	1	1	1	1	1	1	1	.992	1.000	-4.960	-5.000
ZERO SCALE+LSB	0	0	0	0	0	0	0	1	.008	1.984	-.040	-9.920
ZERO SCALE	0	0	0	0	0	0	0	0	.000	1.992	.000	-9.960

(A) Basic unipolar negative operation.

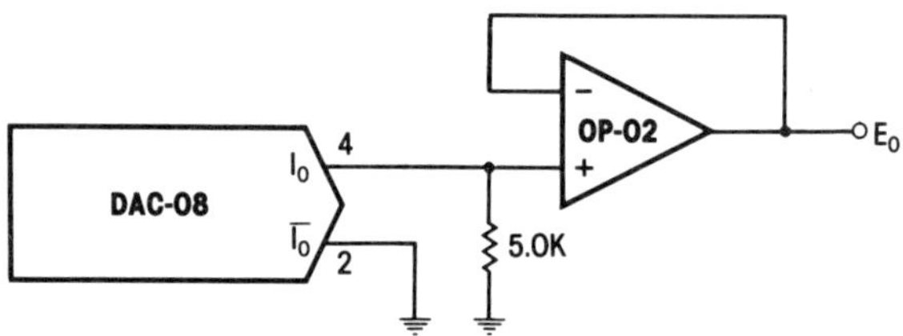

FOR COMPLEMENTARY OUTPUT (OPERATION AS A NEGATIVE LOGIC DAC) CONNECT NON-INVERTING INPUT OF OP-AMP TO $\overline{I_O}$ (PIN 2); CONNECT I_O (PIN 4) TO GROUND.

	B1	B2	B3	B4	B5	B6	B7	B8	I_O mA	E_O
FULL SCALE	1	1	1	1	1	1	1	1	1.992	-9.960
FULL SCALE-LSB	1	1	1	1	1	1	1	0	1.984	-9.920
HALF SCALE+LSB	1	0	0	0	0	0	0	1	1.008	-5.040
HALF SCALE	1	0	0	0	0	0	0	0	1.000	-5.000
HALF SCALE-LSB	0	1	1	1	1	1	1	1	.992	-4.960
ZERO SCALE+LSB	0	0	0	0	0	0	0	1	.008	-.040
ZERO SCALE	0	0	0	0	0	0	0	0	.000	.000

(B) Buffered negative unipolar (single) output operation.

Fig. 6-25. DAC-08 series 8-bit binary,

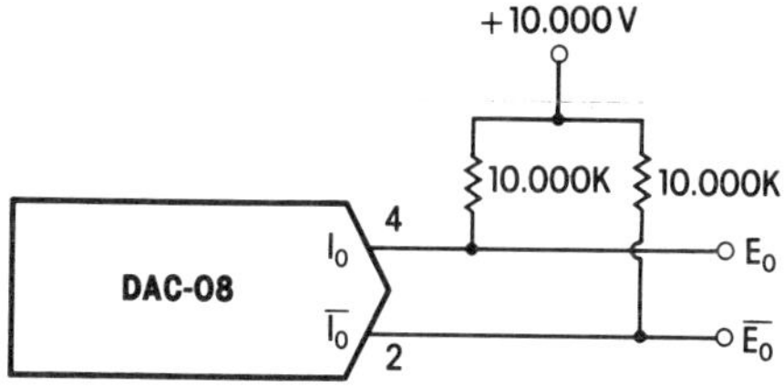

	B1	B2	B3	B4	B5	B6	B7	B8	E_0	$\overline{E_0}$
POS FULL SCALE	1	1	1	1	1	1	1	1	−9.920	+10.000
POS FULL SCALE−LSB	1	1	1	1	1	1	1	0	−9.840	+ 9.920
ZERO SCALE+LSB	1	0	0	0	0	0	0	1	−0.080	+ 0.160
ZERO SCALE	1	0	0	0	0	0	0	0	0.000	+ 0.080
ZERO SCALE−LSB	0	1	1	1	1	1	1	1	+0.080	0.000
NEG FULL SCALE+LSB	0	0	0	0	0	0	0	1	+9.920	− 9.840
NEG FULL SCALE	0	0	0	0	0	0	0	0	+10.000	− 9.920

(C) Basic bipolar operation.

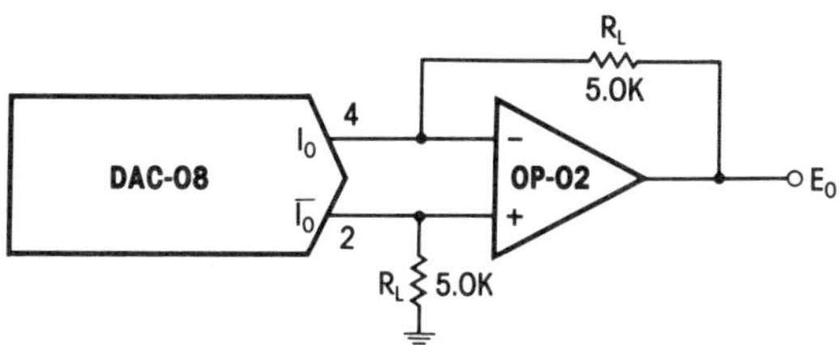

IF $R_L = \overline{R_L}$ WITHIN ±.05%, OUTPUT IS SYMMETRICAL ABOUT GROUND.

	B1	B2	B3	B4	B5	B6	B7	B8	I_O mA	$\overline{I_O}$ mA	E_O(V)
POS. FULL SCALE	1	1	1	1	1	1	1	1	1.992	.000	+9.96
POS. FULL SCALE −LSB	1	1	1	1	1	1	1	0	1.984	.008	+9.88
(+) ZERO SCALE	1	0	0	0	0	0	0	0	1.000	.992	+.040
(−) ZERO SCALE	0	1	1	1	1	1	1	1	.992	1.000	−.040
NEG. FULL SCALE +LSB	0	0	0	0	0	0	0	1	.008	1.984	−9.88
NEG. FULL SCALE	0	0	0	0	0	0	0	0	.000	1.992	−9.96

(D) Differential to single-ended output operation.

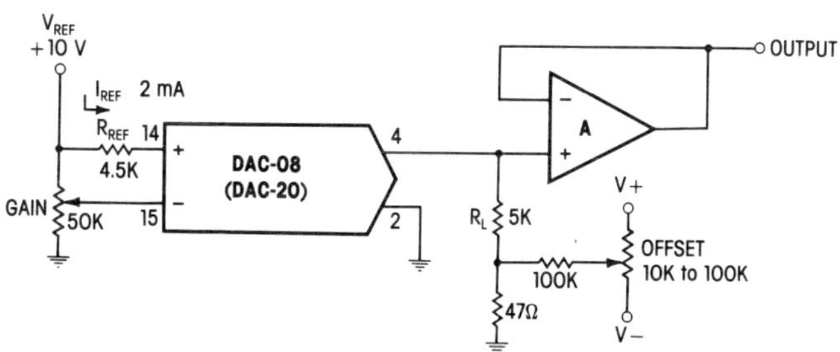

(E) Alternate method of gain, offset adjustment.

Courtesy Precision Monolithics, Inc.

unipolar/bipolar d/a circuits.

For negative output voltages, the circuits of Fig. 6-26B can be used. These are buffered circuits very similar in operation to those in Fig. 6-25B except for their coding of bcd and complementary bcd.

Calibration trim of these circuits can be accomplished by use of Table 6-4, and the general concept of Fig. 6-23 or Fig. 6-25E.

AD7520 Series

The AD7520 series of multiplying d/a converters is illustrated in unipolar and bipolar binary scaled d/a circuits of Fig. 6-27. The AD-7520 series includes the AD7520 and AD7530 (10 bits) and the AD-7521 and AD7531 (12 bits). Comments here on general operation of the AD7520 are applicable to all devices. In addition, comments on the basic d/a operation are also applicable to the AD7522, which is essentially an AD7520, with a double buffered input register.

In Fig. 6-27A the AD7520 is shown in its basic connection as a binary scaled unipolar d/a converter. It is used with an external current-to-voltage converter op amp, which operates in conjunction with the on-chip feedback resistor. Here an AD741 is illustrated for general purpose use. Devices suitable for higher speed are the AD301AL (operated feedforward), the AD518, and the AD509. High speed op amps will require a 10 to 20 pF feedback capacitor for optimum stability (the data sheet includes details).

The Schottky diode clamp at the I_{out} terminal is recommended to prevent possible d/a latch conditions, for all AD7520 type devices (even if not always explicitly shown).

The AD7520 series can accommodate either positive or negative voltage reference sources, and will deliver an output voltage which is the negative of the reference, multiplied by the fraction represented by the digital word. A +10-volt reference, for instance, results in a −9.990-volt full scale output (10 bits), or −9.9976 volts (12 bits). The reference may in fact be any voltage in the range of ±10 volts, and d/a operation is ratiometric for all reference input voltages within this range. Thus, either positive or negative scaled outputs are available by choice of reference polarity, and output range is available by choice of reference voltage level.

Power supply range of the AD7520 is +5 to +15 volts, and this voltage should be selected to match the logic drive to be used. For a +5-volt supply, inputs are TTL and 5-volt CMOS compatible; 15-volt CMOS is accommodated with a +15-volt supply level. Logic coding is straight binary in all devices.

This circuit illustrates (in terms of pin numbering) an AD7520 or AD7530 10-bit device; an AD7521 or AD7531 12-bit unit operates in identical concept, with appropriate changes to pin connections.

Calibration for gain and offset of an AD7520 circuit is somewhat different than the prior discussed devices. Offset should be trimmed to zero by using the output op amp's internal null scheme, for best d/a linearity (as shown in Fig. 6-23A). When the op amp's actual input offset as measured between inputs is zero, d/a linearity will be optimum.

Gain calibration of the series is ±0.3% (typ), which is sufficiently low for many applications without trimming. For exact full scale calibration, the most convenient gain trim is to adjust the reference voltage, when convenient, as applied to the reference input. For a reference voltage of +10 volts as supplied by one of the adjustable reference sources, trim range will usually be sufficient in itself to calibrate the d/a converter.

For fixed level or system references which cannot be altered, the gain trim schemes shown in Fig. 6-27B may be used. In Fig. 6-27B, Part 1 a two-pot gain trim method is shown; R1 will reduce gain to nominal if it is initially high, R2 will increase it, if low. An alternate single control scheme is shown in Part 2 of the figure, which can either increase or decrease gain, as required to accommodate the worst case d/a calibration limits.

Fig. 6-27C shows bipolar operation of the AD7520 series devices. In this application, both the $I_{out\,1}$ and $I_{out\,2}$ output terminals are used in a push pull configuration which provides offset binary operation. With all zeros in, the ladder current is diverted to $I_{out\,2}$, then inverted in A2, and finally reinverted in A1, providing an output equal to the reference voltage. The 10-megohm resistor provides a 1 LSB offset for correctly scaled bipolar operation, in accordance with Table 6-4.

Calibration of an AD7520 series device is somewhat different for bipolar operation, due to the nature of the circuit scheme used. The two step offset trim method described in the figure nulls the individual offsets of amplifiers 1 and 2 (as described above), and will result in both the lowest offset and best d/a linearity simultaneously. Gain trim can then be accomplished for full scale (See Table 6-4) using the method of Fig. 6-27B.

AD7522

The AD7522 d/a converter is quite similar to the AD7520 series in its basic d/a mode of operation, and has many common application considerations. The main difference in the device is its double-buffered input register, and the attendant ability to parallel or serial load input data. The AD7522 as applied to unipolar and bipolar d/a circuits is illustrated in Fig. 6-28.

In Figs. 6-28A and B are shown the basic analog connections for operation of the AD7522 in its basic unipolar and bipolar d/a modes. Digital signal details are omitted here, for clarity.

As can be noted, these circuits are quite similar to the analogous AD-7520 circuits. Exceptions are the inclusion of power-up fault protection diodes CR1 and CR2, which prevent possible failure modes. CR1 is a

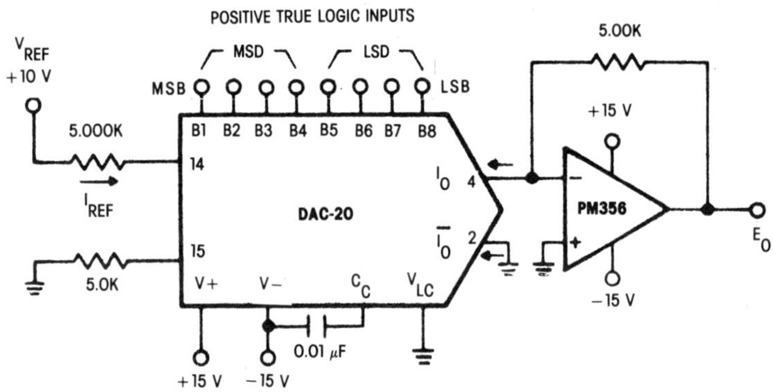

NORMALIZED INPUT	DIGITAL INPUT		I_O	E_O
	MSD	LSD		
0	0000	0000	0	0
10	0001	0000	0.20mA	+1.0V
20	0010	0000	0.40mA	+2.0V
30	0011	0000	0.60mA	+3.0V
40	0100	0000	0.80mA	+4.0V
80	1000	0000	1.60mA	+8.0V
99 (FS)	1001	1001	1.98mA	+9.9V

1. BCD

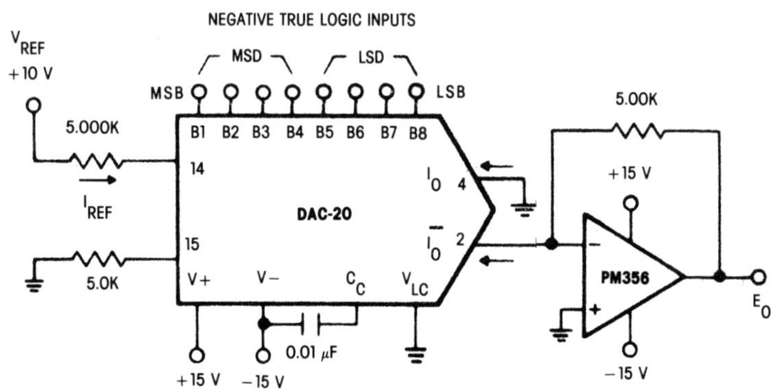

NORMALIZED INPUT	DIGITAL INPUT		$\overline{I_O}$	E_O
	MSD	LSD		
0	1111	1111	0	0
10	1110	1111	0.20mA	+1.0V
20	1101	1111	0.40mA	+2.0V
30	1100	1111	0.60mA	+3.0V
40	1011	1111	0.80mA	+4.0V
80	0111	1111	1.60mA	+8.0V
99 (FS)	0110	0110	1.98mA	+9.9V

2. COMPLEMENTARY BCD

(A) Bcd positive output voltages.

Fig. 6-26. DAC-20 series

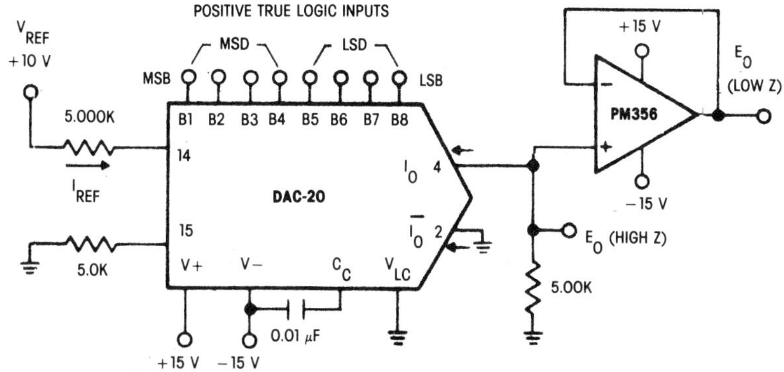

NORMALIZED INPUT	DIGITAL INPUT		I_o	E_O
	MSD	LSD		
0	0000	0000	0	0
10	0001	0000	0.20mA	-1.0V
20	0010	0000	0.40mA	-2.0V
30	0011	0000	0.60mA	-3.0V
40	0100	0000	0.80mA	-4.0V
80	1000	0000	1.60mA	-8.0V
99 (FS)	1001	1001	1.98mA	-9.9V

1. BCD

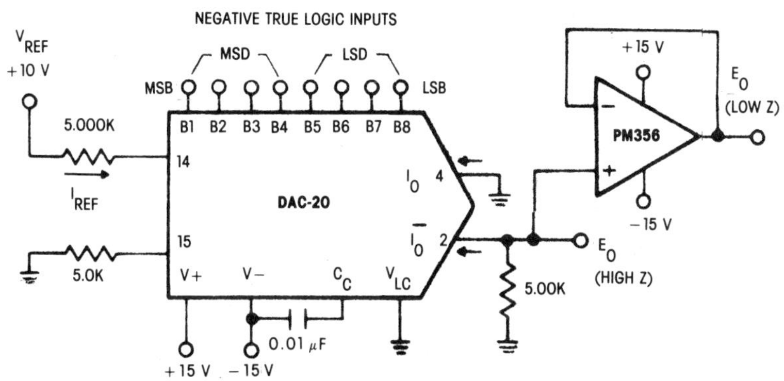

NORMALIZED INPUT	DIGITAL INPUT		$\overline{I_O}$	E_O
	MSD	LSD		
0	1111	1111	0	0
10	1110	1111	0.20mA	-1.0V
20	1101	1111	0.40mA	-2.0V
30	1100	1111	0.60mA	-3.0V
40	1011	1111	0.80mA	-4.0V
80	0111	1111	1.60mA	-8.0V
99 (FS)	0110	0110	1.98mA	-9.9V

2. COMPLEMENTARY BCD

(B) Negative output voltages.

Courtesy Precision Monolithics, Inc.

2 digit bcd d/a circuits.

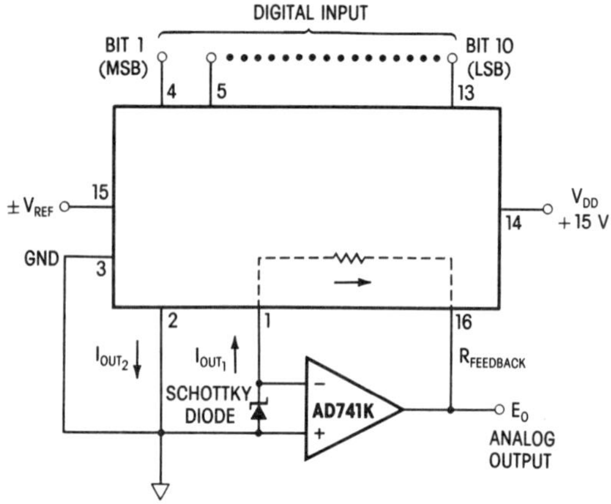

(A) Unipolar operation.

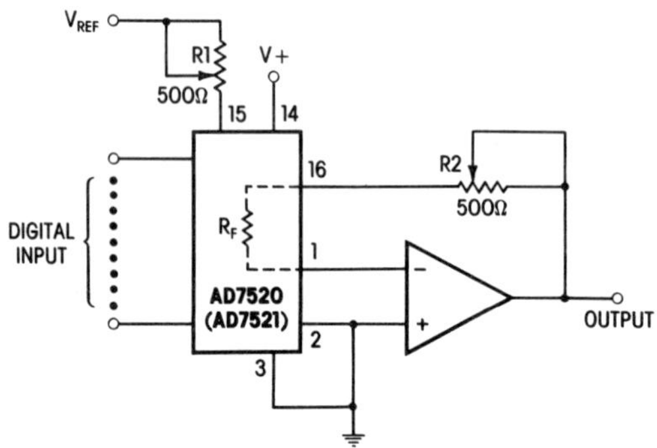

1. TWO POTS

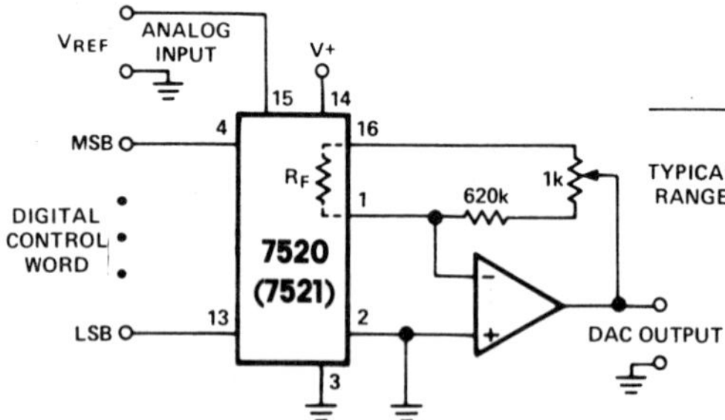

LADDER RESISTANCE		WORST-CASE GAIN ADJUSTMENT RANGE
	5k	-0.7% to +21%
TYPICAL RANGE	8k	-1.1% to +12%
	10k	-1.4% to +9%
	NOMINAL	
	13k	-1.8% to 6%
	20k	-3% to 0.7%

2. SINGLE POT

(B) Calibration trim methods.

Fig. 6-27. AD7520 series

general purpose unit, and CR2 is a Schottky type. CR3 and CR4 perform functions as previously noted.

The AD7522 has gain scaling options, in addition to those of the AD7520, due to two feedback resistor types. With feedback applied to RFB1, gain is unity; applied to RFB2 it is ½. Gain is ¼ with feedback applied to, RFB2, and RFB1 jumpered to $I_{out\ 1}$.

Input data can be entered into the AD7522 in a variety of ways, but the two basic modes are either in parallel or serial fashion. These are illustrated in Figs. 6-28C, D, and E.

Fig. 6-28C shows the simplest method of use, loading of parallel input data in a single 8- to 10-bit byte. For this mode, control inputs SRI, SPC and $\overline{SC8}$ are grounded, and data inputs appear at DB0 and DB9,

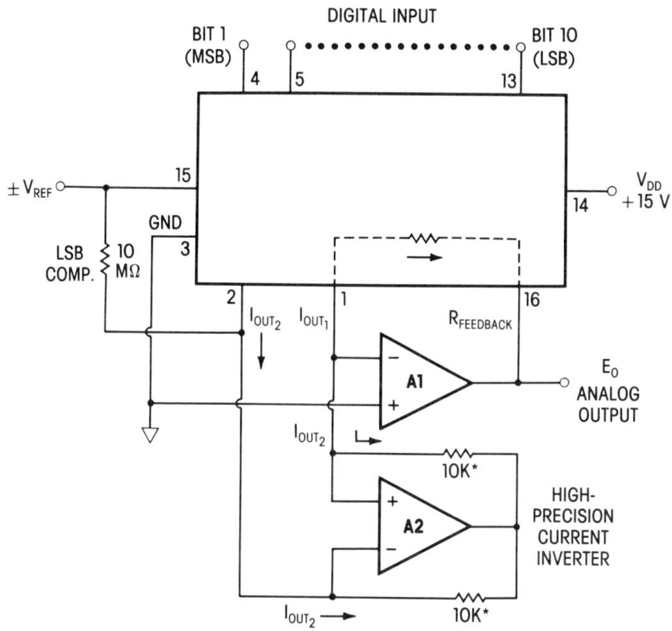

*Absolute accuracy unnecessary. Match and tracking are essential.

BIPOLAR OFFSET TRIM:

1. With V_{REF} = +10 V and all 1's IN, adjust A2 offset for 0.000 V at A2 output.
2. With MSB = 1, other bits = 0, adjust A1 offset for 0.000 V at A1 output.

GAIN TRIM:

Using a method from Fig. 6-27B, adjust GAIN as outlined in Fig. 6-23B.

Note - Repeat if necessary.

(C) Bipolar operation.

Courtesy Analog Devices, Inc.

unipolar/bipolar, d/a circuits.

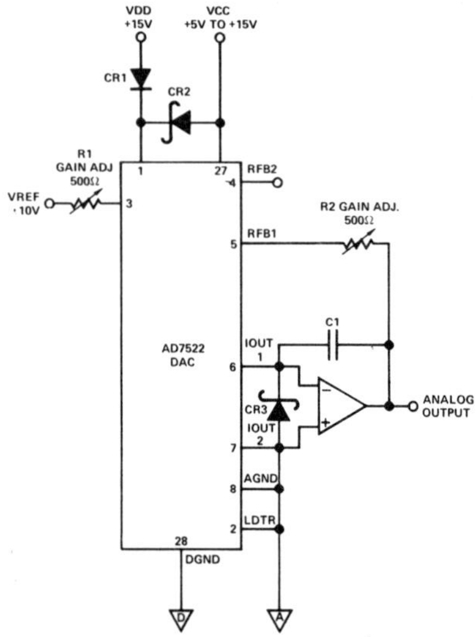

(A) Basic unipolar.

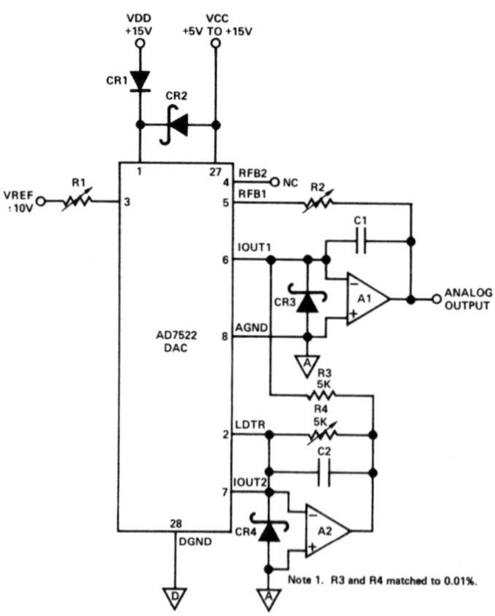

(B) Basic bipolar.

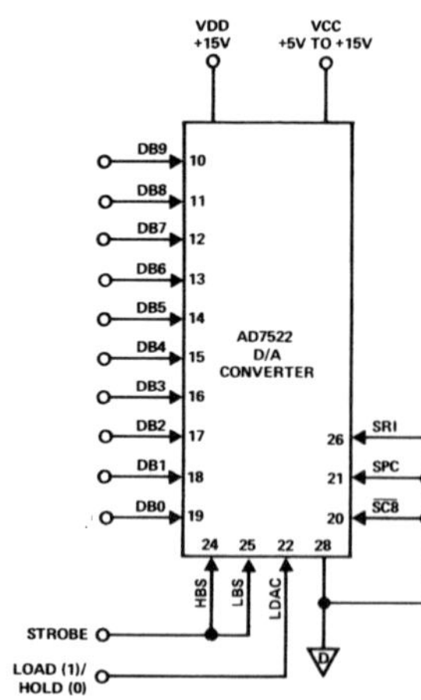

(C) 8- to 10-bit parallel loading.

Fig. 6-28. AD7522 unipolar/bipolar, d/a

the LSB through MSB inputs. Strobe inputs HBS and LBS are operated in common, and when taken high, will clock the data into the input register. The data may then be loaded into the d/a register to update the d/a output, by holding the LDAC control input high. Data will follow the input register when high, and will be locked out when LDAC is low.

Note that various degrees of resolution use are possible, according to the accuracy requirements and the specific AD7522 grade in use. For less than 10-bit resolution operation, the unused data inputs should be grounded.

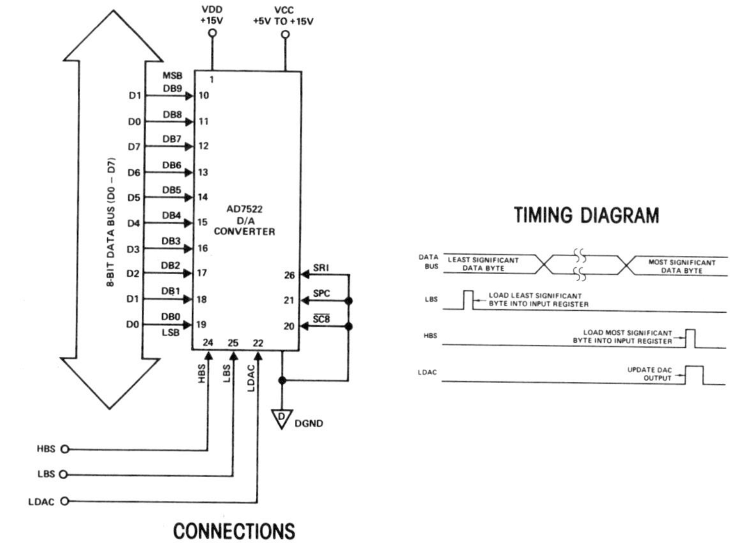

(D) Double byte (10-bit) loading.

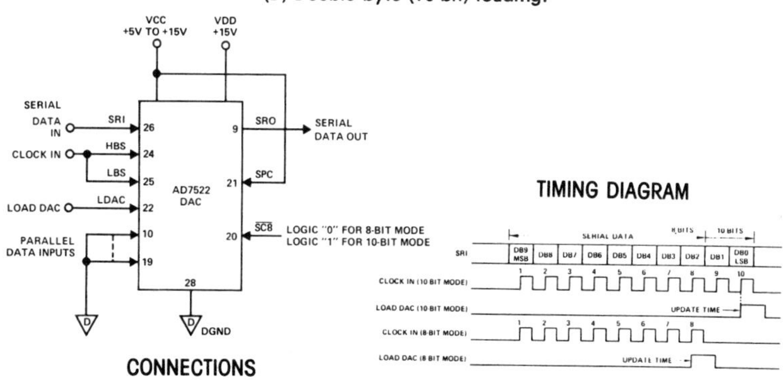

(E) Serial loading.

Courtesy Analog Devices, Inc.

circuits with parallel or serial inputs.

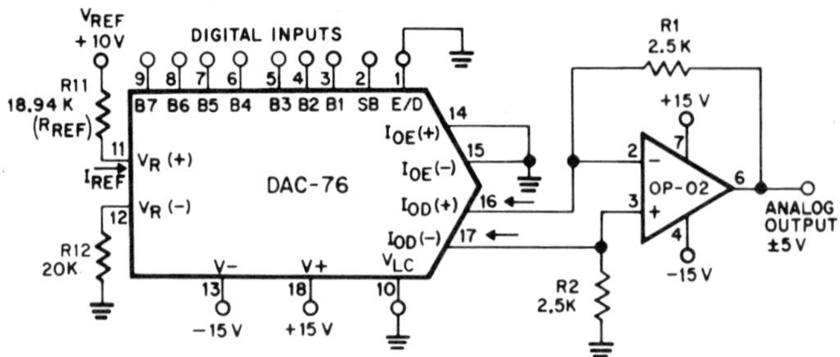

$$I_{REF} = \frac{V_{REF}}{R_{REF}}$$

IDEAL VALUES:

$I_{REF} = 528\ \mu A$

$I_{FS} = 2007.75\ \mu A$

For D/A conversion, or decode mode, output current $= I_{OD}$

$$I_{OD} = \frac{I_{C,S}}{8031} I_{FS}$$

Where, $I_{C,S} = 2[2^C (S + 16.5) - 16.5]$

C = chord 0 through 7

s = step 0 through 15

(see table below)

and $I_{FS} = 3.8\ I_{REF}$

	E/D	SB	B1	B2	B3	B4	B5	B6	B7	E_O
POS FULL SCALE	0	1	1	1	1	1	1	1	1	5.019V
(+) ZERO SCALE +1 STEP	0	1	0	0	0	0	0	0	1	0.0012
(+) ZERO SCALE	0	1	0	0	0	0	0	0	0	0V
(−) ZERO SCALE	0	0	0	0	0	0	0	0	0	0V
(−) ZERO SCALE +1 STEP	0	0	0	0	0	0	0	0	1	-0.0012
NEG FULL SCALE	0	0	1	1	1	1	1	1	1	-5.019V

NORMALIZED DECODE OUTPUT (SIGN BIT EXCLUDED)

STEP \ CHORD		0	1	2	3	4	5	6	7
		000	001	010	011	100	101	110	111
0	0000	0	33	99	231	495	1023	2079	4191
1	0001	2	37	107	247	527	1087	2207	4447
2	0010	4	41	115	263	559	1151	2335	4703
3	0011	6	45	123	279	591	1215	2463	4959
4	0100	8	49	131	295	623	1279	2591	5215
5	0101	10	53	139	311	655	1343	2719	5471
6	0110	12	57	147	327	687	1407	2847	5727
7	0111	14	61	155	343	719	1471	2975	5983
8	1000	16	65	163	359	751	1535	3103	6239
9	1001	18	69	171	375	783	1599	3231	6495
10	1010	20	73	179	391	815	1663	3359	6751
11	1011	22	77	187	407	847	1727	3487	7007
12	1100	24	81	195	423	879	1791	3615	7263
13	1101	26	85	203	439	911	1855	3743	7519
14	1110	28	89	211	455	943	1919	3871	7775
15	1111	30	93	219	471	975	1983	3999	8031
STEP SIZE		2	4	8	16	32	64	128	256

(A) Buffered output.

Fig. 6-29. DAC-76 sign-magnitude

Data and control inputs are TTL compatible with $V_{cc} = +5$ volts and are CMOS compatible with $V_{cc} = +15$ volts. Coding is straight binary.

For compatibility with 8-bit microprocessor data bus applications, a scheme of double byte loading is used, as shown in Fig. 6-28D. In this connection, the DB0 to DB7 inputs connect to bus lines D0 to D7, and the DB8 and DB9 inputs connect to D0 and D1. The data is loaded in a two-byte sequence, the 8 lower bits first, then the 2 highest order bits in a second byte. The LBS and HBS inputs control this sequence. When the complete 10-bit word is loaded, the data can be loaded into the d/a register by the LDAC input. The timing diagram explains this sequence.

For serial data loading, the hookup is as shown in Fig. 6-28E. The SPC input is tied to V_{cc}, and $\overline{SC8}$ connected as appropriate for an 8- or 10-bit mode. All parallel data inputs are grounded, and the serial data is fed to SRI. The clock is applied in common to LBS and HBS, and 8 (or 10) positive edges clock the data into the input register. Thereafter, the d/a register may be loaded by a positive level to LDAC.

The DAC76

The DAC76 companding d/a converter is illustrated in a sign plus magnitude, exponentially coded d/a circuit in Fig. 6-29. The DAC-76 is in some regards similar to the DAC-08 and DAC-20, and features similar logic inputs, programmable thresholds, a similarly operating reference circuit, and current source outputs. Its outstanding operational

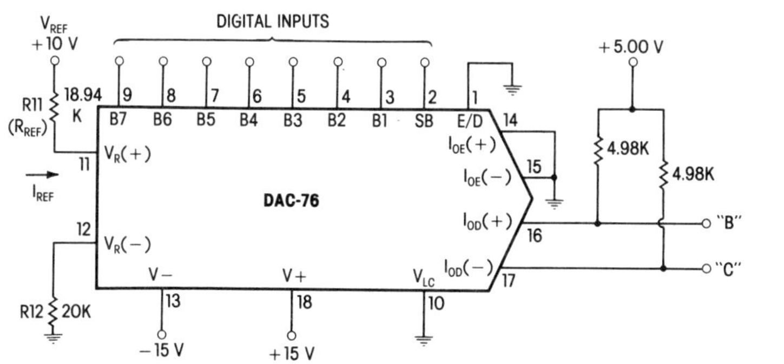

INPUT CODE	OUTPUT VOLTAGE (V) "B"	"C"	DIFF
1111 1111	−5.00	+5.00	−10
1110 1111	+0.02	+5.00	− 4.98
1000 0000	+5.00	+5.00	0
0000 0000	+5.00	+5.00	0
0110 1111	+5.00	+0.02	+ 4.98
0111 1111	+5.00	−5.00	+10

NEGATIVE OUTPUT VOLTAGE COMPLIANCE $V_{OC}(-)$

V− \ I_{FS}	1.0 mA	2.0 mA	4.0 mA
−12 V	−2.8 V	−2.0 V	−0.4 V
−15 V	−5.8 V	−5.0 V	−3.4 V
−18 V	−8.8 V	−8.0 V	−6.4 V

Minimum negative compliance

$V_{OC}-(\text{Min}) = (V-) + (2 I_{REF} \cdot 1.6) + 8.4\ V$

(B) Resistive load output (unbuffered).

Courtesy Precision Monolithics, Inc.

exponential d/a converter.

difference is the nonlinear coding, which is exponential in nature. The device's transfer characteristic is illustrated here in detail, in the light of this application.

In the d/a circuit shown, the DAC-76 is connected in the decode mode, by the use of the I_{OD} (+) and I_{OD} (−) outputs, and a (wired) low logic level to the E/D pin. The unused encode outputs are grounded. This circuit has 7 magnitude bits of control, plus a sign bit.

The voltage output range is summarized in the brief table, and illustrates the effect of the sign and magnitude bits. The output is a symmetric bipolar range of ±5 volts (nominal). The OP-02 device shown is general purpose; high speed operation would imply use of an OP-16 or PM356 type amplifier. The two load resistors used should be matched in value for bipolar symmetry. Gain calibration can be achieved through trim of the reference current (or voltage). As shown, the circuit uses a +10-volt reference source. Reference and full scale current levels are shown in the figure.

This circuit can also be modified for a ±10-volt output range, by changing R1 to 5K and adding an additional 5K resistor from I_{OD} (+) to ground. These two 5K resistors should be matched in value, and twice the 2.5K R2 value.

A general summary of the DAC-76 current output characteristic versus digital input is contained in the notes, and the Normalized Decode Output table. For the active I_{OD} output, either I_{OD} (+) or I_{OD} (−), the current can be generally described by the equation

$$I_{OD} = \frac{I_{c,s}}{8031} I_{FS}$$

where,

$I_{c,s}$ is the solution of the equation

$$I_{c,s} = 2[2^c(s + 16.5) - 16.5]$$

I_{FS} is the full scale output current. Full scale output current for the decode mode is 2007.75 μA, for a 528 μA reference current, a scaling factor of 3.8. As one example, for step 1 of chord 0 (0000001) $I_{c,s} = 2$; therefore, I_{OD} will be 2/8031 times 2007.75 μA, or 0.5 μA. In the d/a circuit shown, an output current of 0.5 μA will result in a voltage of 1.25 mV, as is indicated in the table for E_O.

For calculation of a given code level output, either the equation for $I_{c,s}$ may be used, or the normalized table. The table as noted excludes the sign bit. In actuality this entire table is applicable to which ever of the two outputs (I_{OD} (+) or I_{OD} (−) is active at a given time. It should be noted that the two outputs I_{OD} (+) and I_{OD} (−) are *not* complementary differential, as is the case of the DAC-08. Only the output programmed by the sign bit is active at a given time; the opposite output appears as an open circuit (and a similar comment applies also to the I_{OE} outputs).

Because of its high compliance current source outputs, the DAC-76 can be used as a d/a converter with a simple load to ground, or to a positive voltage source. An example of a sign-magnitude coded d/a converted in unbuffered form illustrates this, as shown in the circuit of Fig. 6-29B.

In this circuit the encode input and outputs are disabled as previously, and the circuit operates generally as before with regard to control inputs.

The outputs are connected to 4.98K load resistors (matched), which are referred to a stable +5-volt source. From the table it can be seen that each output swings a 10-volt range, which is symmetric about zero. Output is subject to loading errors, and may be buffered if desired, similar to Fig. 6-25B.

An application point of interest to circuits of this type is that the DAC-76 negative compliance range must be observed in applications where large output swings are desired. The device limits are a function of the V− supply, and the full-scale current, and are summarized in the figure.

AD562 Series

In Fig. 6-30 is shown a general circuit family applicable to application of the AD562 device series to d/a converters of 12-bit binary or 3 digit bcd operation, operating in both unipolar and bipolar modes. This circuit, as can be noted, is a composite one, and the various coding options are available by jumpering for various ranges and unipolar or bipolar scaling. Binary or bcd operation is chosen by the appropriate part number suffix.

The AD562 contains all the required scaling resistors for reference current, bipolar offset, and gain adjustment. To make the d/a function complete, an external reference source of 10 volts is used, such as an AD581 or AD2700 (see Chapter 3). The output op amp operates in the current-to-voltage mode, and is chosen for the desired speed and accuracy. Suitable device examples are the AD301AL (feedforward) for high dc and ac performance with low cost, the AD504L for high dc accuracy, and the AD509 for high speed and fastest settling time (see Fig. 6-13B).

The AD562 is trimmed for gain and bipolar offset with high resolution 100-ohm trimpots, connected as shown in Fig. 6-30. Unipolar offset may be adjusted via the null terminals of output op amp, or the more general method of Fig. 6-23A. Logic input sensitivity is adjustable between TTL or CMOS levels, as shown in Fig. 6-17B. Logic inputs are straight binary (or bcd), and calibration voltage levels can be found in Table 6-4. A capsule summary of the device's range and code capability is shown in the figure.

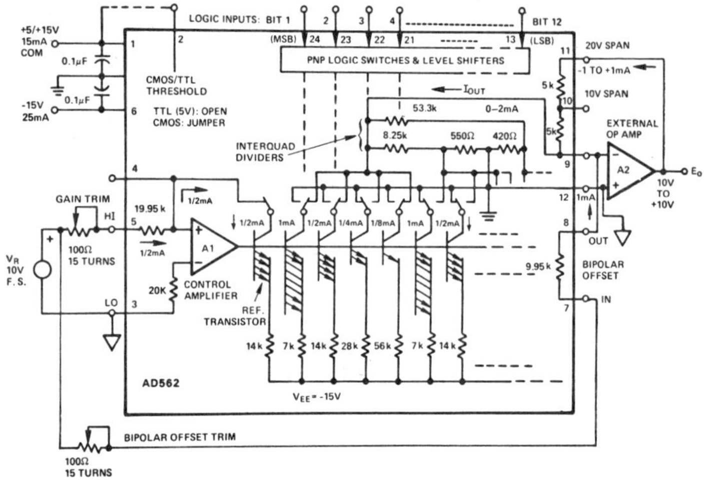

CIRCUIT CONFIGURATION

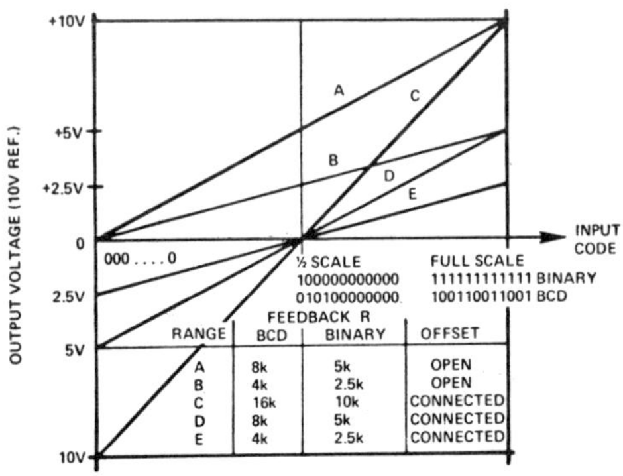

OPERATING RANGE OPTIONS

Courtesy Analog Devices, Inc.

Fig. 6-30. AD562 series 12-bit binary/bcd, unipolar/bipolar d/a circuits.

AC Reference Multiplying Applications

The treatment thus far of the various multiplying d/a converter devices has been in the context of their conventional uses, as unipolar and bipolar scaled d/a converters which use dc (only) references. The section which follows treats these devices in applications which are more closely allied to their generic device class, or true multiplying uses. In these circuits the emphasis is on the handling of ac reference inputs, and digitally controlled multiplication of these signals.

The multiplying d/a converter function is conceptually illustrated in Fig. 6-31. Here is shown a "block box" multiplying d/a converter, which

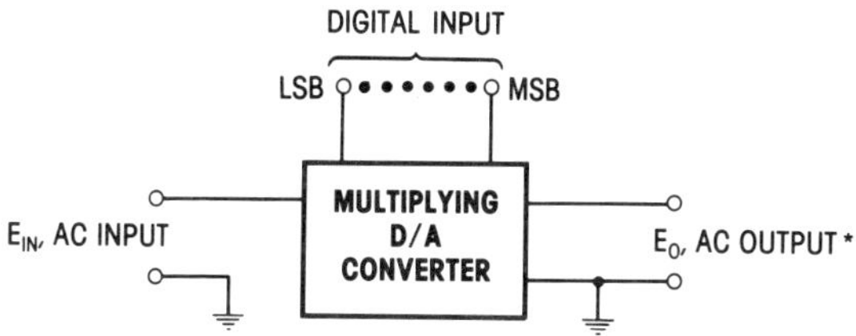

*$E_0 = K E_{IN}$
Where K = fraction represented by digital input.
Two quadrant operation of E_{IN}, E_0 is implied.

Fig. 6-31. Function of a multiplying d/a converter.

accepts an ac input signal (or reference), and a digital input. The output, E_o, is the product of the input signal E_{in} and K, where K is a fraction represented by the digital input. The output is therefore a scaled replica of the input signal.

This function may generally be viewed as a digitally controlled potentiometer, where the digital control replaces the conventional mechanical input (or shaft position). For a given digital input code the output can be seen to follow amplitude or envelope changes in E_{in}, as the transmission characteristics of the reference to output path are linear. Also, bipolar operation of both input and output signals is inherent.

A graphic general representation of multiplying d/a operation is shown in Fig. 6-32. There are two general classes of multiplying d/a converters, two quadrant and four quadrant.

A two quadrant characteristic is shown in Fig. 6-32A. In this type of multiplier the digital code controls only the output magnitude, but not the sign. Maximum output occurs with a full scale digital input of 11 . . . . 11, and minimum at a digital input of 00 . . . 01 (1 LSB). Zero output occurs at a digital code of 00 . . . 00, which is an off state. Control dynamic range is then simply the ratio of digital full scale to 1 LSB. A summary of the dynamic range control relations for binary converters is given in Table 6-5.

As can be noted from this table, greater resolution yields greater dynamic range, control, and selection accuracy. Where high dynamic range, accuracy and selection resolution are required, two converters may be an effective system solution. These cases are also summarized in Table 6-5.

A four quadrant multiplying characteristic is shown in Fig. 6-32B. In this type of converter, both inputs control the output sign. Output is maximum with a negative sign at a digital input of 00 . . . 00, and maximum with a positive sign at a digital input of 11 . . . 11. Zero output occurs at digital half scale, or 100 . . . 00.

Both two and four quadrant types of converters have their uses, depending upon the nature of the application requirements. Two quadrant

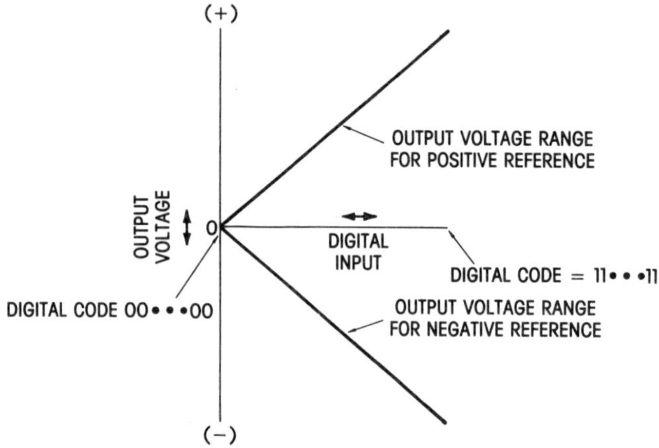

(A) Two quadrant.

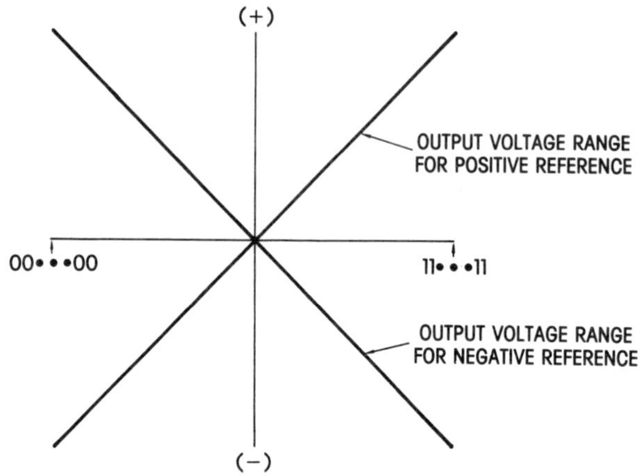

(B) Four quadrant.

Fig. 6-32. Digital/analog multiplying characteristics.

types are analogous to unipolar dc output d/a converters in that the digital input controls magnitude, but not sign. Four quadrant types are analogous to bipolar dc output d/a converters, where the digital input controls magnitude and sign. In both types the input reference sign directly controls the output sign.

Applications considerations appropriate to multiplying d/a converters include all of the general points covered thus far, plus the consideration of reference channel dynamics, as well as the output interface and its dynamics. These device considerations are in addition to the control range dynamic considerations discussed previously.

Multiplying d/a converters find application wherever it is desired to control ac signal levels by digital means. Examples are servo and control systems, signal processing, audio systems, pulse and/or video systems, and programmable instrumentation. Most general purpose d/a

Table 6-5. Theoretical Attenuation Performance, Binary Scaled Multiplying D/A Converters*

Number of bits (N)	Maximum attenuation limit (dB)	Attenuation limit for ±1/2 LSB ±1 dB (dB)	Attenuation limit for ±1 dB resolution (dB)
	Single Binary Coded D/A Converters		
6	36	24	17
7	42	30	23
8	48	36	29
9	54	42	35
10	60	48	41
11	66	54	47
12	72	60	53
	Cascaded Binary Coded D/A Converters		
6	72	34	22
7	84	46	34
8	96	58	46
9	108	70	58
10	120	82	70
11	132	94	82
12	144	106	94

* W. G. Jung, "Applications considerations for IC Data Converters Useful in Audio Signal Processing," **JAES**, December 1977.

converters can handle bandwidths which accommodate the audio range, and more specialized circuits are useful into the megahertz region.

Reference Modulation Techniques

Although there are many IC multiplying d/a converters a great deal of them fall into a category of device which is confined to unipolar operation, insofar as reference input(s) and output(s) are concerned. Representative of this type of converter are devices such as the MC1408 and DAC-08, and the others mentioned within Fig. 6-20. To use these devices for ac inputs and outputs, operation must be set up in such a way that the reference and output current are always unipolar, although they can contain the desired ac components. Thus, such a device may be viewed as having dc input and output currents, which are modulated by ac. It is this manner in which they must be used for multiplying d/a converters, due to the fact they cannot handle true bipolar currents.

Two basic methods of modulating the reference input current of a multiplying d/a converter are shown in Fig. 6-33. Either is suitable, the main selection difference being that the method in Fig. 6-33A has a

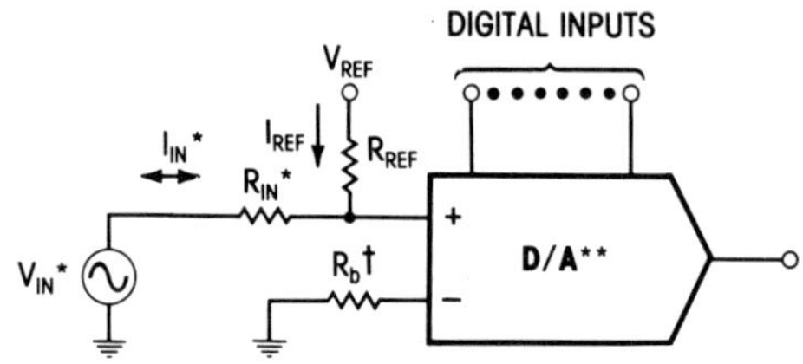

$*I_{IN} = \frac{V_{IN}}{R_{IN}}$, $I_{IN\,(PEAK)} < I_{REF}$

$\dagger R_b = R_{IN} \parallel R_{REF}$

**Applicable to these D/As.

1406	DAC-08
1408	DAC-20
3410	DAC-76
AD562	

(A) Current input.

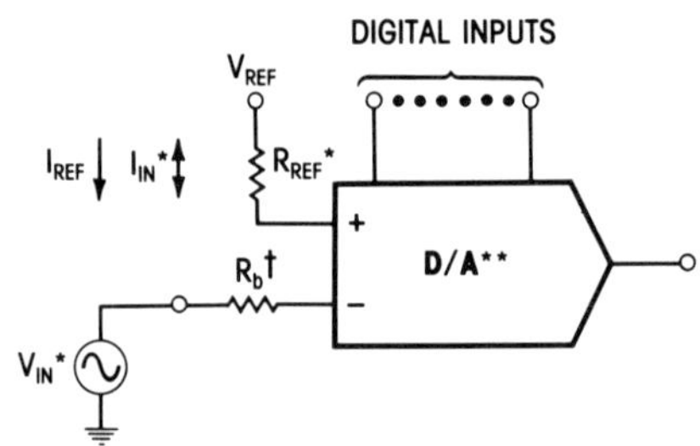

$*I_{IN} = \frac{V_{IN}}{R_{REF}}$, $I_{IN\,(PEAK)} < I_{REF}$ or $V_{IN\,(PEAK)} < V_{REF}$

$\dagger R_b = R_{REF}$

(B) Voltage input.

Fig. 6-33. Reference modulation techniques.

low input impedance (the value of R_{in}), and that in Fig. 633B has a high input impedance.

Both methods vary the current flowing into the d/a (+) reference input, which in turn varies the ladder current (see Fig. 6-19). In Fig. 6-33A the current in input resistor R_{in} (I_{in}) adds (or subtracts) to the dc current, I_{REF}, flowing in R_{REF}. A fundamental constraint is that the peak current in R_{in} must always be less than I_{REF}, so that the net reference input current cannot go to zero. In practice this should be guaranteed by selecting the reference current to be at least 10 to 20 percent greater than the highest peak current in R_{in}

In Fig. 6-33B the current in R_{REF}, and thus that into the (+) input, is varied by modulation of the voltage across R_{REF}. This is due to the fact that the d/a converter (+) and (−) inputs will be maintained at a near zero potential difference. In this circuit, a net I_{REF} greater than zero can be guaranteed by constraining the (+) peak of V_{in} to less than the reference voltage. Also, V_{in} must not exceed the reference input common mode limits.

In both circuits, minimum dc error is achieved when a compensation resistor R_b is used, with a value equal to the equivalent resistance seen at the (+) input. This resistor is optional, and may sometimes be deleted for certain applications.

Choice of reference current for these modulation methods must be done with consideration of the d/a converter rated maximum I_{REF}. For a device I_{REF} rating of 2 mA, for instance, a dc bias of just over 1 mA will allow peak positive current swings of 2 mA, without allowing the current to decrease to zero on negative going peaks. This general procedure will be made more clear by a circuit example.

In Fig. 6-34 is shown a basic two quadrant multiplying d/a circuit which uses the just described biasing. In this case either an MC1408 or DAC-08 device may be used, whichever is suitable.

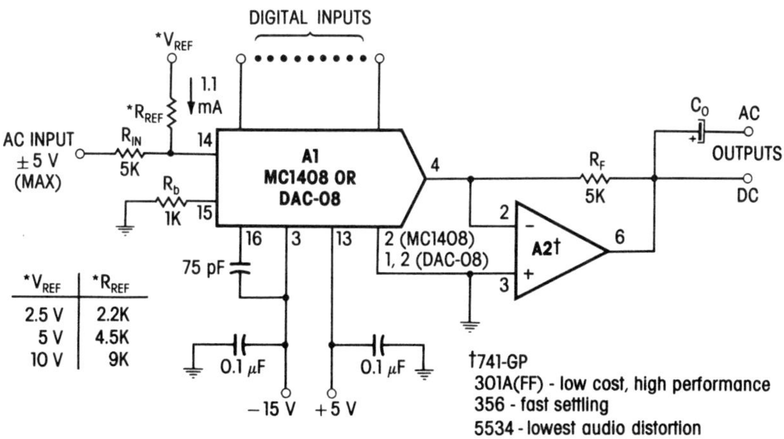

*V_{REF}	*R_{REF}
2.5 V	2.2K
5 V	4.5K
10 V	9K

Fig. 6-34. Basic two quadrant multiplier (ac-coupled).

An I_{REF} of 1.1 mA is established by R_{REF}, in conjunction with V_{REF}. The R_{REF} value is chosen from the table, according to the reference voltage level to be used.

Here an output-current to voltage-converter op amp is used. R_F is chosen so that with the full scale d/a dc output current (1.1 mA in this case) the op amp will rest quiescently at about +5 volts. Here R_F is 5K, which causes the output to rest at a level of +5.5 volts, with a full scale digital input. With the application of reference modulation, the d/a output current can swing ±1 mA, or ±5 volts at the op-amp output. Note that the output is an ac signal atop a dc bias level.

Input current scaling resistor R_{IN} is chosen so that the highest expected input level will cause a peak current of 1 mA (or less). Here a ±5-volt swing is allowed by a 5K input resistor. R_F and R_{IN} of this circuit determine the maximum gain, or the d/a converter overall scaling factor at maximum digital input. Here it is chosen to be unity, in actuality 255/256. In general this gain will be

$$\text{Gain} = \left(\frac{2_n - 1}{2_n}\right)\left(\frac{R_F}{R_{IN}}\right)$$

The implication given by this equation is that a multiplying converter also may be used with a constant scaling factor which is equal to the R_F/R_{IN} ratio. However, in the example just discussed, R_F and R_{IN} are best chosen for close to a unity gain ratio because of dc bias restrictions.

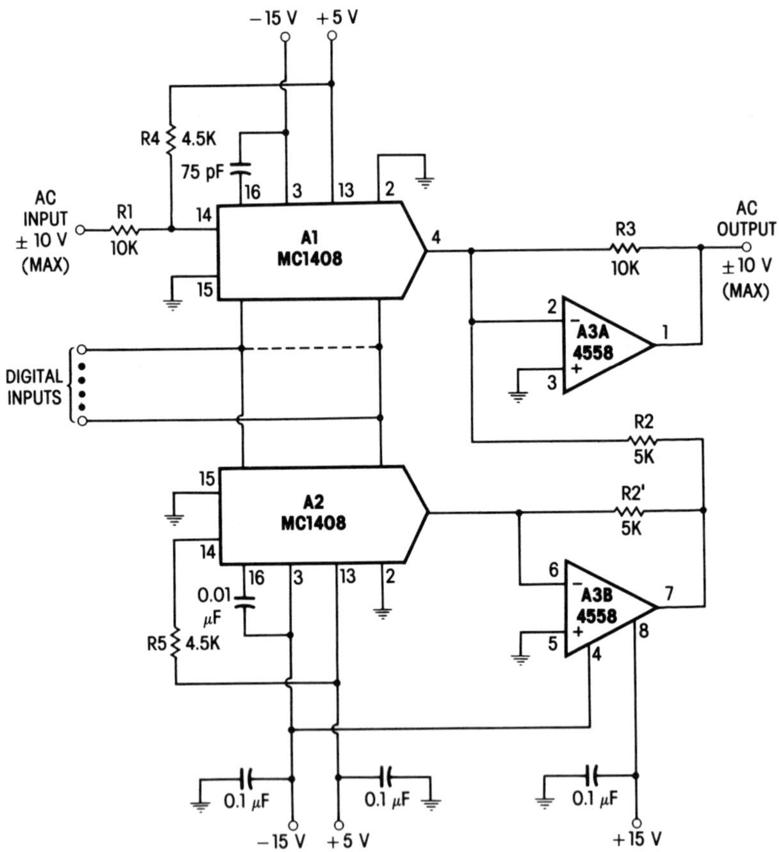

(A) Using the MC1408.

Fig. 6-35. Direct coupled

However, this need not always be so when the bias considerations are dealt with separately, as the next few examples will show.

This circuit output, as has been noted, consists of the desired ac signal which rides on a dc baseline. As the digital input changes, both the ac and dc components of the output change. The effect of the dc component, will create a base-line shift in the output while transmitting the desired variable amplitude ac. However, C_0 will only remove the *static* dc component, and there will still be step changes transmitted to the output for each code change. These transients will also be disturbing to some circuits, so a more complete means of their suppression is desirable.

Wide reference channel bandwidth is achieved in multiplying circuits using d/a converters of this type by using compensation as appropriate to the signal frequencies being handled. For the MC1408 and DAC-08,

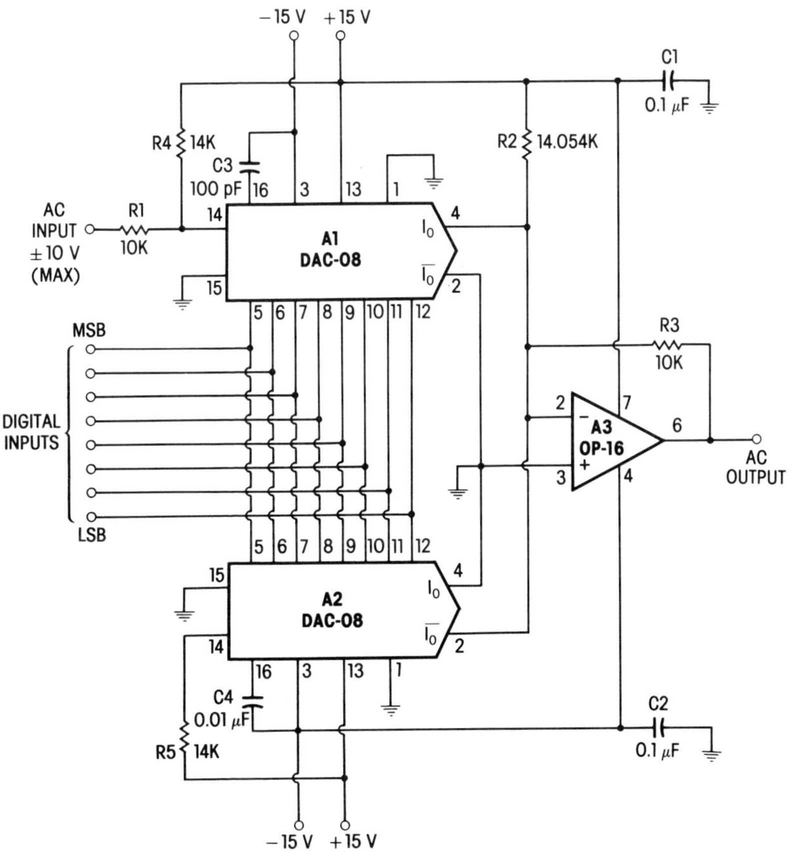

(B) DAC-08 current-to-voltage output.

two quadrant multipliers.

this is accomplished with reduced compensation, such as the 75 pF value shown. Also, the output amplifier used should be consistent with the range of output levels and frequencies in use. A general-purpose 741 type will allow 10-volt peak-to-peak output swings up to approximately 15 kHz before slew limit; higher frequency operation should employ faster amplifiers such as a feedforward-operated 301A, or a 356 which

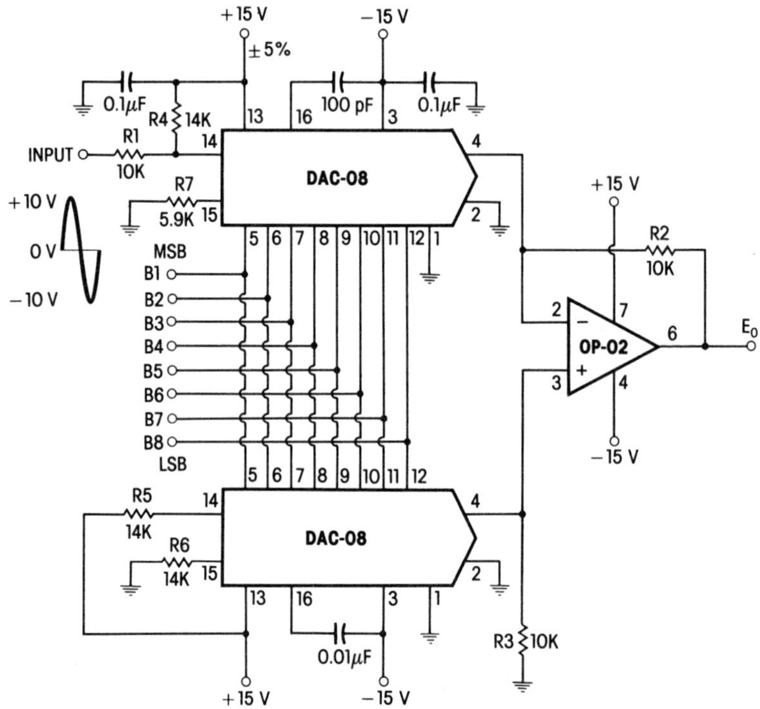

(C) DAC-08 differential to single-ended output.

Fig. 6-35 (cont). Direct coupled two quadrant multipliers.

is also a fast-settling (1.5 μs) type. Both of these devices will provide full power response to beyond 100 kHz. Very low audio distortion will be realized with the 5534 op amp.

This general circuit concept can also be used with the 6-bit MC1406 and 10-bit MC3410 types, with appropriate pin changes.

Direct Coupled Two Quadrant Multipliers

Several schemes exist for suppressing the dc output component in unipolar current multiplying d/a converter types. They are illustrated in Fig. 6-35.

Fig. 6-35A shows a hookup, using two MC1408 devices, which cancels the dc component in the output, allowing transient-free operation and unrestricted gain ranging over the dynamic range.

In this circuit device A1 is operated with a modulated reference, and this portion of the circuit delivers a multiplied dc + ac current to A3A, the output summing amplifier. D/a device A2 operates with an unmodulated reference and delivers an identically scaled, but dc only current to A3B. This current is inverted by A3B and returned to the top summing amplifier via R2, and cancels the dc component from d/a converter A1. With the digital inputs of A1 and A2 driven in parallel, their dc output currents track over the dynamic range, allowing effective cancellation in the output. As a result the output contains only the desired multiplied ac signal.

Overall ac maximum gain is unity, as set by R3, a value of 10K in this example. However, R3 can be scaled for lower (or higher) gains if desired, without interference with biasing levels. Signal handling capability is ±10 volts at input, as set by R1, and ±10 volts at the output, set by R3 and the ±15-volt supplies to A3.

The +5-volt supply is used as a reference in this instance, although in general it is not recommended. In a case such as this it is allowable because variations in the supply are cancelled by the dc current cancellation scheme. Reference currents for both channels should be identical by match of R4 and R5, and also by match of R2 and R2′. These four resistors are preferably matched pairs. Note that with the value ranges used, all resistors can be contained in a tapped 10K array.

The output amplifiers are chosen here as a 4558, a dual type, for single package convenience. This will allow an output full power bandwidth of about 15 kHz. For higher-speed operation, the top device can be speed optimized, using such types as the TL072 or LF353 for high performance audio. Note that the lower channel, which operates only for dc, need not necessarily be speed optimized.

A scheme which uses the differential current output of the DAC-08 device to advantage is shown in Fig. 6-35B. This circuit again uses two d/a devices, and combines their dc output currents in a single summing amplifier, A3.

With the digital inputs driven in parallel, and the reference input of only A1 modulated, the current I_{o1} will consist of scaled dc + ac, while current $\overline{I}_{o2}$ is complementary scaled, but dc only. The current $I_{o1} + I_{o2}$ then consists of both a constant dc current, and also the desired (scaled) ac current.

The dc component, which represents a fixed offset, is cancelled by resistor R2, which sums a current into A3 equal to I_{FS}. For an 8-bit d/a converter, I_{FS} is equal to 255/256 I_{REF}, so R2 is made the value of R_{REF} multiplied by 256/255. Since this R2 value is not standard, it can be a fixed 14K resistor similar to R4 plus a fixed 56-ohm, or a 100-

ohm trimmer. The +15-volt supply is used as a reference, and R4 and R5 should preferably be matched for best dc cancellation.

R3 sets the overall gain, and can be scaled as previously mentioned. Signal handling is ±10 volts at input and output.

The output amplifier, since it operates in the current-to-voltage mode, can be speed optimized. A feedforward type useful here is the OP-01, and the OP-16 allows full power output to beyond 100 kHz.

Another DAC-08 circuit which uses differential cancellation is shown in Fig. 6-35C. This version is similar to that just described in that one d/a converter is modulated, and the second converter has dc only. Here, however, the I_o (only) outputs are used to drive a differential current-to-voltage converter. This requires that resistors R2 and R3 be matched, as well as R4 and R5.

The amplifier shown is a general-purpose unit in terms of speed; faster operation can be achieved with an OP-16 or PM356 type.

All of the just described circuits are capable of 8 bits of control resolution which yield a dynamic range of 48 dB (Table 6-5). Binary scaling in general has a relatively limited dynamic range however, as opposed to exponential scaling. With exponential scaling, equal decibel per step weighting can be achieved, allowing greater ranges of control for a given number of control bits, such as popular 8-bit microprocessor based data systems.

An example of an exponential control characteristic multiplying d/a converter is shown in Fig. 6-36, using the DAC-76 as the d/a converter. This circuit operates similarly to the others in terms of reference input, and is scaled for a maximum gain of unity.

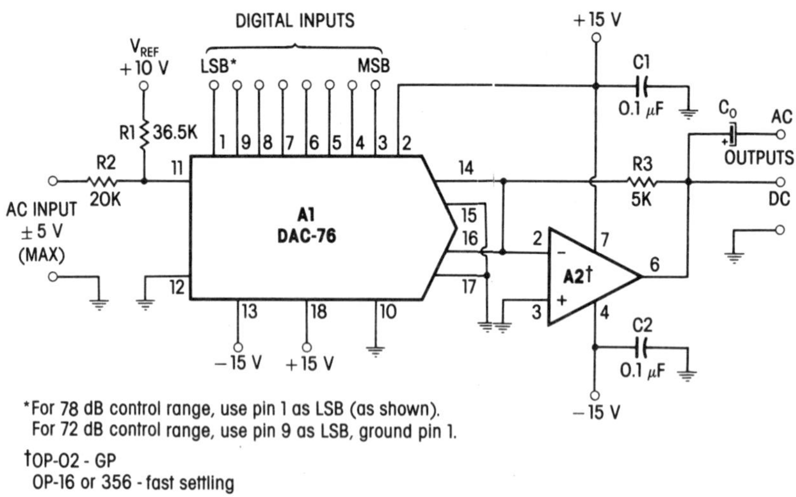

Fig. 6-36. Exponential control characteristic two quadrant multiplier.

The DAC-76 has the ability to operate with either a 78 dB or 72 dB control range, dependent upon the use of the encode/decode outputs. For 78 dB range, the 7 most significant control bits are used as normal, and an 8th bit (LSB) is added by using the E/D input, pin 1. With the encode and decode outputs tied together, this adds additional output steps to the control characteristic, increasing the dynamic range by 6 dB, to 78 dB total. The gain changes are nominally 0.15 dB per step. For a 72 dB dynamic range, the E/D input is grounded, and the gain change is 0.3 dB per step, nominally.

The DAC-76 is also a unipolar output current device, therefore this circuit has a dc output component (similar to Fig. 6-34). The dc component can be removed by using cancellation techniques, such as those that have just been described. Some highly optimized DAC-76 multiplying d/a circuits are described in further detail within Chapter 7.

A most straightforward form of two quadrant multiplying d/a converter is realized by the AD7520 series of converters. This is because these devices achieve their attenuation by a passive R/2R ladder and there is no inherent requirement for bias current in the reference channel. As a consequence, application is considerably simplified, and bipolar (ac) reference inputs are readily accommodated.

In Fig. 6-37 a basic AD7520 type two quadrant multiplier is shown, as suitable for use with AD7520 or AD7530 10-bit device. The ac signal input (reference) is applied to pin 15, the reference input, and may range up to ±10 volts. The internal ladder and switches scale the output to achieve multiplication, and the internal feedback resistor operates in conjunction with a current-to-voltage converter op amp, here shown as an AD301A, operated feedforward.

Maximum gain of the circuit is unity times the multiplication ratio, and calibration can be trimmed (if desired) as shown in Fig. 6-27. A table of convenient values is given for calculation of the required digital fraction input for various stage gains, for both single and two stage systems. Note that this is a general table applicable to binary scaled devices. This general circuit can use 8- through 12-bit resolution devices, as represented by the AD7520 (30) or AD7521 (31). It can also use the AD7522, with the added convenience of digital input registers.

The 301A amplifier type mentioned is capable of high performance with low cost; other amplifier types suitable are the AD518 and AD509.

A valuable asset of the AD7520 series is its very low reference channel feedthrough (off-state attenuation); it is specified as less than ½ LSB to 10 kHz. At lower frequencies it is appreciably better than this figure, approaching 90 dB below 2 kHz.

Four Quadrant Multiplying D/A Circuits

For applications which require that the digital input control both output sign and magnitude, four quadrant multiplying d/a circuits are

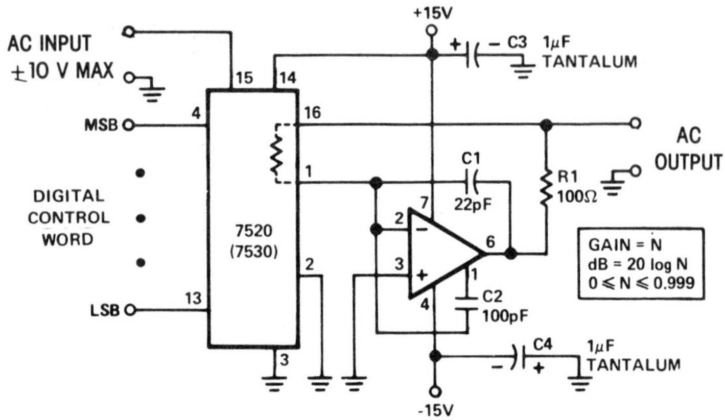

DIGITAL INPUTS FOR TYPICAL ATTENUATIONS

dB ATTEN.	N: INPUT NUMBER – VALUE 1 STAGE	2 STAGES
0	0.999	0.999
-1	0.891	0.944
-2	0.794	0.891
-3	0.708	0.841
-4	0.631	0.794
-5	0.562	0.750
•		
-10	0.316	0.562
-15	0.178	0.422
•		
•		
-30	0.032	0.178
-40	0.010	0.100
-50	0.003	0.056
-60	0.001	0.032
-70	–	0.018
-80	–	0.010
-90		0.006

Courtesy Analog Devices, Inc.

Fig. 6-37. AD7520 series two quadrant multiplier.

used. A DAC-08 four quadrant multiplying d/a circuit is shown in Fig. 6-38. Here devices A1 and A2 are differentially crossconnected and feed an output differential to a single-ended current converter, consisting of op amp A3 and two matched resistors.

In this circuit the high impedance reference input connection to both d/a converters is used, and input ranges are ±5 volts (10-volt differential). Overall gain at either positive or negative full scale limit is nominally unity, with maximum output ±4.98 volts. The table of operating values illustrates the coding arrangement used, which is symmetric about zero. As shown, the circuit uses a medium speed op amp; higher speed can be achieved with an OP-16 or 356 type device for A3.

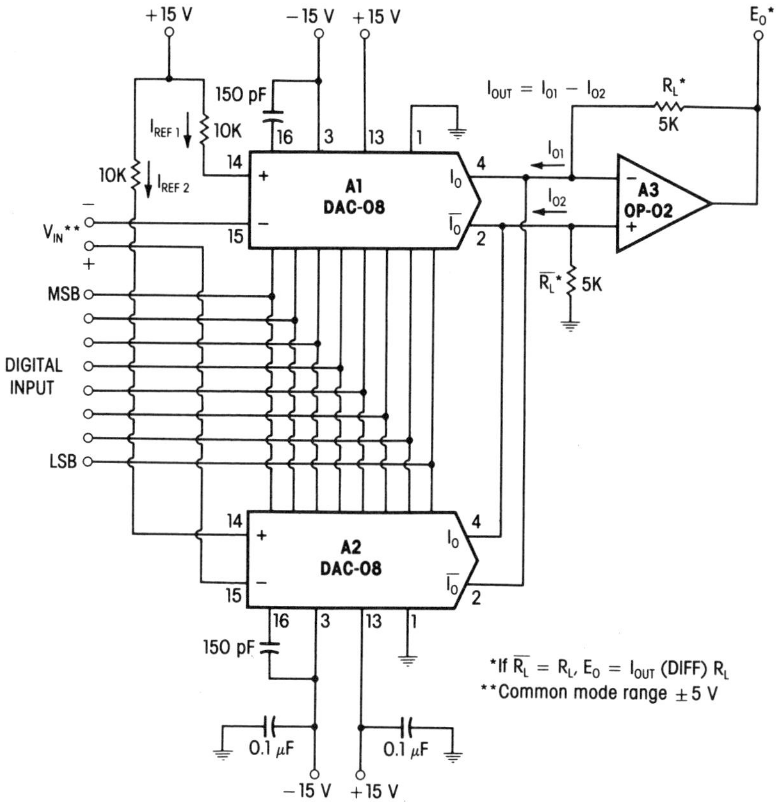

OPERATING VALUES

DIGITAL INPUT	$V_{IN}(+)$**	$V_{IN}(-)$**	V_{IN} DIFF.	$I_{REF\,1}$ (mA)	$I_{REF\,2}$ (mA)	I_{O1} (mA)	$\overline{I_{O2}}$ (mA)	I_{O1} (mA)	I_{O2} (mA)	$\overline{I_{O1}}$ (mA)	I_{O2} (mA)	I_{OUT} DIFF.	E_O* (V)
1 1 1 1 1 1 1 1	+5 V	−5 V	+10 V	2.000	1.000	1.992	0	1.992	0.996	0	0.996	0.996 mA	4.98
1 0 0 0 0 0 0 0	+5 V	−5 V	+10 V	2.000	1.000	1.000	0.496	1.496	0.500	0.992	1.492	0.004 mA	0.02
0 1 1 1 1 1 1 1	+5 V	−5 V	+10 V	2.000	1.000	0.992	0.500	1.492	0.496	1.000	1.496	−0.004 mA	−0.02
0 0 0 0 0 0 0 0	+5 V	−5 V	+10 V	2.000	1.000	0	0.996	0.996	0	1.992	1.992	−0.996 mA	−4.98
1 1 1 1 1 1 1 1	0 V	0 V	0 V	1.500	1.500	1.494	0	1.494	1.494	0	1.494	0.000 mA	0.000
1 0 0 0 0 0 0 0	−5 V	−5 V	0 V	2.000	2.000	1.000	0.992	1.992	1.000	0.992	1.992	0.000 mA	0.000
0 1 1 1 1 1 1 1	+5 V	+5 V	0 V	1.000	1.000	0.496	0.500	0.996	0.496	0.500	0.996	0.000 mA	0.000
0 0 0 0 0 0 0 0	0 V	0 V	0 V	1.500	1.500	0	1.494	1.494	0	1.494	1.494	0.000 mA	0.000
1 1 1 1 1 1 1 1	−5 V	+5 V	−10 V	1.000	2.000	0.996	0	0.996	1.992	0	1.992	−0.996 mA	−4.98
1 0 0 0 0 0 0 0	−5 V	+5 V	−10 V	1.000	2.000	0.500	0.992	1.492	1.000	0.496	1.496	−0.004 mA	−0.02
0 1 1 1 1 1 1 1	−5 V	+5 V	−10 V	1.000	2.000	0.496	1.000	1.496	0.992	0.500	1.492	0.004 mA	0.02
0 0 0 0 0 0 0 0	−5 V	+5 V	−10 V	1.000	2.000	0	1.992	1.992	0	0.996	0.996	0.996 mA	4.98

Fig. 6-38. DAC-08 four quadrant multiplying d/a converter.

The AD7520 series as used in a four quadrant multiplier is shown in Fig. 6-39. As may be noted, this circuit is very similar to the bipolar scaled AD7520 d/a application previously described, which in itself is inherently four quadrant, if the reference is variable in sign.

Here ac reference inputs of up to ±10 volts are applied to the reference input, and the output is controlled in sign and magnitude by the

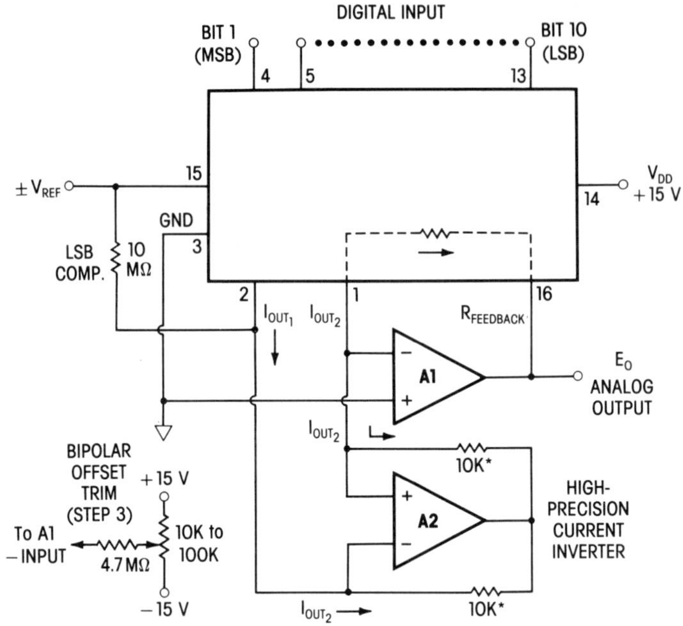

*Absolute accuracy unnecessary. Match and tracking are essential.

DIGITAL INPUT*	V_{REF}	OUTPUT*
1111111111	+10 V	+9.980 V
1000000000	+10 V	0.000 V
0000000000	+10 V	−10.000 V
XXXXXXXXXX	0	0.000 V
1111111111	−10 V	−9.980 V
1000000000	−10 V	0.000 V
0000000000	−10 V	+10.000 V

*10 bit operation illustrated

BIPOLAR OFFSET TRIM:

1. With V_{REF} = +10 V and all 1's IN, adjust A2 offset for 0.000 V between A2 inputs.
2. With MSB = 1, other bits = 0, adjust A1 offset for 0.000 V between A1 inputs.
3. After (2), if D/A output is not zero use trim network to adjust for 0.000 V out.

GAIN TRIM:

Using a method from Fig. 6-27B, adjust GAIN as outlined in Fig. 6-23B.

Note - Repeat if necessary.

Courtesy Analog Devices, Inc.

Fig. 6-39. 7520 series, 10-12 bit, four quadrant multiplying d/a converter.

digital input. Operation follows the code table, which as can be noted, is very similar to offset binary with the exception that the reference now also controls output polarity. Any of the AD7520 series devices can be used, dependent upon the resolution requirements. A 10-bit example is illustrated here.

Output amplifiers should be chosen as described above for the two quadrant AD7520 circuit. For general purpose uses, dual devices may be found useful, making the circuit more simple.

INTERNAL REFERENCE D/A CONVERTERS

The DAC-100

Fig. 6-40 shows the DAC-100 series of 10-bit devices operated in complementary binary coded converters, in both unipolar and bipolar modes.

Although the DAC-100 is a current output, internal reference d/a converter, it must be used with an external current-to-voltage converter op amp. Resistors are included on-chip for the scaling of both unipolar and bipolar modes of operation. The choice between 5- or 10-volt operating ranges is made by a part number suffix.

Fig. 6-40A shows how the DAC-100 is connected in general with regard to power supplies etc; and how unipolar operation is achieved. The on-chip feedback resistor, R_S is either 2.4K (5 volt) or 4.8K (10 volt) and with a nominal 2 mA DAC-100 output current yields a 0 to +5 volts, or 0 to +10 volt range. If a bipolar input stage output amplifier is used, such as the OP-01, it is recommended that R1 be used with a value equal to R_S. For FET input op amps, pin 3 of the amplifier may be grounded directly.

Calibration of the DAC-100 is achieved for full scale (gain) by adjustment of the 200-ohm potentiometer. Offset is adjusted either via the op-amp offset trim, or by the more general method of Fig. 6-23A. For unipolar scaling, the calibration levels of Table 6-4 can be used as appropriate for 10-bit operation. Note that with the DAC-100 complementary coding, full scale is achieved for all 0s in; zero scale is achieved with an all 1s input.

Logic inputs of the DAC-100 are both TTL and CMOS compatible, but logic swings must be confined to +6 volts or less. Thus either TTL or CMOS 5-volt sources will be directly compatible. For less than 10-bit resolution uses, the unused logic inputs should be tied to the +5-volt logic supply voltage.

Bipolar operation of the DAC-100 is achieved as shown in Fig. 6-40B. An external +6.4-volt voltage reference source is used, which sums a ½ full scale (1 mA) current into the sum point by means of bipolar offset resistor R_B. Bipolar ranges available are ±2.5 volts and ±5 volts, with R_S values of 2.4K and 4.8K respectively. Bipolar calibration levels are found in Table 6-4, with the stipulation of complementary coding.

The DAC-100 is a fast settling d/a converter, with a specified settling time of 375 ns (maximum), to ½ LSB accuracy for 10-bit operation. However, since it must be used with an op amp, the overall settling time will by necessity be greater, and typical minimum settling times in practice will be 0.5 to 1 μs, dependent upon the op amp selected. See Table 6-3 for pertinent amplifier choices.

The AD561

Fig. 6-41 shows the AD 561 10-bit internal reference d/a converter in a variety of binary scaled d/a circuits. Shown here are unipolar and bipolar circuits, in both buffered and unbuffered forms.

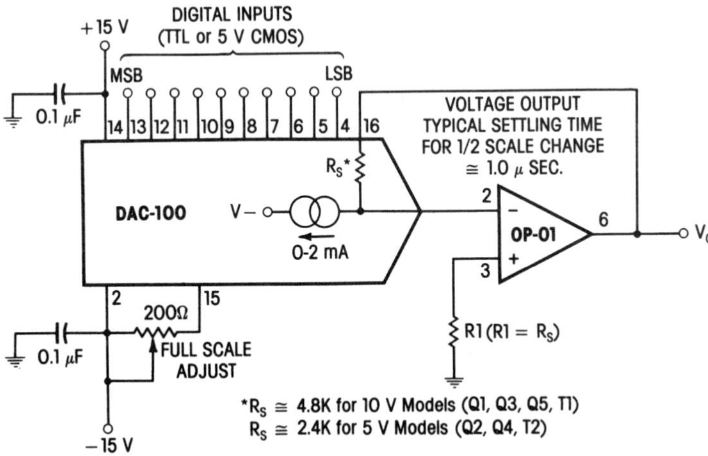

(A) Unipolar, 0 to 5 volts or 0 to 10 volts.

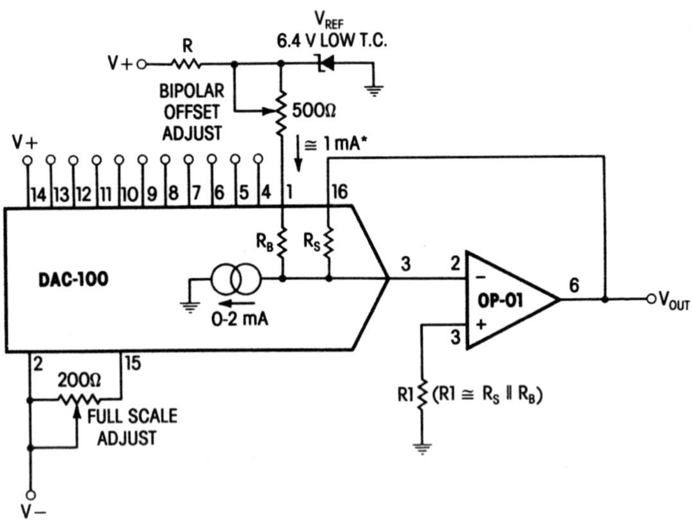

(B) Bipolar, ± 2.5 volts or ± 5 volts.

Fig. 6-40. DAC-100 series 10-bit complementary

In Fig. 6-41A are shown buffered voltage output circuits, using an external op amp in the current-to-voltage conversion mode. In Fig. 6-41A, Part 1, operation is unipolar, with a range of 0 to +10 volts. OFFSET adjustment is accomplished by trim of the op amp, and GAIN by trim of R2 (see Table 6-4). Settling time of this d/a converter plus op-amp combination to ½ LSB is 600 ns, typically. The 10-pF feedback capacitor should be optimized for the particular layout, and the AD509 is compensated for unity gain (see AD509 data sheet).

Digital inputs of the AD561 are programmable for threshold by adjustment of the positive power supply; as shown here they are TTL or 5-volt CMOS compatible by virtue of a +5-volt supply. Operation is straight binary in this circuit. Power supply bypassing is not shown, but should take the form of 0.1-μF capacitors on each supply line.

The bipolar circuit of Fig. 6-41A, Part 2, uses the on-chip 2.5K offset resistor to sum a ½ scale current into the summing point, and thus yield bipolar operation. R3 trims the level of offset current and is used as the OFFSET calibration for −5 volts out with all 0s in. R2 trims the overall gain at ½ scale for an output of 0 volts with the MSB (only) on. Coding is offset binary for this circuit and output voltage range is ±5 volts. The AD509 amplifier allows a settling time of 1.1 μs to be achieved. In both of these circuits a bias compensation resistor is used for minimum offset error.

Because of the output voltage compliance range of −2 to +10 volts and the on-chip scaling resistors, the AD561 can also be used to implement complete converters with unbuffered voltage outputs. This can

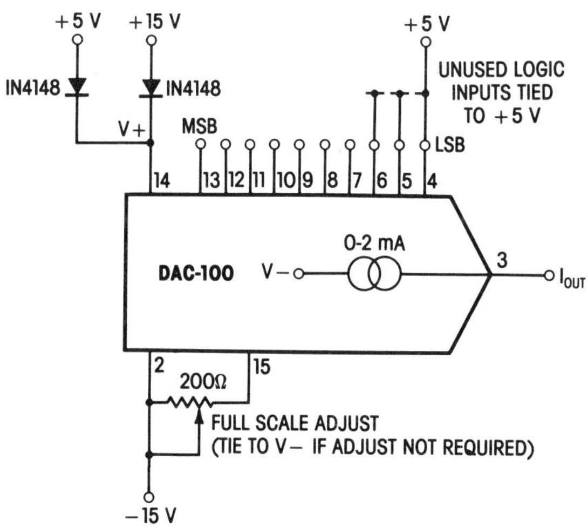

Courtesy Precision Monolithics, Inc.

(C) Power supply sequence protection circuit.

binary, unipolar/bipolar d/a circuits.

be advantageous, both from the standpoint of system simplicity and operating speed, as eliminating the output amplifier almost always allows the lowest settling times to be realized. Two examples of this type of use are shown in Fig. 6-41B.

In Fig. 6-41B, Part 1, a complementary binary unipolar circuit is shown which uses the high voltage compliance to advantage, and also makes double use of the internal reference. The AD741 op amp amplifies the +2.5-volt reference to a +9.99-volt level, which then serves as the referal level for the internal 5K load resistor. With all 0s in, the output will be at +9.99 volts; with all 1s it will drop to 0 volts. The 50-ohm OFFSET pot adjusts the zero scale level with all 1s in, while the 200-ohm pot adjusts the full scale level (GAIN) with all 0s in. Although the output impedance is high as shown (5K) it can be lowered if de-

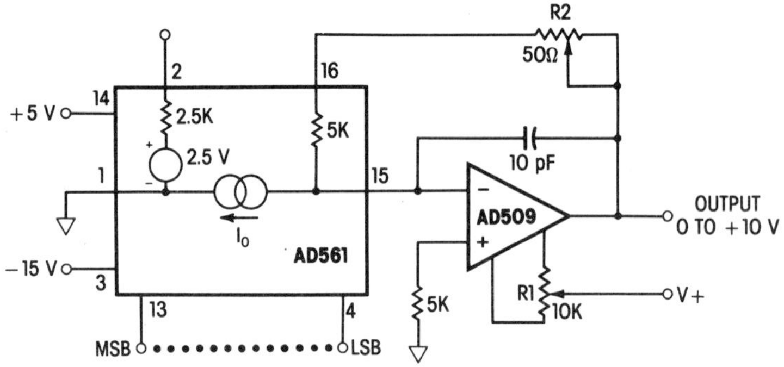

1. UNIPOLAR, 0 TO +10 V

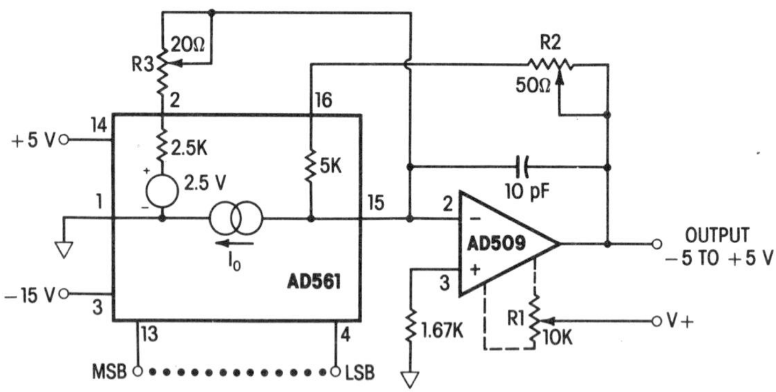

2. BIPOLAR, ±5 V

(A) Buffered voltage outputs.

Fig. 6-41. AD561 10-bit binary

sired, by use of a buffer as shown in Fig. 6-22. Settling time of this circuit will be on the order of 2 μs, assuming a load capacitance $\leqslant$25 pF.

A higher speed configuration is the complementary binary scaled bipolar circuit of Fig. 6-41B, Part 2. This circuit uses the 2.5K internal resistor to provide offset, and the 5K plus an external 2.5K (R_x) to

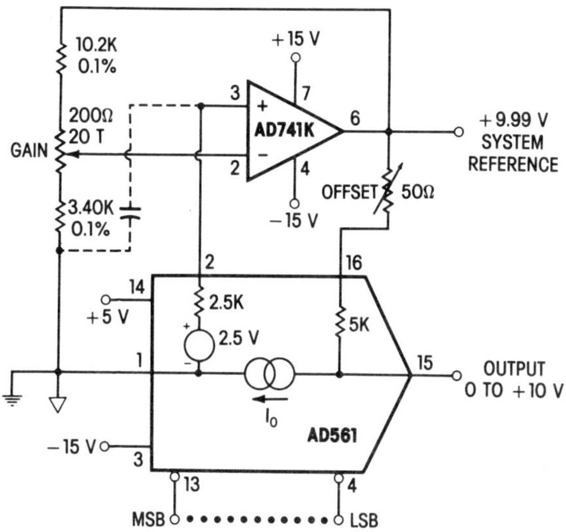

1. COMPLEMENTARY UNIPOLAR, 0 TO +10 V

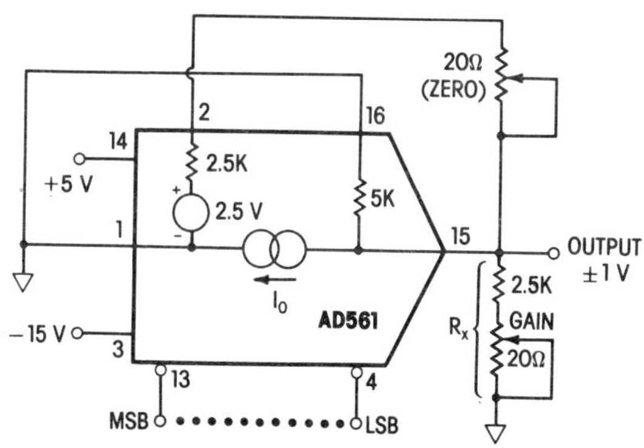

2. COMPLEMENTARY BIPOLAR, ±1 V

DIGITAL INPUT	OUTPUT
0000000000	+1 V
1000000000	0 V
1111111111	−0.998 V

Courtesy Analog Devices, Inc.

(B) Unbuffered voltage outputs.

unipolar/bipolar d/a circuits.

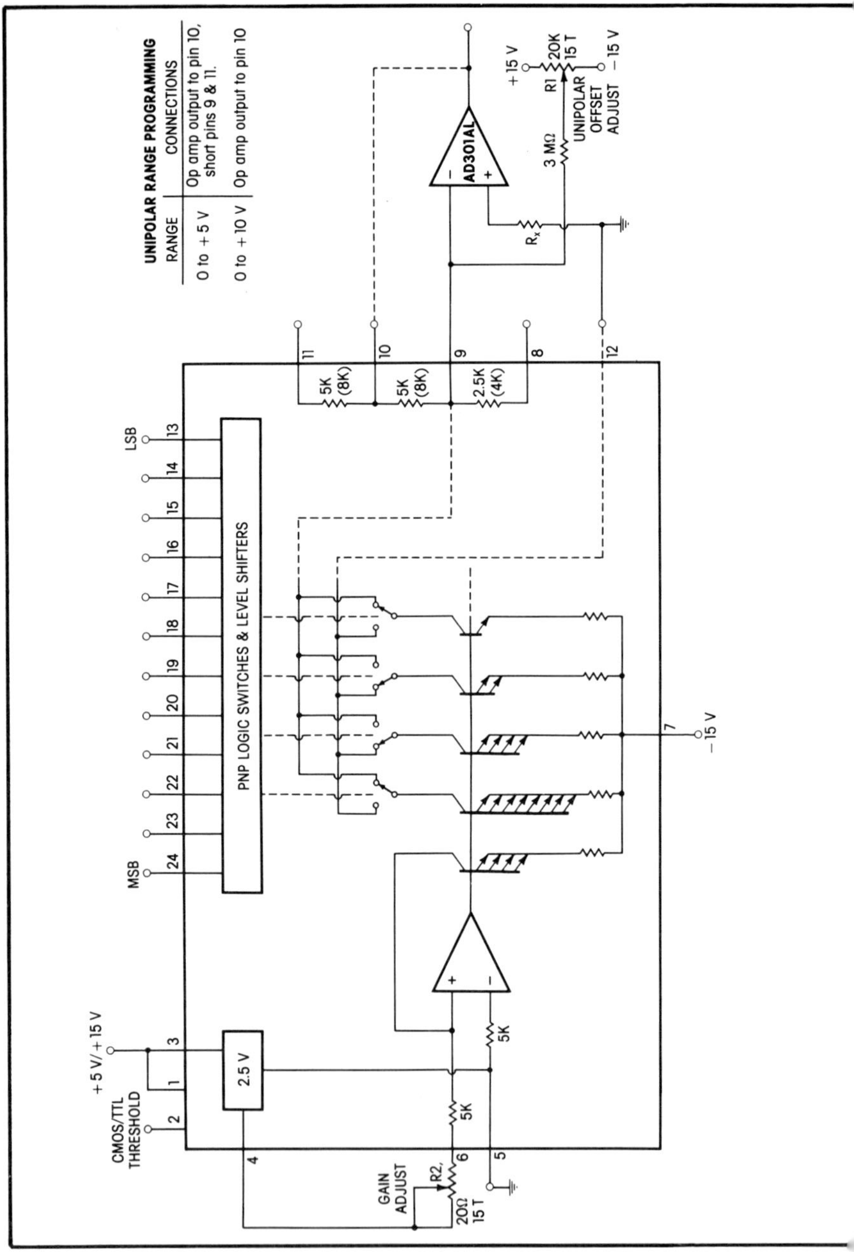

(A) Unipolar 0 to 5 volts, 0 to 10 volts.

Fig. 6-42. AD563 12-bit binary/bcd,

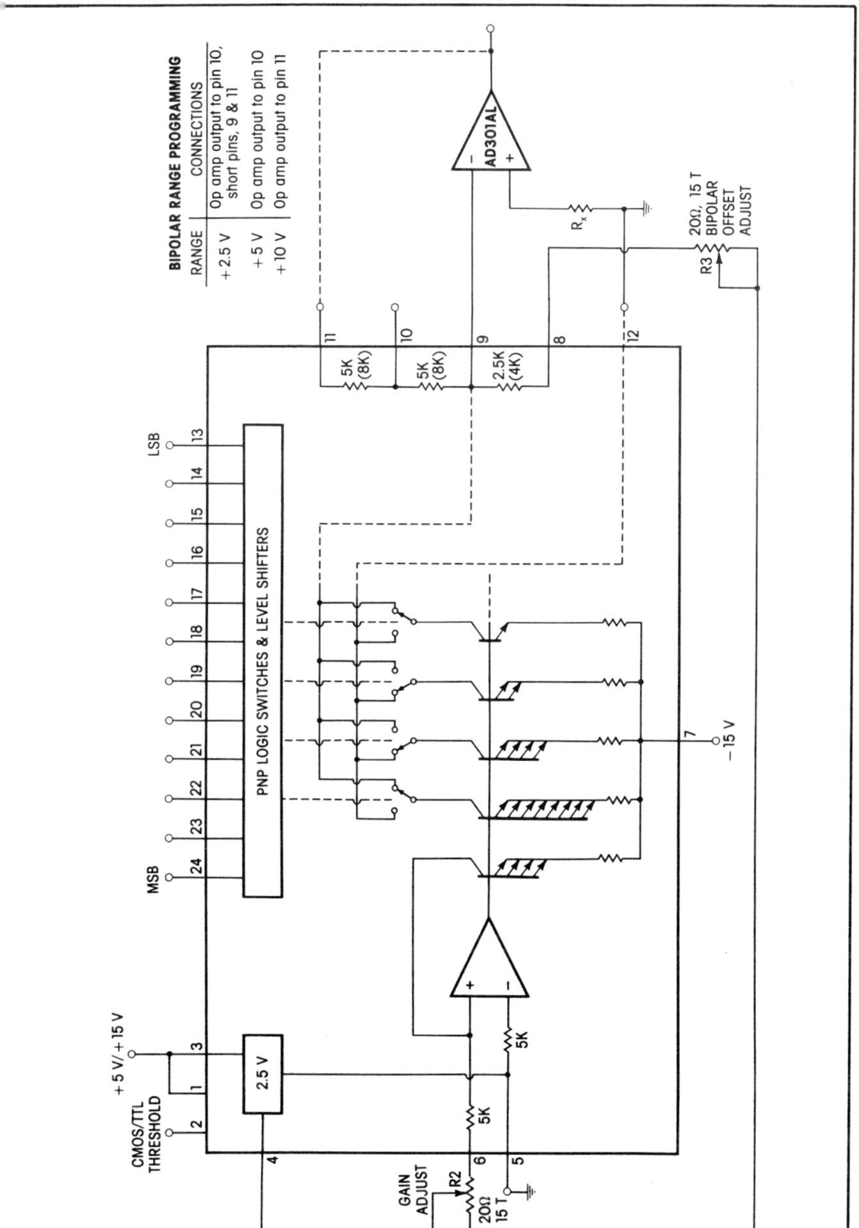

Courtesy Analog Devices, Inc.

(B) Bipolar ± 2.5 volts, ± 5 volts, ± 10 volts.

unipolar/bipolar d/a circuits.

scale operation for ±1 volt. The OFFSET 20-ohm pot is adjusted for 1 volt (all 0s), and the GAIN 20-ohm pot trims R_x for 0 volt out with the MSB (only) on.

The equivalent impedance of the AD561 load for this circuit is 1K, and with very low stray capacitance this will result in a settling time of less than 400 ns, for $C_L \leqslant 25$ pF. This makes the circuit suitable as an integral part of fast a/d converters, video display controls, and other such speed critical uses.

The AD563

Fig. 6-42 illustrates the use of the AD563 series in d/a circuits for either 12-bit binary or 3 digit bcd coding, for either unipolar or bipolar operation. For this device series, like the AD562, various scaling arrangements are chosen by jumper options, while the choice of binary or bcd operation using the AD563 series is made by part number suffix.

Fig. 6-42A shows the AD563 connected for unipolar operation. The output ranges are 0 to +5 volts or 0 to +10 volts, and are programmed as indicated in the table. Logic input threshold is programmed as described in Fig. 6-17. Supplies are +5 to +15 volts, and −15 volts, bypassed with 0.1-μF capacitors (not shown). Zero scale is adjusted by the OFFSET control as shown, or alternately by the method appropriate to the amplifier in use. The GAIN pot adjusts full scale range, as appropriate (see Table 6-4).

The internal reference is powered by a V+ connection to pin 3, and the +2.5-volt output supplies a reference current into pin 6 of the d/a converter.

Choice of output amplifier determines the overall settling time, and use of the AD509 will result in a settling time of 1.2 μs, to ½ LSB of 12 bits. This will be somewhat faster for bcd operation, due to the greater error band.

The resistor values in parentheses are for bcd models; standard binary values are as noted.

The AD563 is operated for bipolar ranges as shown in Fig. 6-42B. The ranges of ±2.5, ±5, and ±10 volts are programmed as per the table, and the trimmers adjust gain and offset as indicated.

INTEGRAL D/A CONVERTERS

Integral d/a converters, as their name implies, are complete and ready for operation without the necessity of additional design or application work. Popular devices of this category are the DAC-80 and DAC-85 series which are available with a variety of coding/scaling arrangements. Since these devices are applied on a relatively straightforward basis, application circuitry for them will not be separately detailed here. Many of their application considerations such as calibration levels, housekeeping details, etc. have been covered already.

REFERENCES

1. Brokaw, A. P. "Analog Signal-Handling for High Speed and Accuracy." *Analog Dialogue,* Vol. 11, No. 2, 1977.
2. ———. "An IC Amplifier User's Guide to Decoupling & Grounding." *Electronic Products,* December 1977.
3. Erdi, G. *The OP-17, OP-16, OP-15 As Output Amplifiers For High Speed D/A Converters.* Precision Monolithics Application Note AN-24, Precision Monolithics, Inc., Santa Clara, California.
4. Gross, A. E. "Join Two Unipolar D/A Converters and Get Bipolar Operation." *Electronic Design,* February 1, 1973.
5. *Interface Data Book.* National Semiconductor Corp., Santa Clara, California.
6. Jung, W. G. "Application Considerations for IC Data Converters Useful in Audio Signal Processing." *Journal of the Audio Engineering Society,* December 1977.
7. ———. "Audio Application Ideas for CMOS DACs." *Analog Dialogue,* Vol. 10, No. 1, 1976.
8. ———. "Convert Unipolar CMOS Signals Into Analog Bipolar Outputs." *Electronic Design,* October 25, 1977.
9. ———. "Digital/Analog Level Control." *db,* September, 1976.
10. National Semiconductor Application Note AN-108. *Transmission Line Characteristics.* National Semiconductor Corp., Santa Clara, California.
11. Percival, R. *Data Bus and Differential Line Drivers and Receivers.* National Semiconductor Application Note AN-83. National Semiconductor Corp., Santa Clara, California.
12. Schoeff, J. A. "Differential and Four Quadrant Applications of the DAC-08." *EE Times Seminar,* February, 1976.
13. Schoeff, J. A.; Soderquist, D. *DAC-08 Application Collection.* Precision Monolithics Application Note AN-17, 1975. Precision Monolithics, Inc., Santa Clara, California.
14. ———. *Differential and Multiplying Digital-to-Analog Converter Applications."* Precision Monolithics Application Note AN-19, 1976, Precision Monolithics, Inc., Santa Clara, California.
15. ———. "Differential and Multiplying Use of Digital-to-Analog Converters." *EE Times,* June 21, 1976.
16. Sheingold, D. H. *Analog-Digital Conversion Notes.* Analog Devices, Inc., Norwood, Massachusetts, 1977.
17. *The TTL Data Book, 2nd Edition.* Texas Instruments, Inc., Dallas, Texas.

18. Manufacturers' Data Sheets:

Analog Devices {AD562, AD563; AD7520, AD7521; AD7530, AD7531; AD7522, AD7523, AD7541}

Harris Semiconductor {HI-562

Motorola Semiconductor Products {MC1406; MC1408; MC3410}

Precision Monolithics {DAC-08; DAC-20; DAC-76; DAC-100}

7

Basic A/D Converter Applications

A/D ANALOG INTERFACES

The type of interfacing which may be necessary at an a/d analog input is typically scaling and/or sign inversion. Usually, these measures can be quite satisfactorily accommodated by a simple op-amp circuit. In brief form, basic suitable op-amp circuits are shown in Fig. 7-1.

In Fig. 7-1A is shown perhaps the most useful a/d analog interface circuit, a unity gain voltage follower buffer. This circuit is capable of the highest precision of the op-amp hookups shown, and requires no precision parts (other than the op amp itself).

Although many op-amp types can be suitable for this function, dependent upon the required accuracy, FET input units are often most attractive, because of their minimal bias currents. The 356 or OP-16 series is a good all around amplifier choice, as it has low drift and fast settling characteristics, and thus will usually minimally degrade the following a/d specifications.

User options for this stage are an offset null (which will not be required if the a/d converter has such an adjustment), and input protection, such as R1 and D1-D2. These latter components are recommended if the analog input line can see "outside world" or other potentially dangerous voltages, which could destroy the buffer (and perhaps the a/d converter following!). General-purpose diodes are indicated, which may be excessively leaky for some applications. Low leakage protection diode types suitable are the Siliconix PAD 1-PAD100 series, and also a great

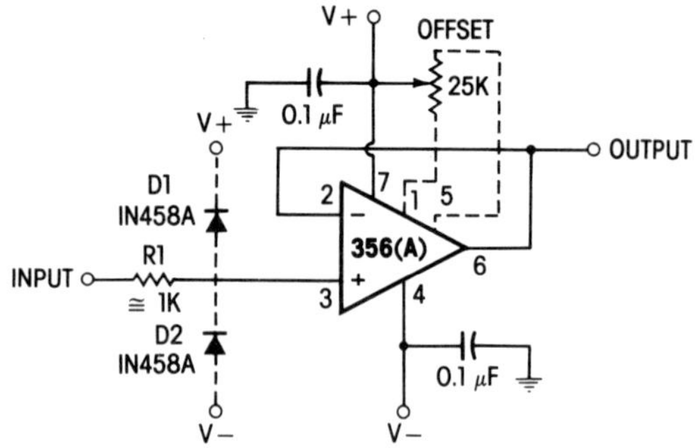

(A) Unity gain follower.

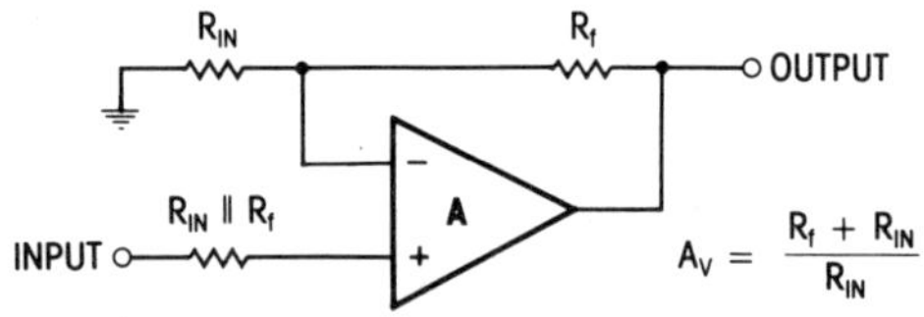

(B) Follower with gain.

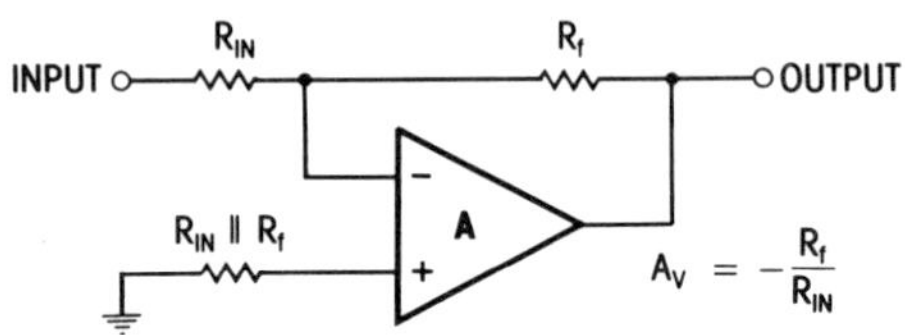

(C) Inverting amplifier.

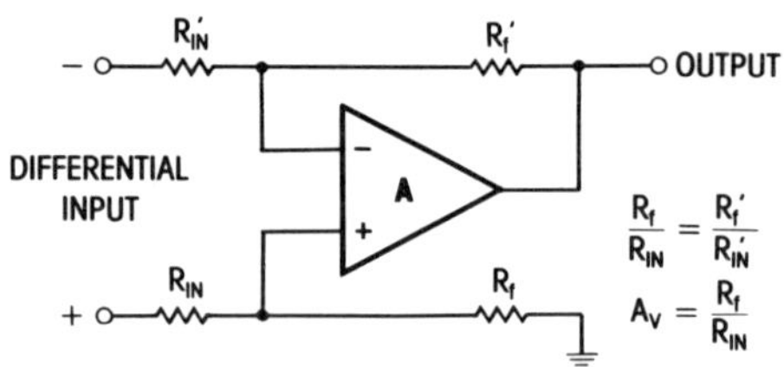

(D) Differential amplifier.

Fig. 7-1. A/d analog buffer, scaling, inversion circuits.

many JFET devices, simply diode connected. Although this protection circuit is only shown in this case as an example, it can be equally applicable to the remaining circuits (and others), for cases where dangerous voltages can exist. Protection against fault conditions should always be considered for good system design.

In general, the choice of an op amp to be used with these circuits should best be based on the required specifications. Fast settling types for instance can be chosen from Table 6-3. Bipolar input stage types should use a bias compensation resistor at the (−) input, for best dc accuracy. Also, offset nulling can be applied (as appropriate for the particular device) if necessary.

The circuits in Fig. 7-1 are generally self-explanatory. Gain is set up by the input and feedback resistors which should be precision types, preferably with low tracking TCs.

A/D LOGIC INTERFACES

Like d/a inputs, a/d logic outputs often require interfacing for optimum marriage into a specific system. The interfacing can be logic buffering and/or inversion, code conversion, data storage, bus or line driving, and change of data format. Interfacing techniques for these problem areas are discussed later, and are sufficiently general to be applied to virtually any a/d converter.

Code Conversion

Most a/d converters are supplied with a good deal of flexibility regarding coding, and may be configured for various codes by simple jumper options. The two most universal codes are offset binary and twos complement, which are very close to one another in terms of hardware conversion. Since the difference between these two codes is only in the sign of the MSB, a simple inverter as illustrated in Fig. 7-2A is all that is necessary to translate one into the other. Very often, the MSB is supplied in both true and complemented form from the a/d converter itself. Thus offset binary a/d converters are easily converted to twos complement. and vice versa.

Bipolar to Sign Magnitude

With appropriate analog plus digital interfacing, an a/d converter that is basically unipolar in its operation can be made to respond to bipolar inputs, with sign-magnitude output coding. Such a scheme is illustrated in Fig. 7-2B.

This circuit (which is an extension of the polarity reversing amplifier described previously with Fig. 5-21) responds to bipolar analog inputs, and automatically transforms them into unipolar dc outputs which can be converted by the a/d converter. A sign bit is derived by the A3 comparator that operates as a zero crossing detector. With this system, the a/d converter responds to magnitude information only, while the sign bit indicates input polarity. The output data has the resolution of the a/d converter used, plus an additional bit for the sign.

This circuit is particularly useful with integrating a/d converters, many of which respond only to unipolar dc inputs. The output can accommodate either unipolar positive or negative a/d converters by rewiring A3 as appropriate (see table).

It should be noted that the basic function of this circuit is that of an absolute value amplifier. As such, there are many op-amp based configurations which can satisfy the general requirement.

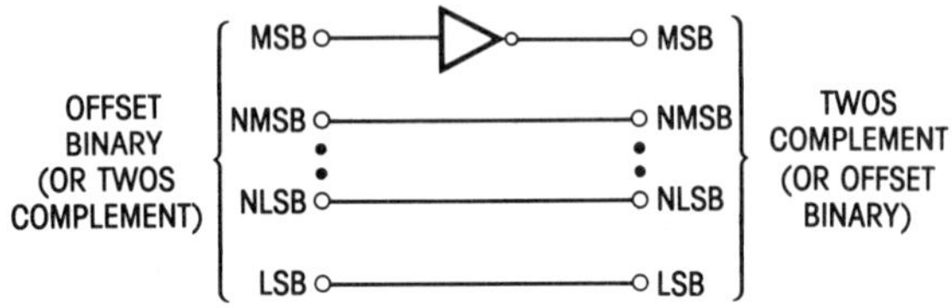

(A) Offset binary to twos complement.

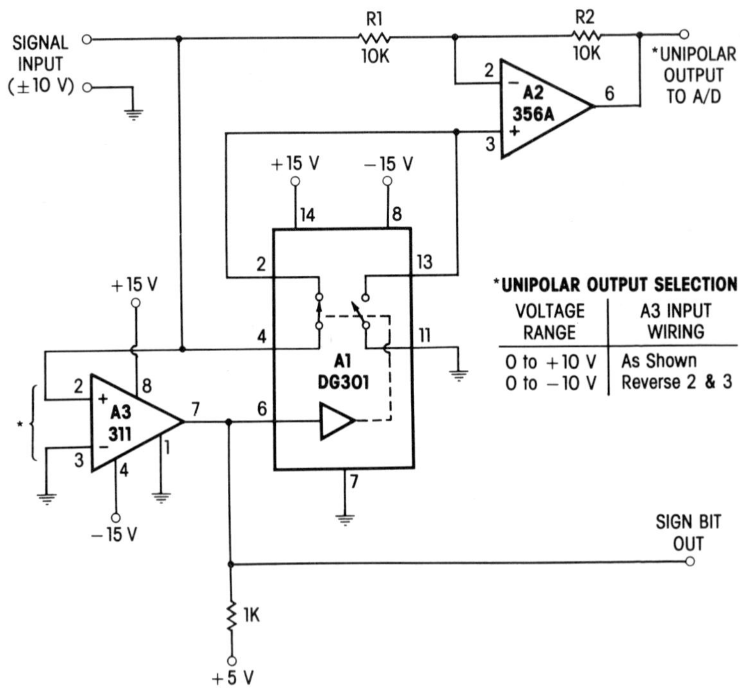

***UNIPOLAR OUTPUT SELECTION**

VOLTAGE RANGE	A3 INPUT WIRING
0 to +10 V	As Shown
0 to −10 V	Reverse 2 & 3

(B) Bipolar to sign magnitude.

Fig. 7-2. Code

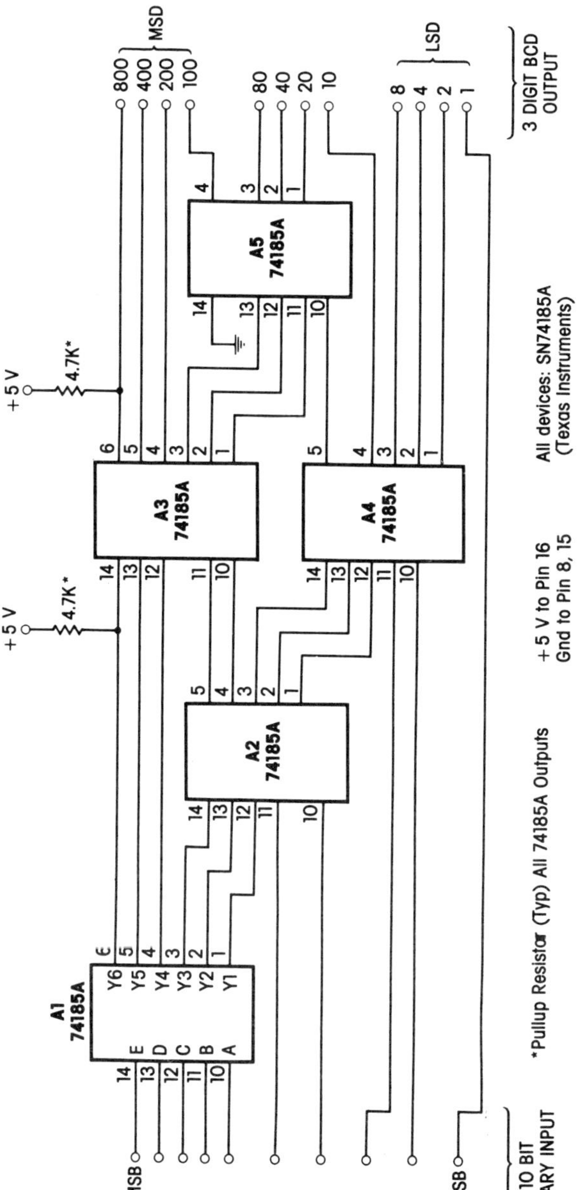

Courtesy Analog Dialogue

(C) Binary to bcd.

conversion techniques.

Binary to BCD

An interesting scheme to translate binary coded a/d outputs into bcd coded data is illustrated in Fig. 7-2C.* This technique, originally described in the reference cited, allows a 10-bit binary coded a/d converter to deliver 3 digit bcd data, with appropriate and sufficient accuracy.

The circuit makes use of the fact that a 10-bit binary converter has 1024 possible codes, while a 3 digit bcd converter requires but 1000. Thus, 3 digit bcd-weighted 12-bit converters actually waste 3096 of their possible 4096 (2^{12}) codes. Therefore, a 10-bit binary converter can conservativly supply the resolution and accuracy for 3 digit bcd operation, given the appropriate logic translation.

Here the conversion is performed by 74185A binary-to-bcd code converters, five devices being required for 10-bit, 3 digit operation. In use the a/d converter must be calibrated for a 3 digit bcd range; i.e., 1 LSB = 10 millivolts and full scale = 9.99 volts (assuming a 10-volt input range).

Although originally illustrated for the Analog Devices ADC-10Z modular a/d converters, the scheme is generally amenable to 10-bit IC converters such as the AD7570, and the AD561 configured within a successive approximation a/d converter (see SA a/d applications section, or AD561 data sheet). Note that the 74185A devices have open collector outputs, so pull-up resistors will typically be required for best noise immunity.

The major advantage of this conversion technique is its speed. The delay through the 74185As will in most cases be less than the actual a/d conversion time, therefore the bcd output will be valid essentially when the (binary) a/d conversion is complete. For other binary/bcd translations, see the 74185A data sheet.

Output Buffers

Quite often a/d outputs may require logic buffering and/or inversion. All of the logic devices previously described with d/a input interfacing are applicable to straightforward logic buffers, with of course suitable consideration as to logic families. With a/d converters however, an additional consideration is compatibility with a data bus. This factor makes 3-state output buffers a desirable and almost standard feature with an a/d converter.

The generalized concept of an a/d output buffer is shown in Fig. 7-3. The a/d converter shown is to be interfaced with the output data requirement. The data buffer can be simply a buffer, one the of 3-state output

---

* J. Fishbeck. "10 Bit Binary to 3 Digit BCD Conversion Using the ADC-10Z As a 3 Digit BCD Converter." *Analog Dialogue,* Vol. 6, No. 2.

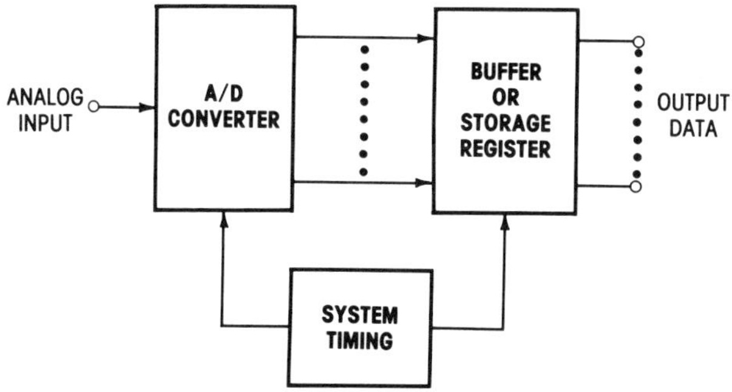

Fig. 7-3. The a/d converter with output buffer/register.

buffers or inverters (one per bit), or it may be a clocked 3-state output register. Both forms of 3-state buffers are illustrated in Fig. 7-4.

The most common and direct form of a 3-state output buffer is a simple inverting or noninverting stage, with a 3-state output enable control. This type of buffer presents the a/d data to the data bus whenever enabled, and goes to a high impedance state when disabled.

Suitable TTL devices for such use are the 8095, 74365A, and 8096, 74366A which are noninverting and inverting hex buffers respectively,

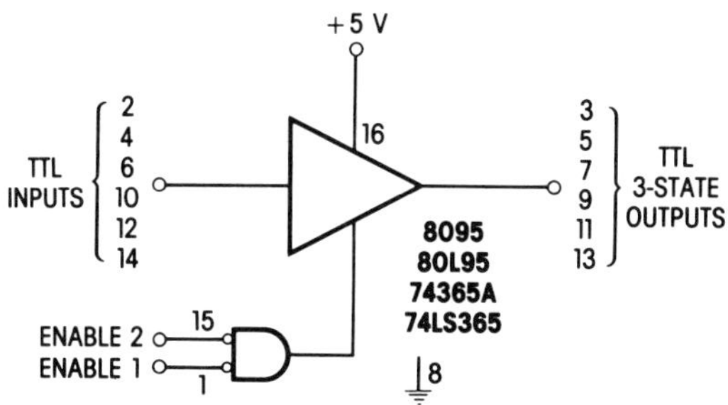

(A) Noninverting.

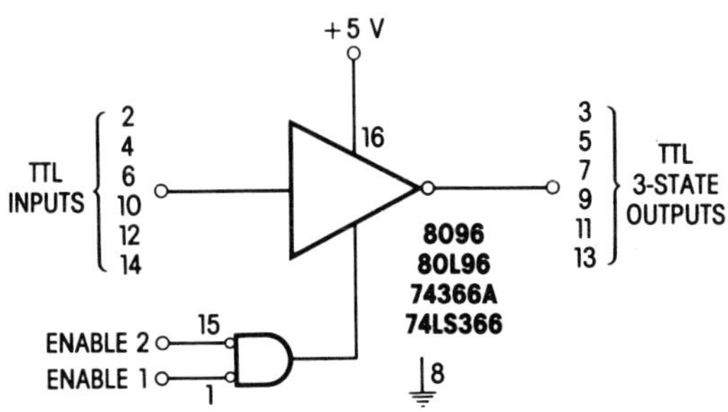

(B) Inverting.

Fig. 7-4. 3-state output buffers.

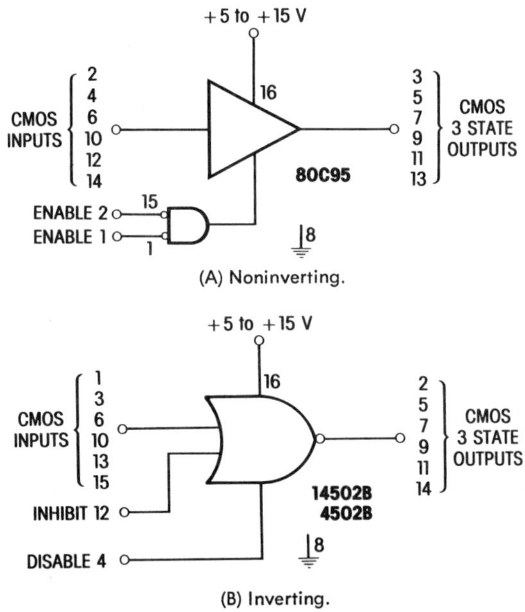

Fig. 7-5. CMOS 3-state output buffers.

shown in Fig. 7-4. The six outputs per device have common 2-input enable controls. Quad devices are also available in the 8000 3-state logic series.

For CMOS applications, a CMOS version of the 8095 noninverting unit is available, the 80C95, shown in Fig. 7-5A. For inverting uses, there is the 14502B, shown in Fig. 7-5B. This is also a hex unit, with a common output disable, and a common input gate or inhibit line. Because of its high drive capability, the 14502B is also useful as a general purpose 5-volt CMOS to TTL buffer.

A generally more useful interface to a 3-state data bus is a clocked D flip-flop register, such as is shown in Fig. 7-6. Here two 8551 or 74173 devices are connected as an 8-bit, clocked, 3-state output register.

In this application the data from the a/d converter is clocked into the register on the positive clock edge. The loaded data will then appear on the data bus when the 3-state outputs are enabled by a low input. Register data is cleared by a positive pulse. Either TTL or CMOS versions of these devices are available, and this register can easily be extended to more bits, if desired.

Serial to Parallel Conversion

It may be desirable in some instances to change the data format, such as between serial and parallel operation. Fig. 7-7 shows an example of

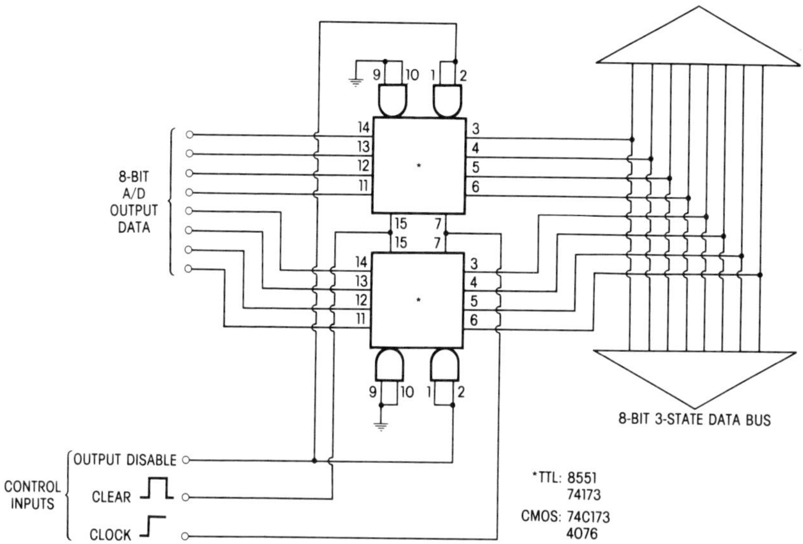

Fig. 7-6. Clocked 3-state output register.

how this can be accomplished using the AD572 a 12-bit a/d converter.

The AD572 has a serial output line; here it feeds the input of a 12-bit shift register composed of 74LS395 devices. These units have 3-state outputs, therefore the loaded data can be fed directly to a 12-bit TTL data bus. The serial output data from the AD572 is clocked into the register on the negative clock edges. The STATUS output of the AD572 is logically ANDed with a data transfer signal. When the conversion is complete and the data transfer line goes high the loaded data then will appear on the data bus.

Parallel to Serial Conversion

Many a/d converters, particularly the successive approximation types, generate a serial data output format during conversion. However, this is not universally true, and there are many instances where a serial format is advantageous; for instance, if the a/d converter is remotely located. Therefore parallel to serial conversion can be useful with virtually any a/d converter.

Fig. 7-8 shows a relatively straightforward means of converting parallel 8-bit data into a serial format, using a single 8-bit register, the 74165. With this device, the parallel data is loaded when the load input is taken low, and clocked out of the register on the positive clock transitions. Both true and complement data outputs are available.

In this particular application the clock inhibit (pin 15) and serial (pin 10) inputs are wired low. For applications greater than 8 bits, two

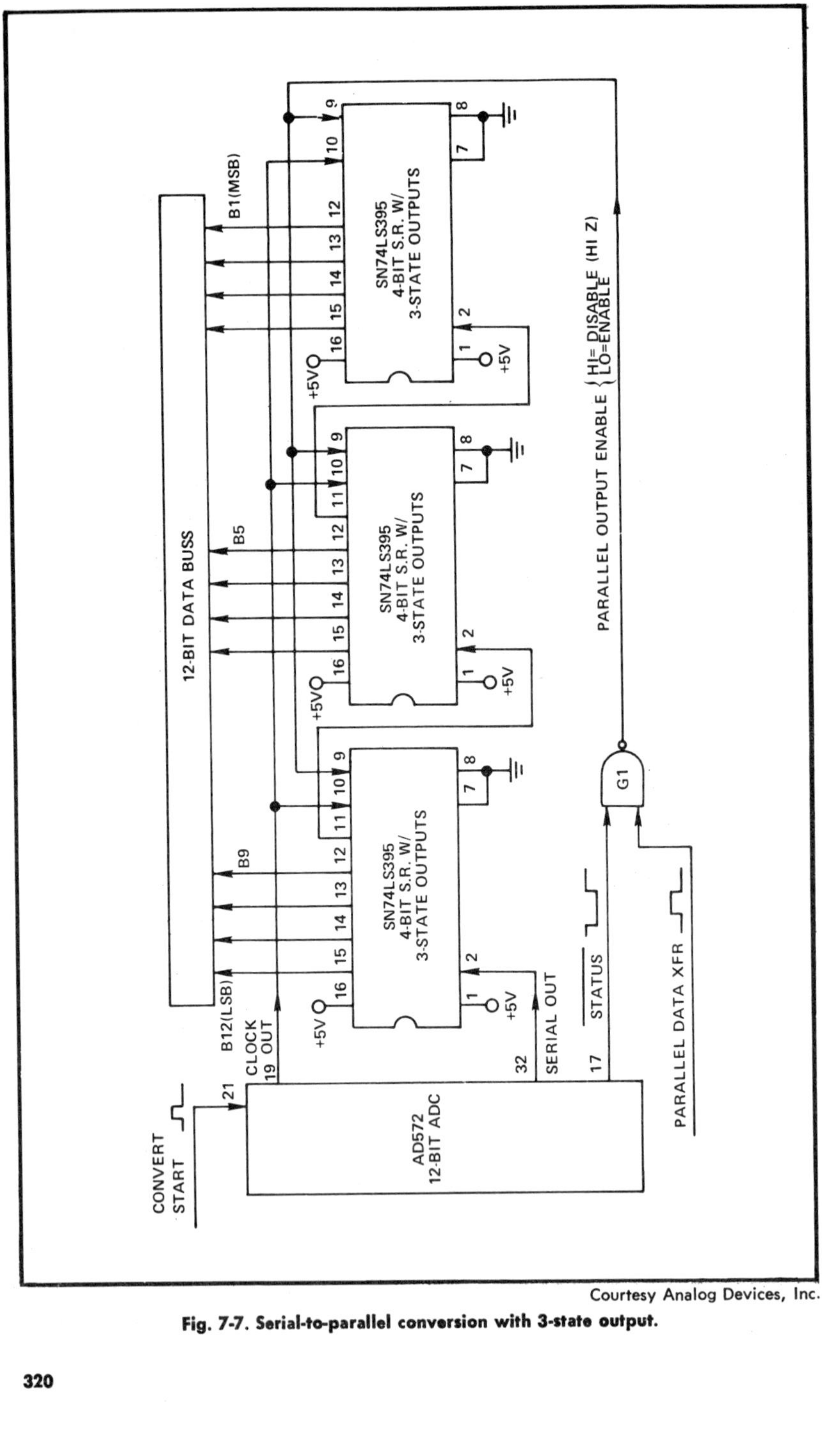

Courtesy Analog Devices, Inc.

Fig. 7-7. Serial-to-parallel conversion with 3-state output.

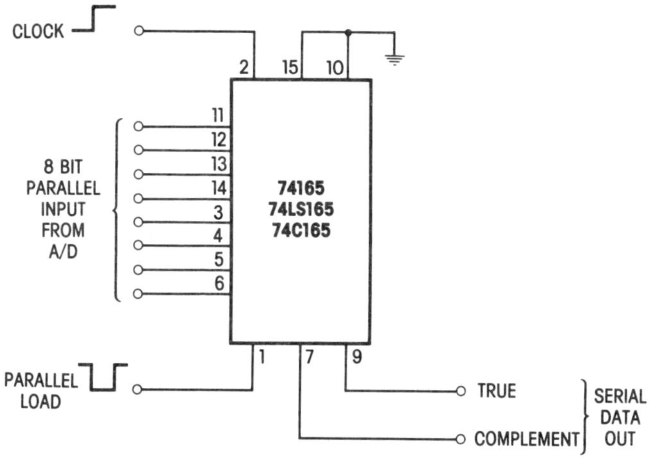

Fig. 7-8. Parallel-to-serial conversion.

74165s can be cascaded by tying the serial input to the output of a preceding stage.

In general, since the speed capability of the 74165 (and others of its type) is far in excess of most a/d conversion times, it should be noted that a serial transmission technique such as this can *multiplex* a number of a/d outputs. The data can then be appropriately demultiplexed at the receiving end, without loss of apparent a/d conversion speed.

CLOCK GENERATORS

There are a number of different areas in data conversion applications which require timing or "clock" signals of some sort. Examples are clocks for successive approximation a/d converters, display multiplex clocks, counter time-base clocks, master system-timing clocks, and others. The range of required frequencies is diverse, and output logic levels can be either TTL or CMOS. Accordingly then, this section briefly discusses a few of the more popular and useful forms of clock generators, and gives guidelines for their effective use.

Astable Multivibrator Clocks

One of the most simple and widely used form of clock generators is the RC astable multivibrator. This type of circuit is based upon charge and discharge of an RC network to form a timing waveform of a defined period and rate. The active components used can range from transistors, logic gates, to op amps and /or comparators. The simple, straightforward and inexpensive types of examples are examined here in Fig. 7-9.

In Fig. 7-9A is shown the absolute minimum of components needed to form a clock generator. This circuit uses a CMOS inverting Schmitt

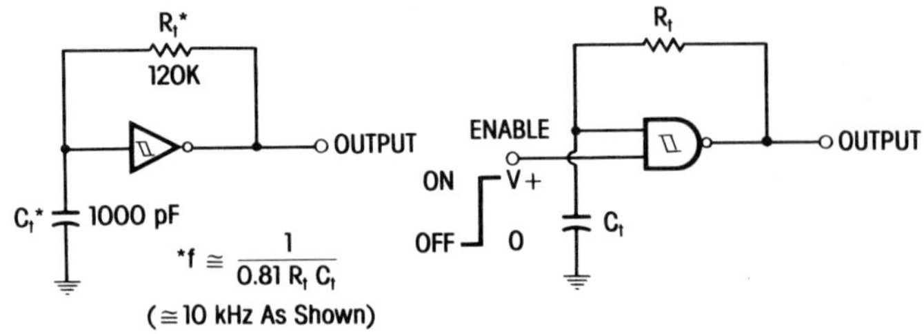

(A) Schmitt trigger astable.

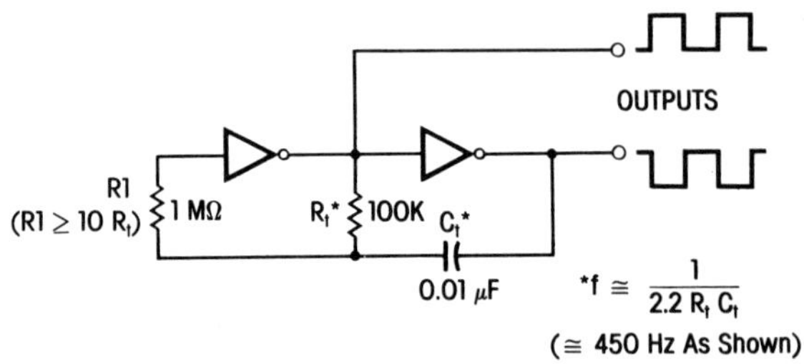

(B) Two inverter astable.

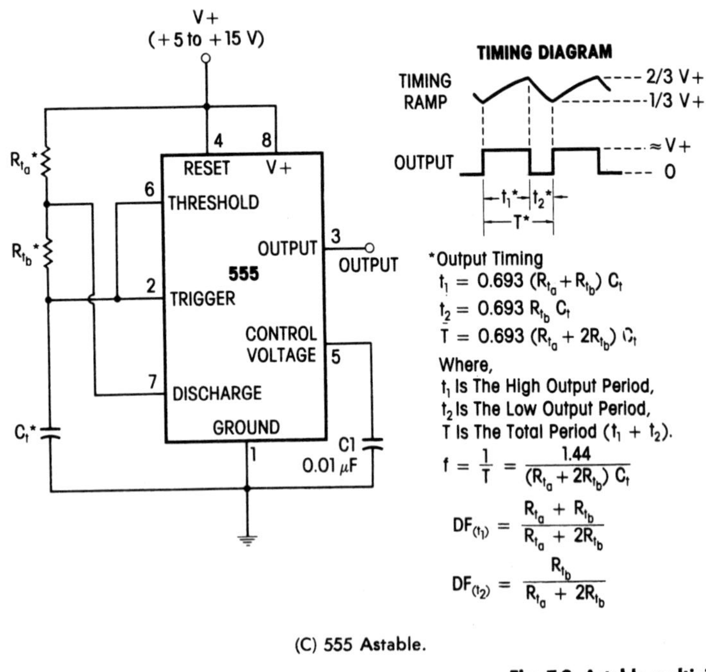

(C) 555 Astable.

Fig. 7-9. Astable multivibrator

trigger such as a 74C14 or 40106, with a timing network composed of R_t and C_t. In operation this circuit oscillates between the upper and lower hysterisis limits of the Schmitt circuit, and produces an output waveform which is nominally a square wave.

This circuit is highly flexible, being capable of a very wide frequency range of operation due to the very low CMOS input currents. It is also a very low power type of oscillator. Operation can be set up for any supply voltage in the CMOS range of +5 to +15 volts. CMOS "B" series devices are recommended, due to their higher gain and greater output drive.

Due to tolerances on device thresholds, the design equations are only shown as approximate, therefore the circuit is not well suited to high

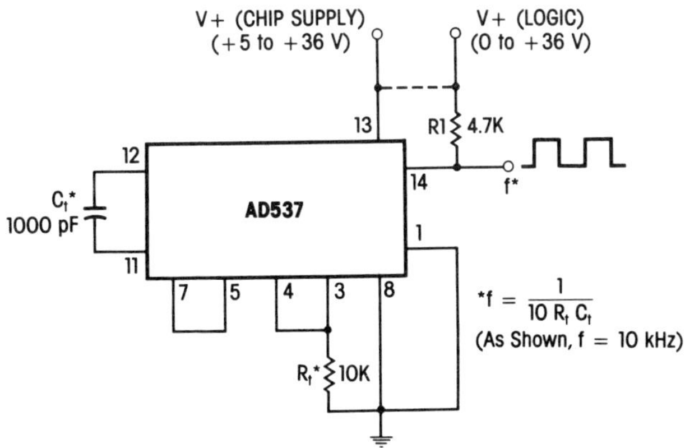

(D) AD537 astable.

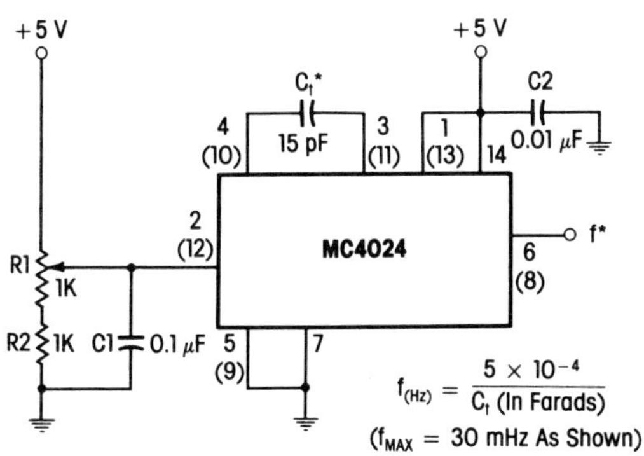

(E) MC4024 dual voltage-controlled multivibrator.

clock generators.

precision use. An alternate version of the circuit is also shown in Fig. 7-9A, using a two input NAND Schmitt, such as a 4093B. This circuit is a gated one, being on when the ENABLE input is high, and held off when it is low.

In Fig. 7-9B is shown a very popular version of a CMOS astable multivibrator, the two inverter circuit. This version is only slightly more complicated but offers good stability over temperature, and the bonus of push-pull outputs.

There is a limit to predictability with this circuit also, due to variations in the thresholds; thus, the equation shown is also approximate. It can use almost any two CMOS inverter pairs, and can also be gated if one of them is a two input gate, such as a 4001 for example. Output waveform is nominally square, but symmetry can vary.

Probably the most popular all around timing circuit device is the 555 timer, shown in Fig. 7-9C. This circuit shows the basic astable hookup of the 555, and how the timing components can readily be selected for specified duty cycles.

The excellent predictability of the 555, typically about 2%, makes it a good choice where the clock waveform must be adjusted rather precisely. It can operate from +5 to +15 volts, is suitable for either TTL or CMOS systems, has high drive capability, and can oscillate up to 100 kHz. There are of course many other variations of astable uses to which it may be applied.

The AD537, although it is basically intended as a V/F converter, is also capable of very effectively being used as a clock source, as in Fig. 7-9D. The attractiveness of the AD537 for this application is in its combination of extreme simplicity, high accuracy and predictability, and application flexibility.

The AD537 can operate at frequencies up to 100 kHz, producing an inherent square-wave output, and can be interfaced with virtually any logic family. The AD537 requires a supply voltage in the range of +5 to +36 volts for operation. Its open collector output stage can be referred to this supply voltage (shown dotted), or a separate logic supply level, up to +36 volts. An example of such a use would be to operate the AD537 from a +15-volt supply, while driving +5-volt logic, with R1 referred to this level.

An IC clock generator chip which operates to appreciably higher frequencies is the Motorola MC4024, shown in Fig. 7-9E. This device is a dual emitter-coupled multivibrator, with a TTL compatible square wave output. Each of the two sections is individually powered through separate pins for system decoupling purposes. The MC4024 operates up to 30 MHz typically, and 25 MHz minimum. It can also be operated in a crystal-controlled mode, for high stability applications. Crystal operation is achieved by replacing the timing capacitor with a series resonant crystal.

In the circuit of Fig. 7-9E the MC4024 is shown operated in a voltage-controlled mode, and is tuned by potentiometer R1. This circuit will operate from below 10 MHz to 30 MHz. Lower frequencies are readily accessible simply by adjusting the value of C_t, using the equation shown. This equation gives the *maximum* frequency of the 3/1 adjustment range. If fixed frequency (nontunable) operation is desired, the tuning network of R1, R2 and C1 can be deleted and the control input (pin 2 or pin 12) wired to +5 volts. This will yield a nominal output frequency according to the equation shown. The dual numbered device pins shown correspond to the two individual device sections.

The MC4024 was one of the first voltage controlled clock generator ICs of its type to be introduced. Other devices now available which operate similarly are the Texas Instrument SN74LS324, single vco with an enable control and complementary outputs; the SN74LS325, a dual vco with complementary outputs; the SN4LS326, a dual vco with complementary outputs and enable control; and the SN74LS327, a pin equivalent of the MC4024.

High Stability Clock Sources

For highly accurate and stable system clock sources, it is often desirable to use a crystal controlled oscillator. There are a large number of ICs now available specifically designed for crystal controlled clock applications, and many others can be readily implemented as crystal oscillators. They are briefly illustrated in Fig. 7-10.

Fig. 7-10A shows a general purpose crystal oscillator, which uses CMOS inverters biased for linear mode operation. The devices used should be one of those listed, or another CMOS unit which is suitable for linear operation in the stage A1A.

This particular application example uses a low frequency watch crystal at 32,768 Hz, a type which is readily available. Other frequencies of operation are easily possible with CMOS stages such as shown, at frequencies up to about 5 MHz or more. A virtue of CMOS device characteristics is the wide supply range of +5 to +15 volts, a feature which can allow ready interfacing to either TTL or CMOS logic.

There are a number of combination oscillator/divider ICs available which provide not only the oscillator circuit, but also a divider chain to subdivide the crystal frequency. This is often desirable on an overall basis, from a specifications and/or economics standpoint.

The National Semiconductor MM5369 is one example of this type device, being composed of an oscillator/buffer stage, plus a divider. With a commonly available 3.58-MHz color tv crystal, it produces a 60-Hz output for time reference uses. It may however, be used as a general purpose oscillator/buffer at frequencies from 10 kHz to 4.5 MHz. A circuit illustrating the use of the 5369 as a 3.58-MHz clock oscillator is shown in Fig. 7-10B. The 5369 operates from supplies of +3 to +15 volts, but will require higher voltages (such as shown) for frequencies above 2 MHz. The 5396 is available with custom programmed frequency division ratios.

Examples of other oscillator/divider ICs which feature crystal stability are the Intersil 7207, 7213, and 7038A, the Motorola MC14521, and the RCA CD4045 and CD4060.

It is often desirable (or necessary) for a clock source to have a 50% duty cycle, sometimes precisely and in complementary form. Although many of the sources mentioned have a nominal square-wave output, it is often not precisely so. Any of the above described oscillators can be used as a *precise* 50% duty cycle square-wave source by following the basic oscillator by a divide-by-two counter. This is simply accomplished by using a 2 × clock, and a type D flip-flop, as illustrated in Fig. 7-11. Logic levels in the range of +5 to +15 volts should use a CMOS 4013 type device, as in Fig. 7-11A. Five-volt logic levels, or high speed circuits (≥10 MHz) should use a TTL type such as a 7474, shown in Fig. 7-11B. Both devices yield complementary square-wave outputs.

VOLTAGE-TO-FREQUENCY AND FREQUENCY-TO-VOLTAGE CONVERTERS

One of the most widespread, low cost, and popular forms of a/d converters has become the voltage-to-frequency converter, or V/F converter

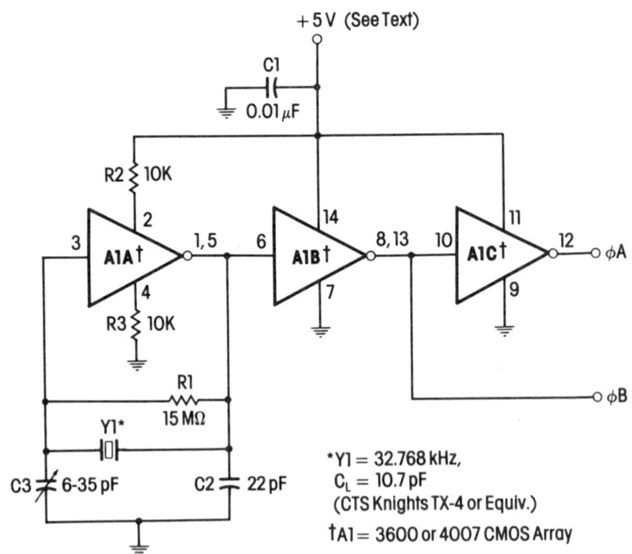

(A) General purpose.

Fig. 7-10. Some types of

as it is popularly called. The V/F converter is basically a highly developed form of voltage-controlled oscillator, where an output frequency is held precisely proportional to an input voltage. Thus, as an a/d converter, the input voltage is converted to an output pulse rate. Typical V/F accuracies are much better than 1% of full scale, with nonlinearity conversion errors of 0.05% or less, often quoted for many types. The V/F converter is a specific class of analog-to-digital converter, as it accepts an analog input voltage and produces an output pulse train. Dynamic ranges of representative devices are 1000/1 to 10,000/1 (or more), with full scale output frequencies usually 10 kHz or 100 kHz. Most of them can also function in an alternate frequency-to-voltage (F/V) mode, described later.

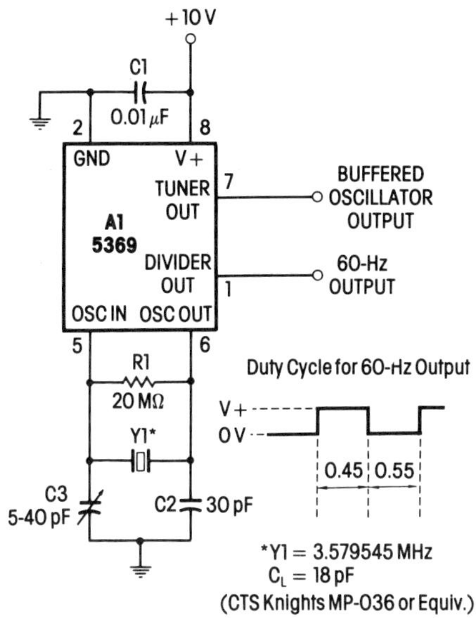

(B) Oscillator/divider combination.

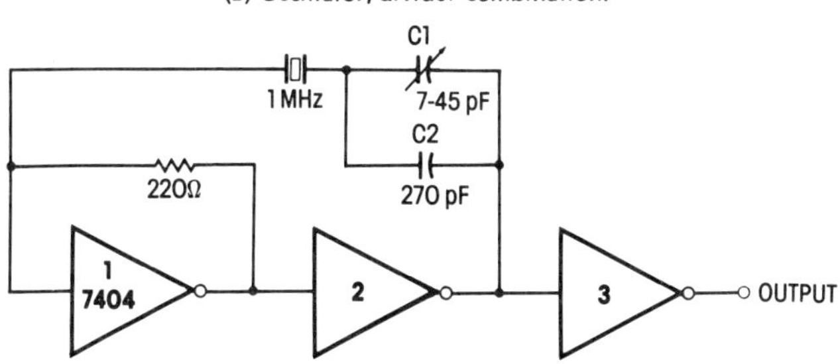

(C) TTL crystal oscillator.

crystal oscillators.

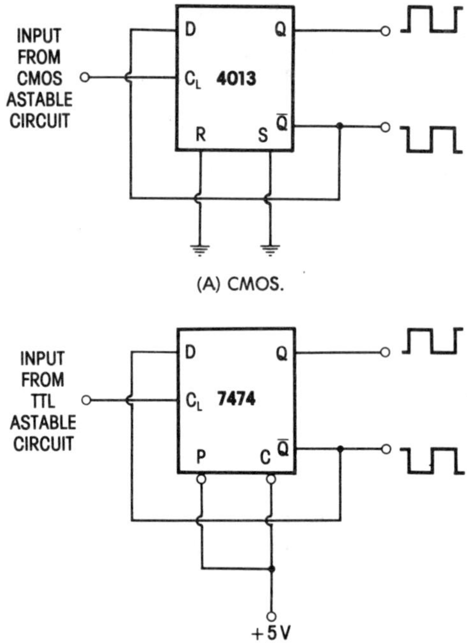

(A) CMOS.

(B) TTL.

Fig. 7-11. Logic dividers for 50 percent duty-cycle clocks.

V/F Converters

Fig. 7-12 illustrates the basic operation of the most popular type of V/F converter, the charge-balancing type. The name "charge balancing" comes from the operation of alternately charging up a capacitor from a reference source, then discharging it with the input. The result is a pulse train with its frequency linearly proportional to the input.

In this circuit C1 is the capacitor that is charged and discharged with a constant current by virtue of the op amp. In general terms, a timing period for a given frequency consists of two periods, t1 and t2. In this type of design, period t1 is fixed in length, and is, in fact, the pulse width of a precision one-shot. Period t2 is variable, corresponding to the input current I1 flowing in resistor R_{IN}. During the time period t1, the gated current source I2 is switched on, and the current which will flow in C1 will be the algebraic sum of currents I2 and I1. I2 must also always be greater than I1 in this circuit. During the t2 time period, the current source I2 is off, and the current in C1 consists of only I1.

The alternate switching of currents I1 and I2 into the integrator causes the output to ramp up and down by a voltage of ΔV, the difference between the amplitude limits. One amplitude peak of ΔV is fixed by the

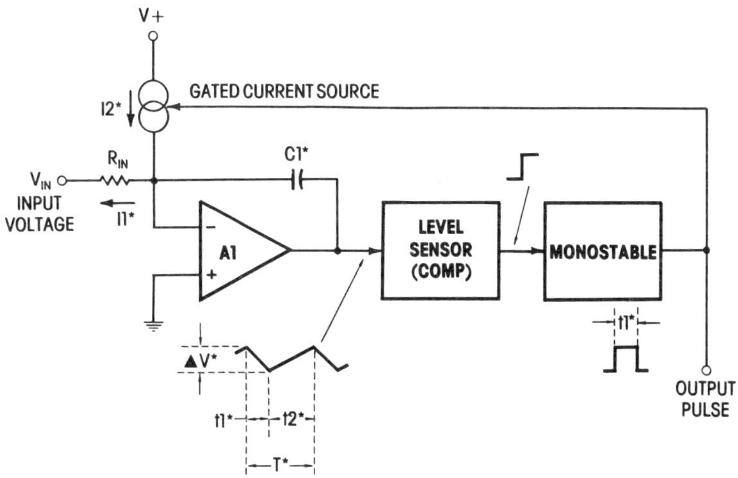

*BASIC V/F RELATIONSHIPS:

$$t1 = \frac{\Delta VC1}{I2\text{-}I1}\text{, and, } t2 = \frac{\Delta VC1}{I1}.$$

Thus, $\Delta VC1 = t1(I2-I1)$, and $\Delta VC1 = t2I1$, So, $t1(I2-I1) = t2I1$.

Then t2 in terms of t1 is:

$$t2 = \frac{t1(I2-I1)}{I1}.$$

Solving For Total Period, T:

$$T = t1 + t2 = t1 + \left(\frac{t1(I2-I1)}{I1}\right).$$

Simplifying, $T = t1\left(\frac{I2}{I1}\right)$.

In Terms of Frequency, $f = \frac{1}{T} = \frac{I1}{t1I2}$.

In Terms of V_{IN}, $f = \frac{V_{IN}}{R_{IN}\,t1\,I2}$.

Fig. 7-12. Block diagram of a charge-balancing type of V/F converter.

comparison threshold established by the level sensor; in the case here it is the positive peak limit. When the ramp reaches this threshold (at the end of t2) the level sensor changes state, which fires the one-shot, producing the output pulse of width t1. The gated current I2 then causes the integrator to ramp down for the t1 period, causing a voltage change of ΔV. ΔV will vary with the magnitude of I1 (and input voltage), being maximum at lower levels of I1.

With the application of some basic charge/voltage relationships and op-amp theory it can be shown that the basic t1 and t2 relationships are

$$t1 = \frac{\Delta VC1}{I2 - I1}, \quad t2 = \frac{\Delta VC1}{I1}$$

These two equations both contain a like term, $\Delta VC1$. This is due to the charge-balancing mechanism. Mathematically, two equations in terms of t1 and t2 can be expressed as being equal to $\Delta VC1$, so they are equal to one another, or:

$$t1\ (I2 - I1) = t2\ I1$$

Then the equation for t2 can be expressed in terms of t1 as

$$t2 = \frac{t1\ (I2 - I1)}{I1}$$

The equation for the total period T is simply the sum of t1 + t2, or

$$\begin{aligned} T &= t1 + t2 \\ &= t1 + \frac{t\ (I2 - I1)}{I1} \end{aligned}$$

after some simplification, this becomes

$$T = t1\,\frac{I2}{I1}$$

or in terms of frequency,

$$\begin{aligned} f &= \frac{1}{T} \\ &= \frac{I1}{I2\ t1} \end{aligned}$$

These two equations are the crux of the entire matter insofar as V/F theory is concerned, and bear some discussion. For instance, from the frequency expression it can be seen that frequency is directly proportional to I1, which is the input current. Therefore conversion sensitivity can be scaled for a given voltage by adjustment of R_{IN}. Frequency is also inversely proportional to t1 and I2. Note that in terms of I1 and I2, the equation is ratiometric. Thus if I2 and I1 are made to track one another in the same proportion there will be no net change in frequency. This point has important practical value, as will be seen in the actual V/F circuits described.

It should also be noted that ΔV and C1 terms have disappeared. The practical implication of this point is that neither C1 nor the comparator threshold need be highly stable to maintain accuracy, as their long term drifts cancel during operation. The required stable elements are only those which effect I1, I2, or t1. In some IC V/F types, most of these components of the block diagram are contained integrally. In others, additional external components, such as the addition of a high performance op amp, allow a high quality V/F converter to be built at very reasonable cost.

RC4151/LM331 V/F Circuits

The RC4151 V/F converter is one of the more simply applied, and one of the most flexible types. It is capable of moderate to high performance, dependent upon the degree of external circuitry used with it. This section illustrates the use of the RC4151 in a variety of applications, from simple to high performance, with a variety of input voltage range/supply voltage options.

The LM331 V/F converter is a circuit patterned after the RC4151, and designed to be functionally interchangeable, with improved performance in many applications. It is pin-compatible with the RC4151, and is an alternate choice for every application circuit. In the circuits that follow, either the RC4151 or LM331 may be used.

Fig. 7-13 shows the 4151 as used in its basic hookup, as a single supply V/F converter. This circuit accepts an analog input in the range of 0 to +10 volts, and converts it to a corresponding pulse train output in

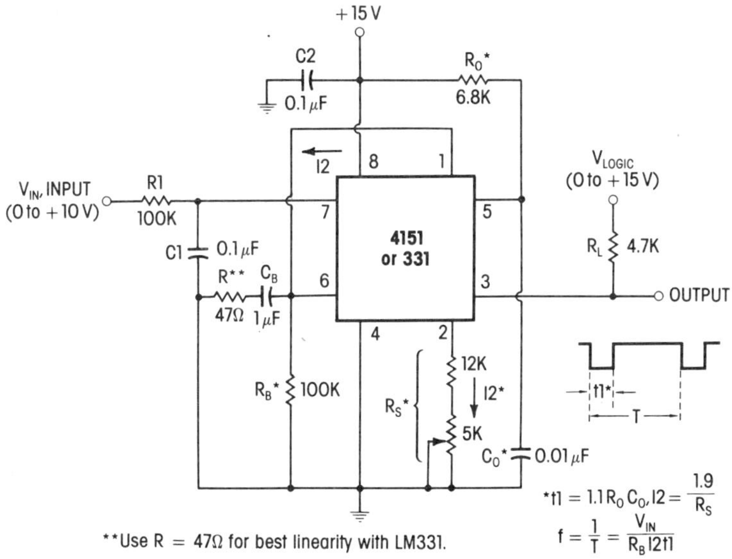

Fig. 7-13. RC4151/LM331 basic V/F converter circuit.

the range of 0 to 10 kHz. Output pulses are negative going, as generally shown in the waveform sketch, and may be interfaced with virtually any logic family by choice of the reference voltage for pullup resistor, R_L. The negative output pulses are of a fixed width t1, where t1 is determined by R_o and C_o by the relationship shown. This characteristics of negative going pulses of width t1 is common to all 4151 circuits described here. Output current sink drive is two TTL loads (3 mA).

In operation, an internal reference current I2 is generated by the 4151, and is gated on during the time period t1. This current charges capacitor C_B, and the network C_B-R_B forms a (simplified) charge-balancing integration network. The average voltage across C_B follows the input voltage applied to the opposite comparator input, pin 7, through the RC filter R1-C1.

In this basic V/F circuit, linearity is limited to about 1% due to the 4151 internal current source voltage compliance errors, comparator common mode errors, and R_B-C_B integration ripple. Also, the configuration is not readily trimmed for offset voltage errors; thus, for best results operation is best confined to about one decade below full scale range. If the LM331 is used in this circuit, the 47 ohm resistor in series with C_B is used and linearity is better than 0.05%, typically.

For a given design, I2 is set up to a nominal level of 135 μA by means of R_S, across which appears the 4151 1.9-volt reference voltage. R_S therefore is a convenient full-scale calibration adjustment; alternately R_o can be trimmed, which adjusts t1.

To set the circuit up for a given full-scale frequency and input voltage range, t1 is first chosen to be less than the full-scale frequency period, T. Here, t1 is chosen as 75 μs, as the full-scale period is 100 μs. The specific value for t1 is not critical, and can be in the range of 0.5 to 0.8T. It must necessarily be less than T, to provide overrange capability. R_o and C_o should be stable types for best temperature performance, preferably with low (or complementary) TCs. C_o should be a low dielectric absorption type, such as polystyrene. R_o can range in value from 6.8K to 680K, while C_o can range from 1000 pF to 1μF. R_S should also be a stable resistor, as it directly affects the reference current I2. I2 is recommended to be in the range of 100 to 160 μA for best results.

R_B can be chosen to scale the input voltage range, as V_{IN} effectively appears across it. The full-scale input voltage is selected by choosing the value of R_B such that with 100 μA flowing, it will drop a voltage equal to the maximum input. Here the range is 10 volts, thus R_B is 100K. Similarly, a 10K R_B would allow a 0 to 1-volt input range for a 0 to 10-kHz output. Where possible, R1 should be made equal to R_B. Note that with this configuration the maximum input range is +10 volts, but higher input voltages can be accommodated by using an input attenuator to reduce the voltage applied to the 4151 to +10 volts maximum.

Precision 4151/331 V/F Converters

Much higher overall accuracy can be achieved with the 4151 or 331 when it is used with an external op amp to form a "complete" charge-balancing V/F converter as seen in Fig. 7-12. Several examples of this type of circuit are discussed, with different input options.

The circuit of Fig. 7-14 uses the 4151 in conjunction with a 356 type op amp in a configuration that features vastly extended dynamic range

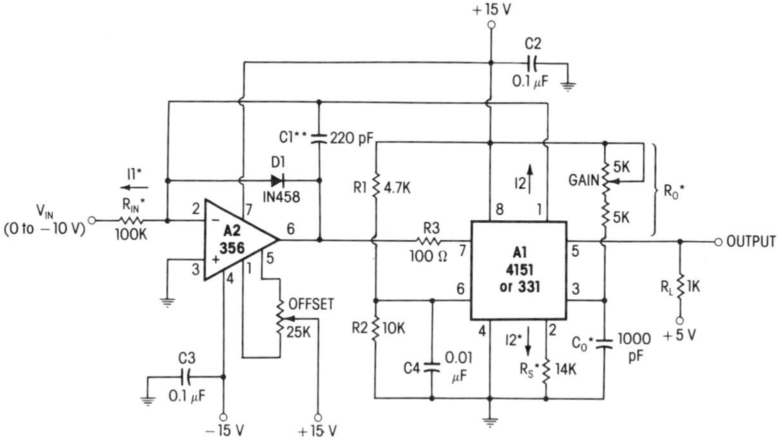

$*t1 = 1.1\, R_O C_O,\ I2 = \frac{1.9}{R_S},\ I1 = \frac{V_{IN}}{R_{IN}}$

Note: $I1 < I2$

$f_O = \frac{1}{T} = \frac{I1}{I2\, t1} = \frac{V_{IN}}{R_{IN}\, I2\, t1}$

$**C1 \geq 2.7 \times 10^{-5}\, t1$

Choose A2 SR $\geq \frac{2.5 \times 10^{-5}}{t1}$

Fig. 7-14. Precision RC4151/LM331 V/F converter.

and linearity. In this circuit A2 functions as the integrator, while the 4151 performs the functions of level comparison and precision monostable. Pin 6 of the 4151, one comparator input, is biased to +10 volts by R1-R2. The op amp feeds the other input, and the comparator will fire the one-shot when the op-amp output reaches +10 volts. In this circuit the op-amp output voltage in operation is quite similar in duration and levels to the waveform shown in Fig. 7-12. In fact, the only practical difference is that the 4151 produces negative going output pulses—otherwise, this circuit follows the block diagram almost precisely.

In this circuit t1 and R_S are chosen as previously in setting up operation for given conditions. R_{IN} is chosen for a full-scale input current I1 which, at maximum V_{IN}, is less than the reference current I2. This is necessary to provide an overrange safety factor; in this example, the 135-μA I2, and 100-μA I1 provide for the overrange, as does the selection of t1 < T.

The op amp operating in the virtual ground mode removes the 4151 current source compliance errors by clamping pin 1 at a fixed level, which increases linearity. Dynamic range of this circuit is maximized by two means; one is the operation of the 4151 at 100 kHz full scale, and the other is using a low input current op amp with offset null.

With an op-amp integrator such as used here, the low end of the dynamic range is limited either by the op amp input bias current, current source leakage, or the amplifier residual offset voltage. If excessive, any of these parameters can ultimately cause the V/F converter to stop functioning at some small input level. With a full scale I1 of 100 μA, a dynamic range of 10000/1 will require a leakage current to be less than 10 nA. The 356, an FET input unit, has a bias current much lower than this, satisfying this standpoint. If a wide dynamic range (10,000/1) is required, a low-leakage V/F device, such as the 331 should be used, with its off-state current source leakage of 1 nA (LM331A). Similarly, a 10-volt input voltage with a dynamic range of 10,000/1 will require an offset voltage of less than 1 mV. This is achieved by nulling the offset voltage of the op amp, which serves as the low-frequency calibration control.

The op amp used must also satisfy a slew rate requirement, to track the highest voltage rate of changes required at its output. This slew rate is dictated by the value of C1 and t1, and varies with frequency. Simple equations for selecting C1, and a conservative slew rate for the op amp

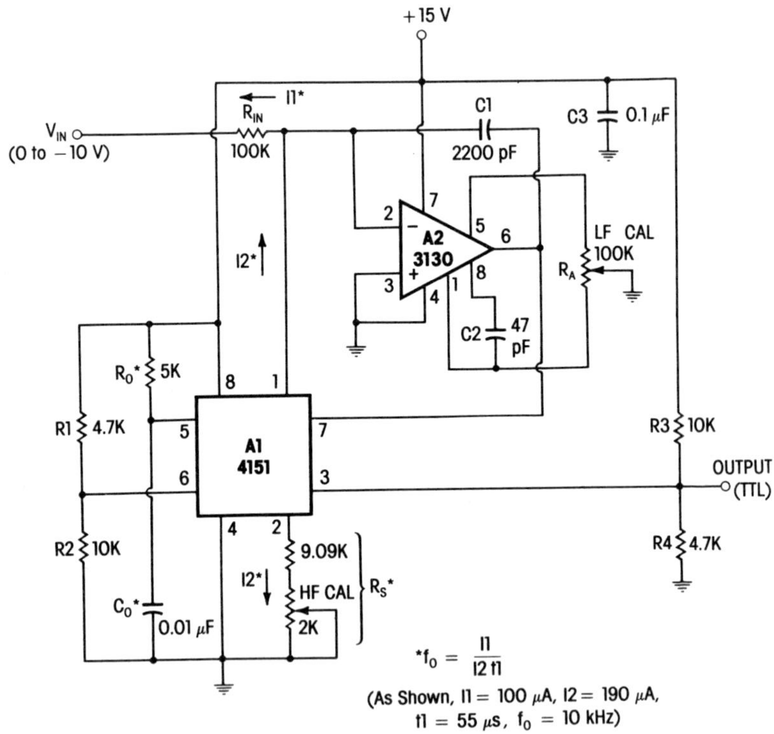

Courtesy Electron Design

(A) Negative input.

Fig. 7-15. Single supply precision

are given. Note that C1 value, tolerance, and TC are not overly critical, but the type used must have low dielectric absorption and current leakage. Similarly, a low leakage (<1nA) type should be considered for D1, if ultrawide dynamic range is desired. Performancewise, this circuit has a nonlinearity that is less than 0.1% of full scale.

While the above comments on op-amp parameters and component selection are presented here in the light of this specific application, they are also generally applicable to most other V/F circuits, using the 4151 and other V/F devices.

Single Supply Precision 4151/331 V/F Circuits

It is often desired to operate a precision V/F from a single supply voltage. Generally, popular supply levels are in the range of +12 to +18 volts, with +15 volts standard. Fig. 7-15 illustrates two circuits that utilize the 4151 operating from a single +15-volt supply, and handling positive or negative input voltages.

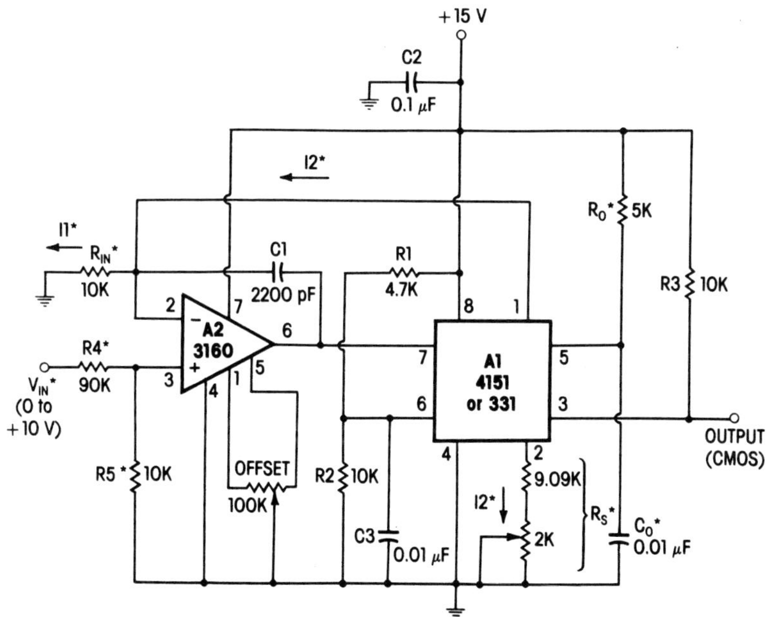

$$*f_O = \frac{I1}{I2\,t1} \text{ (General) In Terms of } V_{IN}\text{, With } \frac{R5}{R4 + R5} = \frac{1}{10}$$

$$f_O = \frac{V_{IN}}{10\,R_{IN}\,I2\,t1}$$

(B) Positive input.

RC4151/LM331 V/F converters.

Fig. 7-15A* is a single supply V/F converter capable of very high linearity, due to the use of an op-amp integrator, as described above. As in Fig. 7-14, this holds the 4151 current output pin 1 at a constant voltage, thus eliminating nonlinearity which is due to current source compliance errors.

The op amp used in this circuit must be capable of operating with its inputs at ground potential, while operating from a single supply. There are several types which can accomplish this, but only a few also have offset null provisions. The 3130, 3160, and 3140 MOSFET input units all possess these two features, and have very low input bias currents, as well as high slew rates. The latter two mentioned can also be used, as well as the 3130 illustrated here.

Aside from the single supply features, this circuit operates similarly to that in Fig. 7-14, and has the same component selection criteria. As shown here, operation is for a −10-volt full scale input, which results in a 10-kHz output. With the offset null of A2 adjusted for calibration at a low frequency, operation can span three or more decades. Linearity errors are less than 0.05% typically, which is appreciably less than the previous circuit. This decrease is due mostly to the lower operating frequency.

For different input voltage operating ranges, R_{IN} is selected for 100 μA of input current, at the desired full scale input voltages. The output is TTL and 5-volt CMOS compatible, as shown.

A single supply precision V/F circuit which operates with positive input voltage is shown in Fig. 7-15B. This circuit is similar to the one just described, with the exceptions of the connections to the integrator A2.

In this circuit, A2 is connected as an integrator also, but with a variable dc level applied to pin 3, which is in fact the input voltage. This causes a voltage to appear across R_{IN} proportional to the input voltage, and thus causes current I1 to flow in the integrator feedback path, as before.

This scheme works well, but is somewhat limited in linearity, due to the variation of voltage at pin 1 of the 4151. If a full (unattenuated) input voltage of 0 to +10 volts is applied to A2, this nonlinearity will be as high as 0.1% or more. With the technique used here, the input is scaled down to 1 volt at the op amp, and the nonlinearity error is reduced by proportion. Linearity error in the circuit as shown is less than 0.05%, which is quite reasonable for its simplicity. Input range is 0 to +10 volts with R4 and R5, or 0 to +1 volt without the attenuation network. These resistors should be stable types.

The circuit uses a 3160 op amp, an internally compensated version of the 3130. Output pulses are 15-volt CMOS compatible.

---

* W. G. Jung. "Precision Voltage-to-Frequency Converter Uses Only Single Supply Voltage." *Electronic Design,* October 11, 1976.

The 9400 V/F Converter

The 9400 is a charge-balancing V/F converter type, and is illustrated in Fig. 7-16. Although it operates on a charge-balancing principle, the 9400 uses a form of *voltage-based* charge balancing rather than current based, as was described in Fig. 7-12. As a result, its hookup configuration is slightly different, but operation is similar from a system viewpoint.

Pins 2 and 3 of the 9400 are the inputs of an MOS input op amp, which operates as an integrator. Pin 2 is the (+) input, and is connected either to ground, or to a small variable offset voltage, as shown here. The 9400 is designed to operate with input currents of 0 to +10 μA, applied to the (−) input; an external resistance such as R_{IN} is used to scale voltage inputs for 10 μA at the full scale voltage-input level. As shown, the voltage input range is 0 to +10 volts for full scale; other ranges are accommodated by appropriate choice of R_{IN}. R_{IN} is also made variable ±10% about nominal, for the purpose of full-scale calibration.

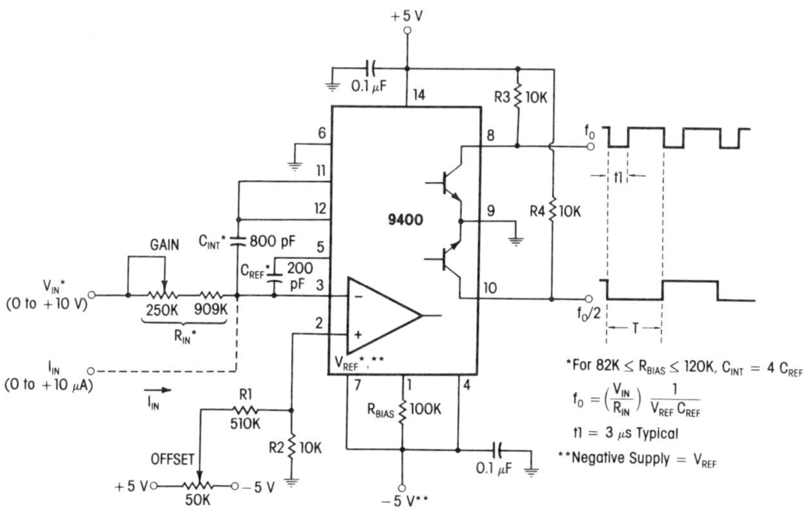

Fig. 7-16. 9400 V/F converter circuit.

The 9400 uses two capacitors in operation, C_{INT} and C_{REF}. Of these C_{REF} is the most critical, and directly affects the transfer characteristic according to the equation shown. C_{REF} may also be used as an alternate calibration means, and should be a stable, low TC, low dielectric absorption type. C_{INT} is selected to be in the range of 3 to 10 times C_{REF}. Here, it is 4 times C_{REF}; R_{BIAS} is 100K +20%.

As may be noted from the figure, the configuration shown uses the negative supply of −5 volts as the reference voltage, which is applied to pin 7. Since V_{REF} directly affects converter accuracy and stability, this voltage should be stable. Also, low impedance ac-bypassing is recommended for V_{REF}, as the pulse currents are appreciable.

Output from the 9400 is in two forms, a negative-going pulse at pin 8 of a nominal 3-μs width, and a half-frequency square wave at pin 10. Both of these outputs come from bipolar npn open collector sources, and must use external load resistances, such as is shown. The reference voltage for these output loads need not be the same level as the 9400 positive supply, as long as it is no greater than 18 volts more positive than pin 4 (negative supply). Also, the emitter of these transistors can be tied to ground (as shown), or to the negative supply. This flexibility allows both outputs to be readily interfaced with most logic families.

With values as noted in the figure, operation is for a 0 to 10 kHz V/F. Over this range nonlinearity is 0.05% (maximum). For 0 to 100 kHz operation, C_{REF} is changed to 15 pF and C_{INT} to 75 pF, and nonlinearity is 0.25% (maximum). Gain TC of the 9400 is typically 25 ppm/°C, exclusive of the reference voltage.

The AD537 V/F Converter

The AD537 V/F converter is unique in that it employs an operating principle based on a *current controlled emitter-coupled multivibrator,* as opposed to the more common charge-balancing V/F types. Like the

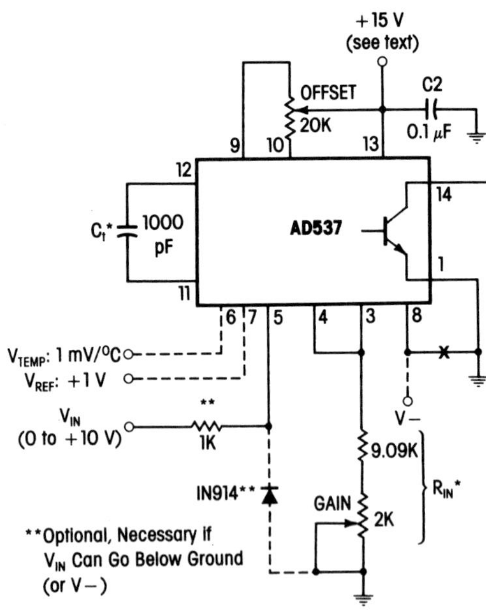

$$^*f_O = \frac{1}{T} = \frac{V_{IN}}{10\,R_{IN}\,C_t}$$

(As Shown, f_O = 100 kHz With V_{IN} = +10 V)

R_{IN} SELECTION

INPUT RANGE	R_{IN}
10 V	10K
1 V	1K
0.1 V	100Ω

(A) Positive inputs.

Fig. 7-17. AD537 V/F

charge-balancing method, the current-controlled emitter-coupled multivibrator is monotonic, but it has the added advantage of an inherent square-wave output. The technique is however not an integrating one (as is the charge balancing method), and may be more noise susceptible in some applications.

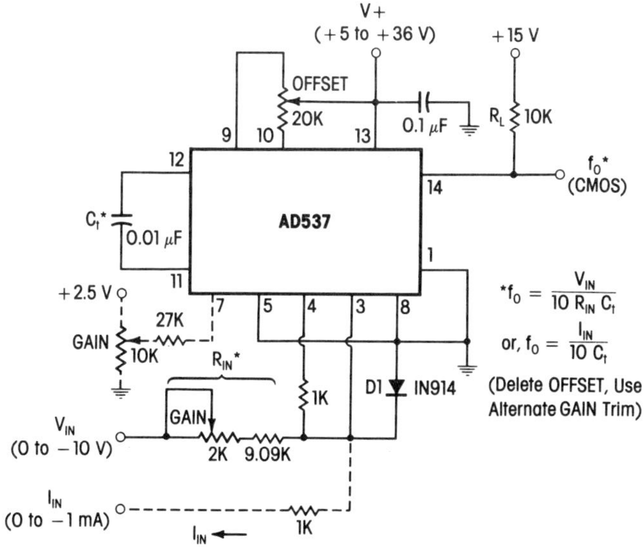

(B) Negative inputs (voltage or current).

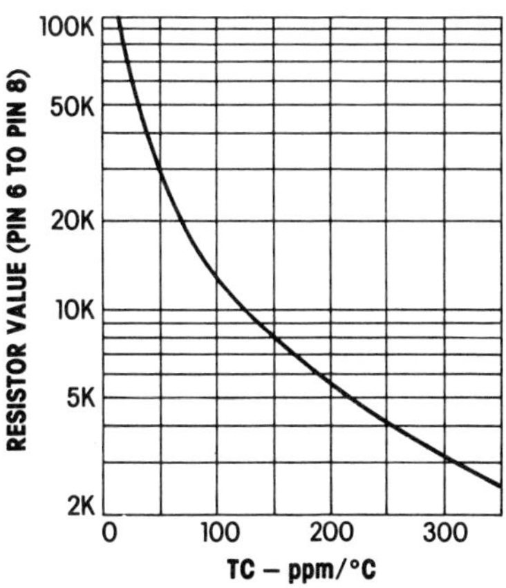

1. TC COMPENSATION

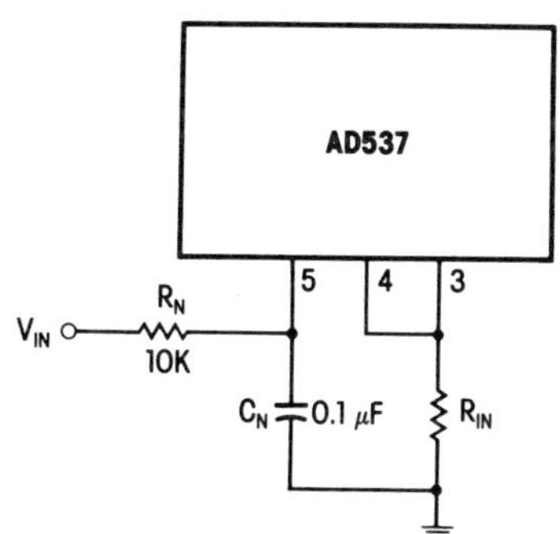

2. NOISE FILTERING

Courtesy Analog Devices

(C) Performance modification options.

converter circuits.

The AD537 is available in various accuracy grades, with nonlinearities as low as 0.05% of full-scale range. It operates to frequencies of 100 kHz or more, and is capable of dynamic ranges of from 10^3 to 10^4. It can be easily configured to operate from either positive or negative voltage sources, as well as current sources. The AD537 is illustrated in its two most basic types of V/F use, in Fig. 7-17.

In Fig. 7-17A is shown a positive input voltage V/F circuit. There are only two critical components used with the AD537 to determine operating frequency and input range, R_{IN} and C_t. R_{IN} is selected as appropriate to the voltage input range to be handled, as per the table. This allows 1 mA of current to flow in R_{IN} at the full scale input voltage level, which is the AD537 recommended operating current level. With R_{IN} as dictated by the desired V_{IN} range, Ct is then selected for the desired full scale frequency, according to the equation shown. For instance, the circuit example shown is scaled for 100 kHz with a +10-volt input, for full-scale operation.

The useful dynamic range of the AD537 can reach as high 10,000/1, for high input voltages such as 10 volts, and low (or trimmed) offset voltages. For maximum dynamic range and best lower scale accuracy, the AD537 input offset voltage should be trimmed by means of the optional OFFSET control shown. The general calibration procedure is for offset to be trimmed at an input level of 10 mV for correct output, then GAIN is adjusted for the correct full scale output with a full scale input voltage. Although offset trim can generally be considered optional on 10-volt and perhaps 1-volt ranges, it should be seriously considered for a range of 0.1-volt full scale, as a 1-millivolt offset represents a 1% error for these conditions.

The circuit accuracy is dependent upon the AD537 device option used, in terms of both nonlinearity and TC (the major errors). Linearity error is 0.15% (typical) for the AD537K, and 0.2% (typical) for the AD537J. These two devices have TCs of 50 and 150 ppm/°C(max), respectively. Lowest TC will be realized for low TC timing components, metal film resistances for R_{IN}, and NPO types for C_t. C_t should also be a low dielectric absorption unit.

As shown in this hookup the AD537 is operated in a single supply mode, with the negative supply terminal (pin 8) grounded. For dual supplies, this pin is returned to the negative supply (shown dotted). For positive input voltages, the positive supply level must be at least 4 volts greater than the highest input level; thus, a +15-volt supply is indicated for this +10-volt input range example.

The output of the AD537 is a square wave from a floating npn transistor (see symbolic sketch). Normally, the emitter of this transistor (pin 1) is tied to logic ground, but it may also be referred to other levels. The open collector may be referred by an external load resistor to any level, up to +36 volts. Here it is +5 volts, compatible with TTL circuitry. The

AD537 (J and K suffixes) will sink 20 mA of current driving TTL low state loads.

As can also be noted from this figure, the AD537 has two outputs which may optionally be used. A 1-volt reference voltage is available at pin 7, V_{REF}. Also, a voltage output which has a linear coefficient of 1mV/°K is available at pin 6, V_{TEMP}. Both outputs are from relatively high impedance sources and should feed high impedance loads, for best accuracy.

Fig. 7-17B shows how the AD537 is used for negative input voltages, or current inputs. For voltage inputs, the general AD537 pinout is the same, but the signal is applied to pins 4 and 3, with pin 5 grounded. This operates the internal op amp in its virtual ground mode, where pin 3 will rest at ground potential. The input resistance R_{IN} is then selected to scale the current into pin 3, for 1 mA full scale. R1 and D1 protect the internal circuitry against possible fault conditions. It should be noted that this method of operation can accommodate negative voltages of theoretically unlimited magnitude, as long as the current is limited to 1 mA full scale. Therefore voltage ranges of −100 or −1000 volts are perfectly feasible, with suitable power derating of R_{IN} of course.

Current inputs with a 0 to −1 mA range are accommodated in this configuration, by use of the alternate (dotted) hookup. The 10K trimmer functions as the full scale calibration, and the design equation is as noted. Use of D1 and R1 is still appropriate. For true current source inputs, error due to AD537 input offset voltage is relatively unimportant, therefore the offset adjustment can be deleted.

In this application example, the circuit is set up for a 10-kHz full scale output, with −10V (or −1 mA) input. Linearity for a 10-kHz range is appreciably better than the positive input circuit counterpart, being 0.15% (maximum) for the AD537J, 0.07% (maximum) for the AD537K.

Power supply ranges for this hookup are not affected by the input voltage, as the inputs operate at ground potential. Therefore the supply may be any voltage in the range of +5 to +36 (single) or ±5 to ±18 volts (dual). The output may be referred to independent levels; here it is shown as +15-volt CMOS compatible.

The AD537 is quite a flexible device, and several of its application options have already been described. Two further options which are generally applicable to the device are illustrated in Fig. 7-17C.

A feature which is unique to the AD537 is the ability to operate with a defined nonzero TC, which is both predictable and programmable, as shown in Part 1 of Fig. 7-17C. By connecting an external resistor from the V_{TEMP} pin (6) to pin 8, a positive TC is introduced. By proper choice of this resistor, cancellation of timing component TC can be realized. For instance, polystyrene capacitors are readily available and desirable for timing use, because of their low dielectric absorption. The characteristic −150 ppm/°C TC of polystyrene can be cancelled by use of an 8.2K resistor from pin 6 to 8, which causes the AD537 to exhibit a like positive TC.

The AD537 is not an integrating type of V/F converter, as are charge-balancing units. Therefore, it will respond to noise or instantaneous input voltage changes without the benefit of low-pass filtering characteristic of integrating type V/F converters. This can give rise to undesirable spurious outputs.

An input filter can be added to reduce this effect, as shown in Part 2 of Fig. 7-17C, using RC components R_N and C_N. These types are not critical and have a negligible effect on dc transfer accuracy. R_N can, in fact, be made nominally equal to R_{IN} (as shown) to cancel bias current error. C_N should be a low-leakage, low-dielectric absorption type, but otherwise it is not critical.

As shown the input filter is appropriate to positive voltage input operation (Fig. 7-17A). For negative voltage inputs, as generally outlined in Fig. 7-17B, R_{IN} may be split into two equal value pairs, with C_N going from midpoint to common.

The A-8400

The A-8400 V/F converter is one of the more "complete" charge-balancing V/F types, capable of operation with relatively few external components, and with high precision. It is illustrated in Fig. 7-18.

The A-8400 contains all of the functional components shown with the general charge-balancing V/F converter of Fig. 7-12. For operation, it requires external (precision) components to define the output pulse width t1, and thus the full-scale operating range. Input resistor R_{IN} is included on the chip, which defines the input current as 0 to +1 mA, with 0 to +10 volts to the 10K (nominal) resistor value. Reference current I2 is set up by an internal regulator as 1.63 mA. The I1 and I2 currents are as just described for minimum component, pin programmed use. Optionally, input voltage range and I2 can be altered if desired, with somewhat increased circuit complexity.

In the circuit shown, the input voltage range is 0 to +10 volts, as applied to R_{IN} (internal) at pin 4. Output pulses of a width, t1 are positive going from an open collector npn stage at pin 11. Here t1 is set up for a pulse width of 6 μs, which yields a 100 kHz full scale output frequency. In general, the pulse width t1 is equal to R_{ext} C_{ext}, and t1 is less than T, the (minimum) output pulse period, which occurs at f_o (maximum).

Offset trim for low-frequency calibration is accomplished by the OFFSET control, for best low range accuracy (optional). Full-scale calibration is accomplished by variation of t1; here this is shown as adjustment of R_{ext}. Alternately, C_{ext} can be trimmed. These components should be stable, low TC types, and C_{ext} is a low dielectric absorption unit. C1 is

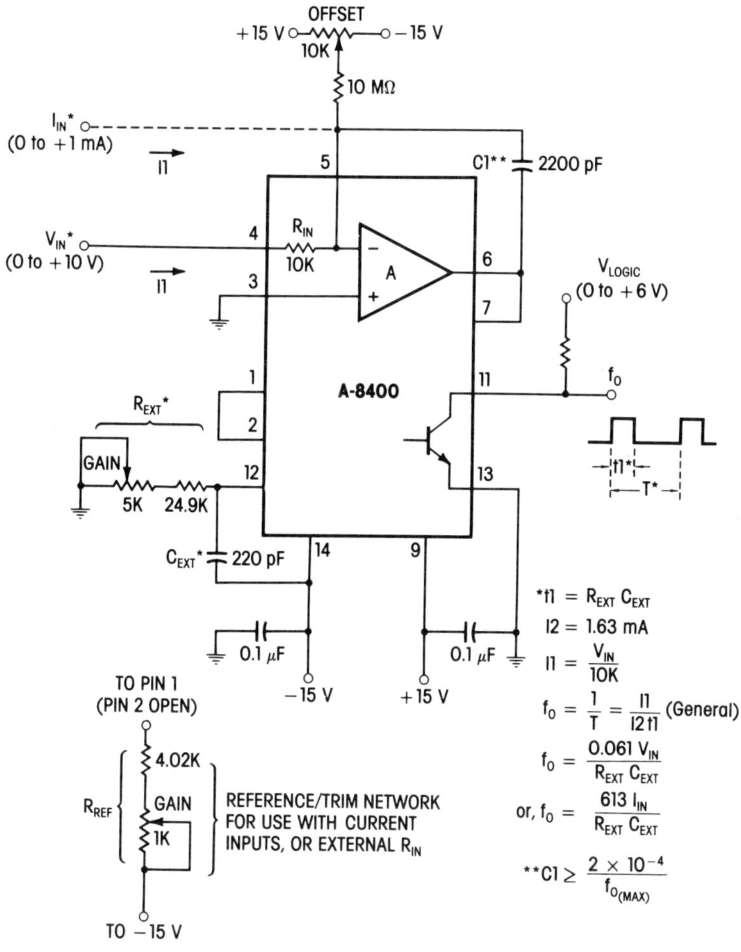

Fig. 7-18. A-8400 V/F converter circuit.

less critical, and may be generally chosen for value by the equation shown.

The A-8400 is operated as described for 0 to +10-volt input voltage ranges, and as can be noted, uses relatively few external precision components. Alternately, greater or smaller input voltage ranges can be accommodated by using an external R_{IN}, connected to the op-amp summing point, pin 5. This resistor is chosen for 1 mA of current, at the desired full scale input voltage; i.e., $R_{IN} = 1K$ for 1-volt full scale. Also, current inputs in the range of 0 to +1 mA are accommodated by applying the input direct to pin 5 (shown dotted).

With an external R_{IN}, or current source uses, the on-chip scaling resistance used to set reference current I2 will generally not yield optimum performance from a tolerance and TC standpoint. For such applications, R_{REF}, an external reference resistance of 4.5K can be used, as indicated in the inset. This resistance can also be trimmed for gain adjustment, as shown. If possible, it is desirable that R_{IN} and R_{REF} have low tracking TCs. The two different design equations for voltage and current input operation are noted in the figure.

The A-8400, as noted above, is capable of low nonlinearities. This version, which operates to 100 kHz, is specified for a 0.05% (maximum) nonlinearity, while 10-kHz range is capable of 0.025% (maximum) nonlinearity. TC is 100 ppm (maximum) for a 10-kHz range. A frequency range of 10 kHz may be achieved by upward adjustment of C1 and (either) R_{ext} or C_{ext}, by a factor of 10, in each case.

The VFC-32

The VFC-32 V/F converter is also one of the more complete charge-balancing V/F types, and capable of high precision operation with few external components. It is illustrated in Fig. 7-19.

The VFC-32 contains all of the functional components that comprise a charge-balancing V/F converter. For use, it requires external (precision) components to define the output pulse width (and frequency range), and the operating input range. Input current range (I1) is 0 to +0.25 mA, and reference current (I2) is 1 mA.

Fig. 7-19A shows the VFC-32 as it is connected for positive range voltage (or current) inputs. For a voltage range of 0 to +10 volts, R_{IN} is selected as a 40K nominal value, with a ±10% trim range for full-scale calibration.

To define operation for a given frequency range, C_t is chosen from the relation shown. This will yield a positive output pulse width, t1, which is related to C_t as shown. With the reference current I2 internally fixed by design at 1 mA, the V/F design equation for f_o is as shown in the figure.

To accommodate input voltage ranges other than 10 volts, R_{IN} is adjusted in proportion for I1 = 0.25 mA at the desired full-scale voltage. For example, for a 1-volt range, R_{IN} would be 4K.

For current inputs in the range of 0 to +0.25 mA, the alternate (dotted) connection is used, with the current I1 applied directly to the summing point, pin 1. The design equation for current input use is noted in the figure, and the C_t value is trimmed for full-scale calibration.

In this example, the full-scale output frequency is 100 kHz, with either a +10 volt (voltage) or +0.25 mA (current) input. Over its dynamic range the VFC-32 is capable of 0.05% (maximum) nonlinearity, for 100-kHz operation. This improves to 0.01% (maximum) for 10 kHz operation. The VFC-32 can also be operated to 500 kHz, with nonlinear-

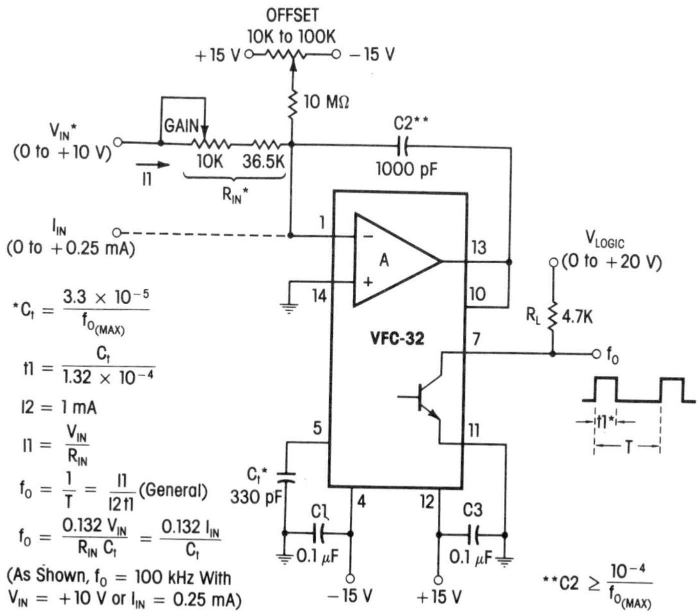

(A) Positive inputs (voltage or current).

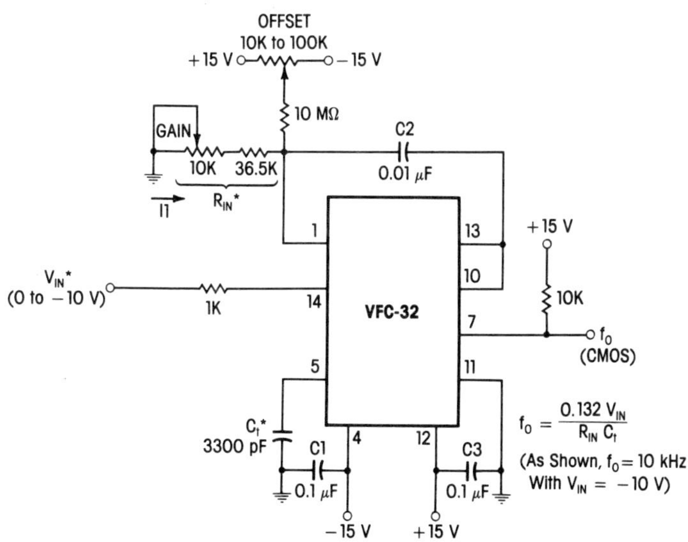

(B) Negative voltage inputs.

Fig. 7-19. VFC-32 V/F converter circuit.

ity of 0.2% (maximum). Gain drift (TC) is 100 ppm/°C (maximum) for the VFC-32KP illustrated (14 pin).

The VFC-32 may also be used with negative voltage inputs which range to −10 volts full scale, as shown in Fig. 7-19B. The general design equation and component selection criteria are the same for this hookup as the positive range version, thus they are not repeated here. This mode of operation cannot accommodate current inputs.

In this circuit, the current I1 appears in R_{IN}, by virtue of the op amp zero differential input voltage, which indirectly impresses V_{IN} across R_{IN}. R_{IN} is adjusted to suit various full-scale input levels of −10 volts or less in magnitude. Voltages greater than −10 volts (more negative) can be accommodated by using a scaling attenuator prior to the VFC-32 input.

The general performance of this circuit is similar to the positive input version, but there is a slight degradation in linearity, to 0.015% (maximum) for a 10-kHz range (as shown here). In this circuit example the output is 15-volt CMOS compatible.

F/V Converters

With few exceptions, V/F converter devices of the charge-balancing type are also usable as F/V converters. In this mode they perform an operation which is an inverse to that of the V/F, converting an input frequency to a proportional voltage. Once the basic function of a charge-balancing V/F is understood, the operation of an F/V follows naturally.

A block diagram of a charge-balancing F/V converter is illustrated in Fig. 7-20. If this figure is compared with that of the V/F converter (Fig. 7-12) it will be noted that many of the elements are identical. In reality, a single circuit may be configured to perform either function, V/F or F/V.

In Fig. 7-20 the input frequency is applied to a level sensor or comparator, which detects when the input signal crosses a defined level. This comparator can be either single ended or differential, as shown.

The comparator fires the one-shot, which produces a pulse of width t1, every time the input crosses the comparator threshold. This pulse gates on a current source, which allows a current I2 to flow into the summing point of the op amp for the interval t1.

In the feedback path is a resistor R_f and capacitor C_f, which serve to scale the output to the desired dc value, and smooth the ripple produced by the input pulses. The output voltage produced is V_o, which is the product of R_f and the average current, I1. Note that R_f performs a scaling function, analogues to the input resistor R_{IN} of the V/F connection.

The basic operating equation of a F/V converter of this type is related to its use as a V/F type, therefore it is derived by manipulation of the V/F defining equation. Here the input variable is f_{in} and the output is V_o, and the equation for V_o is

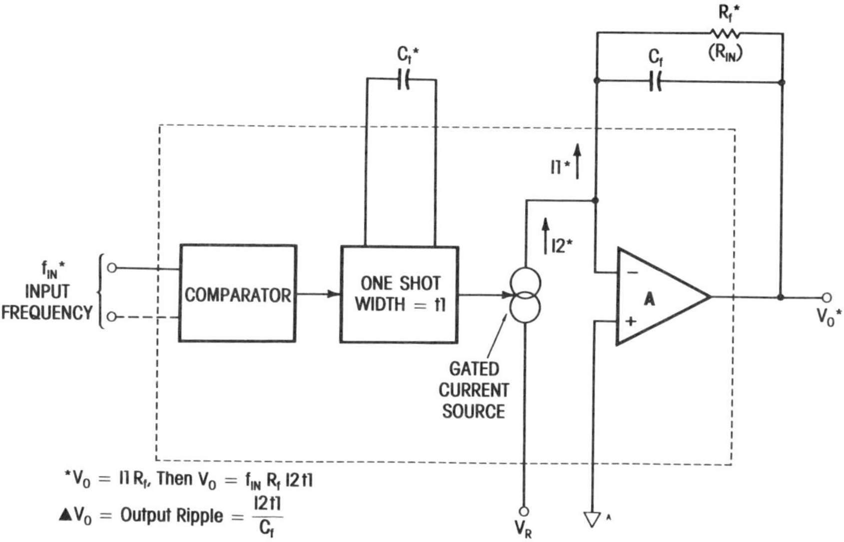

$$*V_O = I1\,R_f, \text{ Then } V_O = f_{IN}\,R_f\,I2t1$$

$$\blacktriangle V_O = \text{Output Ripple} = \frac{I2t1}{C_f}$$

$$\tau = \text{Response Time} = R_f\,C_f$$

Fig. 7-20. Block diagram of a charge-balancing type of F/V converter.

$$V_o = f_{in}\,R_f I2t1$$

To operate at a given frequency the time t1 is scaled, usually via C_t. Output voltage is scaled (or calibrated) via R_f, or in some cases I2 (when accessible). Output voltages can be up to 10 volts, positive or negative (dependent upon device), and are in buffered form due to the op amp. This voltage will be linearly proportional to f_{in}, over a dynamic range of 10,000/1 or more, dependent upon the device. While the basic output range of a V/F converter is unipolar due to fact that the current I2 never changes sign, a bipolar voltage range can be produced by offset of V_o, if desired. This is accomplished by feeding a fixed (reference) current into the op-amp sum point.

Although the integrator capacitor C_f does not appear in the basic design equation, it does have a secondary effect on output in two regards. Since the input is in the form of pulses, the output V_o will exhibit a ripple of an amplitude ΔV_o, with an amplitude as shown. C_f can be increased to reduce this ripple as desired, but as it increases, the response time of the F/V converter, which is $R_f\,C_f$ also increases. This is a basic tradeoff of the F/V converter.

Before proceeding with application circuits using F/V converters, it should be pointed out that minor interfacing may be required with some types, as shown in Fig. 7-21. Perhaps the most common need for interfacing is the necessity for input pulse shaping or differentiation, as in Fig. 7-21A. Here the RC time constant is chosen to be less than the basic pulse width, t1. Also, a voltage divider such as R1-R2 may be needed to establish a static bias above (or below) the F/V input threshold, which is variable dependent upon the device.

If the input signal is a high amplitude, voltage limiting may be required, as in Fig. 7-21B.

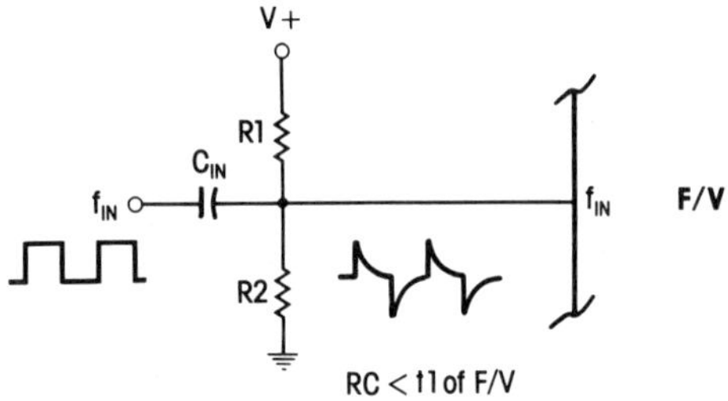

(A) Input differentiator.

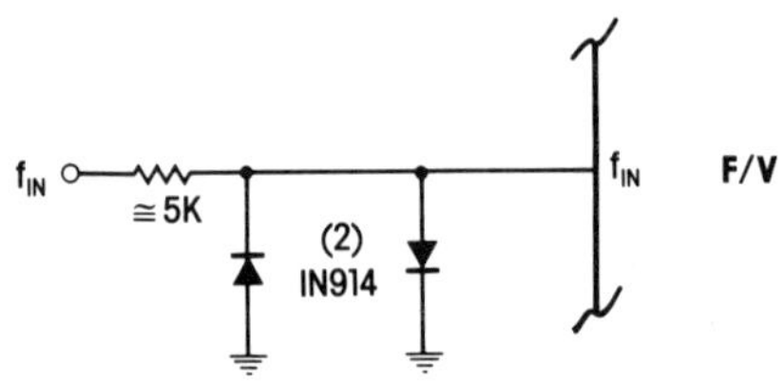

(B) Voltage limiting.

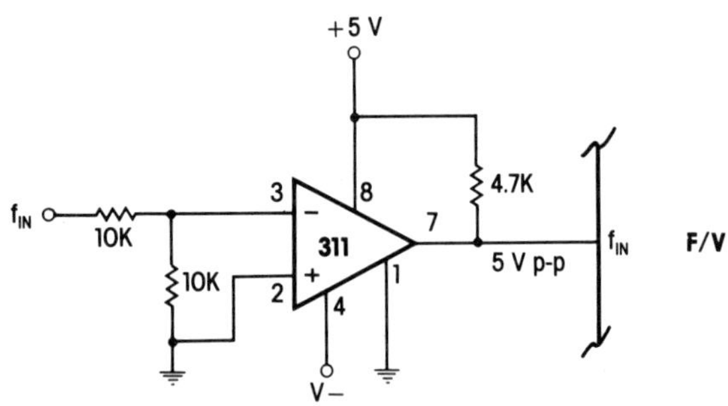

(C) Zero crossing detector.

Fig. 7-21. F/V input interfaces.

Finally, if the input is not rectangular in shape, a high gain comparator may be required to raise the level up to the amplitude needed by the F/V converter, and provide fast edges for triggering. A comparator such as the 311 is useful for this, and is shown in Fig. 7-21C in typical use as a zero crossing detector.

F/V Application Circuits

In Fig. 7-22 are illustrated F/V circuits using the RC4151 in the F/V mode. These circuits use the device in either a single or dual supply configuration.

In Fig. 7-22A is shown a basic 4151 circuit, operating from a +15-volt supply. This circuit does not use an op-amp integrator and is therefore limited in linearity to about 1%. Its virtues are simplicity and modest cost.

The comparator of the 4151 has differential inputs, pins 6 and 7. For F/V use, one is biased for dc (pin 7), and a differentiated pulse applied to the other (pin 6). The equation shown summarizes the operation of this circuit, which is scaled for 0 to +10-volt output, with a 0 to 10-kHz input. Note that either R_S, R_B, R_o or C_o can be used to trim calibration. Output impedance is the value of R_B, and C_B determines output ripple.

The circuit of Fig. 7-22B is capable of much higher precision, due to the use of the op-amp integrator. The basic operating equation of this circuit is the same as shown in Fig. 7-22A and it uses ±15-volt supplies.

The circuit can be trimmed most easily for gain via R_S, and for offset via the op-amp null terminals, as shown. C_1 is the integrating capacitor in this circuit, and with a value of 0.1 μF, ripple is about 100 mV. Linearity over the output voltage range of 0 to −10 volts for this circuit is on the order of 0.05%.

The 9400 is illustrated in its use as a F/V converter in Fig. 7-23. This circuit is a dual (±5-volt) supply configuration, and may be set up for operation for either a 10-kHz or 100-kHz full-scale range.

This circuit is fairly straightforward, and with f_{IN} signals of less than 2 volts, the frequency input can be connected directly to pin 11. For a 10-kHz input scale R_{INT} is 1 meg; for 100 kHz it should be reduced to 100K. Since C_{REF} appears in the output equation and is relatively small, calibration may be sensitive to layout and stray capacitance. C_{REF} should therefore be variable for calibration, and lead lengths as well as stray capacitance minimized.

This circuit has a nonlinearity of less than 0.1% to 100 kHz, and will typically be better for 10-kHz ranges. Output ripple can be minimized by adjusting the value of C_{INT}. Note that this should be a low-leakage, mica or polystyrene capacitor.

The A-8400 is illustrated as an F/V converter in Fig. 7-24. This device features low nonlinearities over the ranges of 0 to 10 kHz and 0 to 100 kHz, 0.025 and 0.05%, respectively (maximum).

As can be noted from the figure, most of the required elements for the F/V function are contained on the chip. Timing components R_{EXT} and C_{EXT} determine the input operating frequency range, and C_{INT} determines the output ripple. Recommended values are noted in the figure. With the respective frequency ranges, this circuit produces a 0 to +10-

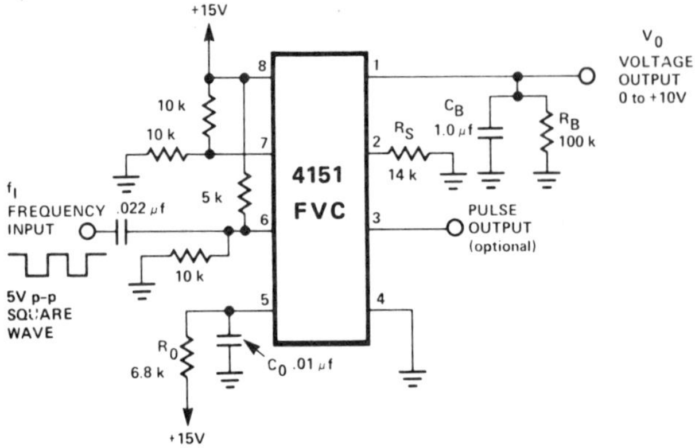

$$V_0 = f_{IN} R_B I_2 t_1$$

$$\text{Where } I_2 = \frac{1.9}{R_S}, \; t_1 = 1.1 R_0 C_0, \; I_2 = 140\ \mu A$$

(As Shown, $V_0 = 10$ V With $f_{IN} = 10$ kHz)

(A) Basic, single supply.

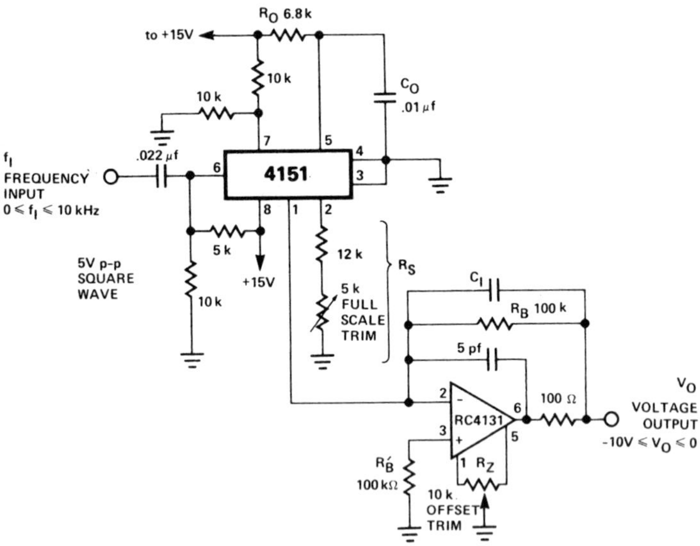

(B) Precision, dual supply.

Courtesy Raytheon Semiconductor Division

Fig. 7-22. RC4151 F/V converter circuits.

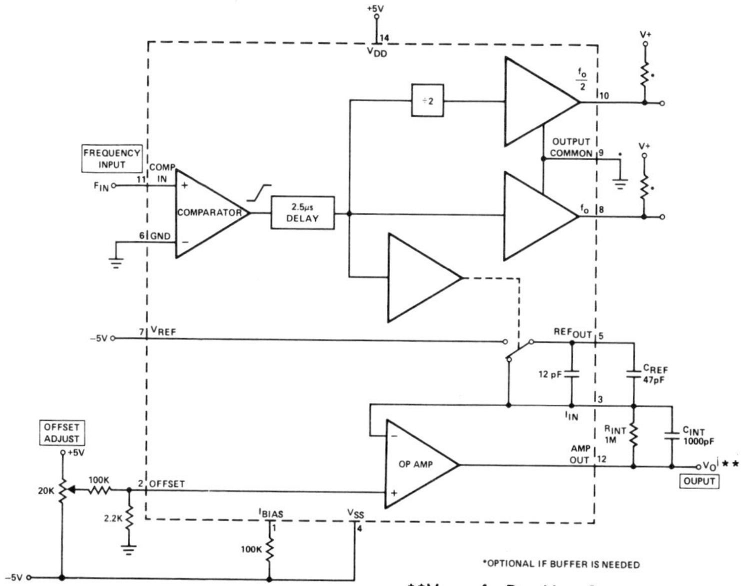

*OPTIONAL IF BUFFER IS NEEDED

$$^{**}V_O = f_{IN}\,R_{INT}\,V_{REF}\,C_{REF}$$

(As Shown, $V_O = 2$ V With $f_{IN} = 10$ kHz)

Courtesy Teledyne Semiconductor

Fig. 7-23. 9400 F/V converter circuit.

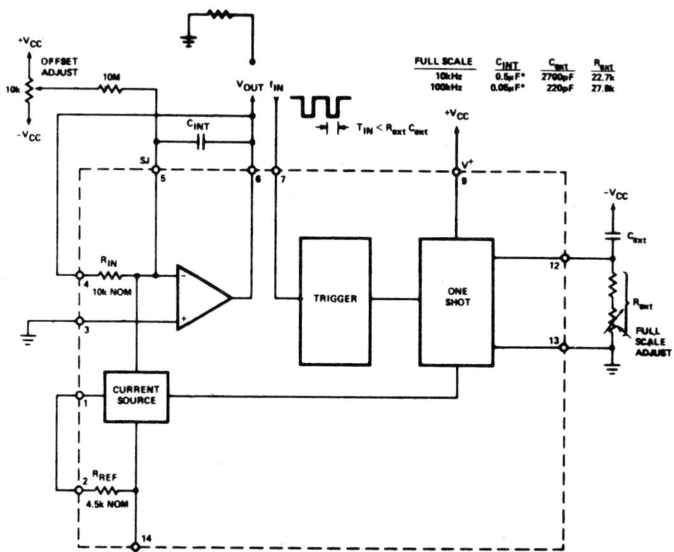

$$V_{OUT} = 16.4\,f_{IN}\,R_{EXT}\,C_{EXT}$$

Courtesy Intech-Function Modules

Fig. 7-24. A-8400 F/V converter circuit.

volt output range. Calibration is achieved via R_{EXT} for gain, and the offset pot.

The VFC-32 is illustrated as an F/V converter in Fig. 7-25. This device also features very low nonlinearities up to 100 kHz, and is usable to 500 kHz with good accuracy. The nonlinearities for 10 and 100 kHz inputs are 0.01% and 0.05%, respectively (maximum).

The VFC-32 uses two external components to set it up for F/V operation, R_{IN} and C_t. C_t is selected as noted for the basic range desired, and a portion of the R_{IN} resistance is used to calibrate for gain adjustment. Offset is trimmed with the network shown. The VFC-32 produces a 0 to +10-volt output voltage for F/V use.

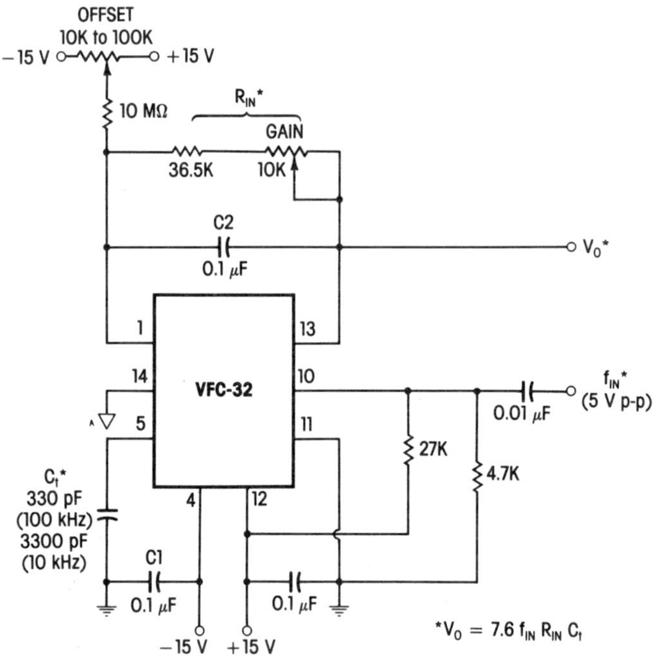

Fig. 7-25. VFC-32 F/V converter circuit.

An input bias and differentiator network is shown as appropriate for 5-volt p-p input levels (such as from TTL drive sources).

The above discussion has outlined the fundamental operating considerations for the various V/F converter devices, operating in both their V/F and F/V modes. Further applications discussions on V/Fs and F/Vs are contained in Chapter 8, and are of a general systems nature.

INTEGRATING A/D CONVERTERS

The area of the integrating a/d converter is quite an active one and there is a wide variety of different IC devices available, with various resolutions, accuracies, etc. We shall discuss a number of representative devices within the context of their intended applications. The discussions are not intended to be an exhaustive treatise on the working of individual devices, as such a treatment is beyond the bounds of reasonable length. For those who are interested, further information of this specific nature can be found in the list of references at the end of the chapter.

General Application Considerations

Already discussed within this chapter have been analog and digital interfacing considerations as applied to a/d converters, and clock generators. Any or all of these details may be necessary in a given integrating a/d circuit requirement, with the a/d function considered as a system. These general requirements are illustrated in Fig. 7-26.

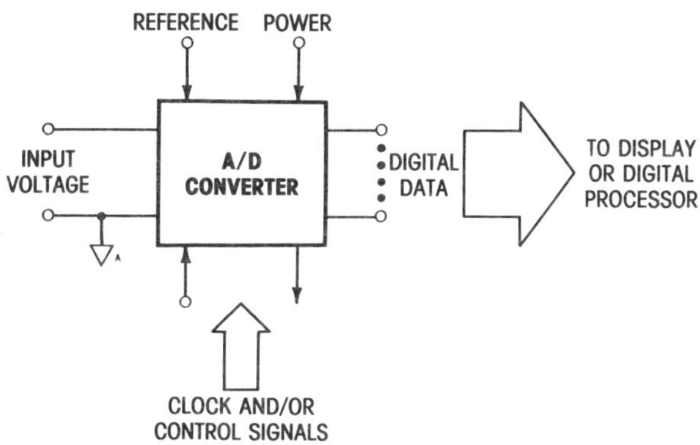

Fig. 7-26. The a/d converter system interface.

As typically used, an integrating type (and others also, such as SA types) a/d converter will require the signals defined within this figure. Clock and timing signals may be internal or external, but are essential to operation of the a/d converter. The analog input is almost always in the form of a voltage. Interfacing here may be in the form of buffering or range scaling to match the voltage range of the source to that of the a/d converter to be used. Power and reference voltages must be applied to the a/d converter, and the reference voltage in particular is almost always one of the prime limits to overall accuracy. The reference techniques discussed in Chapter 3 come into play here, as many a/d converters do not provide an internal reference. Also, in many cases where a reference is provided, it is often inconsistent with the converter's potential accuracy (usually from a drift standpoint).

The output of the a/d converter is either bcd or binary digital data, and it may be in either parallel or multiplexed form. Generally, integrating a/d converters fall into either of two categories of intended applications—display/instrumentation oriented, or data acquisition/processing oriented uses. Instrumentation or display oriented a/d applications are almost universally bcd outputs, such as used in digital panel meters (dpms) digital voltmeters (dvms) or digital multimeters (dmms). The most popular data format here is multiplexed bcd, which minimizes the number of device output pins, and where the required multidigit scan time is not an application disadvantage. When the output data is to be used further "downstream" by a computer or other digital processing, the data can be either bcd or binary, but usually it is desired to be in a parallel format. Some integrating a/d converters provide such outputs directly, and often also in 3-state form, for data bus uses.

Overall, available integrating a/d converters tend to be largely oriented towards instrumentation/display applications, such as dvms and multimeters. There are natural reasons for this, among them are the relatively slow conversion speed (no disadvantage to a human observer), the high accuracy capability (a prerequisite for instrumentation), and not insignificantly, the low cost or high performance/price ratio.

Quite appropriately then, it is this class of applications upon which this section places the most emphasis. Examples of circuitry will be presented for integrating a/d converters of up to 4¾ digits, and a full host of multimeter peripheral circuits such as current, ohms, and ac voltage measurement, as well as the basic dc voltage measurement capabilities of the a/d converter itself. Also covered will be auto-polarity and autoranging considerations.

BCD Scaling/Coding

One of the sometimes confusing applications points of a bcd a/d converter is the concept of the fractional digit, as represented by a "3½" or "4½" digit meter. This is so because the a/d devices as presently in practice do not literally deliver the extra ½ digit of range, as the connotation would seem to imply. A 3½ digit range instrument would seemingly provide a range midway between 1,000 and 10,000, or 3 and 4 digits which would be 3162, or $10^{3.5}$.

In practice, the ½ digit has come to mean simply an *overrange* capability, which usually doubles the basic 3- or 4-digit range. This is illustrated in Table 7-1.

As can be noted, a 3-digit a/d converter provides a basic scale of 999, while a 3½-digit unit ranges to 1999. Similarly, a 4-digit unit provides a scale of 9999; 4½ a scale of 19999. Also shown in this table are some voltage ranges as typically used in popular converters. The basic ranges are, of course, extendable (upward) by external range dividers. The

Table 7-1. Popular Scaling Arrangements for BCD-Coded A/D Converters

Resolution (digits)	Basic Scale	Accuracy (potential, or implied)	Typical Ranges
3	0 to 999	0.05%	±0.999 V or ±9.99 V
3 1/2	0 to 1999	0.05%	±1.999 V or ±199.9 mV
4	0 to 9999	0.005%	±9.999 V
4 1/2	0 to 19999	0.005%	±1.9999 V

"½" digit may then be best understood if it is considered just an over-range digit, which in fact it is.

A very important factor to be considered with an a/d converter is its real accuracy, which is dependent upon many factors. Column three of this table shows the accuracy consistent to the a/d scale, which is ±½ LSD of the basic range. This accuracy is here termed a *potential* or *implied* accuracy, as a particular a/d converter may or may not have a nonlinearity sufficiently low to achieve this accuracy, and other error sources can degrade operation as well.

The most important overall factor to be considered about accuracy is that the theoretical number as listed here will in practice usually be down graded by a number of additional errors. Most a/d devices specify a non-linearity error or "accuracy" figure which is a measure of the devices basic conversion error. However, this is almost always given in terms that exclude the effects of the reference voltage (usually external). Also, any analog interfacing or scaling can add additional error factors, which must be considered into the overall error allocation. This becomes increasingly important for 4 or more digit circuits, and/or low scale ranges, where noise and drift become larger errors.

Most display-oriented integrating a/d converters have an auto-polarity feature which allows the device to achieve the full rated resolution for analog inputs of either polarity. Thus the output of such an a/d converter is in the form of sign-magnitude coding—for example, 3½ digits plus sign.

Also, most display-oriented a/d converters have an *auto-zero* feature which eliminates the need for a user-adjusted zero control. The zero function is achieved electronically by on-chip switches which periodically ground the input, and null the offsets of the analog loop to a zero digital output. This procedure corrects for both the initial offset of the amplifiers, and their drift with time and temperature, and can achieve typical corrected drift figures on the order of 1 μV/°C, or less.

DVM Application—General Considerations

The most simple and straightforward application of an integrating a/d converter is within a dvm, (or dpm). Typically this application is a "committed" one, that is, it is intended to measure and display only a single specific range of voltage, such as 0 to ±1.99 volts, for example. A simplified block diagram of such a use is shown in Fig. 7-27.

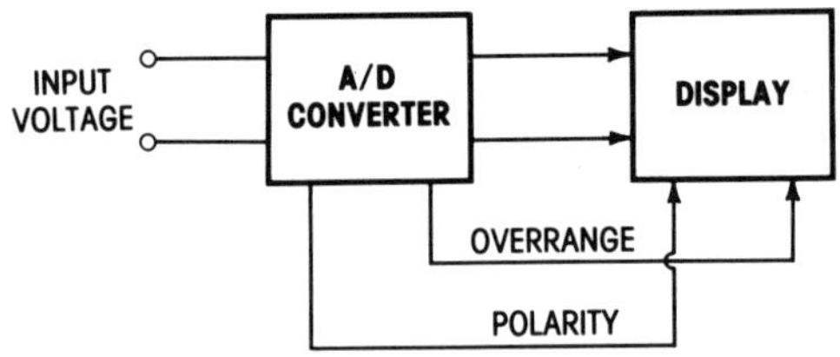

Fig. 7-27. Block diagram of a dvm application of an integrating a/d converter.

Here the a/d converter accepts the input analog voltage and converts it to bcd data which drives the display data inputs. Digit drivers, required for some a/d devices, are not shown in this generalized diagram. The a/d converter output polarity bit drives the (±) indicator of the display, while the overrange bit either blanks or (in some cases) causes the display to flash, indicating an overrange condition. The display used can take on a variety of forms, including LED, liquid crystal, or gas discharge 7-segment indicators. Examples of all are included in the circuits which follow.

Digital Multimeter Application—General Considerations

A more developed form of use for the a/d converter is the complete digital multimeter, which performs all of the functions of the classic analog multimeter—ac or dc volts, ohms, and current measurement. In addition, it can (optionally) have functions of auto ranging in addition to the basic (to the a/d device itself) operation of auto zero and auto polarity. A block diagram of such a use for an a/d converter is shown in Fig. 7-28.

In this meter, the a/d converter always performs as a basic voltage-measurement device, and the other ranges and functions serve to convert higher voltages or other parameters to the basic voltage measurement range of the a/d converter—such as 0 to ±2 volts for example.

For ac voltage measurements, an ac-to-dc converter rectifies incoming signals which have been ranged by the range switch, and converts them to an averaged dc voltage calibrated in terms of ac rms input. A more accurate option for this function is a true rms-to-dc converter, which reads the dc rms equivalent voltage with relative independence of input waveform, allowing higher conversion accuracy for nonsinusoidal waveshapes.

The ohms-to-volts converter applies a calibrated current to the external unknown resistor, and the resulting voltage drop is read by the a/d converter. By Ohm's law this voltage will be directly proportional to the

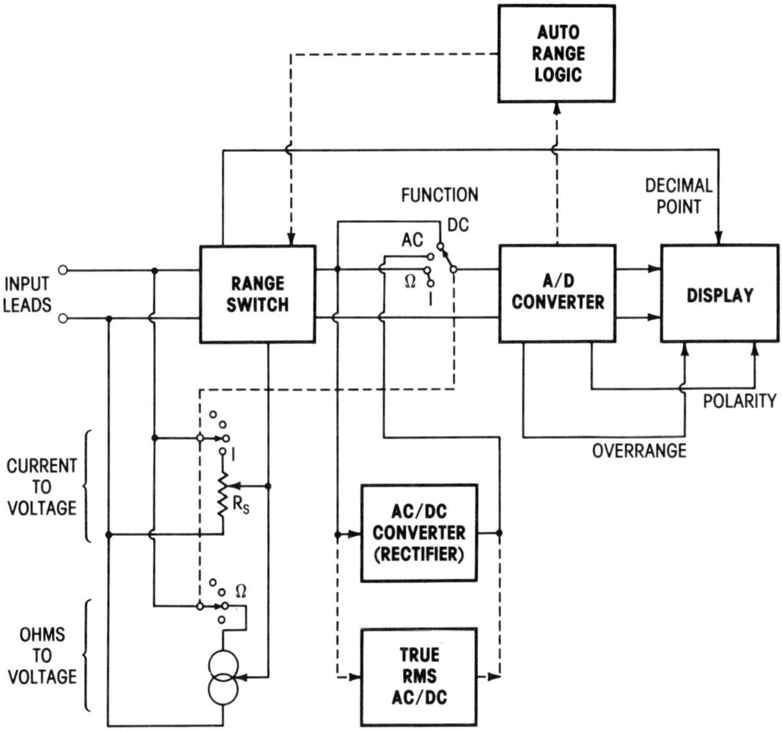

Fig. 7-28. Block diagram of multimeter application of an integrating a/d converter.

resistance. The current level is changed to accommodate various full scale resistance ranges.

The current-to-voltage converter passes an unknown current through a precision sample resistor, R_S. The voltage will be proportional to the external resistance. The range resistor R_S is changed for various full-scale current ranges.

A useful option for a multimeter is an auto-ranging circuit, which automatically positions the range switch (and display decimal point) for an on-scale and most accurate reading. This circuit makes use of a/d under/over range output signals, and drives either relays or analog switches to achieve the appropriate input ranging.

Examples of all of the above described concepts are included in the applications which follow.

Integrating A/D Circuits

This section deals with application circuits which illustrate the use of integrating a/d converters in circuits suitable for DVMs or DPMs, digital multimeters, and data acquisition applications, with a wide range of flexibility and accuracy capabilities. Virtually any diverse requirement can be met using one of the following circuits, either in its presented form, or via minor modification(s). Examples are given for differing analog input ranges, power supplies, display formats, clock frequencies, as well as the fundamental considerations of accuracy and resolution.

The LD110/LD111(A)/LD114 Series

The Siliconix "LD" series of integrating a/d converters are capable of a wide range of flexibility. Basically these chip sets—the LD110/LD111 the LD110/LD111A and the LD111/LD114—are 3½ digit converters which accept analog inputs of either 0 to ±199.9 millivolts or ±1.999 volts. They require an external reference and clock source, and have multiplexed bcd data outputs. The devices themselves require ±12-volt power supplies for operation, and as typically used with TTL logic to drive LED displays, a +5-volt supply also is needed. All devices of this series feature both auto-zero and auto-polarity operation.

As presented here, the format is first in terms of the basic digital voltmeter circuits themselves, then the additional peripheral circuits required to implement the complete multimeter function.

LD110/LD111(A) Circuits

A basic circuit which illustrates the function of the various components used with these a/d devices is the LD110/LD111 circuit of Fig. 7-29. As shown here, this circuit is appropriate for analog inputs of ±199.9 millivolts, and drives a common-anode LED display.

To set up the LD110/LD111 devices for operation, R2 is selected for the desired nominal input range, either 10K for 200 millivolts or 100K for 2 volts. R1 plus the series trim resistor is selected to allow a nominal 82μA of current with the reference voltage used applied across it. For a +6.8-volt reference as shown here, this makes the resistance 82K, realized by a 75K fixed resistor plus a 20K trimmer, which serves as a full-scale calibration adjustment. The R1 and R2 resistors should be high stability types, such as metal film units, as their relative drift rates will affect the gain TC of the a/d converter.

The exact reference voltage level used is not highly critical and can be in the range of +2 to +11 volts. Higher range voltages are preferred, such as the 6.8-volt level shown. The 2N4274 (National Semiconductor) transistor operates as a zener reference device, and has a typical TC of 50ppm/°C; for high stability a more tightly specified unit can be used such as the LM329 (see Chapter 3). The E507 is a constant-current diode which regulates the zener current at 1.8 mA, or in general near the minimum TC current of the reference diode used.

The passive components used with the LD111 function with the internal integrator and auto-zero system, and are generally not highly critical to operation. Table 7-2 shows a list of recommended values for these

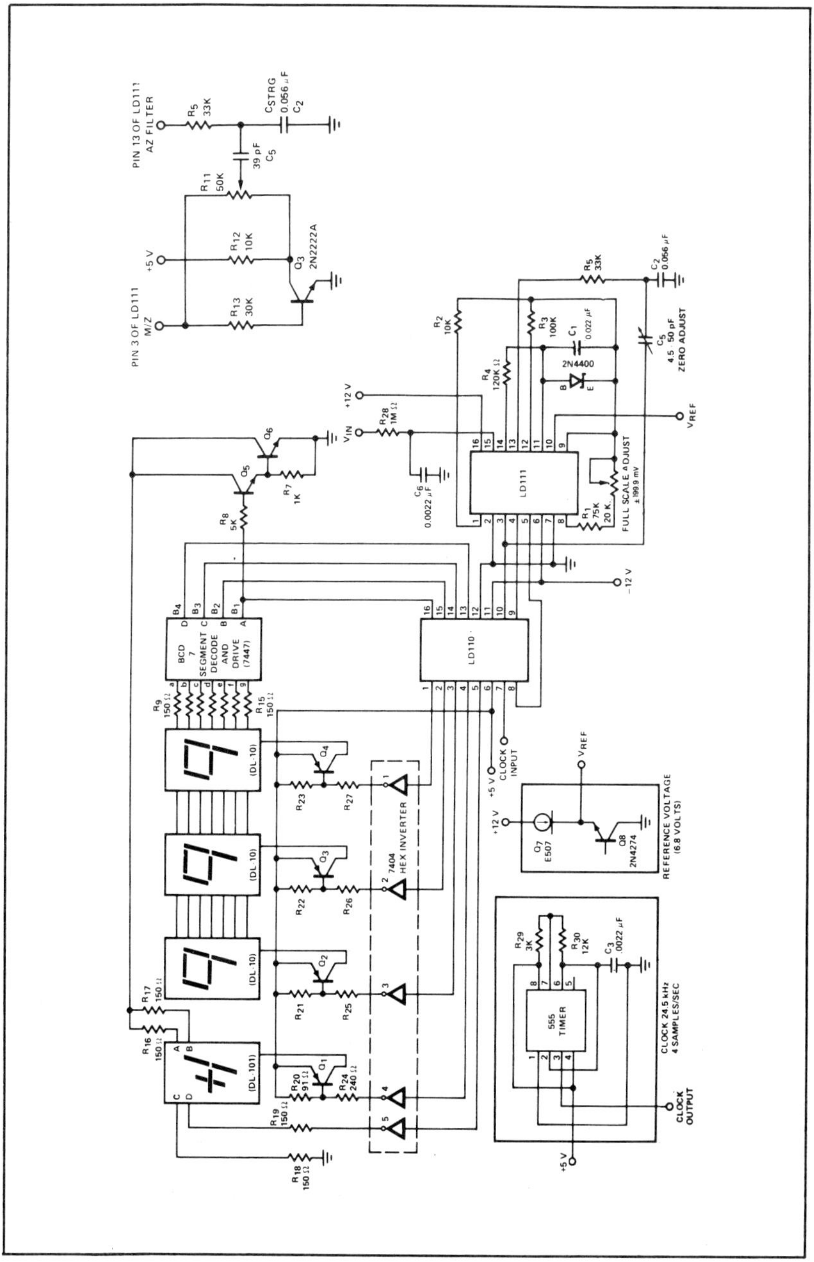

Courtesy Siliconix, Inc.

Fig. 7-29. LD110/LD111 ± 199.9-millivolt common-anode LED display.

Table 7-2. LD110/LD111 Auto-Zero Filter Values

f_{IN} (kHz)	C_{INT} (μF)	C_{strg} (μF)	R_4 (KΩ)	R_5 (KΩ)
2 to 10	0.1	1.0	68	15
10 to 20	0.039	0.1	240	47
20 to 40	0.022	0.056	120	33
40 to 75	0.01	0.056	82	18

units, as a function of different clock frequencies. It is recommended that C1 (C_{INT}) and C2 (C_{STRG}) both be film dielectric types, such as Mylar. The 2N4400 emitter-base junction serves as an overvoltage protection clamp for the integrator (note—not required with LD111*A*).

The clock frequency used here is at 24.5 kHz, and is supplied by the 555 timer astable oscillator in the inset, resulting in 4 conversions per second. In general, the conversion rate will be the clock frequency divided by 6144. For other clock drive sources to be used, the LD110 clock input is TTL compatible.

In an integrating converter, rejection of normal mode ac noise (ac components on the dc to be measured) is maximized when the signal integration period is an exact multiple of the ac line period. In the U.S., this implies the reciprocal of a harmonic of 60 Hz; in Europe 50 Hz. Note that the exact frequency itself is not that important; what is important is that it be an *integral* multiple of the line frequency.

Trimmer C5 of this circuit serves to provide small variable offset voltage, which may be necessary due to noise induced in a specific layout. As shown in the basic figure it is a unipolar type of compensation; the alternate offset adjustment in the inset provides bipolar offset compensation, and may be preferable. This trim circuit can generally be used with any of the LD series converters (with the exception of the LD130).

This dvm circuit uses 3 Litronix DL-10 common-anode 7 segment displays, plus a DL-101 polarity/overflow display. These particular types are 0.270 inch character, DIP style packaged devices. There is of course no reason why this circuit cannot be used with other LED display devices, with similar electric characteristics, but other mechanical configurations. To this end, a short summary of popular 7-segment LED single-digit display is given in Table 7-3. All types within a given format are electrically and mechanically equivalent, and interchangeable. All types listed are red in color, but many are also available in green or yellow formats in the same package style. The individual manufacturers should be consulted for further details on the respective devices (see Appendix).

Drive for the display is via a BCD to seven-segment decoder, the 7447, and discrete pnp digit-select drivers. The 150-ohm resistors determine the segment brightness. The pnp drivers can be 2N5139 or

Table 7-3. Popular Single Digit LED Displays

	Fairchild	Hewlett-Packard	Monsanto	Litronix	T.I.
0.270" CC Characters, CA 14 Pin DIP ±1			MAN1 MAN101	DL-10 DL-101	TIL302 TIL304
0.300" CC Characters, CA 14 Pin DIP ±1		5082-7740 5082-7731 5082-7732	 MAN71 MAN73	DL-704 DL707 DL-701	TIL313 TIL312 TIL327
0.500" CC Characters CA ±1	FND500 FND507			DL-750 DL-747 DL-61	TIL322 TIL321 TIL330

similar types; the npn units can be 2N2222 or 2N3904s. These devices are not highly critical.

In this basic circuit no decimal point positioning is shown, and for a ±199.9-millivolt display the decimal point for the second LSD would be active. If range switching is used in conjunction with a circuit such as this, the decimal point should follow the range in use, of course.

It can be noted that the analog input to this circuit is applied through a series 1-megohm resistor to the input, pin 15. This resistor provides a/d overvoltage protection, and its importance cannot be overemphasized. The use as shown here is typical, and should be considered mandatory for any a/d circuit used for metering purposes, where potentially catastrophic voltages can be common. The resistor introduces no dc error as the LD111 (and all a/d circuits described here) has FET inputs, with currents in the low pA range. The bypass capacitor provides noise filtering; its leakage current should be very low, such as is typical of polystyrene or Mylar dielectrics.

LD110/LD111 ±1.999-Volt DVM Circuit

In Fig. 7-30, the LD110/LD111 pair is shown in a 3½ digit, ±1.999-volt dvm application. This circuit is similar in many respects to Fig. 7-29, therefore only the essential differences are discussed.

For ±1.999-volt scaling, R2 is changed to a 100K value. The R1 value remains the same, except for this case it adjusts full-scale calibration for ±1.999 volts.

Aside from the scaling change, the other major difference of this circuit is in the mode of display operation. Common-cathode LED display digits are used; Fairchild FND-357s and a FND-358. Digit-select cathode drive is from the 9667 inverter. The anode driver used is the Fairchild 9368, a device which provides a constant 20 mA segment drive current without the need for current setting or limiting resistors, thus allowing considerable simplification. Sign drive is via the 2N5139 discrete transistors with 150-ohm current-limit resistors.

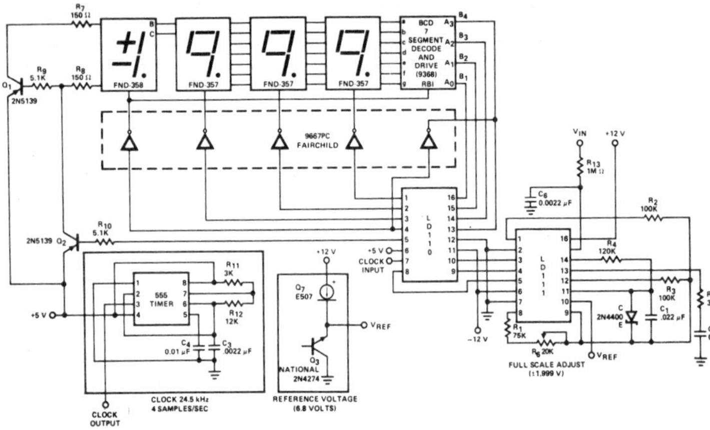

Courtesy Siliconix, Inc.

Fig. 7-30. LD110/LD111 ± 1.999-volt common-cathode LED display.

"Housekeeping" circuit details pertinent to these circuits are relatively important, and in general applicable mostly to other dvm circuits as well. High current paths, such as used by digital portions and the LED displays, should be wired separately from the analog portions, with the digital grounds returned to the +5-volt supply. The analog ground should be kept separate, and returned to digital circuitry at one point only, to prevent spurious coupling of noise components.

The digital supply should be heavily bypassed with several microfarads of capacitance and the analog supplies (±12 volts here) bypassed at the a/d device pins. Also, it is preferable to keep high impedance and/or sensitive components associated with the a/d conversion process close to the device socket by using minimum lead lengths. Outside foil of capacitors such as C1, C2, and C6 should be connected to circuit common, or the lower impedance node of the two available.

LD110/LD111A ±199.9-mV Circuit

Fig. 7-31 illustrates the use of the LD110/LD111A pair as a 3½ digit, ±199.9-millivolt range dvm. This circuit features improved linearity, drift, and overall accuracy over the two just described, due to the improved performance of the LD111A.

Functionally, this circuit is similar to the LD110/LD111 circuit just described, and scaling/range selection is performed by the selection of R2 (±199.9 millivolts here). It also uses a similar reference scheme and clock circuit (24.5 kHz), and performs 4 conversions per second.

The LED display in this circuit illustrates a further option available to the user, utilizing Monsanto MAN4610 and MAN4630 common anode units. Segment drive is provided to the cathodes by the Fair-

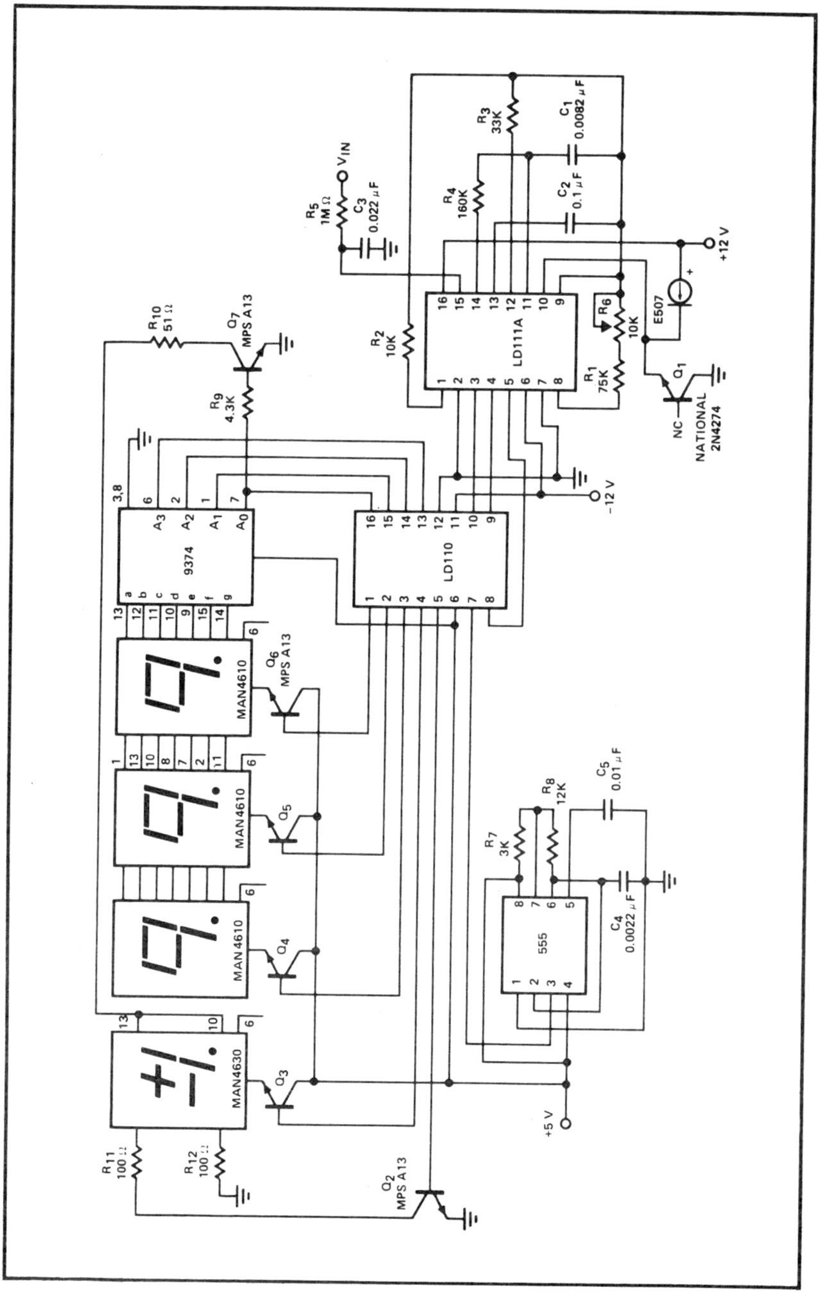

Courtesy Siliconix, Inc.

Fig. 7-31. LD110/LD111A ± 199.9-millivolt common-anode LED display.

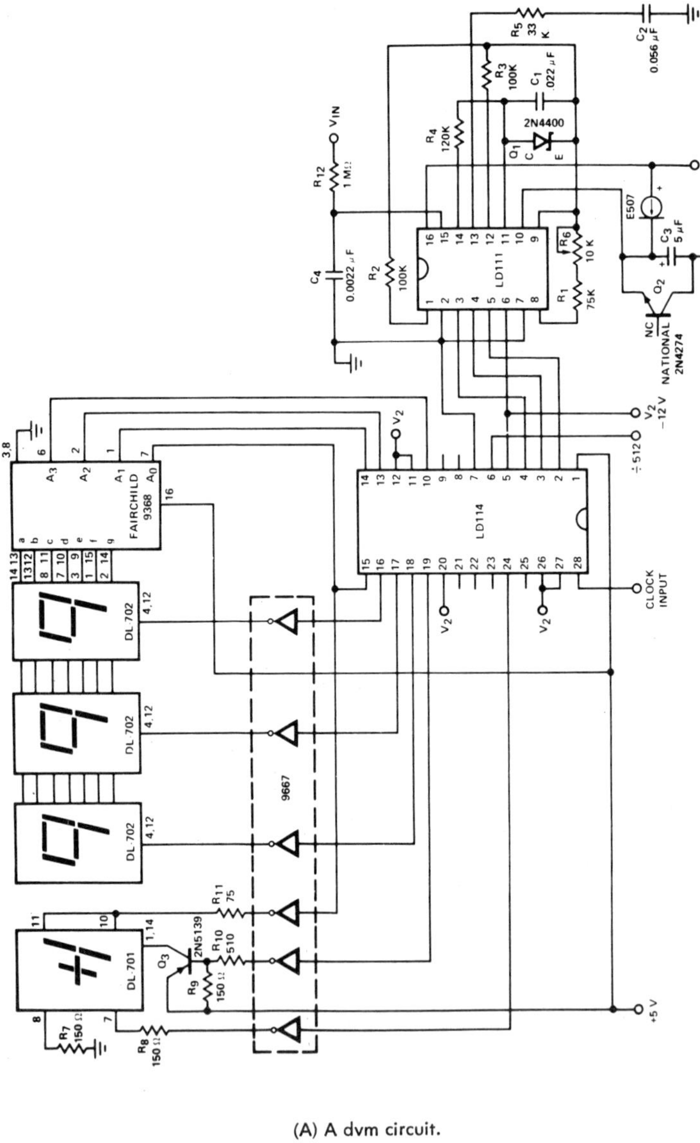

(A) A dvm circuit.

Fig. 7-32. LD114/LD111 ±1.999-volt,

child 9374, a bcd to seven-segment decoder/driver with a current-limited output, which eliminates the need for resistors. Digit drive is provided by the discrete MPSA13 npn transistors.

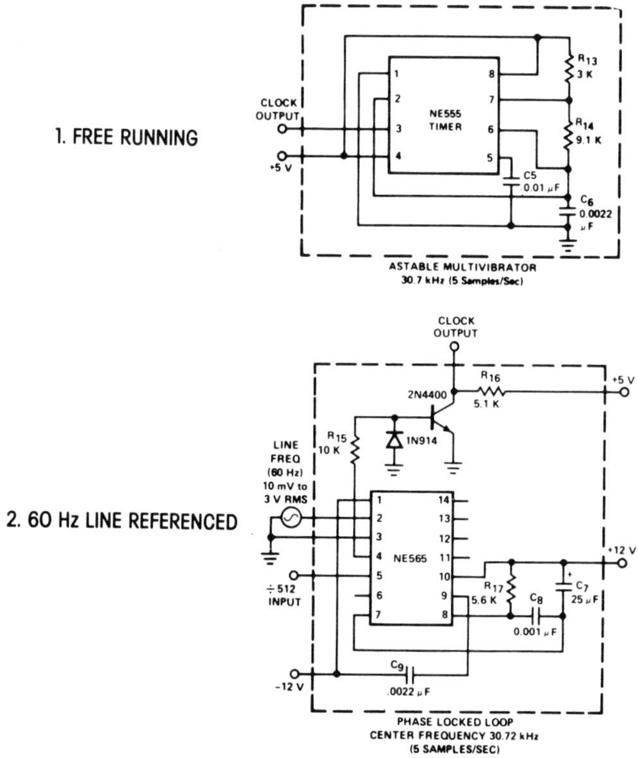

(B) Clock source options.

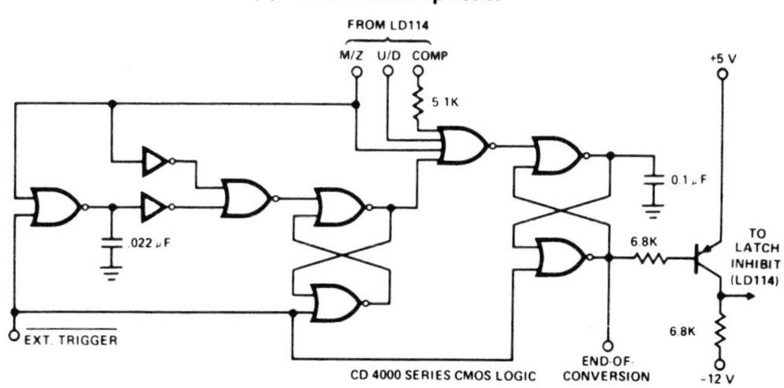

(C) External trigger circuit option.

3½ digit LED display. Courtesy Siliconix, Inc.

LD114/LD111 ±1.999-Volt DVM Circuit

The LD114/LD111 chip pair is illustrated in Fig. 7-32 as a ±1.999-volt 3½ digit dvm with an LED display. The analog portions of this circuit are similar to Figs. 7-29 and 7-30, and R2 can be adjusted to 10K for a ±199.9-millivolt application. The main functional difference is in the increased flexibility provided by the LD114 digital processor.

As noted previously, line referenced noise rejection in an integrating dvm is maximum when the clock is adjusted to an exact multiple of the noise frequency. The LD114 provides an optimum means of realizing this feature, as it incorporates an on-chip divide-by-512 clock counter, which can be used with external phase-locked loop (PLL) to provide a line-referenced clock source. When operated in this manner, normal mode rejection is increased to 80 dB, whereas it is only 40 dB when used without the phase-lock option.

For the LD114, the clock is operated at 30.72 kHz, which results in a 5 per second conversion rate. In Fig. 7-32B Part 1 is shown a simple 555 timer based clock; Part 2 is a clock circuit using the LD114 divide-by-512 output. This circuit produces a 60-Hz referenced 30.72 kHz clock to drive the LD114. Resistor R17 sets the free-running frequency, and should be changed to 4.7K for 50-Hz reference frequencies.

With the exception of the actual display units used, the LED display drive method is similar to that described with Fig. 7-30.

Another option available with the LD114 is a latch-inhibit function, which is activated when pin 27 is connected to −12 volts. With pin 27 at the +5-volt supply level, the normal conversion process occurs. An external circuit can be used with the LD114 to operate it in a convert-on-command mode, as shown in Fig. 7-32C. This circuit, composed of CD4000 series CMOS logic operated on the +5-volt supply, will cycle the LD114/LD111 through a single conversion after the EXT TRIGGER input is taken low. The end-of-conversion output goes high when the conversion is complete, and the data valid at this time remains in the latches.

Driving Other Displays

Thus far the LD series has been shown in applications using only LED displays. Two other display types which are also popular are gas discharge and liquid crystal (LCD) types. These are illustrated in Fig. 7-33.

In Fig. 7-33A is shown a method of driving the Beckman SP752 type gas discharge displays with high voltage DD700 drivers (Beckman). These drivers perform the bcd to seven-segment decode function, but with 80-volt outputs which can drive the displays directly. A positive 180-volt source is required for these displays, and a discrete decimal point driver using npn high voltage types such as 2N3440 (typical) is

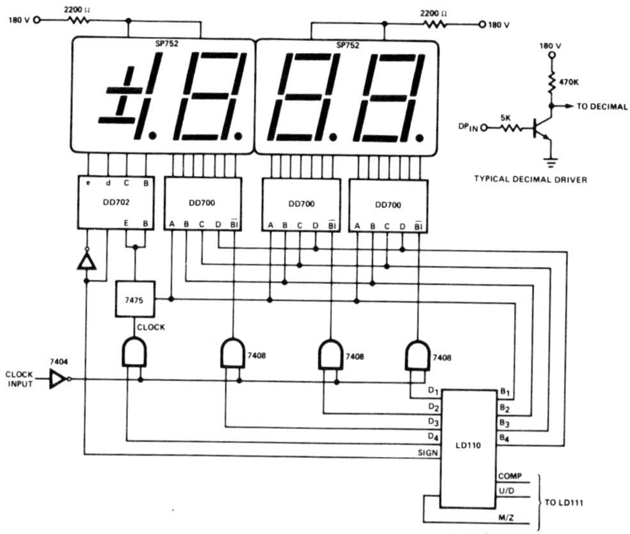

(A) Gas discharge.

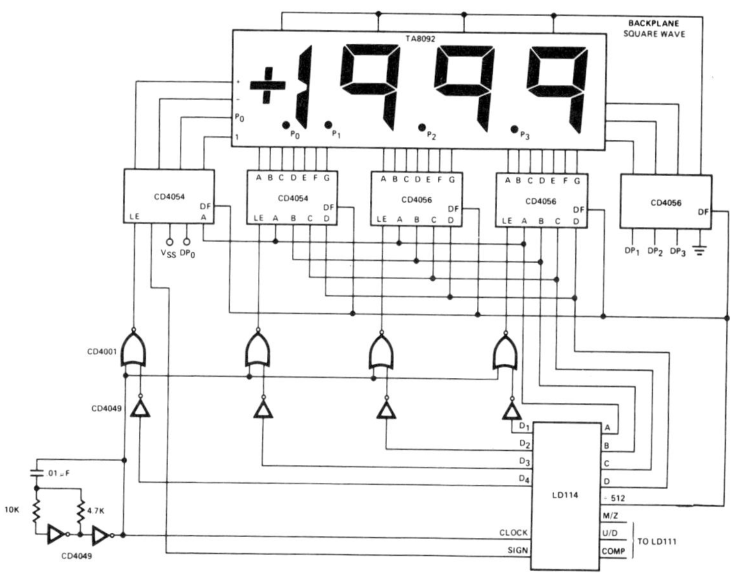

(B) Liquid crystal.

Courtesy Siliconix, Inc.

Fig. 7-33. Driving other display types.

shown. 7400 series TTL logic is shown driving the DD700 inputs. This display interface is a general one, and may be used with any of the previously described LD110/LD111 circuits.

Fig. 7-33B shows an LCD display interface, using the characteristics of the LD114 to advantage. Here the LD114 divide-by-512 output provides the 30-Hz signal required for backplane drive, in conjunction with the CD4056 drivers. A 15.4-kHz clock is used, as provided by the CD4049 astable multivibrator. All logic is CD4000 series, operated from the +5-volt supply. This circuit can be used in conjunction with the analog section described in Fig. 7-32, for instance, to provide a ±199.9-millivolt or ±1.999-volt LCD dvm.

Multimeter Interfaces

To make a complete digital multimeter based on one of the already described basic dvm circuits, various interfaces can be added. Several such circuit possibilities are shown in Fig. 7-34, and are compatible with the LD series device circuits.

Fig. 7-34A shows an auto-ranging circuit to be used in concert with the LD110/LD111, operating with a basic ±199.9-millivolt range. Resistors R1 through R5 form an input attenuator which is switched to reduce the input always to a 199.9-millivolt range. TTL logic is used, driven by the LD110 digit outputs which will clock the 7496 register when all digit outputs are low, representing an overrange condition. Underrange is decoded by gating of D4 and B4, which presets the 7496 to all 1s. For this state, the DG201 switches will be off and the circuit is in a 199.9-millivolt scale condition. Should the voltage input be excessive, the overrange state will persist and the 7496 will be clocked up range, until the desired range is reached.

Current-to-Voltage Converter

Most dvm schemes to measure current employ a precision shunt resistor, across which is measured a voltage proportional to the unknown current. It is desirable in such a technique to minimize the voltage dropped across the sense resistor, both from the standpoint of minimizing change in the current level itself, and minimizing power dissipated in the sense resistor, which can alter its value.

The current-to-voltage converter circuit shown in Fig. 7-34B can measure accurately currents over an 8-decade range, from 200 nA to 2 A full scale, due to the use of a low input current op amp, the LF356.

In this circuit, on the lowest ranges of 200 nA to 200 μA, the voltage dropped across the I_{IN} terminals is close to zero, due to the op amp. For these ranges the output voltage is $2I_{IN}R1$, and the scaling shown provides 200 millivolts of output for the respective ranges. For the higher ranges shunt R4 is used, and the op amp operates as a gain-of-10 amplifier.

The offset control should be adjusted for zero output with the leads shorted, and R1 through R4 values should be precision types, as is consistent with the desired overall accuracy.

Ohms-to-Volts Converter

A simple circuit for realizing an ohmmeter function with LD111 converters is shown in Fig. 7-34C. Here a section of a TL084 quad FET input op amp is used as a floating current source which is programmed by ohms range resistor R_A. The voltage across R_A, which must be held constant, is established by CRO33 and R1-R2. The CRO33 current regulator diode sets up a constant current in R1-R2 which has a nominal TC of zero. R1 is adjusted for a precise 1 volt across R_A, which calibrates the circuit. This can most conveniently be accomplished with a precision reference resistor for R_X, such as 1K.

The circuit can accommodate full scale resistance ranges of 2K to 20 megohms, which are read on the 2-volt scale of an LD111 converter circuit.

The TL084 is shown here from the standpoint of efficiency, as the next circuit employs the remainder of the four devices. Single devices for A1 could be a TL081, CA3140 or LF356.

AC-DC Converter (Averaging, Rms, Sine Wave Calibrated)

An ac-to-dc converter of the averaging type is illustrated in Fig. 7-34D. This circuit uses the remaining four sections of the TL084 FET input op amp, and provides an output which is an averaged dc level equivalent to the rms input.

It accepts a 2-volt rms full-scale ac level from the range switch, and buffers it in the A2 stage. This stage is ac-coupled to remove any dc component, and features a high input impedance of 22 megohms. Stages A3 and A4 form a full-wave rectifier and averaging filter. R8 calibrates the average dc output to a level equal to the rms level of a sine wave input. R9 provides a dc offset trim for the op amps, allowing optimum low-scale accuracy (optional).

With the TL084 high slew rate op-amp devices, operation is possible to 100 kHz with low error, and accuracy is better than 1% to below 10 Hz.

True RMS-to-DC Converter

A form of ac-to-dc converter which is generally much more useful than the just described average responding type is a *true rms*-to-dc converter. A true rms converter produces a more accurate dc output for nonsinusoidal inputs such as pulse or rectangular waveform, including dc component(s).

The Analog Devices AD536 rms-to-dc converter chip is used in this circuit, and is a highly accurate, completely pretrimmed device. An

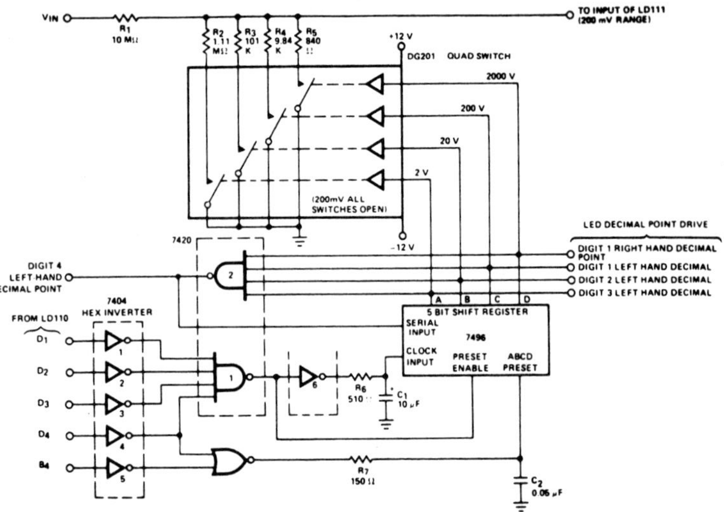

Courtesy Siliconix, Inc.

(A) Auto-ranging circuit.

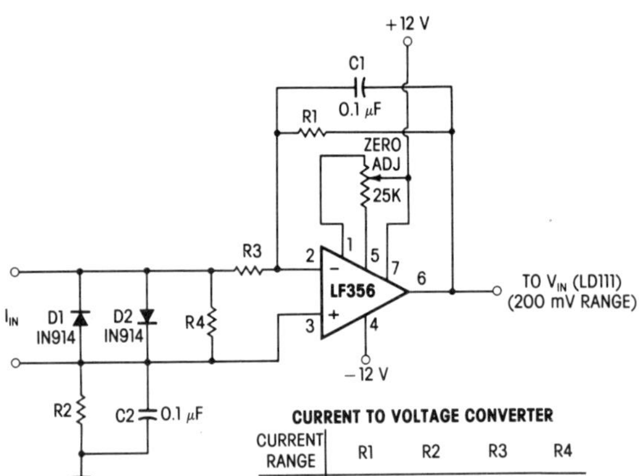

CURRENT TO VOLTAGE CONVERTER

CURRENT RANGE	R1	R2	R3	R4
200 nA	500K	500K	0	∞
2 μA	50K	50K	0	∞
20 μA	5K	5K	0	∞
200 μA	1K	0	0	∞
2 mA	50K	0	5.0K	10.0
20 mA	50K	0	5.0K	1.0
200 mA	50K	0	5.0K	0.1
2 A	50K	0	5.0K	0.01

Courtesy Siliconix, Inc.

(B) Current-to-voltage converter.

Fig. 7-34. Different types

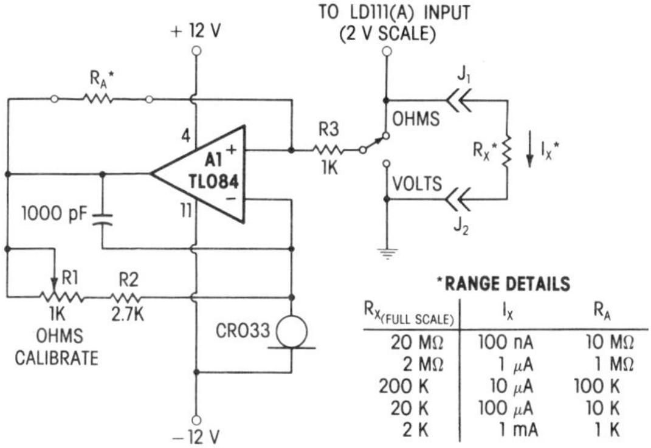

*RANGE DETAILS

$R_{X(FULL\ SCALE)}$	I_X	R_A
20 MΩ	100 nA	10 MΩ
2 MΩ	1 μA	1 MΩ
200 K	10 μA	100 K
20 K	100 μA	10 K
2 K	1 mA	1 K

(C) Ohms-to-volts converter.

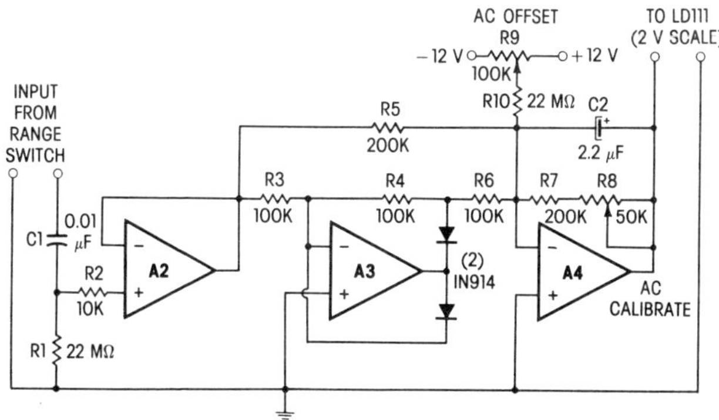

(D) Ac-to-dc converter.

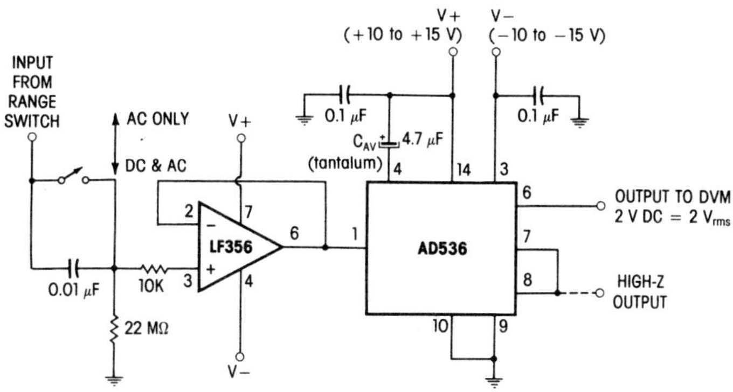

(E) Rms-to-dc converter.

of multimeter interfaces.

AD536J has a basic accuracy of 0.5%, while the AD536K is accurate to 0.2%.

As shown in this circuit, the AD536 provides this accuracy for inputs in the range of 0 to 2 volts rms. A high input impedance is provided here by the LF356 op-amp buffer, which "unloads" the range switch. For circuits that can drive the 16.7K AD536 input impedance at pin 1, this buffer stage can be eliminated. An input mode switch selects either a dc-coupled input, or ac coupling only. This can conveniently allow measurement of ac, or dc + ac components of a signal.

The AD536 has a conversion scale factor of unity for an rms input, therefore it will read 2 volts rms as a 2-volt dc output. Dc + ac components are read as their correct rms equivalent, or

$$V_{O(dc)} = \sqrt{V_{in(dc)}^2 + V_{in(ac)}^2}$$

The circuit has a frequency dependent error of 1% or less, with 2 volts rms in, up to 100 kHz, and down to 10 Hz. Settling time of the rectifier filter is determined by C_{AV}, which is connected across an internal 25K resistance. As shown here it is 0.5 second.

This circuit can use power supplies in the range of +10 to +15 volts, and is thus compatible with not only "LD" series converters but many others as well. For more general purpose on ±15-volt supplies, it can handle inputs up to 7 volts rms, with accuracy as previously quoted.

8052(A)/8053(A) + Logic DVM Circuits

One of the most flexible performance combinations available in chip set integrating a/d converters is a composite of the 8052 (or 8052A) plus the 8053 (or 8053A) and discrete logic devices. These devices, with appropriate logic, can be used to implement 3½ to 4¾ digit bcd, or up to 16-bit binary dvm a/d circuitry, with analog input ranges of from ± 199.9 millivolts up to ±4.0000 volts.

A representative "family" circuit for these devices used in such a manner is shown in Fig. 7-35. It may be used with minor modification to form any of the dvm scale/function combinations shown in the table.

In this circuit, the control of which dvm family "member" is realized is governed by connections to IC5, a 4-bit counter, termed the *state* flip-flop. This counter drives two points, termed MSB and MSB-1, respectively. Connection of these points to the counter outputs shown implements the appropriate timing for a given range.

The decade counters of this circuit are contained within the TIL306 (Texas Instruments) counter/latch display devices. The clock inputs are driven by a (external) 120-kHz, TTL-compatible source.

Operation of this dvm family may be more readily appreciated by reference to a simplified functional diagram, Fig. 7-35B. For the connections as specifically noted in Fig. 7-35A, the range is 20000 counts.

This is as would be used for a 4½ digit, ±1.9999 volt dvm, with a +1.0000-volt reference (as adjusted by the 1K pot). In general the analog range of the 8052/8053 is ±2 V_{REF}, where V_{REF} is the voltage applied to the 8053(A) pin 2. The reference can be derived from the +1.85-volt output of the 8052, or it can be furnished by an external reference source.

Logic used in this dvm is generally CMOS, plus one 74L04 hex inverter, and uses a +5-volt supply. The data valid output goes high at the end of conversion, and parallel bcd outputs are available (if desired) from the TIL306s. The circuit will free run at 3 conversions per second, if the sync input is held at +5 volts. With this line low, the circuitry is held in auto zero, and will commence a conversion on a rising edge. An overrange condition is indicated by a blinking display of 19999, and a high overrange output.

Although a 4½ digit application is shown here, operation can be modified for 3½ digit use by deleting the center TIL306, changing the clock to 12 kHz, and making the (noted) changes to the MSB and MSB-1 connections.

Some specific examples of other family options are best explained in the context of specific figures. For example, the 16-bit binary scaled converter of Fig. 7-36 operates as a 15-bit + sign converter, scaled for a sensitivity of 0.1 mV/LSB. This yields a ±3.2768-volt analog range, with output data appearing in 15-bit complemented latched format from the CD4042 "D" latches. A clock frequency of 136 kHz provides a nominal 3 conversion per second rate, and the reference voltage is adjusted to +1.6384 volts for scale colibration. Otherwise, the circuit considerations are generally as previously described.

Certain high analog ranges, such as ±4 volts, are best accommodated with the 8053(A) by using an external *negative* reference voltage, with the 8052(A) reference unconnected. This type of operation is illustrated in the ±3.9999-volt range, 4¾ digit bcd dvm circuit shown in Fig. 7-37.

To accommodate a negative reference, the drive to the 8053(A) pins 6 and 10 are interchanged, otherwise the circuit generally operates as the previous bcd circuits. With a range of 4¾ digits, a high-stability low TC reference is desirable, such as an LM399 diode operated in a negative fashion, with an output scaling potentiometer to yield −2.0000 volts.

Application points which are common to all 8052/8053/7101/7103 a/d devices are proper treatment of grounds, supply, and signal path wiring. Also, the integrating and reference capacitors used should be high quality, low dielectric absorption units. Polypropylene is preferred, with polystyrene as an alternate choice. Layout of analog portions of the circuitry should be compact and direct, with ground planes and screening or guarding recommended.

FULL SCALE	V_{REF}	TOTAL NUMBER OF DECADE COUNTERS	CONNECT MSB-1 to	CONNECT MSB to
+200.0 mV	+0.1000 V	4	Q_A	Q_B
+2.000 V	+1.000 V	4	Q_A	Q_B
+400.0 mV	+0.2000 V	4	Q_B	Q_C
+4.000 V	−2.000 V	4	Q_B	Q_C
+800.0 mV	+0.4000 V	4	Q_C	Q_D
+2.0000 V	+1.0000 V	5	Q_A	Q_B
+4.0000 V	−2.0000 V	5	Q_B	Q_C
+3.2768 V	+1.6384 V	4*	Q_C	Q_D

*Number of 4-Bit Binary Counters

PACKAGE DESIGNATION
1 CD4027 - DUAL J K MASTER SLAVE FLIP FLOP
2 CD4025 - TRIPLE 3-INPUT NOR GATE
3 CD4073 - TRIPLE 3 INPUT AND GATE
4 CD4070 - QUAD EXCLUSIVE-OR GATE
5 74C161 4-BIT BINARY COUNTER
6 74L04 HEX INVERTER

NOTE FOR 3½ DIGIT USE 16KHz CLOCK FREQUENCY AND DELETE TIL306 FROM MIDDLE.

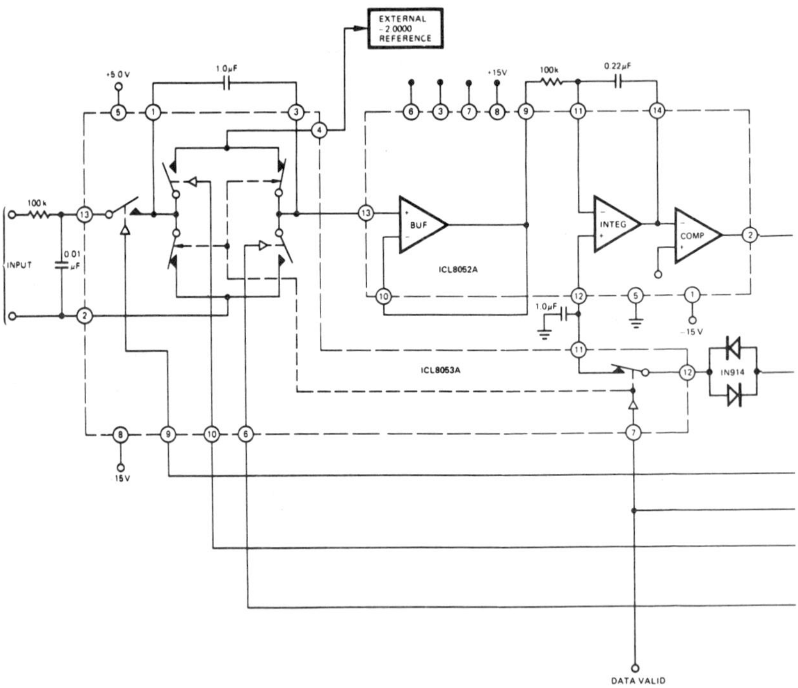

(A) Circuit

Fig. 7-35. The 8052(A)/8053(A)

8052/7101 DVM Circuits

Considerable overall circuit simplification is possible in 3½ digit dvm applications by using the 8052 in concert with the 7101 digital processor. The 7101 provides the digital logic and analog switching, and furnishes a parallel latched 3½ digit output, along with sign and overrange signals. It has an internal clock, requires a start input signal to initiate conversion, and furnishes a busy output signal.

A simplified diagram of a 8052/7101 3½ digit dvm circuit is shown in Fig. 7-38. This circuit can be used for both ±199.9-millivolt and ±1.999-volt ranges, as selected by the dpdt switch which selects either a 0.1- or 1-volt reference, and changes the integrator resistor from 10K to 100K. The 1K pot is adjusted to calibrate the ±1.999-volt range; then if the 9K and 1K resistors provide a precise 9/1 ratio this also

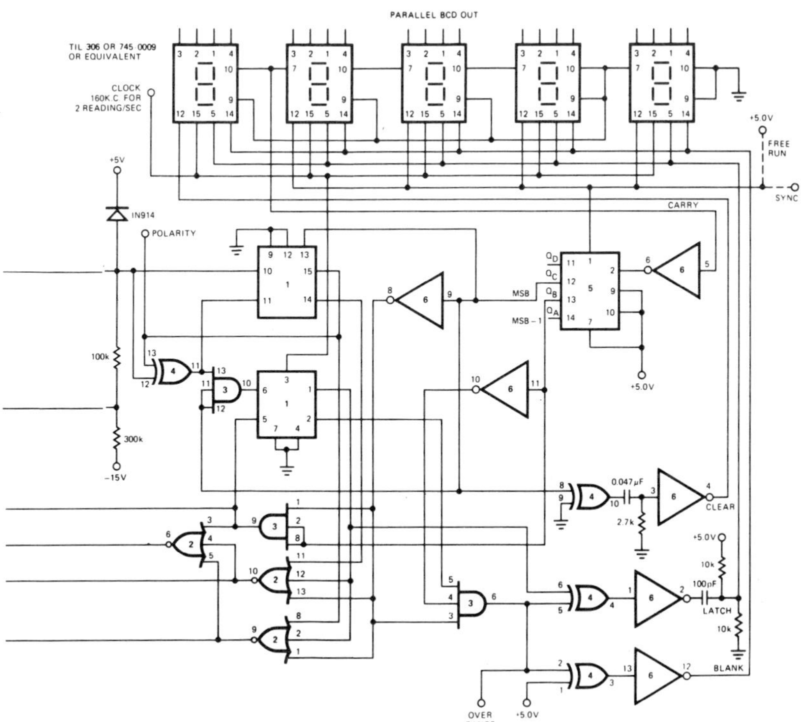

Courtesy Intersil, Inc.

schematic.

family of dvm circuits.

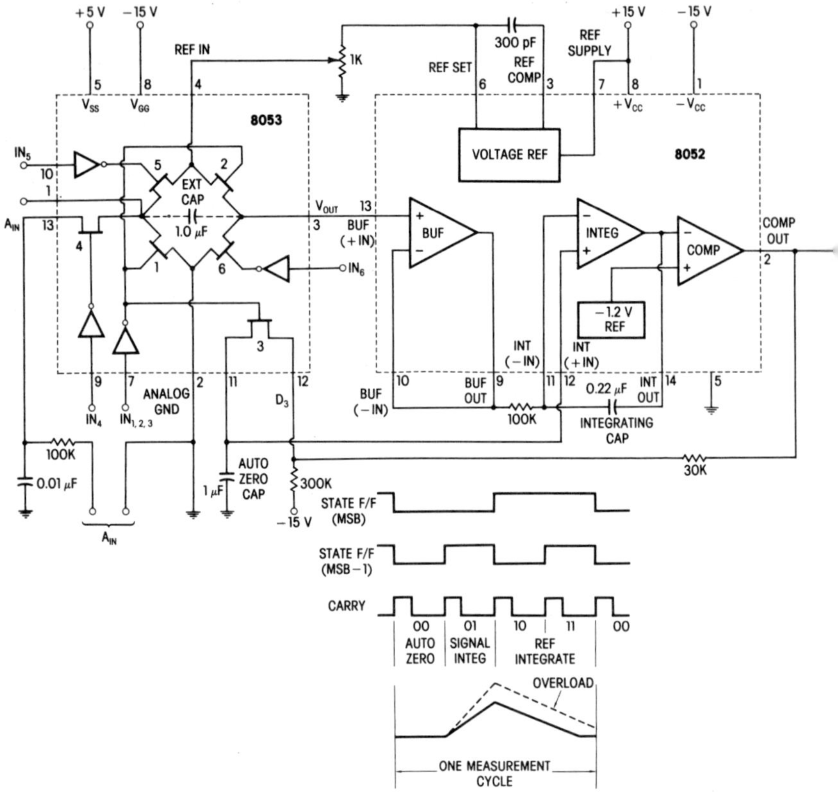

(B) Simplified

Fig. 7-35 (cont). The 8052(A)/8053(A)

calibrates the ±199.9-millivolt range. Auto-zero, reference and integrator capacitors are selected as previously described for the 8052.

Outputs of the 7101 are in the form of latched parallel data, which can be used as shown to drive an LED display. The display drive details are not all shown here, but can be similar to those previously described, using three 7447s for instance, plus 150-ohm segment current-limit resistors, and common anode displays.

The clock frequency of the 7101 is determined by the capacitor between pins 25 and 28, and the value shown provides a 3 conversion per second rate. The 7101 must have the start-reset (pin 36) held low prior to conversion, to enable the auto zero cycle. When taken high, a conversion is started, and this pin must return low prior to the end of conversion.

The out-of-range signal at pin 23 is low for counts of 2000 or more. The bcd data outputs are all high state true, with the exception of the

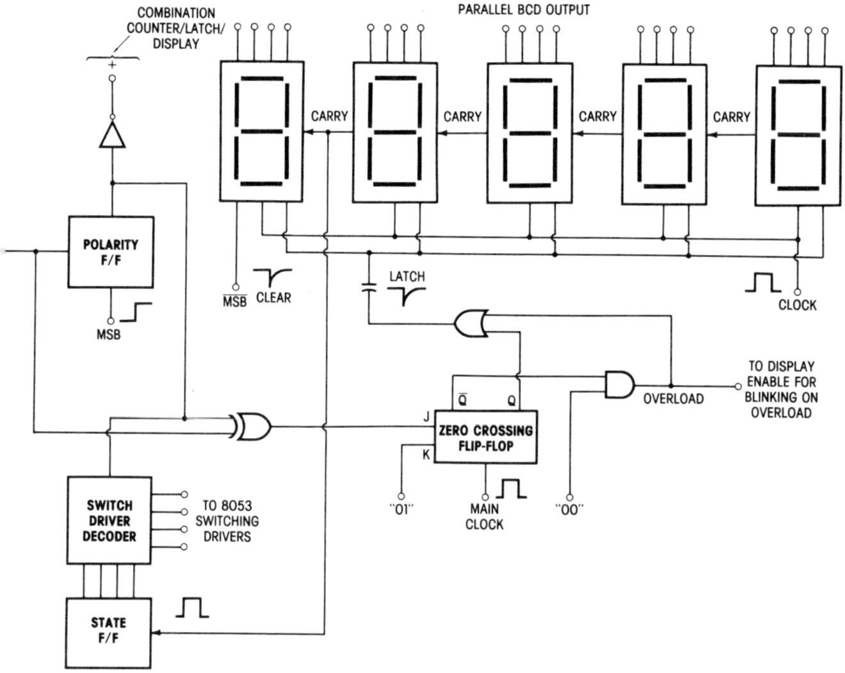

functional diagram.

Courtesy Intersil, Inc.

family of dvm circuits.

1000 bit, which is low for counts over 999. These outputs are TTL compatible.

8052A/7103A

Very high performance with relatively low component count is possible using the 8052A/7103A a/d chip pair. These devices feature 4½ digit operation with multiplexed bcd outputs, and are shown in a typical application in Fig. 7-39.

Functionally, this circuit is similar to an 8052/7101 combination, with the exception that the 7103A has multiplexed 4½ digit bcd outputs. The 7103A uses an external 120-kHz clock, providing 3 conversions per second.

In this circuit, a common anode LED display is illustrated, such as MAN1s and a MAN101 (see Table 7-3), driven by a 7447 and discrete npn anode drivers. The transistors are general purpose types such as 2N3904 or 2N2222. Note that the common anode units can be contained in a single array, such as the CA3082 (RCA).

As shown here this circuit is suited for ±1.9999-volt use, with the reference voltage adjusted for +1.0000 volt. A high resolution trimmer is recommended for this adjustment, and the reference divider should use stable components. The 8052A reference is capable of a 40 ppm/°C (typical) TC; for lower TCs an external reference can be used, in which case pins 3 and 6 are open, pin 7 grounded, and the 300-pF capacitor removed, (see Chapter 3 for low TC reference options). The reference is the primary determinant of stability and accuracy in this circuit, and the remaining components are of lesser (but not trivial) importance. A low dielectric absorption type integrator capacitor should be used, as noted previously.

The 8052/7103 combination (no "A" suffix) can also be used for 3½ digit applications with relatively minor changes. For such a use the 7103 pin 2 is grounded, placing it in the 3½ digit mode, and a 12-kHz clock is used. An 8052/7103 device pair used in this manner is still the same basic pair; the difference is that the "(A)" suffix devices have been tested for 4½ digit accuracy.

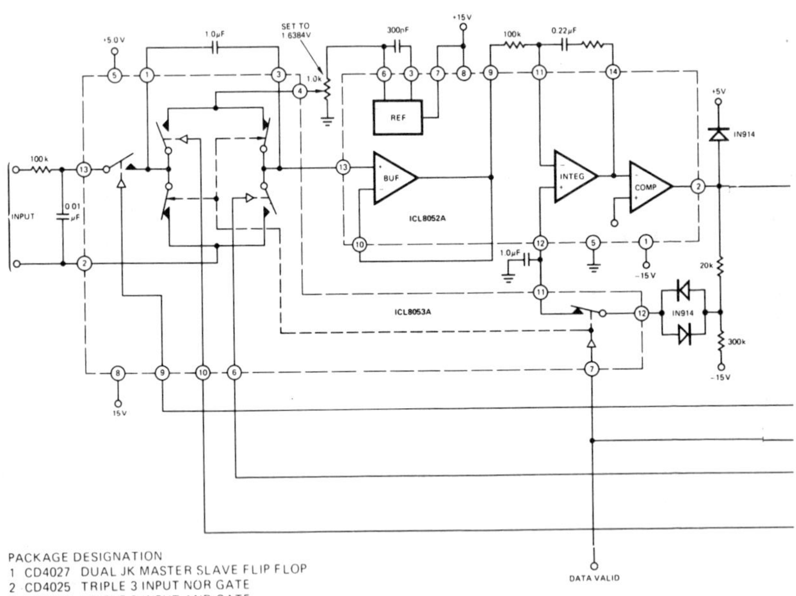

PACKAGE DESIGNATION
1 CD4027 DUAL JK MASTER SLAVE FLIP FLOP
2 CD4025 TRIPLE 3 INPUT NOR GATE
3 CD4073 TRIPLE 3 INPUT AND GATE
4 CD4070 QUAD EXCLUSIVE OR GATE
5 74L04 HEX INVERTER
6 10 74C161 4 BIT BINARY COUNTER
11 14 CD4042 QUAD CLOCKED 'D' LATCH

Fig. 7-36. A ± 3.2768-volt,

The run/hold input, pin 4, will allow the 7103(A) to free run when high (or open). When taken low, an in-process conversion will complete, then be held as long as the low state persists. Triggered conversions can be initiated by pulsing this pin from low to high.

The STROBE output is a series of five negative going pulses, which occur sequentially, each successive pulse at the midpoint of the respective D5, D4, etc. outputs. These pulses may be used to demultiplex the output data to a parallel bcd format in a fairly simple manner.

Single-Chip Integrating A/D Converters

This section of the chapter deals with IC integrating a/d converters which are in single-chip form, and so require relatively few additional components for operation. Although there are few general differences of operation between most of these types and the chip set types already discussed, there are practical differences which may favor their use for certain applications. These are the low-component count, and also low-power and/or low-voltage operation, and direct-display drive. As previously noted, these types of a/d devices are largely oriented towards display or instrument uses, but there are two devices notable for their data-acquisition capabilities.

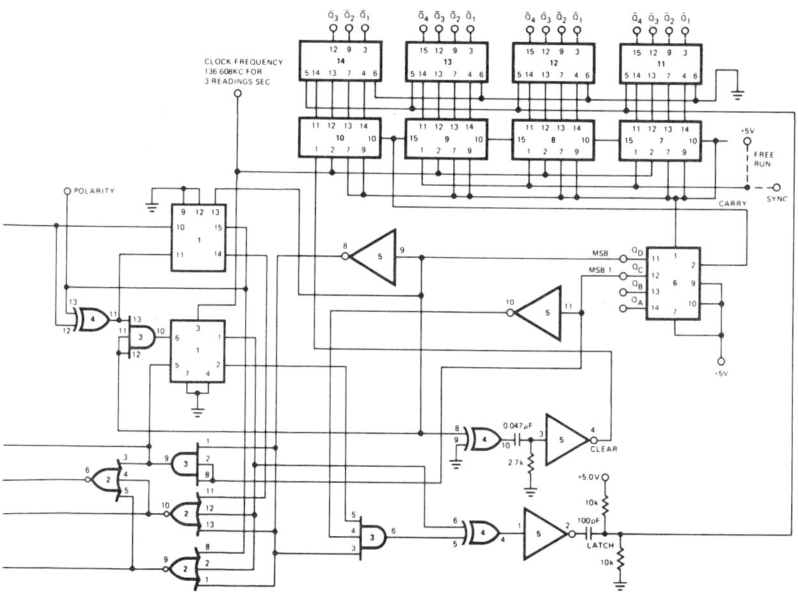

Courtesy Intersil, Inc.

16-bit binary converter.

LD130/LD131 3 and 3⅓ Digit DVM Circuits

The LD130 and LD131 a/d chips are designed for 3 and 3⅓ digit applications, respectively. Both devices employ ±5-volt power supplies, have internal clock oscillators, and use an external 2-volt reference voltage. They have fixed internal range scaling; for 0 to ±0.999 volt or 0 to ±1.499 volts, respectively. They require only a few external passive components to implement the a/d function.

The LD130 is illustrated in a basic ±0.999-volt 3 digit dvm circuit in Fig. 7-40. In Fig. 7-40A is shown a version of the LD130 circuit which uses an LED display, the monolithic NSN-33 (National Semiconductor). The NSN-33, a common-cathode unit, is driven by a CA3083 npn array digit driver, with the segment drive provided by the 4511, a CMOS bcd-to-7 segment decoder. The segment resistors R7 through R13 are 470-ohm units. For use of the monolithic display mentioned, a pnp transistor driver is required, the 2N4403. The DP output goes to the decimal point input of the display; the D_n input to the appropriate digit output of the CA3083.

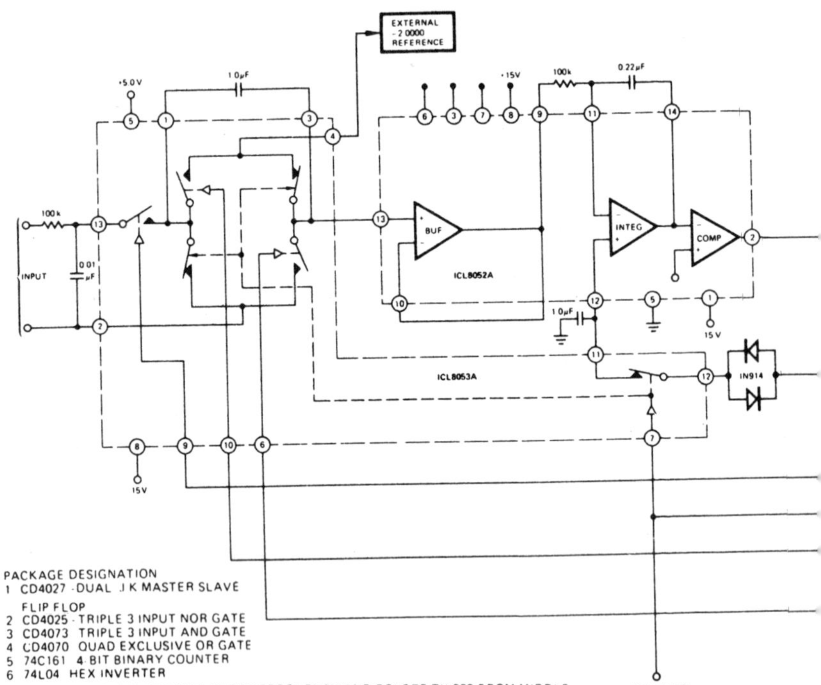

Fig. 7-37. A ± 3.9999-volt,

The LD130 and LD131 use a low voltage reference of 2 volts provided here by the constant current through the CRO33 (a low TC regulator diode) and R1-R2. R2 is adjusted to the output reference level required for full-scale calibration of 0.999 volt (LD130) or 1.499 volt (LD131).

Capacitors C1 and C2 perform functions within the a/d converters analog subsystem, and should be selected for low dielectric absorption properties. The values shown are valid for a clock frequency of 20 kHz, which results in a conversion rate of 3 per second. Capacitor C3 sets the oscillator frequency.

Other circuit details include an FET, the E204, which prevents a possible power-on lockup condition of the LD130. The diode and capacitor in the +5-volt line to the LD130 reduce display-current induced supply modulation, which might cause measurement errors.

This circuit can be used as the basis for a complete dmm, with functions of auto ranging, dc and ac volts, and ohms measurement. Complete details of construction (including pc layouts, stuffing chart, and parts list) are included in Siliconix DA76-3 (see chapter references).

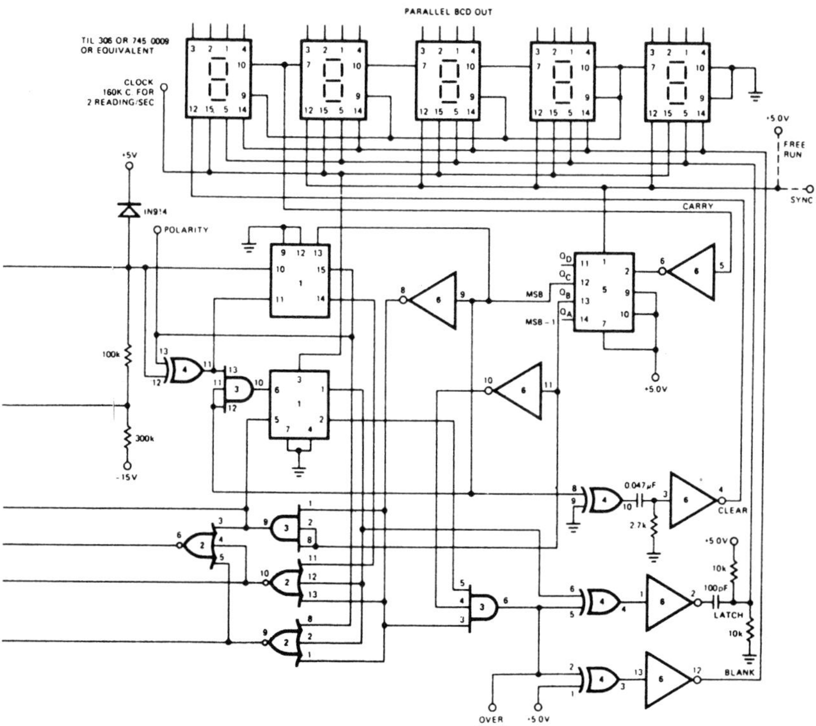

Courtesy Intersil, Inc.

4¾ digit dvm circuit.

The lowest power consumption in an LD130 dvm circuit is realized when low power displays are used, such as the LCD display dvm circuit of Fig. 7-40B. In this circuit the CD4024 divides the D3 output, which occurs at about a 1-kHz rate, down to a 30-Hz square-wave frequency suitable for backplane drive of the TA8092 LC display. The CD4056 function is to demultiplex and latch the bcd data, and drive the display. The CD4055 provides polarity and DP drive.

The circuit can be used with supplies of ±5 volts, either fixed station or portable. For portable use, the battery hookup in the inset is useful, and can provide 150 or more hours of continuous duty with No. 216 batteries.

Note that this circuit is somewhat general in that an input filter/protection RC network is not shown, nor is the antilockup FET. Nevertheless, both are still appropriate and recommended for reliable use.

LD131 Circuit

A dvm circuit based on the LD131 and suitable for +1.499-volt, 3⅓ digit applications is shown in Fig. 7-41. Much of this circuit is

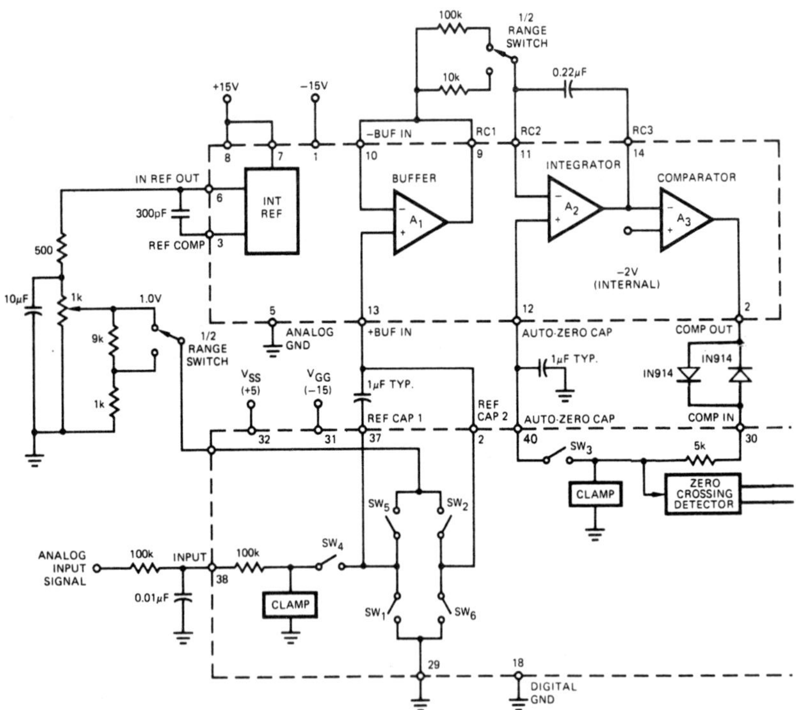

Fig. 7-38. 8052/7101 ± 199.9-millivolt/

similar in concept to the LD130 circuit of Fig. 7-40A, so only important differences are discussed.

LED displays are used, FND-357s and a FND-358 (Fairchild), driven by a 74C48 bcd-to-7 segment driver and discrete npn cathode drivers. The CMOS logic circuitry to the left forms an overflow decoder/one-shot, which drives the overflow display for inputs greater than 0.999 volt. Output point "b" of this circuit is also used for overrange and underrange applications.

The CMOS/discrete circuit to the right forms an (optional) overrange indicator which blanks the 3 LSDs of the display for inputs greater than 1.5 volts, via the blanking input of the 74C48. Output point "a" of this circuit is also used for overrange and underrange applications.

Auto-Ranging Circuits

Earlier in the chapter a variety of dmm peripheral circuits were discussed. Many of these are applicable as well to use with the LD130 or LD131, as long as consideration is given to proper function at the re-

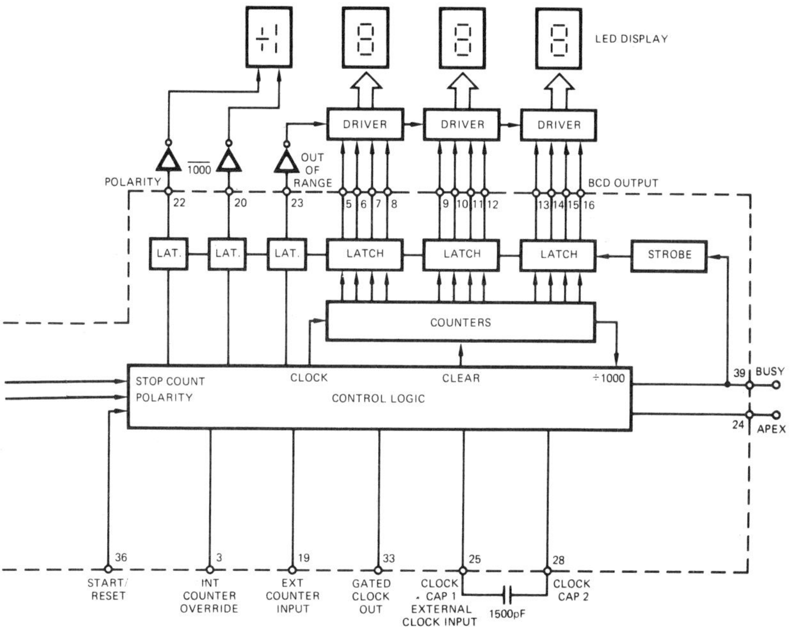

Courtesy Intersil, Inc.

±1.999-volt, 3½ digit dvm.

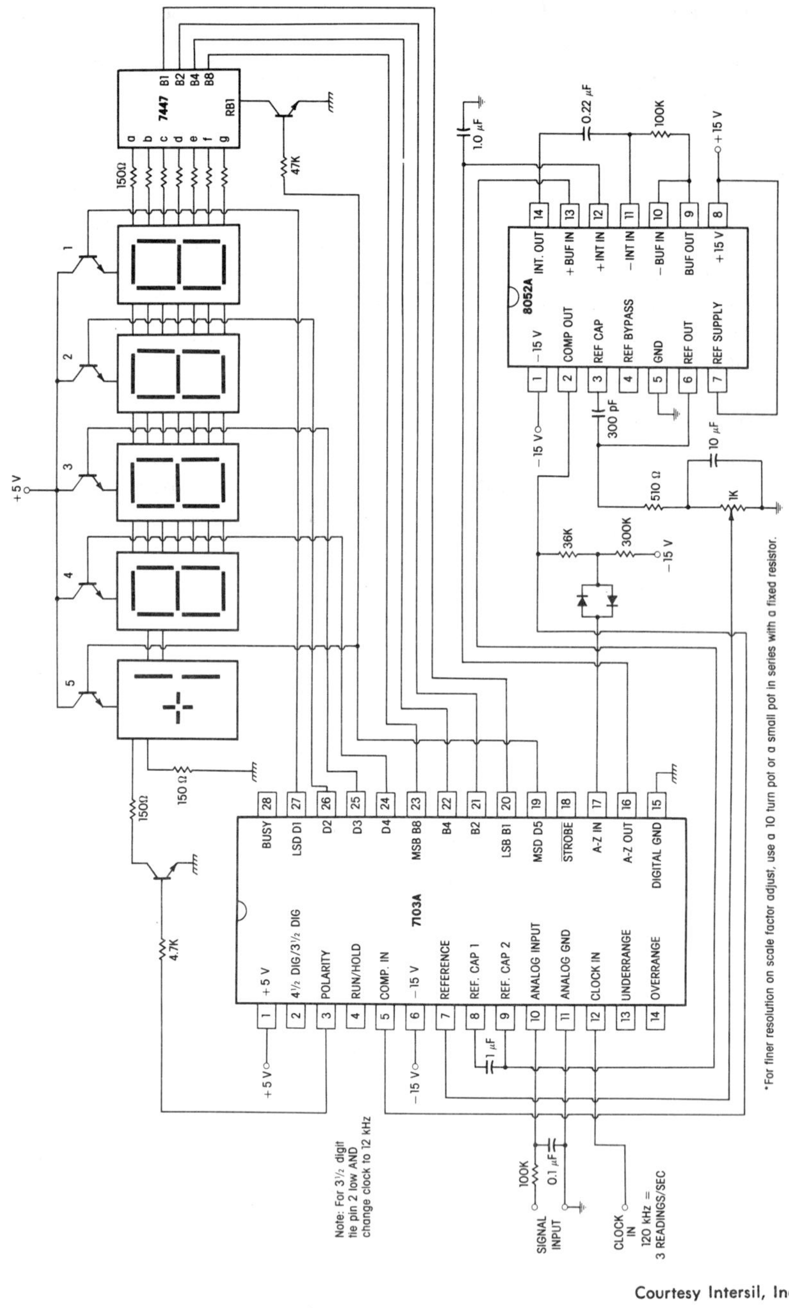

Courtesy Intersil, Inc.

Fig. 7-39. 8052A/7103A ± 1.9999-volt, 4½ digit dvm.

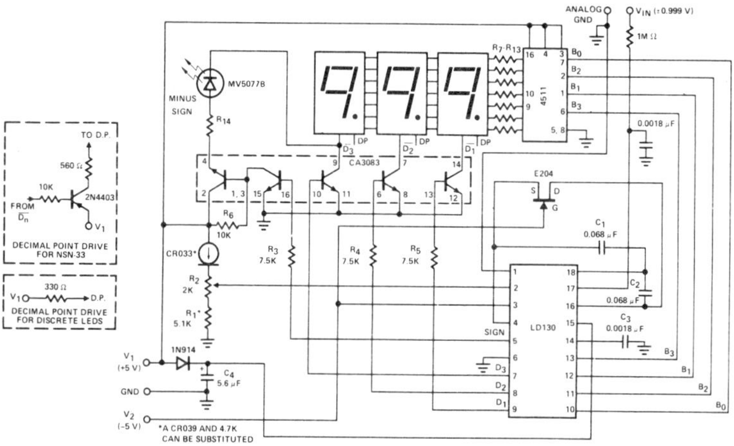

(A) LED display.

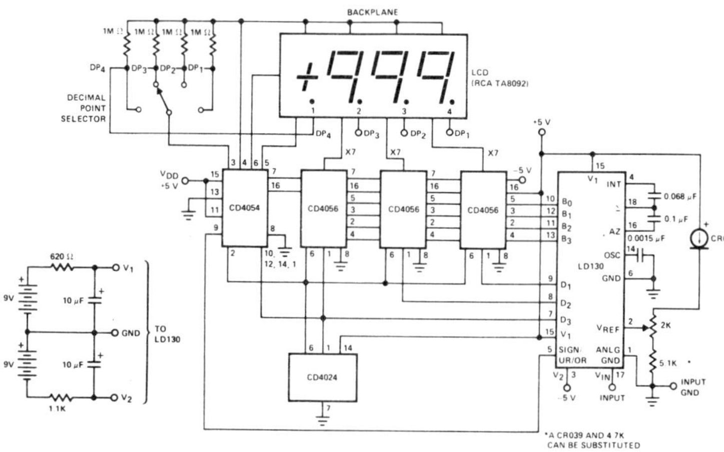

(B) LC display.

Courtesy Siliconix, Inc.

Fig. 7-40. LD130 ± 0.999-volt, 3 digit dvm circuits.

duced (±5-volt) power supply levels, and suitable output voltage scaling. The ohms-to-volts converter and ac-to-dc converter of Fig. 7-34 for instance, can be used with ±5-volt supplies.

Unique to the LD130 and LD131 of course, are considerations for their operation in an auto-range mode. Circuitry for this function is shown in Fig. 7-42.

The LD130 circuit shown in Fig. 7-42A is similar to the scheme used with the LD110/LD111 devices, a shunt attenuator. With the LD130, the underrange and overrange condition is multiplexed and can be de-

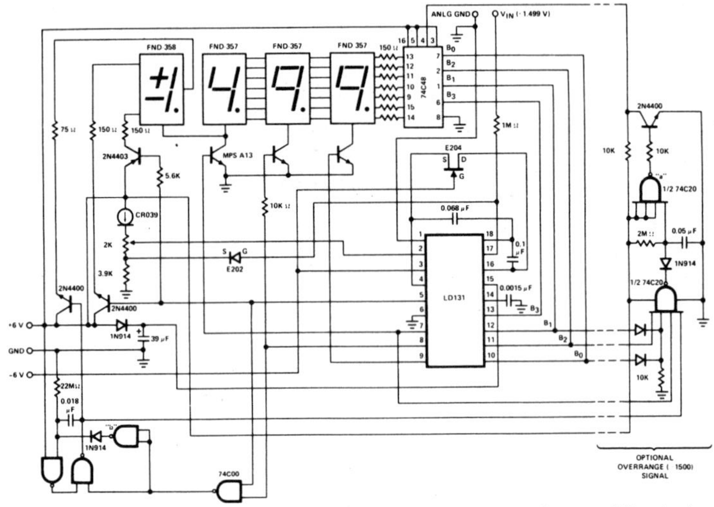

Courtesy Siliconix, Inc.

Fig. 7-41. LD131, ± 1.499-volt, 3⅓ digit dvm.

coded by ANDing the SIGN/UR/OR output with D1 (and D2), as shown. For an underrange and overrange condition, a pulse will occur to step the shift register, until a proper scale has been found. The attenuator shown provides input ranges of ±0.999 volt through ±9990 volts (5 decades) with a 10-megohm input resistance on the four higher ranges.

The LD131 circuit is similar in terms of range shifting, but the logic decoding is more complex. The upper gate ANDs the same two pin outputs of the LD131, as for the LD130. The lower gate ANDs the "a" and "b" signals as derived by the logic circuit shown in Fig. 7-41. As with the LD130, the basic range of the LD131 is multiplied by scales of 1 through 10,000.

Further information on dvm support circuitry for the LD130 and LD131 can be found in the references at the end of the chapter.

MC14433 DVM Circuits

The MC14433 a/d converter is suitable for 3½ digit dvm applications, with basic analog input ranges of ±199.9 millivolts or ±1.999 volts. It uses low voltage power supplies of ±5 volts with low current drain, has an internal clock, and requires an external reference voltage source. Digital output is in the form of multiplexed bcd data, with overrange and end of conversion signals also provided. Basic dvm circuits suitable for use with either LED or liquid crystal displays using the MC14433 are shown in Fig. 7-43.

The LED display circuit of Fig. 7-43A uses a common-anode LED display (see Table 7-3), driven by discrete npn digit drivers from the

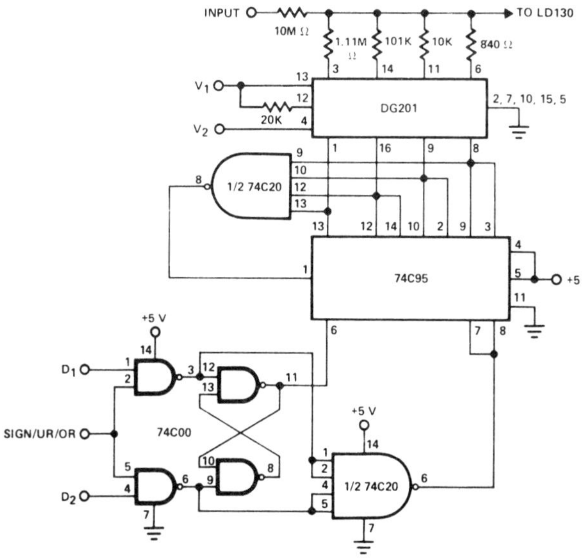

(A) LD130.

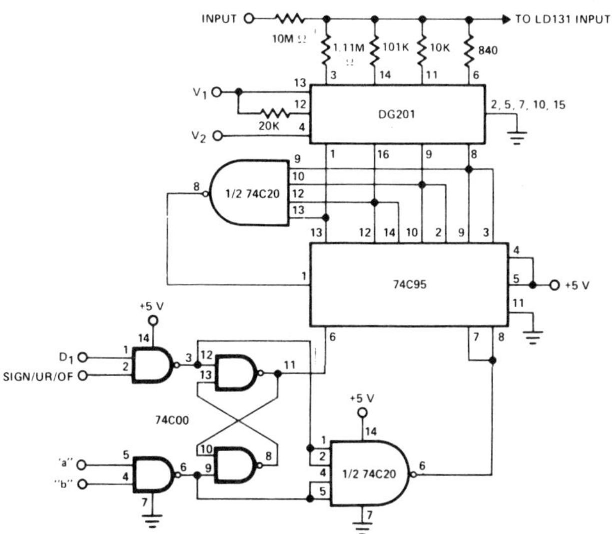

(B) LD131.

Courtesy Siliconix, Inc.

Fig. 7-42. Auto-ranging circuits.

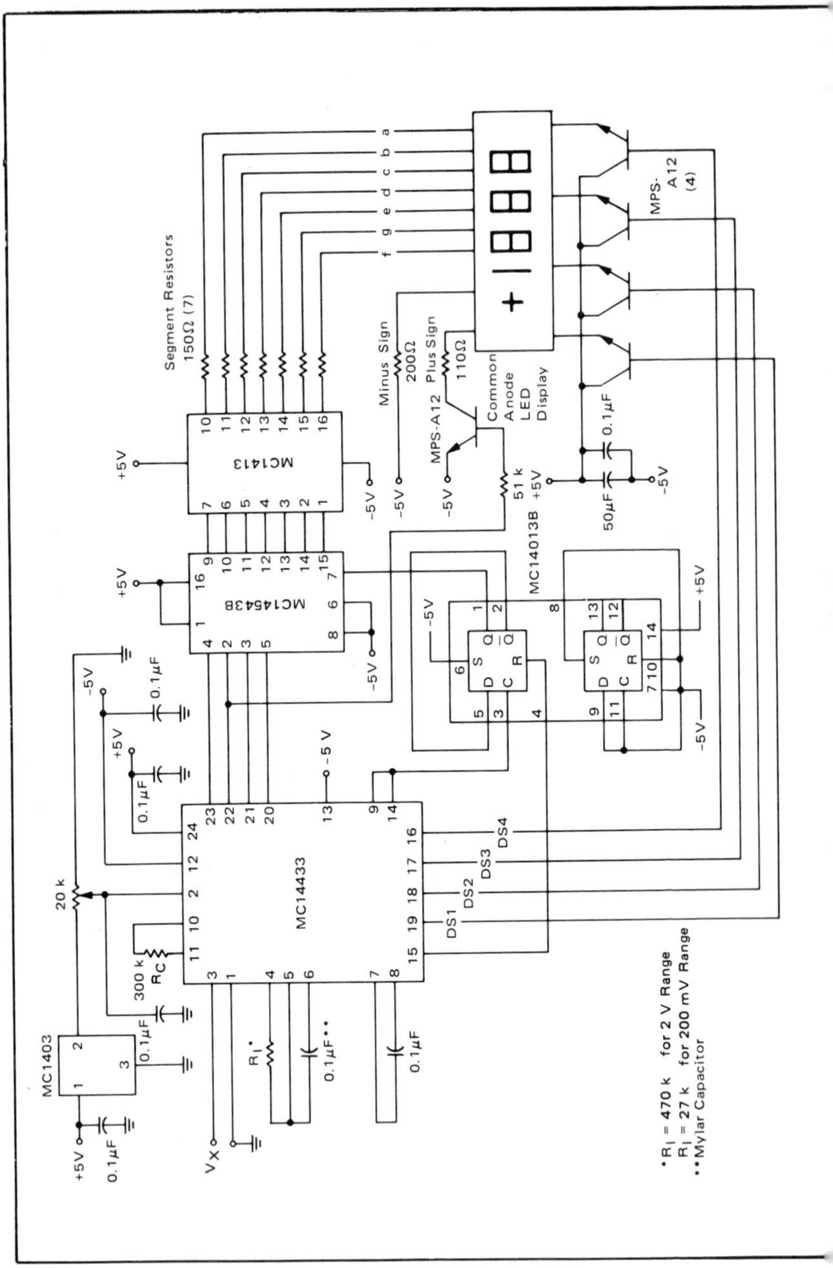

(A) LED display.

Fig. 7-43. MC14433 ± 199.9 millivolt/

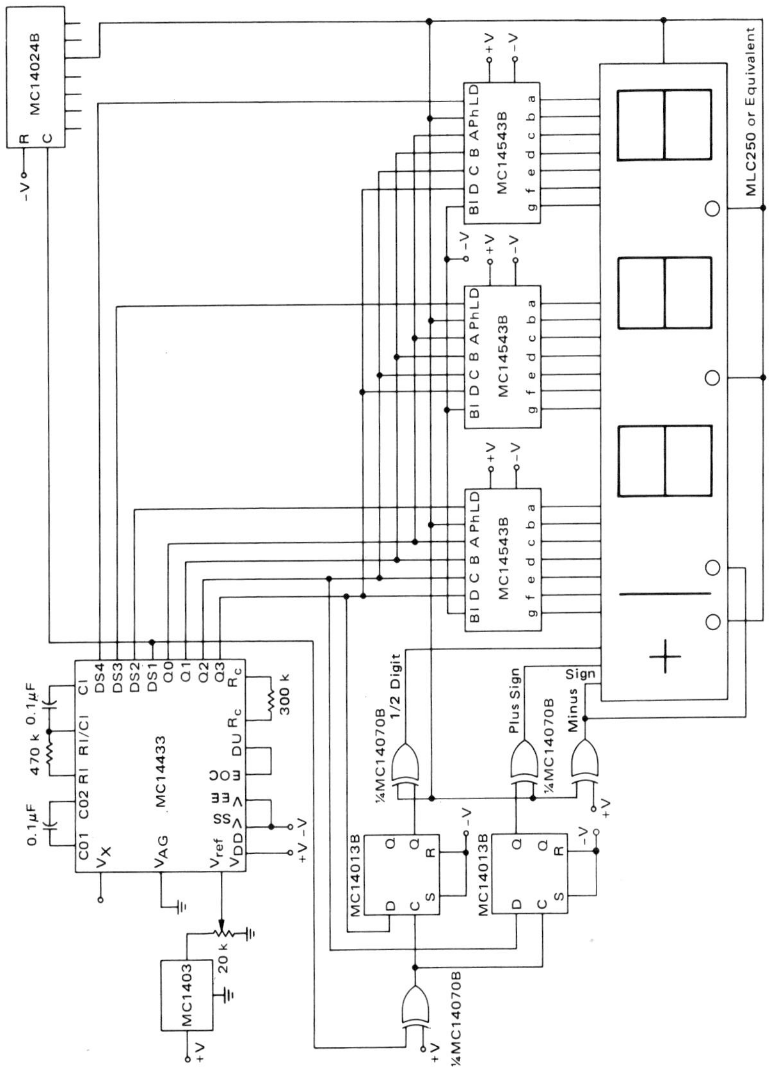

(B) LC display.

Courtesy Motorola Semiconductor Products, Inc.

±1.999 volt, 3½ digit dvm circuits.

digit outputs, and a combination MC14543B bcd-to-segment decoder, plus an MC1413 segment driver. The single D flip-flop, ½ of the MC14013, provides a display overrange indicator, causing it to flash on overrange inputs by blanking the MC14543B, via pin 7.

The operating clock frequency of the MC14433 is set by R_c; with the 300K value shown it is 66 kHz, resulting in 4 conversions per second. In general, the conversion rate of the MC14433 is the clock frequency divided by 16,400.

The reference voltage used with the MC14433 determines the analog input range, which is essentially equal to the reference applied, but bipolar. Thus for a +200-millivolt reference, the input range is ±199.9 millivolts; with a +2-volt reference it is ±1.999 volts. In the circuit shown an MC1403 supplies a regulated +2.5 volts, which is then adjusted to the desired level for full-scale calibration, by the 20K pot. The bypass capacitor reduces noise susceptibility. The other necessary adaptation to change between nominal 200-millivolt and 2-volt scales is an adjustment of integrator resistor R1. This resistor should be 470K for a ±1.999-volt scale; 27 K for a ±199.9-millivolt scale.

The integrator capacitor is recommended as a 0.1-μF value for both scales, and should have low dielectric absorption. The offset capacitor is a 0.1-μF value, and is connected between pins 7 and 8. It should also (preferably) be a quality, low-leakage type.

The analog input, here termed V_X, is applied between pins 3 and 1. Pin 3 is the analog (high) input, pin 1 the analog ground, termed V_{AG}. Both of these input pins are high impedance, with typical input currents of 20 pA. In executing a conversion, the MC14433 operates in response to the voltage differential between these two pins, and analog ground (pin 1) need not necessarily be tied directly to circuit ground, as is shown here. The input common-mode range is to within 3 volts of the power supplies, or effectively ±2 volts. Common-mode rejection is typically 65 dB over this range. Although not specifically shown here, some input series resistor protection is appropriate for both of these inputs, or at least the analog input, if potentially dangerous voltages (> ±5 volts) can be seen. A 100K series resistor and a low-leakage noise-bypass capacitor are recommended, as generally described earlier.

To control large supply current transients on the ±5 volt supplies, a heavy bypass is used (with LED displays), and should be wired between the digit driver collectors and the −5-volt pin of the MC1413. Smaller 0.1-μF capacitors are used at the MC14433 supply pins.

For low-power applications, the liquid-crystal display dvm circuit of Fig. 7-43B is appropriate. This circuit operates similarly to the just described LED circuit with regard to analog inputs, scaling, etc., and the essential differences are in the display.

Here an MLC250 (Motorola) display is used, driven by three MC14543B decoders for digit drive, with the MC14013s and MC14070

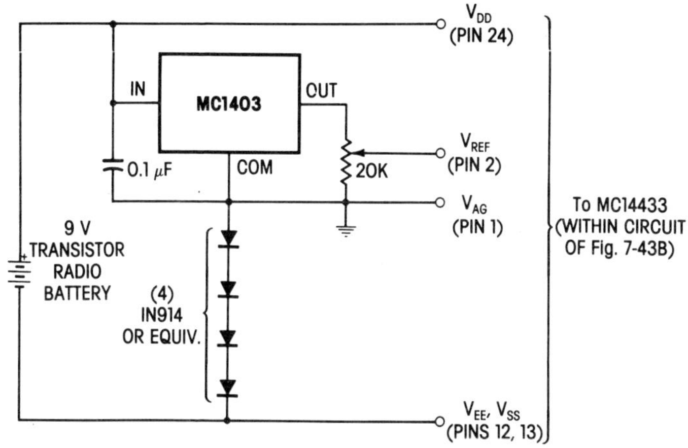

(A) V_{REF}, V_{AG} generation for single supply use.

(B) Crystal oscillator clock.

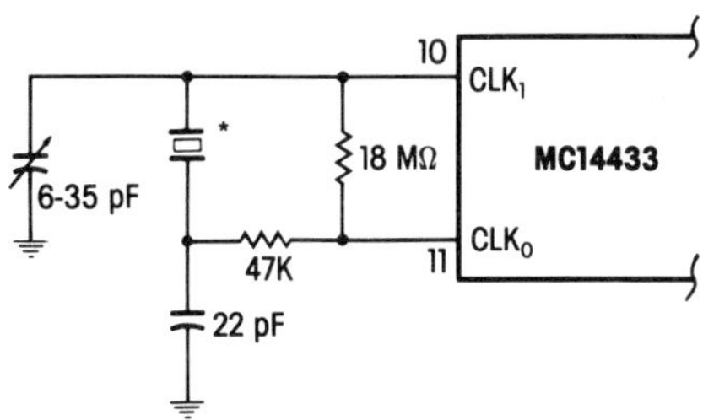

*MTQ32A, 32.768 kHz Watch Crystal
(Motorola Quartz Timepiece Crystal Group)

Courtesy Motorola Semiconductor Products, Inc.

Fig. 7-44. MC14433 operational options.

driving the sign and ½ digit inputs. A low-frequency square-wave drive for the backplane is derived from the DS1 output, after division by the MC14024 counter. As shown the system operates from ±5-volt power supplies, with a typical drain of about 2 mA. The low power consumption of this circuit makes it attractive for battery use.

There are two particularly useful operating options which can be used with the MC14433 to increase utility and effectiveness. These are shown in Fig. 7-44.

In Fig. 7-44A is shown a scheme to operate the MC14433 from a single 9-volt battery, for use with the just described LCD dvm. This circuit takes advantage of the differential input capability and high CM input range of the MC14433.

Here the MC1403 reference voltage source is "stacked" atop a string of silicon diodes, which are configured to drop approximately 2.8 volts, a level sufficient to bias V_{AG} to a proper operating range. The low 1 mA quiescent current of the MC1403 biases the diodes, without increasing the overall power consumption of the circuit, as would be true with conventional low TC reference diodes.

In Fig. 7-44B is shown a particularly useful clock option. The oscillator of the MC14433 can be used for crystal-controlled operation with the general connections shown, for crystal frequencies of 30 to 40 kHz.

As noted earlier in this chapter, rejection of 60-Hz input noise components (normal mode noise) is maximum when the reciprocal of conversion interval is an exact multiple of 60 Hz. For the MC14433, a crystal frequency of 32.768 kHz is fairly close to this optimum relationship, and this also happens to be a frequency which is readily available for watch crystal use.

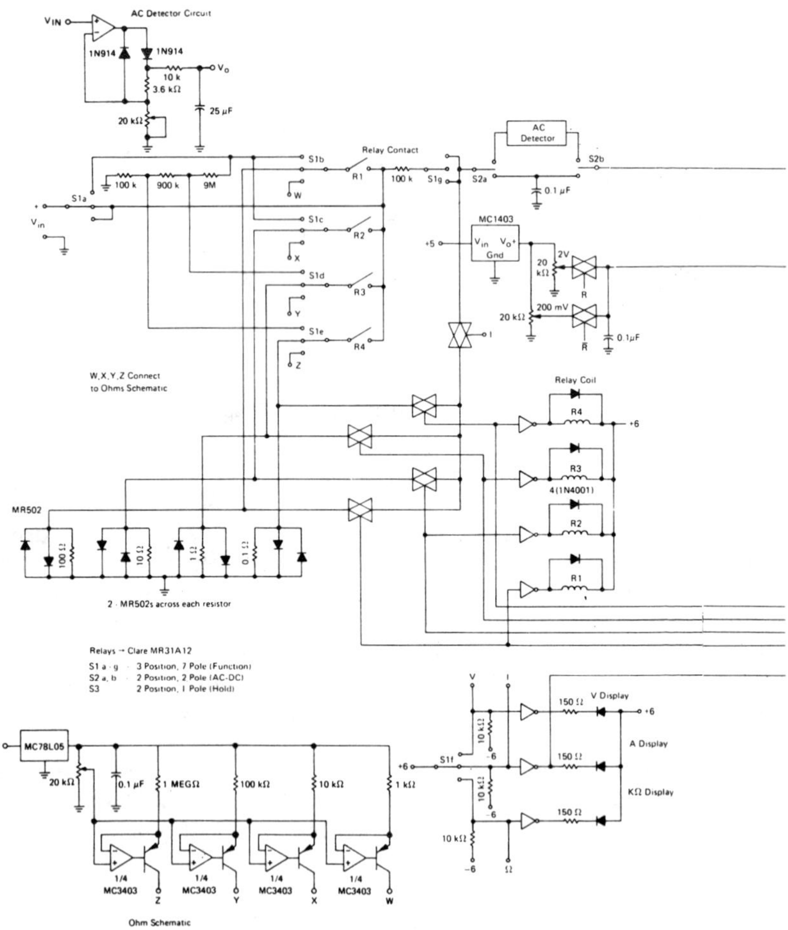

Fig. 7-45. The MC14433

The crystal can be trimmed to frequency if desired, or the trimmer can be replaced by a fixed 22-pF capacitor. This circuit can be used with any of the MC14433 dvm circuits to improve noise rejection quite simply.

MC14433 Auto-Ranging DMM

A complete auto-ranging dmm using the MC14433 is illustrated in Fig. 7-45, with capability of ac and dc voltage measurement from ±199.9 millivolts to ±199.9 volts, ac and dc current from 1.999 mA to 1.999 A, and resistance from 1.999K to 1.999 megohms. It uses the MC1433 plus CMOS logic, relays, and some op amps to accomplish these functions.

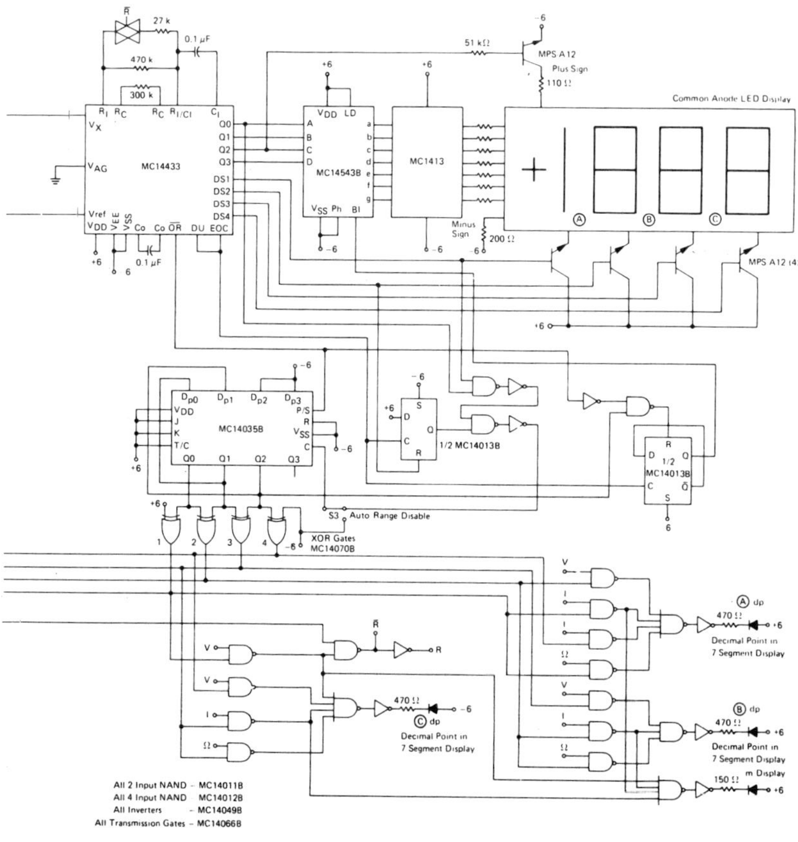

auto-ranging dmm circuit.

The basic dvm circuit is an LED display circuit, similar to Fig. 7-43. The auto-range logic selection is accomplished by the MC14035B, a right-left shift register, which cycles the input range either up or down as appropriate, until a correct scale is found. Automatic display decimal point logic is included, as is function display for the V, I, or ohms function. Selection between ac or dc operation is manual, by switch S2.

Since both the 199.9-millivolt and 1.999-volt dvm scales are used, these should be calibrated separately, initially. Ac voltage measurements are calibrated by the 20K pot, which adjusts the average reading detector to the rms level of a sine wave. No separate calibration is provided by the voltage scales or current scales; these are accurate to the degree of precision afforded by the scaling resistors. All ohms ranges are calibrated by a single pot which sets the voltage across the ohms range emitter resistors to 1 volt. The current sources are 2N3906 transistors, and are driven by a MC3403 low-power op amp. The ohms reference voltage is provided by an MC78L05 5-volt regulator, which should be powered from the unregulated input. The dmm itself uses ±6-volt power supplies, which can be derived with MC7806 (+6 volt) and MC7906 (−6 volt) three-terminal regulators. Further details on the operation of this circuit can be found in Motorola Application Note AN-769 (see chapter references).

The 7106/7107

The 7106 and 7107 are high-performance, low-power, single-chip, 3½ digit a/d converters usable on ±199.9-millivolt and ±1.999-volt ranges. All the necessary a/d functions are contained in a single IC, including seven-segment decoders, display drivers, reference, and a clock. The 7106 is designed to interface with a liquid crystal display (LCD) and includes a backplane drive, while the 7107 will directly drive common-anode LED displays.

These devices are by intent committed to display/instrumentation uses, and as such offer no major advantages for data processing uses. For display uses however, they offer many advantages, such as simplified application with few peripheral components, as well as good electrical characteristics.

The 7106 and 7107 are illustrated in two basic dvm connections in Fig. 7-46. These two circuits as shown are suited for ±199.9-millivolt scale uses, for either device. The differences between the two circuits lies basically in the display and power supplies used with the respective a/d devices.

The 7106 LCD circuit is shown in Fig. 7-46B, and uses a single power supply of 9 volts. With the low chip current drain of less than 1 mA, this circuit is highly suitable for battery operation. Input signals are applied to the 7106 between the IN HI and IN LO terminals, after passing through an input noise filter/protection network R5-C5. The internal

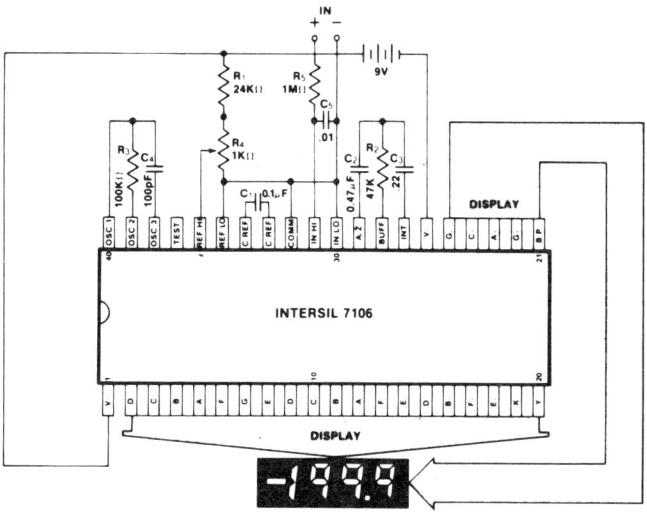

(A) 7106 LC display.

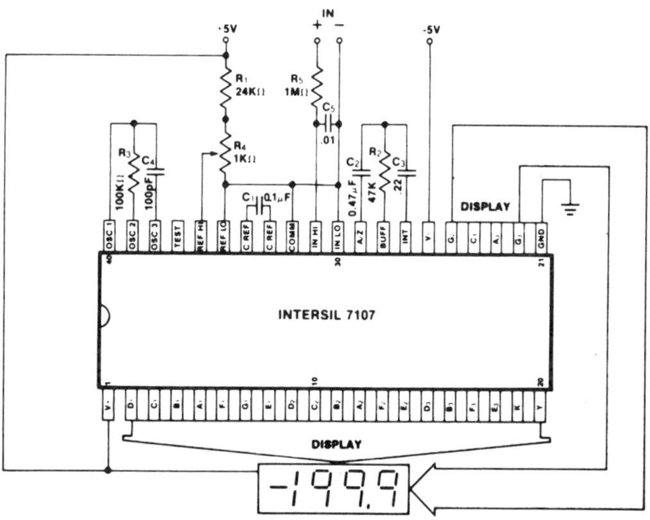

(B) 7107 LED display.

Courtesy Intersil, Inc.

Fig. 7-46. Basic 7106/7107 3½ digit dvm connections.

reference voltage of the 7106 and 7107 appears between the COMMON pin and V+, and is regulated at a level of 2.8 volts. As can be noted here, resistors R1 and R4 divide this voltage down to a lower level, either 100 millivolts (as true here), or 1 volt. The reference voltage of the 7106 (or 7107) determines the full scale analog input range, which is in general $\pm 2\ V_{REF}$. R4 serves as the FS calibration control for the circuit, and C1 is a reference voltage storage capacitor, across which the voltage applied to REF HI and REF LO will appear.

R3 and C4 are the clock circuit frequency-determining components, and as shown result in about a 45-kHz clock frequency.

R2, C2, and C3 function within the a/d analog subsystem, as integrating resistor, auto zero, and integrator capacitors, respectively. Of these, R2 and C2 vary for input range selection (for other than the ±199.9-millivolt scale shown here).

As can be noted from the battery connection between V+ and V−, the 7106 requires only a single supply potential for operation. The analog inputs are differential in their mode of operation, thus this circuit can be considered to be truly floating in its method of functioning. In the circuit as shown, the IN LO terminal is referenced within the common mode range of the 7106 converter by means of its connection to COMMON, pin 32.

The 7107, shown in Fig. 7-46B, is similar in many operational regards to the 7106, so its differences of detail will be discussed here. Two power supplies are used with the 7107, ±5 volts, which are applied with the common point to the P.S. ground, pin 21. The input signal is applied between IN HI and IN LO as before, but the IN LO terminal can be referenced to either COMMON or GND as may be appropriate to the source being measured.

Operationally, both circuits have both auto-zero and auto-polarity operation, with negative inputs displaying a (−) sign. Overrange inputs are displayed as a ±1XXX indication (3 LSDs blanked).

Further insight into the operation of the 7106 and 7107 devices and the details of different modes are best explained by regarding the functions of various subsystems. These are shown in Fig. 7-47.

In Fig. 7-47A are shown the circuit connections and options for ±0.2 and ±2-volt nominal ranges, and for various common-anode inputs.

As can be noted from this circuit, the 7106/7107 generates a 2.8-volt reference which appears between V+ (pin 1) and COMMON (pin 32). This voltage is used as the internal reference, after appropriate division for the scale in use. R1 and R2 comprise a reference voltage divider, with values selected to provide a nominal 0.1 or 1.0-volt output, with a 100-μA current in the divider. Suggested values are shown in the table, and R2 should (preferably) be a high resolution trimmer. This internal reference connection is applicable to both the 7106 and 7107, and has a TC of less than 100 ppm/°C. External reference voltages can also be used for lower TCs if desired, and will be described subsequently. They are most generally applicable to the 7107, which can thermally modulate the internal reference under certain conditions, due to the relatively high LED currents in use.

Aside from the change of reference voltage between scales, the other necessary change is adjustment of the integrator resistor and auto-zero capacitor, with appropriate values shown in the table. The integrator capacitor remains fixed at 0.22 μF, and a polypropylene type is preferred here. The values as shown are for a clock frequency of 48 kHz; they should be proportionally scaled for arriving at higher (or lower) clock rates.

In general ±199.9-millivolt scaling of the 7106 and 7107 is preferred, because it allows the greatest range of input common-anode voltage, and it is the test condition for which both devices are specified.

Common-mode voltages are developed within the 7106 or 7107 input stages when the IN LO terminal is not connected to COMMON. This terminal is the common reference for the analog subsystem, and is a low impedance sink for currents of up to 30 mA. Its relationship to the remaining circuits of the a/d converter may be better appreciated with reference to Fig. 7-47B.

With regard to operation of the 7106, the IN LO terminal is strapped to COMMON and the converter will then see no common-mode voltage. With the 7107, this is also true with IN LO tied to COMMON, but this type of operation is best suited to floating sources, or those not directly referred to the ±5-volt supply common (GND). For sources with the analog common tied to GND (±5-volt supply common) the 7107 IN LO terminal can be tied to this same GND, pin 21. This mode of operation does create a common-mode input voltage, which in general will be on the order of 2.2 volts (V+ −2.8 volts). This in itself will not necessarily be a problem, as the 7107 has a typical 86-dB common-anode rejection. It can however give rise to internal voltage swing limiting for a ±2-volt scale; therefore 7107 operation with IN LO connected to GND is best used with ±199.9-millivolt ranges for highest accuracy (and also for the considerations mentioned above).

Other Optional Operating Configurations

As mentioned above, in certain instances it may be desirable to improve upon the basic operation of the 7106 or 7107 by one of the operating options. These are illustrated in Fig. 7-48.

In Fig. 7-48A are shown two schemes for using external references with the 7106 or 7107. As previously noted, the 7107 is subject to self heating under certain conditions, which can cause a shift in the displayed voltage. When an external reference is used, the thermal path which causes this is broken, and the effect is eliminated. Therefore, although both schemes shown here will work with both the 7106 and 7107, they are most desirable with the 7107 where they will have decided benefit, but will also not cause a large percentage increase in power drain. Addition of one of these circuits, while possible with a 7106 will approximately double the supply drain.

Reference diodes in the range of 6 to 7 volts can be connected as in Fig. 7-48A Part 1, with excitation from both supply voltages. A desirable diode type here is one of the synthesized units, such as the LM329, which can operate at relatively low currents (1 to 2 mA), and the low dynamic impedance provides immunity to supply changes.

In Fig. 7-48A Part 2 is shown a low voltage 1.22-volt reference diode, which can be preregulated by the 7106/7107 2.8-volt reference, providing good supply immunity.

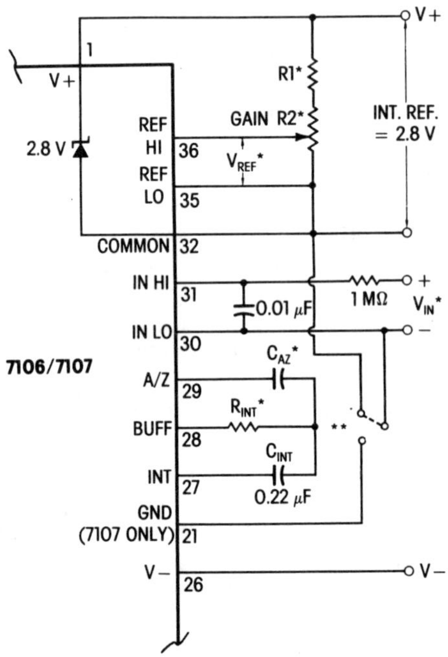

***INPUT RANGE/OPERATING VALUES**

V_{IN}	V_{REF}	R1	R2	R_{INT}	C_{AZ}
±199.9 mV	+100 mV	24K	1K	47K	0.47 μF
±1.999 V	+1.000 V	15K	10K	470K	0.047 μF

(Values Shown For f_{CLK} = 48 kHz)

****COMMON MODE INPUT OPTIONS**

7106: Tie V_{IN}−/IN LO Terminal To COMMON, Use 9 V Supply

7107: Tie V_{IN}−/IN LO Terminal To COMMON For Floating Sources, 0.2 or 2 V Nominal Range.
Tie V_{IN}−/IN LO Terminal To GND For Sources Common To ±5 V, 0.2 Nominal Range.

(A) Circuit connections.

Fig. 7-47. The 7106/7107

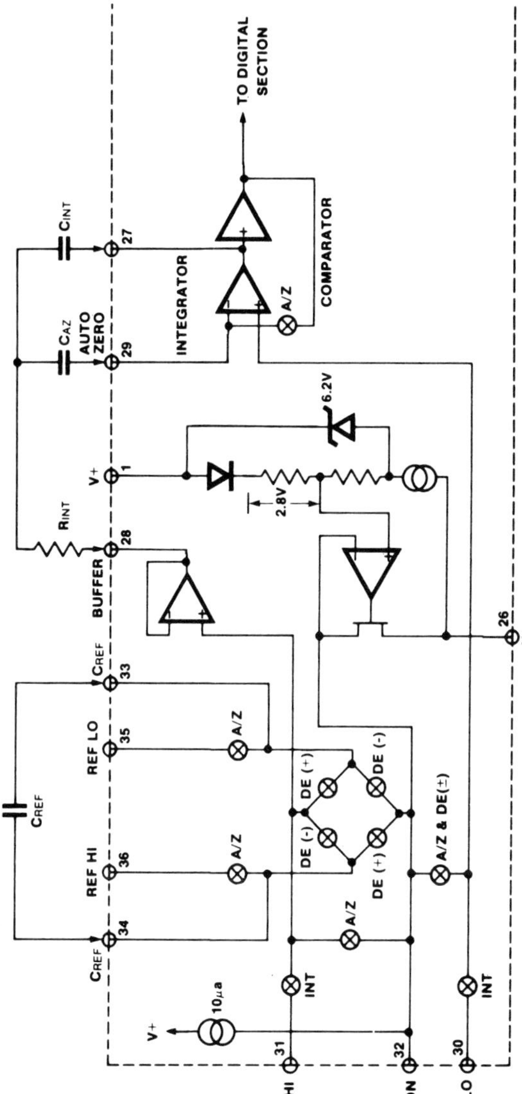

(B) 7106/7107 analog subsystem.

Courtesy Intersil, Inc.

operating range input options.

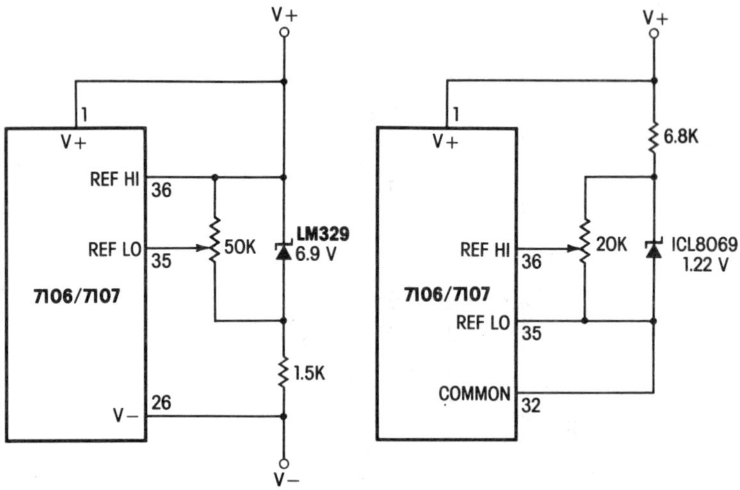

1. USING 6-7 V REFERENCE

2. USING 1.2 V REFERENCE

(A) External references.

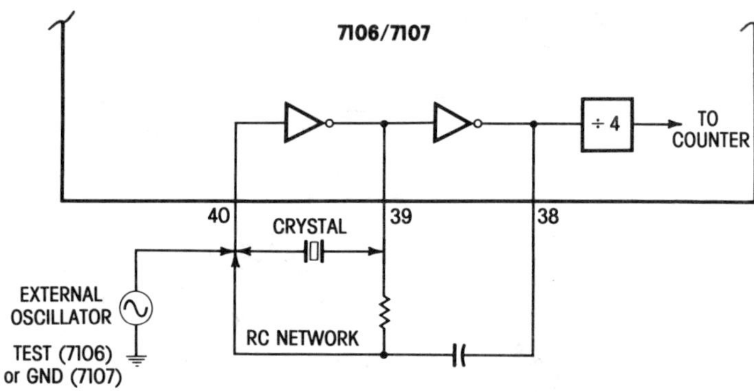

CLOCK OPTIONS

CLOCK MODE	CONNECTIONS TO PINS: 40	39	38
EXT	CLOCK SOURCE	–	–
RC*	RC NODE	R	C
XTAL**	XTAL, R	XTAL, R	–

$*f_{CLK} = \frac{1}{2.2\,RC}$

$**f_{CLK} = f_{XTAL}, 10\ M\Omega \leq R \leq 20\ M\Omega$

OPTIMUM CLOCK FREQUENCIES

f_{OPT} (kHz) 60 Hz	50 Hz	CONV/SEC
	28.571	1.79
34.285		2.14
	33⅓	2.08
40	40	2.5
48		3
	50	3.13
60		3.75
	66⅔	4.17
80		5

(B) Clock options.

Fig. 7-48. Other 7106/7107 optional configurations.

It can be noted from these two circuits that the reference inputs (REF HI, REF LO) can be rather arbitrarily located between the supply limits of V+ and V−. This is true in practice, due to the 7106/7107 common-mode rejection capability. The main criterion is just that REF HI be the more positive of the two inputs, and the applied voltage be stable.

The reference capacitor, not shown here, is generally a value of 0.1 μF, between pins 33 and 34. It may be desirable to increase this to 1 μF, for large reference common-mode voltages. A common-mode voltage will be generated by the reference when the REF LO terminal is *not* tied to COMMON.

The most simple method of clock generation with the 7106/7107 is to use the on-chip CMOS inverters as a simple RC astable multivibrator. It is this type of circuit which is used in Fig. 7-46. However, other possibilities also exist, and are illustrated in Fig. 7-48B.

The three basic clock modes are EXT, RC, and XTAL and they are set up by the connection options shown in the figure. External clock signals should be 5 volts p-p in amplitude, with the source common terminal as indicated. RC operation is achieved by the simple RC connection, with $f_{clk} = 1/2.2RC$. Crystal operation is achieved by connecting a crystal of the desired frequency between 39 and 40, with a parallel high-resistance bias resistor.

For the 7106/7107, there exists a series of optimum clock frequencies which will result in the highest normal-mode rejection. These frequencies are shown for both 60- and 50-Hz power frequencies; also shown are the resulting conversion rates. These frequencies are of course applicable to any mode of clock operation; however, best noise rejection will result when the frequency is well maintained and predictable, as typified by crystal use.

8700 Series A/D Applications

8700 series of a/d converters are different from most integrator a/d types in a number of regards. A major difference is operation based on a form of V/F conversion as opposed to the dual slope (or modified dual slope) schemes discussed thus far. One important practical implication of this point is that faster conversion times are possible, to as low as 1.2 ms for an 8 bit device. The 8750 3½ digit device converts in 10 ms which is about an order of magnitude faster than a typical 3½ digit dual slope unit.

Other technical differences include a current-mode input (as opposed to the other devices thus far discussed, which have high impedance voltage inputs) and operation that is basically unipolar (as opposed to bipolar), and not zero corrected.

Dependent upon the specific device used, an 8700 series a/d converter can be suited for both display/instrumentation or data acquisition/

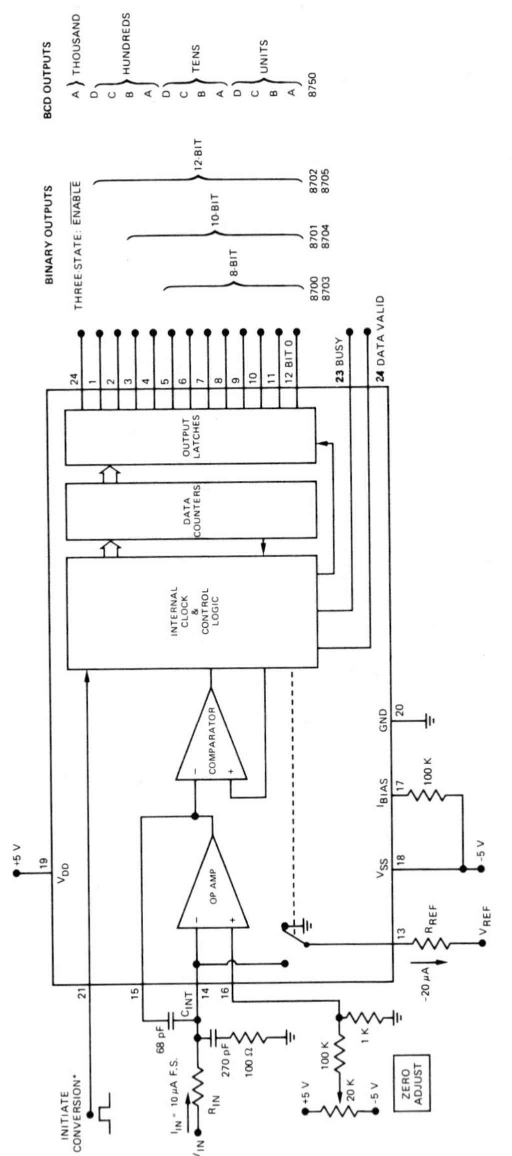

(A) Connection diagram.

Fig. 7-49. The 8700 series

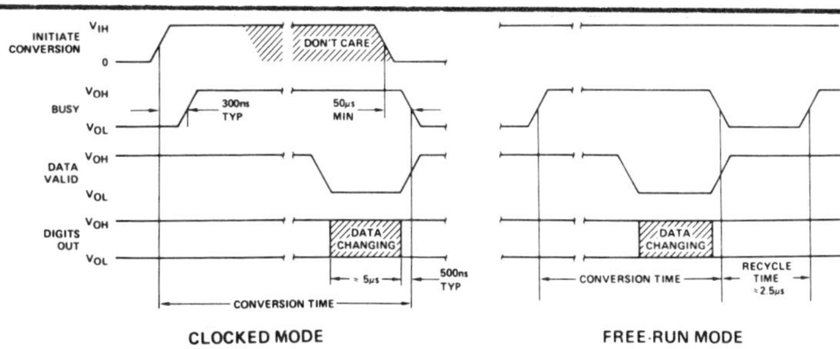

Courtesy Teledyne Semiconductor

(B) Timing diagram.

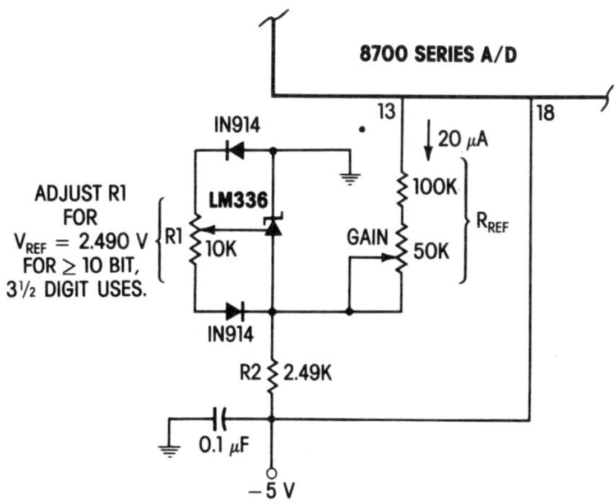

(C) Low-voltage reference.

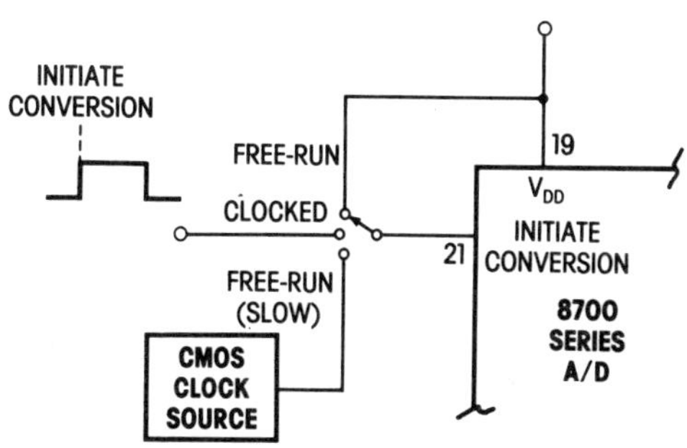

(D) Conversion-rate control.

Courtesy Teledyne Semiconductor

general operation.

processing uses. There is a good deal of flexibility in the options available; 8-, 10-, and 12-bit binary units with normal and 3-state outputs, plus the mentioned 3½ digit bcd unit.

A general connection diagram of the 8700 series units is shown in Fig. 7-49A, and a timing diagram in Fig. 7-49B. The 8700 series (all devices) uses an externally supplied reference current of 20 μA, and an analog input of 10 μA for full scale. The input resistor R_{IN} is selected for the input full-scale voltage, to allow 10 μA into the summing junction; typically it will be 1 megohm for a 10-volt range. Full-scale calibration is achieved by variation of either R_{IN}, R_{REF}, or V_{REF}, as may be convenient to the particular use. Both of these resistors should be precise, stable components, as should V_{REF}. The use of the devices' −5-volt power supply potential is not generally recommended, unless it is sufficiently stable to qualify as a reference source.

The requirement for a 20-μA current reference is somewhat unique and a simple scheme for deriving it from the −5-volt supply is shown in Fig. 7-49C. The LM336 is a 2.49-volt reference diode, with a trim terminal. When R1 and the diodes are used and V_{REF} is trimmed via R1 to 2.490 volts, the resulting TC is typically 10ppm/°C over a 70°C range, which is sufficient for 3½ digit accuracy. For 8-bit or limited-range applications, the trim pot and diodes can be eliminated, and R2 should be a 5% type. The GAIN pot will adjust for full-scale tolerance of the remaining components, and should be a multiturn unit. For sufficient gain calibration range, C_{INT} should be a 5% unit, and R_{IN} a 1% unit. The remaining circuit components of Fig. 7-49A are less critical. The zero network applies voltage to the 8700 input op amp to cancel its residual offset, and should be used for best accuracy.

Operation of the 8700 series devices can be either free running or externally triggered. The timing diagrams for these two basic modes are shown in Fig. 7-49B, and connections are shown in Fig. 7-49D. The 8700 input levels are 5-volt CMOS compatible, and a complete conversion is signaled by a low to high DATA VALID output transition, or a high to low BUSY transition (see timing sequence).

8750 3½ Digit LCD DVM

A useful circuit example of the 8750 3½ digit a/d converter is within an LCD dvm. This application is illustrated in Fig. 7-50. In this circuit the 8750 operates from ±5-volt supplies, and directly drives +5-volt operated CMOS logic. The 4543 devices function as bcd-to-7 segment decoders and drivers for the LCD. A simple and effective 50-Hz square wave drive is derived for the LCD backplane by dividing the BUSY signal, which occurs at a 100-Hz rate. This is accomplished by the CD4013 flip-flop.

Input scaling is here shown for a full-scale range of 19.99 volts, which includes the 10-μA nominal range (or 10 volts), plus overrange.

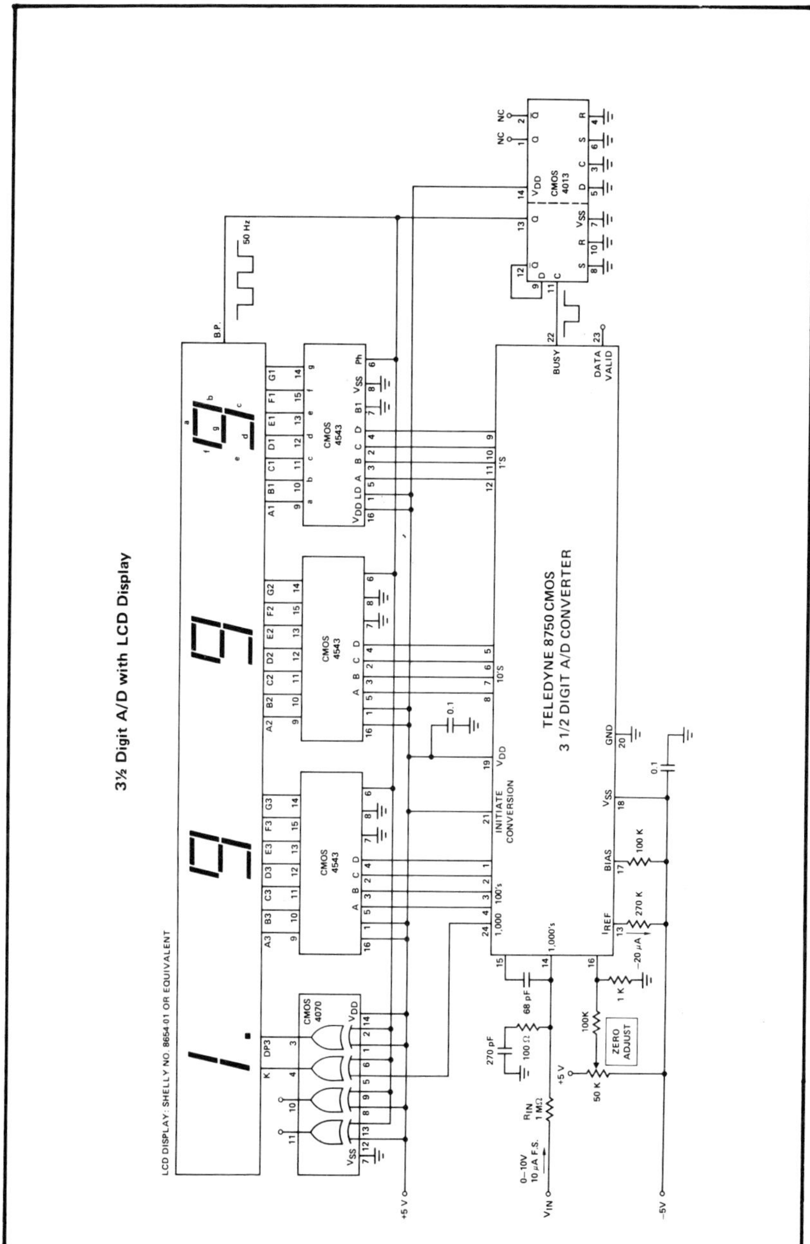

Courtesy Teledyne Semiconductor

Fig. 7-50. The 8750 3½ digit LCD DVM.

As shown, operation is unipolar positive range, but bipolar operation in a sign-plus-magnitude mode can be achieved by preceding the circuit by a sign-changer circuit, such as that of Fig. 7-2B.

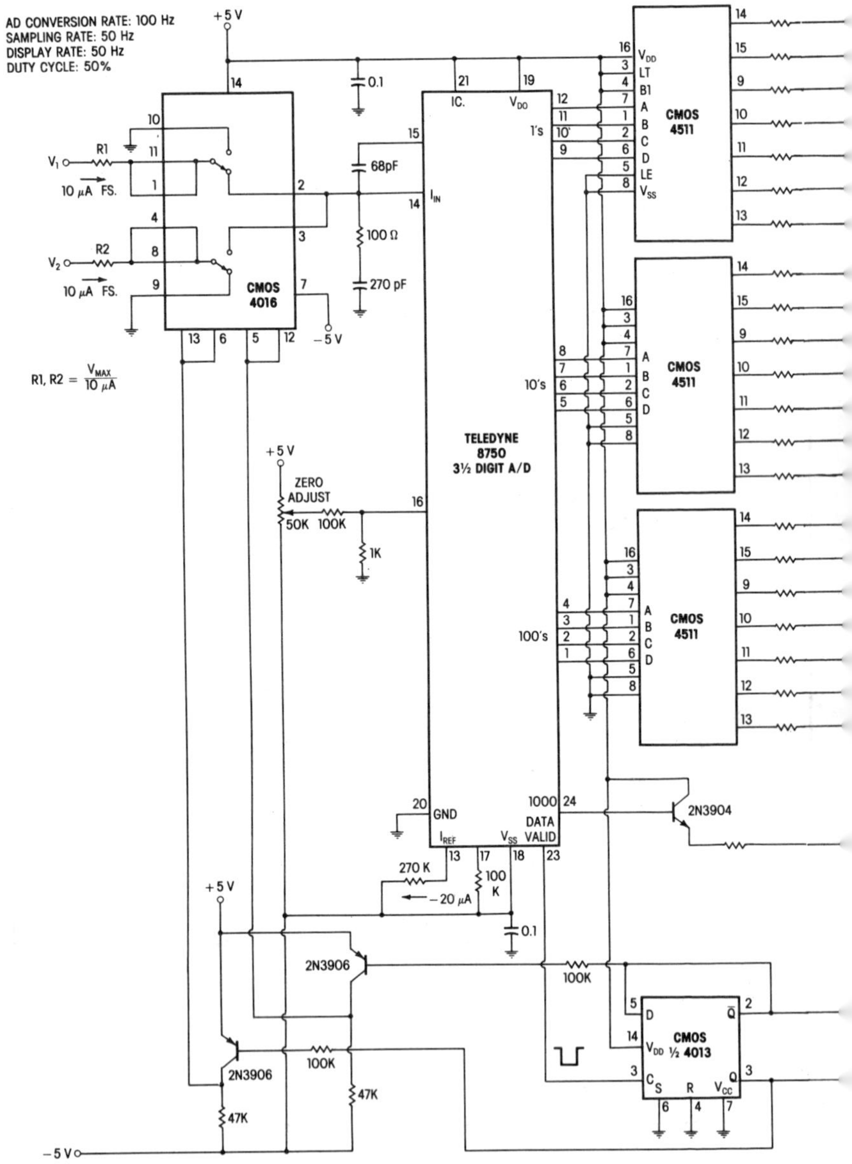

Fig. 7-51. The 8750 dual

8750 Dual Input/Display 3½ Digit DVM

LED displays can also be used with the 8750, as in the example shown in Fig. 7-51.

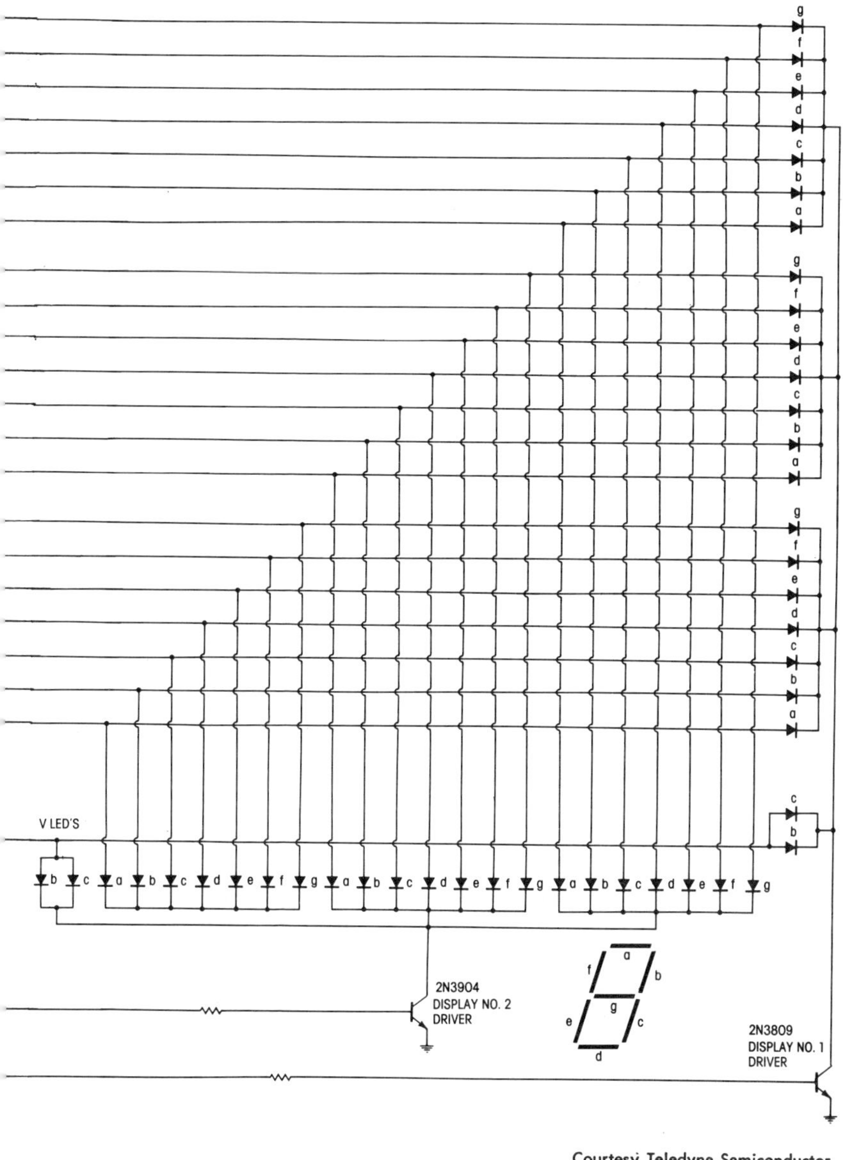

input/display 3½ digit DVM.

This circuit is unusual in the sense that it takes advantage of the fast conversion rate of the 8750 to multiplex two inputs, V1 and V2, and display the respective bcd conversions on two different 3½ digit LED displays. Thus it is effectively two dvms simultaneously, and can efficiently monitor two variables while displaying both.

Multiplexing is achieved by the CD4013 flip-flop, which divides the 8750 DATA VALID output by two. The 50-Hz square wave at the Q and $\overline{Q}$ outputs drives the npn common-cathode LED drivers in complementary fashion, while also driving the pnp level shift transistors. The pnp devices (2N3906 or equivalent) drive the control inputs of the CD4016 analog switch, which selects either V1 or V2. CMOS bcd-to-seven segment decoders are used; 4511s and the 21 output series resistors determine the segment current. Typically the resistors will be on the order of 400 ohms, or less.

Note that the reference and offset voltages are common to *both* inputs in this circuit, so individual full-scale adjustment should be done by individual trim of R1 and R2.

Auto-Ranging Circuit

A useful auto-ranging system which can be used with the 8750 a/d converter is shown in Fig. 7-52. The circuit uses three CMOS logic packages, and operates from ±5 volts.

In operation, the logic detects two digital scale conditions. For counts of 1999 or more, the CD4068 NAND gate will go low and set the RS flip-flop to a 1 at the HIGH RANGE output. This will cause the input analog switch to close the R1 path (switch state shown), and cause the HIGH indicator to light (if used). When the count drops below 40, the CD4078 NOR gate will reset the RS flip-flop to a LOW RANGE output high, and close the R2 input path (and light the LOW indicator, if used).

Note that although this scheme is shown here for specific count conditions with the 8750, it can also be adapted to other ranges (and devices) if desired, as the logic is sufficiently general.

The binary-scaled devices of the 8700 series consist of the 8700, 8701, 8702 for 8-, 10-, and 12-bit uses, and the 8703, 8704, and 8705 are their counterparts with 3-state outputs. All of these devices are applied very similarly to the just discussed 8750, with the exception of the binary scaling/coding.

The most frequent applications of these devices is within data acquisition systems where the digital output data is transmitted for further processing. In applying any of the devices the general connection diagram is just as shown in Fig. 7-49A. For the 8703, 8704, 8705 devices, pin 24 becomes the ENABLE input for the output 3 stage latch. The data is enabled (on) when this pin is low; when it is high the outputs are in the 3rd (off) state.

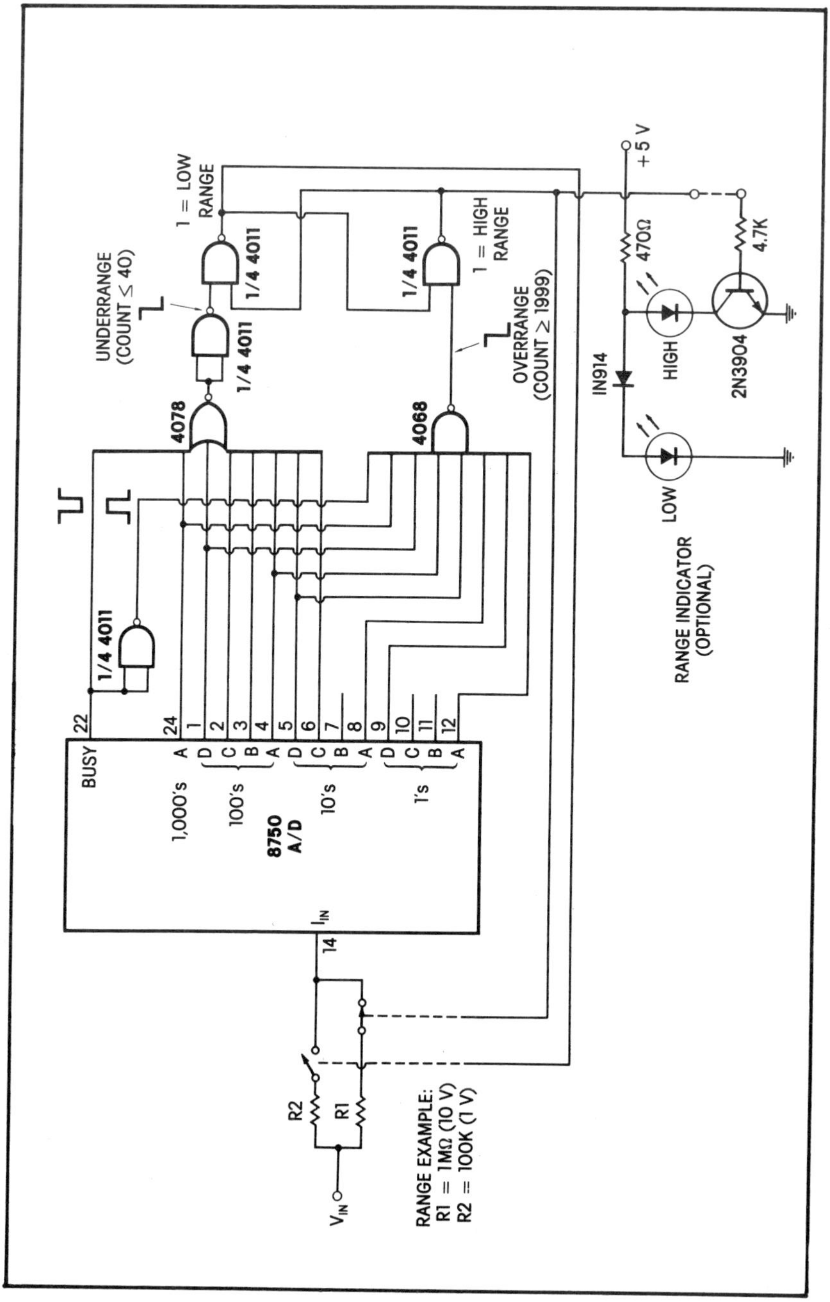

Fig. 7-52. Auto-ranging system.

To set up an 8700 series device the input and reference are selected as was generally described. A common use for this type of converter is as a unipolar range a/d converter. R_{IN} is selected for the full scale range in use to allow 10 μA of input current; for example 10 V/1 meg, 5 V/500K, and so forth.

For bipolar range use, the offset circuit of Fig. 7-53 can be added to allow bipolar input voltages to be accomplished. This general circuit can be used with a variety of input voltages and reference voltages, as can be noted from the table. The main criterion is that the offset current must be +5 μA (½ scale) and the full scale input current is +5 μA. For calibration levels for this or 8/10/12 bit unipolar use, the levels of Table 6-4 can be used.

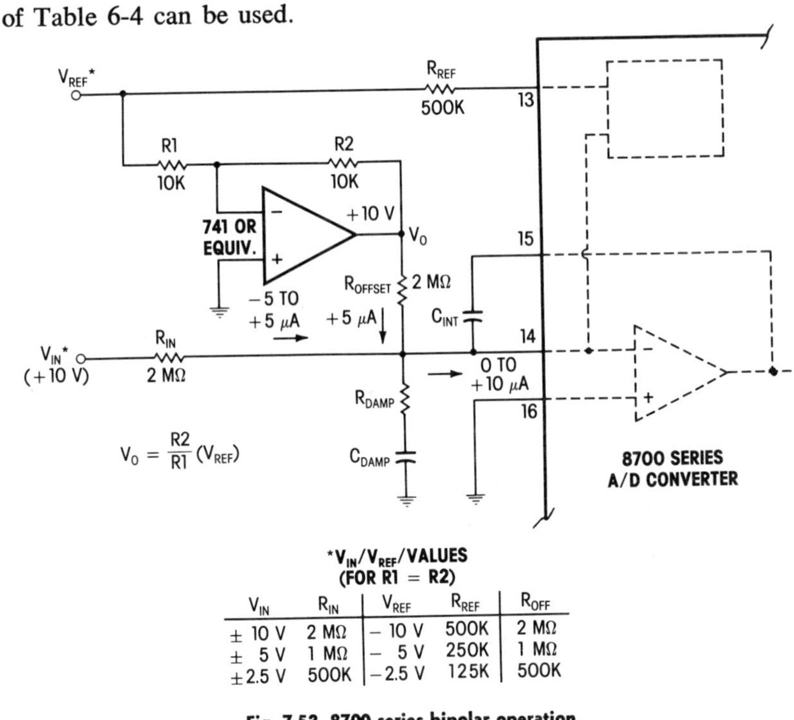

***V_{IN}/V_{REF}/VALUES**
(FOR R1 = R2)

V_{IN}	R_{IN}	V_{REF}	R_{REF}	R_{OFF}
± 10 V	2 MΩ	− 10 V	500K	2 MΩ
± 5 V	1 MΩ	− 5 V	250K	1 MΩ
±2.5 V	500K	−2.5 V	125K	500K

Fig. 7-53. 8700 series bipolar operation.

The outputs of an 8700 series a/d converter can be used directly in parallel form, and the 8703/8704/8705 have 3-state parallel outputs. In some instances however it may be desirable to transmit output data in serial form. An interface to accomplish this is shown in Fig. 7-54.

Here a set of three 74C195 shift registers are added to an 8702 12-bit a/d converter. Data can be read out either in a 12-bit serial fashion with timing as noted, or may also be read out as 3 × 4-bit parallel words, during clock pulses, 1, 5, and 9. Although shown with the 8702 12-bit

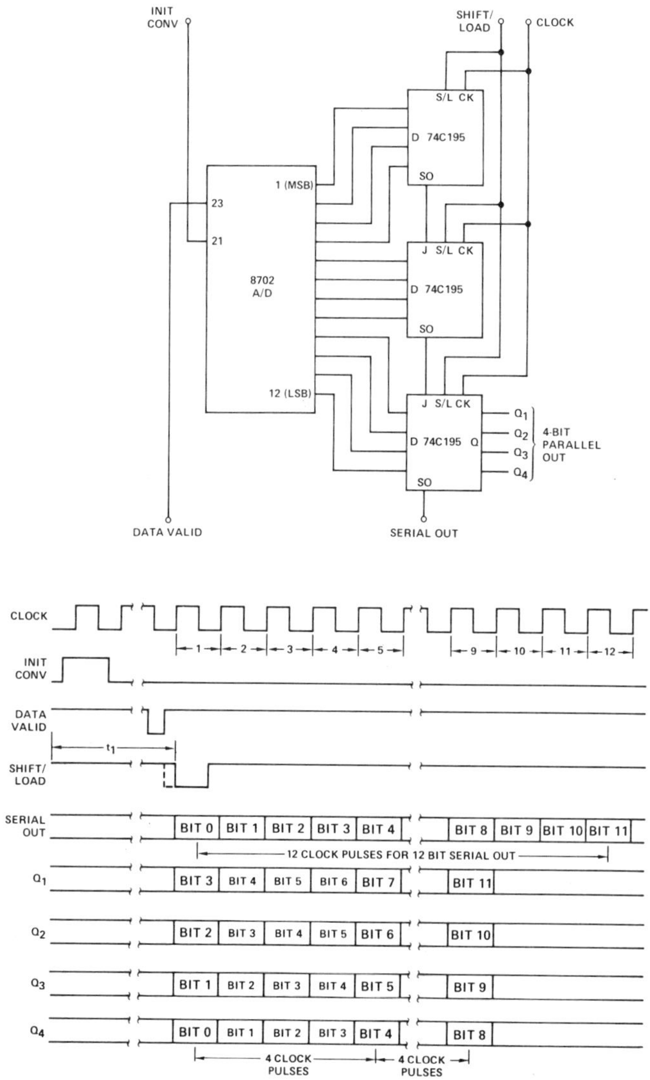

System reads parallel outputs at clock pulse 1, 5, 9

Shift/load may be taken low when Data Valid goes high or at fixed time t_1, after INIT CONV ($t_1 \geq 24$ ms for 8702). Shift/load then must return high before clock pulse 2 and must remain high until all data is read out. Recommended clock frequency 1 kHz min for serial, 750 Hz min for parallel output.

Courtesy Teledyne Semiconductor

Fig. 7-54. 8700 series 12-bit serial/3 × 4-bit parallel output interface.

devices the general scheme is also applicable to the 8700 and 8701 for 8- and 10-bit uses.

Microprocessor Interfaces

One of the most useful features of the 8700 series is the ability to be readily interfaced with the microprocessor, which is becoming predominant in digital processing. The 8703 8-bit device is readily used for such applications, as it can be communicated via its INIT CONV, BUSY, and DATA VALID outputs, and output data requested via the ENABLE input. Its 8-bit data output is directly compatible with 8-bit microprocessor data buses.

Fig. 7-55 illustrates the use of the 8-bit 8700 a/d converter with the 8080A microprocessor and typically used hardware. In this application 80L98 3-state output buffers are used with the 8700, which provides inverted data to the bus, and the inverting 8228 bus receiver/driver.

The 8080A starts an 8700 conversion by the INIT CONV input. When the conversion is complete, the a/d converter signals the 8080A of data availability, via the 74C74 flip-flop and a INT REQ (interrupt request). The 8080A then will cycle through its service routine, and read the data from the 8700 output by addressing the 80L98 3-state buffer. Further details of this, including software, are contained in the Teledyne Application Note AN-8 (see chapter references).

With care in logic design and consideration as to the nature of the analog data being processed, the effective a/d speed can be measurably increased in certain instances. The interface of Fig. 7-56 for instance, will only initiate a new a/d conversion when the a/d data has changed 1 LSB (or more) from the last conversion. In this manner slowly changing input data need not unnecessarily be cycled through a new a/d conversion. The 1.2-ms conversion time of the 8700 8-bit device is also an aid to increase data processing speed, as it is relatively fast for an integrating a/d converter.

The AD7550

The AD7550 13-bit binary a/d converter is a device most generally suited to data-acquisition applications, where the resulting digital output is intended for further digital processing, such as microprocessor-based systems. It is a highly accurate device, with very low TCs for both gain and offset. Outputs are in 3-state form for data bus use, conversions can be externally started, it has an internal clock, and busy and overrange output signals are provided.

Fig. 7-57 illustrates a basic AD7550 operational hookup and the external components required. R1-C1 are the integrator components, and the resistive divider from V_{REF} to analog ground develops a voltage of $V_{REF}/2$ at pin 7. For free-running, self-clocked uses, capacitors C2 and C3 are connected to the STRT (start) and CLK (clock) inputs.

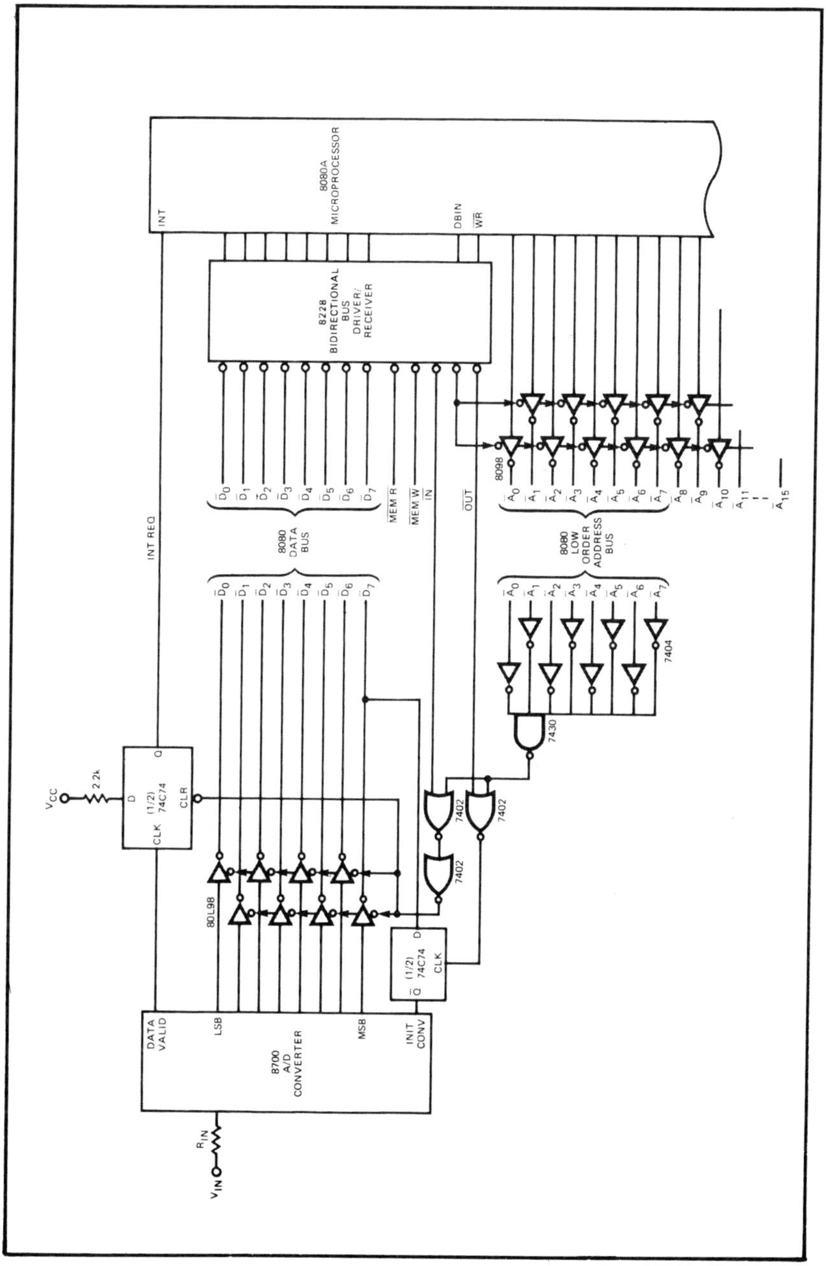

Courtesy Teledyne Semiconductor

Fig. 7-55. 8700 microprocessor interface.

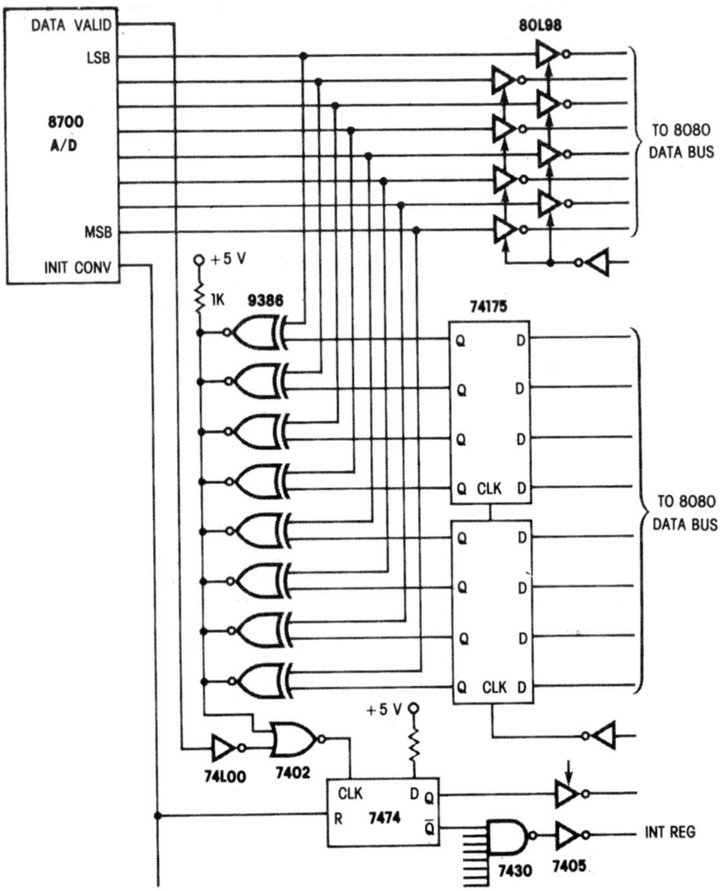

Courtesy Teledyne Semiconductor

Fig. 7-56. A/d charge-interrupt interface.

The 3-state control inputs are STEN (status enable), HBEN (high byte enable) and LBEN (low byte enable). The former input controls outputs BUSY, $\overline{\text{BUSY}}$ and OVRG (overrange) which are active when it is high, 3-state (off) when low. HBEN and LBEN control data outputs DB8 through DB12 and DB0 through DB7 respectively, with a similar sense of control. When not used, the 3-state control inputs are tied to VCC, and data will then continuously appear at the outputs, being updated with each conversion.

Digital levels are a function of VCC, which has a range of +5 to +12 volts. At +5 volts the levels are in TTL form; at +12 volts they are CMOS compatible. Supplies of +12 volts and −5 volts are used for the analog portions of the AD7550, VDD and VSS.

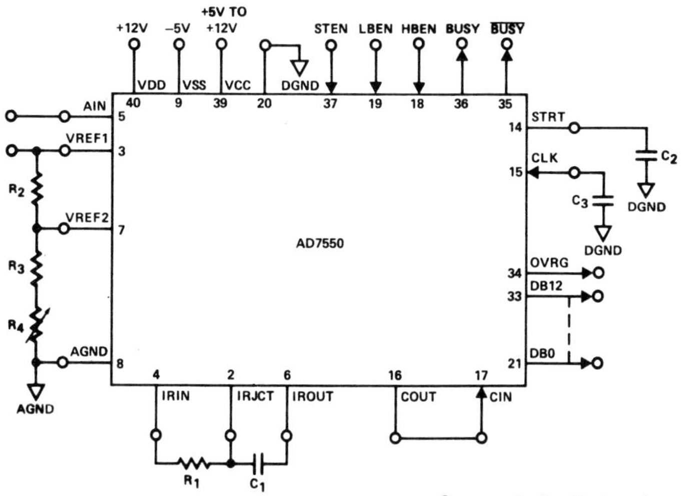

Courtesy Analog Devices, Inc.

Fig. 7-57. AD7550 operational diagram.

Fig. 7-58 is a representative circuit use of the AD7550, showing all details of connections for typical input ranges.

The AD7550 requires an external reference voltage which is related to the analog input V_{IN}, as $V_{REF} = 2.125\ V_{IN}$. This factor indirectly defines the analog input range, which must also be within the range of VDD or VSS in magnitude (whichever is lower). The analog input is bipolar, and is applied between the AIN and AGND terminals.

The table in this figure gives a convenient summary of V_{REF} and V_{IN}, and the passive component values required for several popular bipolar input ranges. For any range, the input impedance will be essentially the value of R1.

The value of C3 determines the internal clock frequency, which will be approximately 200 kHz for the value shown. Conversion time is on the order of 200 ms, but will vary somewhat with different R1-C1 values. An external clock, if used, is applied to the clock input (with C3 deleted).

With C2 connected to the start input, the AD7550 will self start and provide free-running conversions of about 5 per second as shown. For externally triggered conversions, a positive-going start pulse is applied to the input.

The BUSY and $\overline{\text{BUSY}}$ outputs provide high and low state output levels during the conversion interval, and the trailing edge may be used as a data valid signal, at which time new data has appeared at the outputs. The OVRG output provides a high state level for inputs outside (above or below) the analog range. The data outputs are shown here in the normal twos complement form, which provides bipolar range scaling.

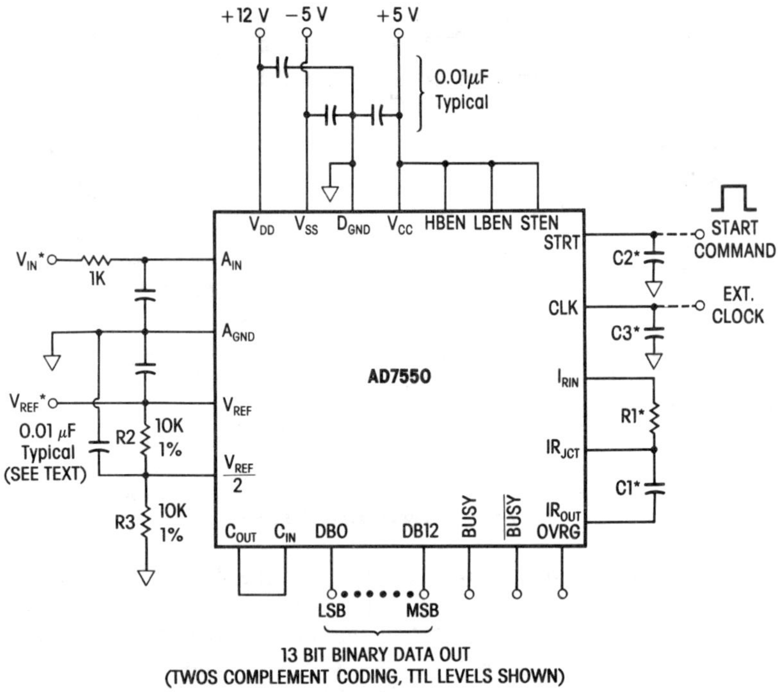

***TABLE OF OPERATING VALUES**

V_{IN}	V_{REF}	R1	C1	C2	C3
±2.000 V	+4.2500 V	2.7 MΩ	0.01 μF	50 pF	220 pF
±2.500 V	+5.3125 V	3 MΩ	0.01 μF	50 pF	220 pF
±4.096 V	+8.7040 V	5 MΩ	0.01 μF	50 pF	220 pF
±5.000 V	+10.6250 V	6.2 MΩ	0.01 μF	50 pF	220 pF

For $f_{CLK} \cong 200$ kHz, $V_{DD} = +12$ V, $V_{CC} = +5$ V, $V_{SS} = -5$ V

Fig. 7-58. AD7550 13-bit binary (twos complement) a/d converter.

Straight offset binary operation can be realized by complementing the MSB (DB12).

Several "housekeeping" type details are necessary if the full potential accuracy of the AD7550 is to be realized. The supply lines should be bypassed with low inductance capacitors to the digital ground pin as shown. This digital ground is returned to the analog ground at one point only. The analog inputs are also bypassed with low-leakage capacitors (such as polystyrenes) to analog ground. The integrator capacitor should be a low dielectric absorption type such as polypropylene or polystyrene, with its outside foil to IROUT. R1 can be a standard resistor type.

Reference voltage considerations for the AD7550 are somewhat unique in that the voltage required for the typically used input ranges are unusual. Two methods of achieving these voltages are shown in Fig. 7-59.

In Fig. 7-59A is shown a method of deriving reference voltages of less than 10 volts, simply by dividing the output of a 10-volt AD581 reference source. The reference TC should be minimized by using a low TC device such as the AD581M, and low TC divider resistors such as metal film or (preferably) wirewound resistors. A portion of the series resistance R_S is trimmed to provide full-scale calibration.

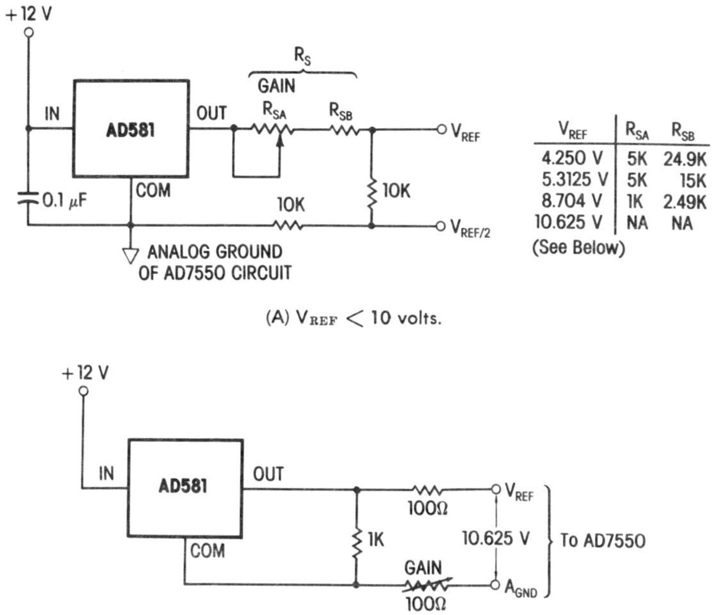

V_{REF}	R_{SA}	R_{SB}
4.250 V	5K	24.9K
5.3125 V	5K	15K
8.704 V	1K	2.49K
10.625 V	NA	NA
(See Below)		

(A) $V_{REF} < 10$ volts.

(B) $V_{REF} > 10$ volts.

Fig. 7-59. AD7550 reference circuits.

For voltages higher than +10 volts, the AD581 can be bootstrapped as shown in Fig. 7-59B. In this case the 11-mA current in the GAIN control provides a 0.625-volt pedestal, yielding a +10.625-volt output.

The connection of the AD7550 to an 8-bit microprocessor data bus is illustrated in Fig. 7-60. This connection places in parallel with the bus (a) DB0-DB7 (b) DB8-DB12 and (c) BUSY, OVRG. These three sets of data can be interrogated by the microprocessor system by selecting LBEN, HBEN, and STEN, respectively. A conversion is started by a pulse to the AD7550 start terminal.

For 3-state data bus use as shown here, the coding must be twos complement. The circuit configuration can be used with any of the AD7550 circuit configurations detailed in Fig. 7-58.

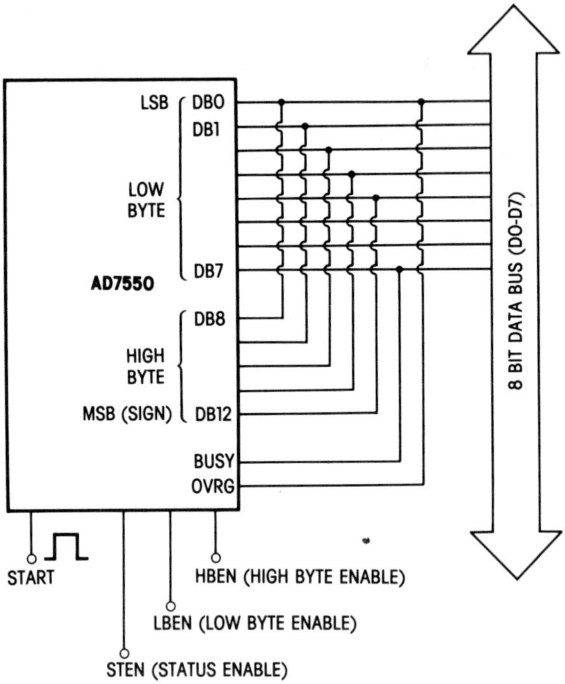

Courtesy Analog Devices, Inc.

Fig. 7-60. AD7550 8-bit microprocessor data bus connections.

SUCCESSIVE APPROXIMATION A/D CONVERTERS

The successive approximation a/d converter (referred to as SA for brevity) is one of the more popular types of a/d converters, because of its relatively high speed and accuracy, with reasonable cost and complexity. For most forms of circuit realization, a SA a/d converter will complete a conversion in n + 1 clock cycles. With the exception of parallel or instantaneous a/d types, this speed is the highest of the techniques in popular use. With the current state of the art in circuit components, the highest practical SA a/d converter speeds are on the order of 100 ns per bit, and a number of commonly available devices can achieve speeds of 1 μs per bit or less, to rated (½ LSB) accuracy. A wide gamut of devices are available for circuit selection, with varying speed, resolutions, accuracies, ranges, etc.

This part of the chapter discusses SA a/d converters with resolutions of 6 to 12 bits with down to 1 μs conversion times, and with a variety of different basic input ranges. The format of discussion will first be in a context of the fundamental SA components which make up such an a/d converter, then with fully developed circuit examples, using both d/a converters and associated hardware to implement the SA function, and also fully committed *integral* SA a/d devices.

The Basic IC SA Converter

A basic (and much simplified) diagram of a SA a/d converter is illustrated in Fig. 7-61. As can be noted, this a/d converter consists of a SA register (SAR), a d/a converter, and a comparator. A fourth function which may or may not be considered as integral to this a/d converter is a clock source. The fundamental details of SA operation have been discussed earlier in Chapter 2, and will not be repeated here, except as pertinent to specific considerations.

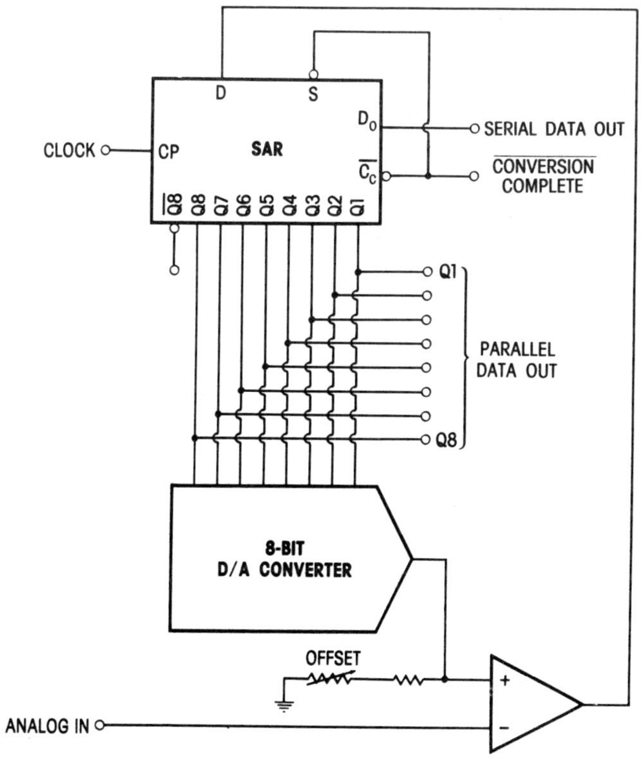

Fig. 7-61. General concept of successive approximation a/d converter.

The key element of a SA type a/d converter is the SAR, which is a special purpose shift register designed to effectively implement the SA a/d function. A very popular series of SARs is the Advanced Micro Devices 2502, 2503, and 2504 devices. These units are TTL designs, and are available in the basic part numbers mentioned for 8-bit (2502,

2503) and 12-bit uses (2504). They are also available in low power versions, i.e., "25L04" (vis-a-vis 2504).

In the circuit as shown, an 8-bit SA a/d converter design is indicated, and the 2502 provides an 8-bit binary output. An MSB complement is commonly provided from all SAR devices, for twos complement uses. Scaling for unipolar or bipolar ranges is determined by the analog circuitry associated with the comparator. The 2502 and 2504 are cascadable with other like units, for resolutions greater than their inherent 8 or 12 bits. All devices require an external signal to start a conversion (START) and they provide a conversion complete output ($\overline{CC}$) signal. As mentioned, the 2500 series SARs are TTL devices, and so use +5 volt supplies. There are also other analogously operating SAR units which use CMOS technology, at some sacrifice in speed, but with greater power supply flexibility. Two examples of these are the Motorola MC14549B and MC14559B, and the National Semiconductor MM74C905.

While many of the finer points of SA a/d operation are best explained in the context of specific circuit examples, there are a few general considerations appropriate to this basic circuit. The comparator, which makes a series of trial bit decisions beginning with the MSB, feeds the D input of the SAR. This comparator output data will then appear, delayed one clock cycle, at the SAR serial output (for the 2502 and 2504), with the MSB first. This sequence of operation will be more apparent with examination of the typical timing diagrams which will follow.

Fig. 7-62 is an illustration of how the 8-bit 2503 and 2502 devices are used in general, and also in a truncated mode. Truncation is simply operation of the SAR for a *fewer* number of bits per conversion than its inherent limit; it is often also referred to as "short cycling." As should be obvious, it greatly increases the utility of a given SAR device, or circuit using the SAR.

This circuit illustrates the basics of a continuous conversion method of operation. Without truncation, continuous conversion is achieved simply by tying the SAR $\overline{CC}$ output directly back to the start input (S) (in a manner shown in Fig. 7-61). With truncation in use, the conversion complete signal is derived from the next unused bit. For example, for a 7-bit converter, the normal LSB (for 8 bits) is used as the conversion complete signal (see diagram). The register can also be truncated to any fewer number of bits, if so desired.

When operating in the truncated mode with continuous conversion, a power-on lockup possibly exists. This is eliminated by the logic gate shown, which can be either type of logic element, as is convenient. If the SAR is to be truncated but externally triggered, the logic is not necessary, and the appropriate low order bit is then used directly as the $\overline{CC}$ signal.

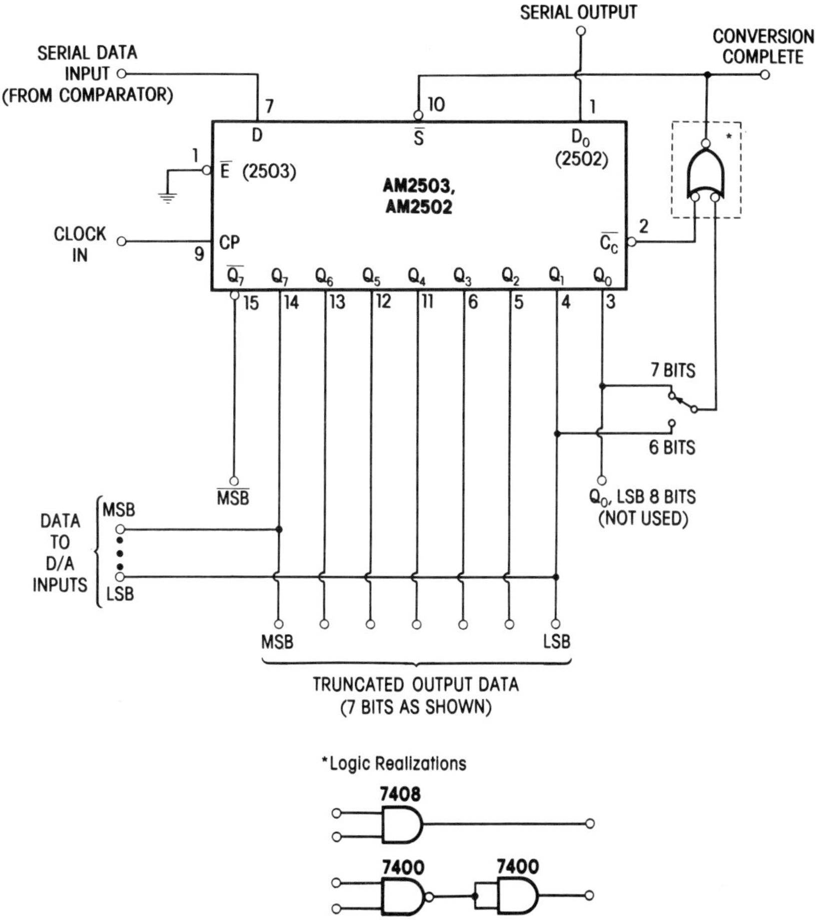

Fig. 7-62. Use of 2502, 2503 SAR in truncated modes.

As mentioned above, SARs can be cascaded or expanded for a/d conversions, with resolutions greater than their inherent number of bits. The general technique is shown in Fig. 7-63.

Here a pair of 2503 devices are operated in a mode which is expanded by driving the $\overline{EN}$ (enable) input of A2 from the $\overline{CC}$ output of A1. This successively enables devices A1 and A2, in order, to perform the complete conversion.

In just a straightforward method of expansion, two 2503s would yield a 16-bit SAR. However many practical applications will require fewer bits, perhaps 13 or 14. In such cases the second SAR is truncated by using the appropriate low order bit as a $\overline{CC}$ signal, such as A2-Q3 as shown here for a 14-bit application. This application shows an ex-

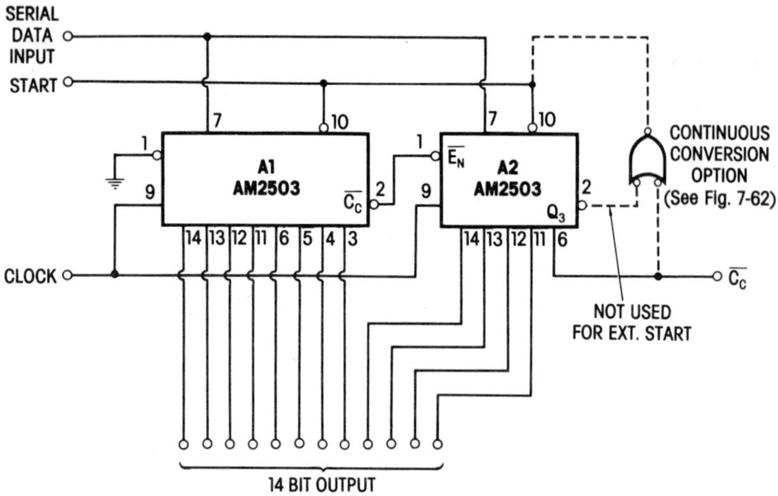

Fig. 7-63. SAR expansion technique.

ternally triggered use; for continuous conversion the antilockup gating is appropriate, as previously described. It should be noted that with the use of 2503 devices, there is no serial output available.

Expanded SAR uses which also require a serial output capability should therefore use the 2504 (or 25L04). The general principles of expansion and truncation are the same as shown here.

While the above generally describes the basic application considerations of SARs, there is another point appropriate to their use, which is output data buffering. This is illustrated in Fig. 7-64.

In Fig. 7-64A is shown an (simplified) 8-bit SA a/d converter using the 2503. With the exception of A2, this diagram is generally quite similar to Fig. 7-61. A2 is an 8-bit type-D register, with 3-state outputs. With the appearance of a $\overline{CC}$ signal from the SAR, the SAR valid output data is clocked into the holding register. Data is held between conversions by A2, being updated with each successive $\overline{CC}$ signal. The 25LS374 is a 3-state output device, making this configuration 8-bit bus compatible. The output data is available with a low output enable signal; and the outputs are in the third state with this input high. An alternate device which can be used for the A2 function is the AM25-LS2520, which also has a register clear input.

As most SA a/d converters have a serial output capability this feature can be very advantageous when data is to be transmitted over appreciable distances, due simply to the necessity for only two output lines (data and clock), rather than "n" parallel outputs. A serial a/d transmitter/receiver scheme illustrating this is shown in Fig. 7-64B.

Here A1, a 2502 performs as an 8-bit SA a/d converter with serial output. A2, a 2503, is an 8-bit serial-to-parallel data converter. In-

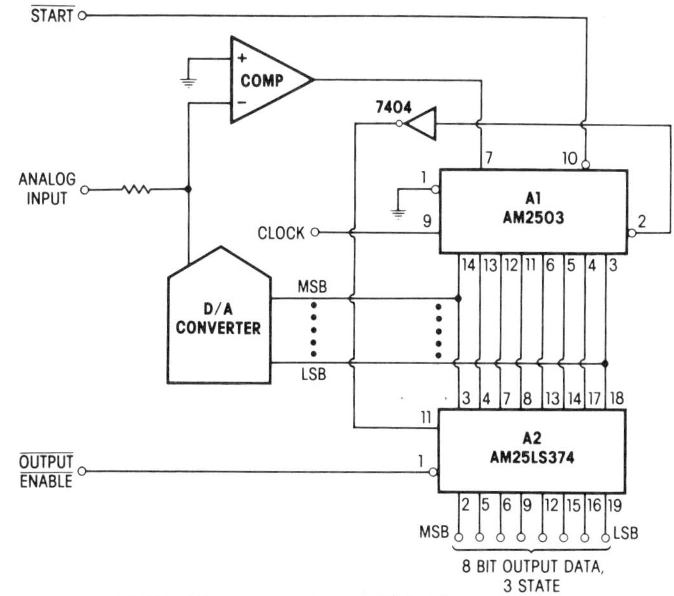

(A) SA a/d converter with parallel holding register.

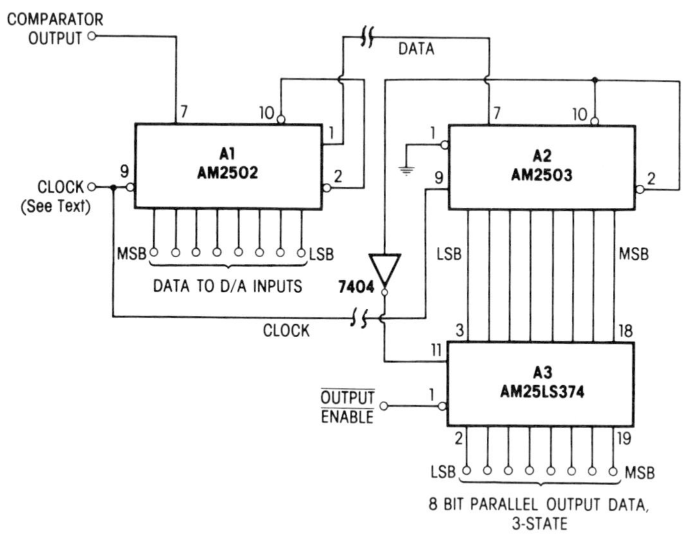

(B) A/d converter with serial output transmitter and serial-to-parallel receiver.

Fig. 7-64. SA a/d buffer techniques.

coming serial data appearing at the A2 input is converted into a parallel form, with valid output data being signaled by a $\overline{CC}$ signal. This $\overline{CC}$ signal, after inversion, clocks the data into A3, an 8-bit holding register (similar to Fig. 7-64A). Output is in 8-bit parallel, 3-state form. A requirement for the proper functioning of this scheme is a gated clock which runs only during conversion, to preserve the proper bit framing in the serial-to-parallel converter and holding register.

SA A/D System Considerations

Before taking up detailed design considerations of SA a/d converters, it is first appropriate to consider the use of the a/d converter within a given system.

Many a/d converters (and SA types in particular) must be used with a preceding S/H (sample/hold amplifier) to function properly. A functional block diagram of a typical hookup would appear as shown in Fig. 7-65.

The reason why an a/d converter must be preceded by a sample hold is a very fundamental one, and is directly related to the speed, or more specifically the *rate of change* of the input signal. If the input to an a/d converter is allowed to change during the conversion interval, the output will by necessity be in error. In practice, the criterion of restraint is that the input must change by no more than ½ LSB during the conversion interval to achieve rated accuracy. This is perhaps a simplistic (or

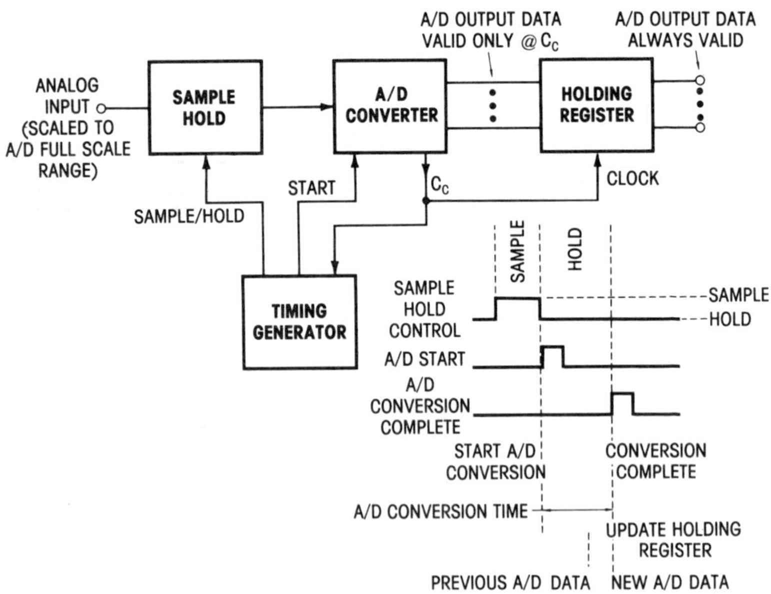

Fig. 7-65. Sample-hold and a/d system.

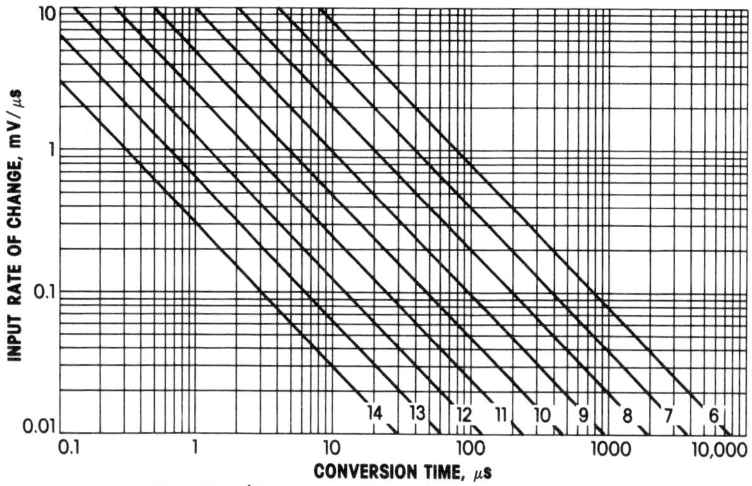

Fig. 7-66. Allowable a/d input signal rate-of-change versus conversion time for various system accuracies.

optimistic) view, as there are many other error sources which are also desired to be held to less than ½ LSB. Nevertheless, it shall be used here to demonstrate its effect. In practice a properly operated (and sufficiently fast) S/H can hold the error due to input signal slope to smaller than ½ LSB errors.

As an illustration of what would result without an S/H amplifier, Fig. 7-66 is a graph of the maximum allowable input rate of change for various a/d system accuracies, for a 10-volt signal level. As an example, an 8-bit system with a 5 μs conversion time will allow only a 3.9 mV/μs input rate-of-change, equivalent to a sine wave at a frequency of 124 Hz. This discussion is, of course, just a different point view with regard to the system's aperture time.

Effectively, what the S/H does for the system is to reduce its aperture time to much more useful levels. The same 5-μs conversion time a/d converter, when used with a 100-ns aperture time S/H amplifier, can track a 7-kHz signal accurately (assuming other errors not to be significant).

Within a system, the S/H and the a/d converter must then be operated in harmony, and the timing of drive signals must allow for sufficient S/H settling to rated accuracy, and a/d conversion time. The simplified timing diagram of Fig. 7-65 illustrates these basic interrelationships.

While the S/H is in its sample (or track) mode, the a/d converter is at rest (not converting). It may however, at the option of the system designer, be statically outputing data from a previous conversion, by means of a holding register (as shown).

When the S/H switches to its hold mode, the a/d converter is given a start convert command, and begins a new conversion. The S/H must remain in hold for the duration of this conversion interval. With a conversion-complete signal, the new a/d output data is valid, and is then shifted into the holding register. At this point the sample/hold would be switched back to the sample mode, to again begin a new conversion cycle.

It should be noted from the timing diagram and this discussion that a practical system will be limited in its output data update rate, or *throughput* rate, by both the S/H and a/d timing constraints. Either of these may be dominant, and they need not necessarily be similar times. The key criteria are the settling or conversion times to the desired system accuracy. The designer can trade off one time against the other to minimize the total conversion interval, and thus maximize system speed, and lower cost.

In this simple example the S/H and a/d converter are shown being driven by separate signals, with the a/d $\overline{CC}$ signal being returned to the timing generator. In practice it is not so much necessary that the signals be separate (they can be the same in some cases, for example with complementary S/H and a/d input sensitivities), but that the timing constraints of the S/H and a/d converter be *mutually satisfied.* In general this will mean a S/H sample drive interval at (or slightly longer) than its required settling time, and a new conversion cycle (S/H returning again to sample) to occur no sooner than the time required for the a/d converter to convert to rated accuracy. Logic transfers, such as loading of the holding register, will in practice usually be appreciably shorter than either of these two major time intervals.

In many of the circuits which follow, all of the details of old conversion and S/H timing (and associated hardware) are not always shown, but they are nevertheless fundamentally required.

IC D/A-PLUS-SCR SA A/D DESIGNS

One of the more simple and cost effective ways to implement a SA a/d converter is to build it up with an IC d/a converter and comparator which work in conjunction with a SAR device. This section discusses a number of examples of this type of SA a/d converter along with the pertinent design criteria.

Basic Design Guidelines

In Fig. 7-67 is shown a circuit for an 8-bit SA a/d converter using the DAC-08, the 2502 SAR, and a CMP-01 comparator (Precision Monolithics). This a/d converter will be used as an illustrative example of several design techniques, which can then be applied in general fashion.

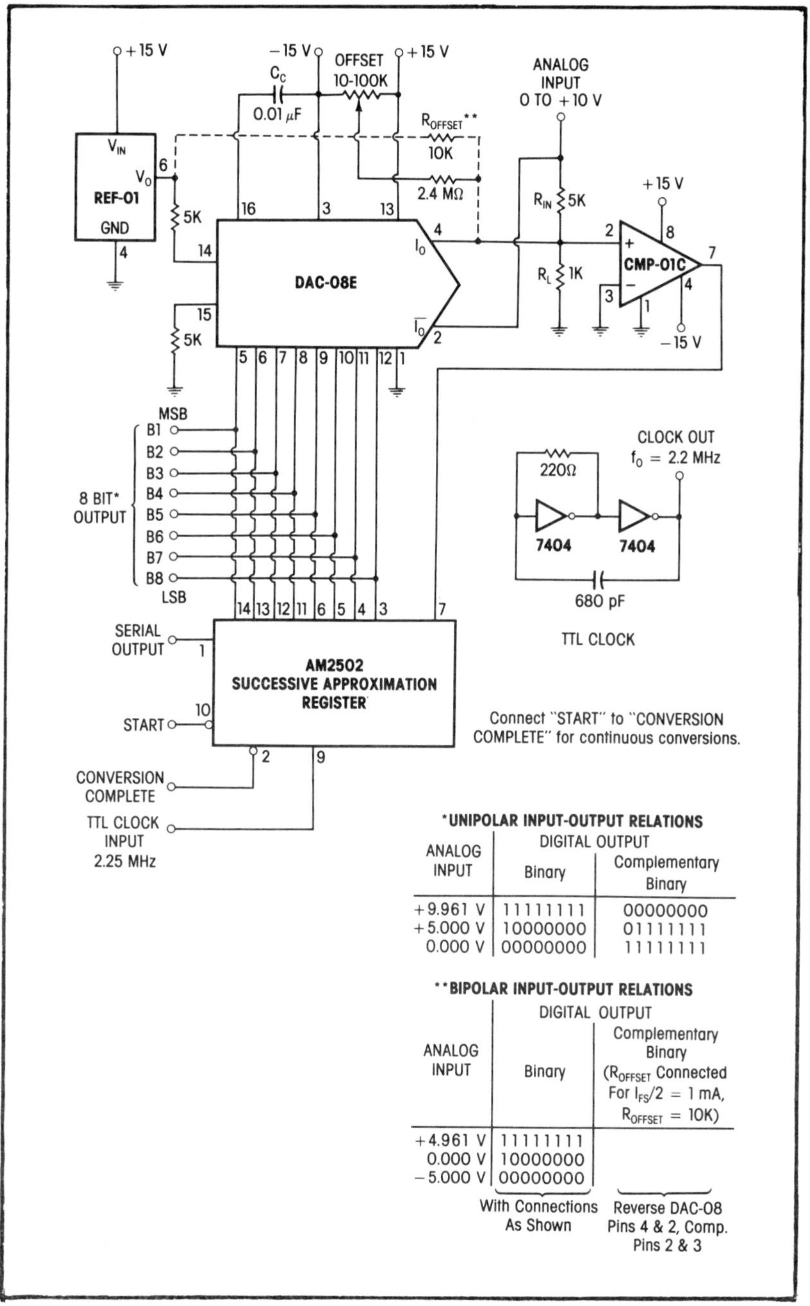

***UNIPOLAR INPUT-OUTPUT RELATIONS**

ANALOG INPUT	DIGITAL OUTPUT Binary	Complementary Binary
+9.961 V	11111111	00000000
+5.000 V	10000000	01111111
0.000 V	00000000	11111111

****BIPOLAR INPUT-OUTPUT RELATIONS**

ANALOG INPUT	DIGITAL OUTPUT Binary	Complementary Binary (R_{OFFSET} Connected For $I_{FS}/2$ = 1 mA, R_{OFFSET} = 10K)
+4.961 V	11111111	
0.000 V	10000000	
−5.000 V	00000000	
	With Connections As Shown	Reverse DAC-08 Pins 4 & 2, Comp. Pins 2 & 3

Fig. 7-67. A DAC-08 4-μs, 8-bit binary a/d converter.

As can be noted from the circuit, the DAC-08 is operated as a current output d/a converter, rather than voltage output to optimize speed. Were the d/a converter of a SA a/d converter such as this to be operated in a voltage output mode, it would severely compromise conversion speed due to the relatively long settling time of the op amp which would be used—as much as a microsecond or more, even for high-speed devices. By contrast, many current output d/a converters can settle in 100 to 300 ns for 8- and 10-bit accuracy; the DAC-08 used here settles in 85 ns.

The analog input voltage is scaled for the desired input range, by selecting the input resistor to drop the full scale input voltage at the d/a converter full-scale output current. The 2-mA d/a converter output here then yields a 10-volt input scale with 5K input resistor. This resistor should be a precision type, and should track the reference input resistor.

The 1K resistor shown as the d/a load serves to reduce the d/a converter overall settling time, by minimizing the RC time constant at this node. It does not directly affect analog input range, and need not be a precision type. This resistor is, however, relatively important, and cannot be arbitrarily selected. While it is desirable to make it as low as possible from the standpoint of speed, it cannot be too low or it will excessively reduce the error input voltage to the comparator, and so lengthen the response time of the comparator.

Comparators respond in general in the shortest times to high overdrive signals, with response time lengthening for lower levels of input overdrive. This effect can be seen in Fig. 7-68, which shows the positive and negative going responses for several popular comparators useful in SA a/d converters.

The load resistor must then allow sufficient error voltage drive to the comparator, with a ½ LSB error voltage signal developed across the equivalent parallel resistance of R_L and R_{IN}. In the case here this resistance is 830 ohms, and ½ LSB (8 bits) of 2 mA is 3.9 μA, yielding a 3.2-millivolt error, or overdrive. From Fig. 7-68B, 2-millivolt and 5-millivolt overdrives are charted for the CMP-01, so the 3-millivolt condition here will have a response time less than the 2-millivolt case, or 100 ns.

The d/a settling time must also be obtained, either from data sheet information or lab data. For the DAC-08, the external RC circuit becomes the dominant factor of setling time for loads greater than 500 ohms. Assuming a minimal capacitance of 25 pF, RC then equals 20 ns. For 8-bit accuracy, 6.2 time constants must be allowed, which gives a settling time of 130 ns. To the two figures of comparator response and d/a settling must then be added an allowance for logic delay. For the 2502, this is on the order of 25 ns, making a total time (per bit) of 250 ns. This time is the maximum speed for rated accuracy, and is

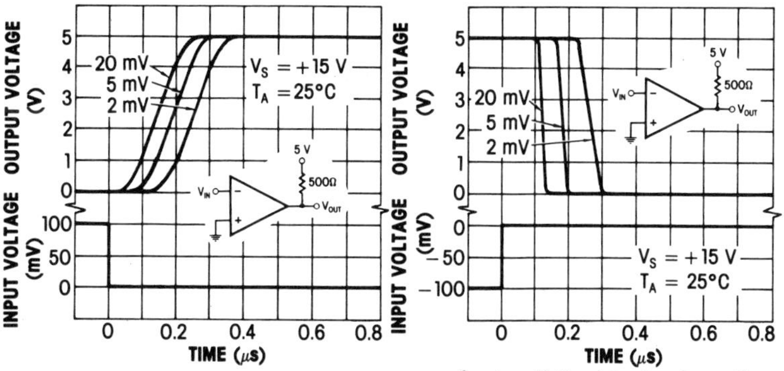

Courtesy National Semiconductor Corp.

(A) LM311.

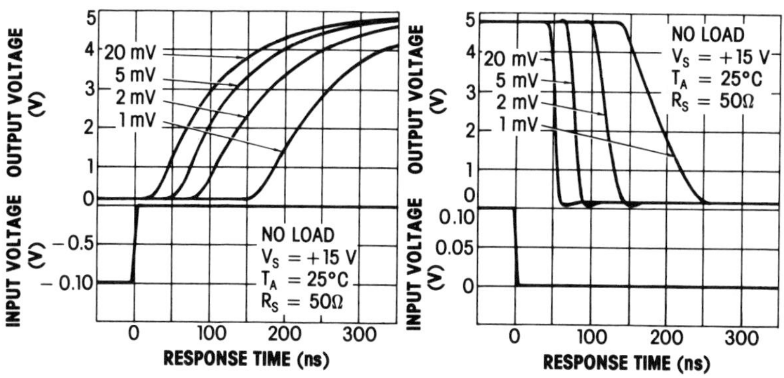

Courtesy Precision Monolithics, Inc.

(B) CMP-01.

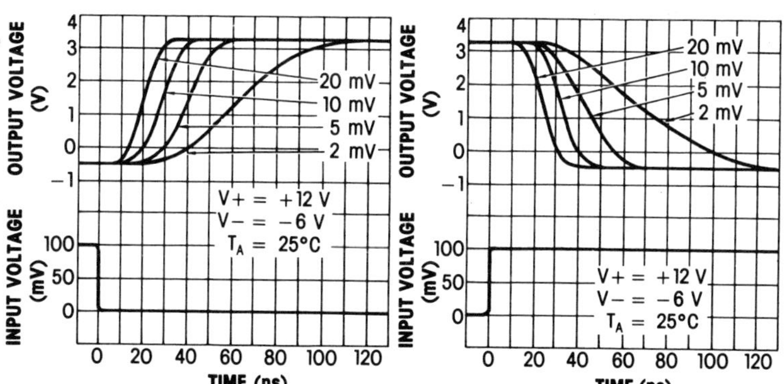

Courtesy Fairchild Semiconductor

(C) μA710.

Fig. 7-68. Comparator response times.

the minimum which can be allowed for a single clock cycle to yield accurate (½ LSB error) SA decisions. For the case here, it implies a maximum clock frequency of 3.9 MHz. The minimum (total) conversion time is 250 ns × (n + 1) = 2.3 μs.

From this discussion it should be obvious that there is a very definite and fixed relationship between the three speed components (d/a settling, comparator response, logic delay), and maximum conversion time and clock rate. Once the speed-per-bit has been determined, the conversion time is then simply this time multiplied by (n + 1). This then fixes the maximum clock frequency as

$$f_{clock} = \frac{n + 1}{\text{conversion time}}$$

The relationship between clock frequency and conversion time for various numbers of bits is summarized in the graph of Fig. 7-69.

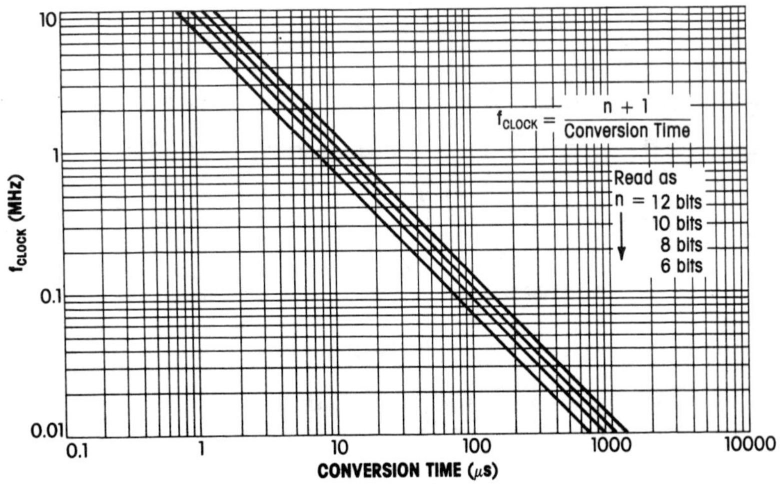

Fig. 7-69. S/A a/d clock rate versus conversion time for various resolutions.

For the example under discussion, the clock frequency is conservatively specified at 2.25 MHz, which yields a 4-μs, 8-bit conversion time (note from Fig. 7-69, the circuit could be clocked as high as 3.8 MHz). From the above calculations it can be seen that a lower comparator response time, if available, would result in an appreciably faster conversion time. If an AM686 comparator is used in this circuit (in a similar fashion) it can allow conversions as fast as 2 μs, with a 4.5-MHz clock.

With an 8-bit, 10-volt full-scale circuit such as this, the LSB value is 39 mV; therefore, no nulling of the comparator is required for offset

Table 7-4. Input Calibration Voltages for Popular Unipolar A/D Converter Scalings

BINARY						BCD				
Bits	Voltage Range	1 LSB (00 . . . 01)	Full Scale (11 . . . 11)	+ 1/2 LSB[1] Zero Scale	Full Scale − 1/2 LSB[1]	Digits	1 LSB (0000 . . . 0001)	Full Scale (1001 . . . 1001)	Zero Scale + 1/2 LSB[1]	Full Scale − 1/2 LSB[1]
6	2.5	0.039	2.461	0.039	2.441					
	5	0.078	4.922	0.020	4.883					
	10	0.156	9.844	0.078	9.766					
8	2.5	0.010	2.490	0.005	2.485	2	0.025	2.475	0.013	2.463
	5	0.020	4.980	0.010	4.971		0.05	4.950	0.025	4.925
	10	0.039	9.961	0.020	9.941		0.1	9.900	0.05	9.850
10	2.5	0.0024	2.4976	0.0012	2.4963					
	5	0.0049	4.9951	0.0024	4.9927					
	10	0.0098	9.9902	0.0049	9.9854					
12	2.5	610 μV	2.4994	305 μV	2.4991	3	0.0025	2.4975	0.0013	2.4963
	5	0.0012	4.9988	610 μV	4.9982		0.005	4.9950	0.0025	4.9925
	10	0.0024	9.9976	0.0012	9.9963		0.010	9.9900	0.005	9.9850
14	2.5	153 μV	2.49985	76 μV	2.49977					
	5	305 μV	4.99969	153 μV	4.99954					
	10	610 μV	9.99939	305 μV	9.99908					
16	2.5	38 μV	2.49996	19 μV	2.49994	4	250 μV	2.49975	125 μV	2.49963
	5	76 μV	4.99992	38 μV	4.99988		500 μV	4.99950	250 μV	4.99925
	10	153 μV	9.99985	76 μV	9.99977		0.001	9.99900	500 μV	9.99850

Notes: Zero Scale = 00 . . . 00 = 0.000 Vin.
Negative scale (or sign-magnitude) circuits use same levels, with appropriate sign.

1 Cal levels are 1/2 LSB from end limits.

a. Adjust OFFSET control for alternate display, 00 . . . 00 → 00 . . . 01.
b. Adjust GAIN control for alternate displays, 1 . . . 10 → 11 . . . 11.

adjustment. A ½ LSB current offset is provided here by the 2.4-megohm resistor to the offset voltage pot. In general, this resistor biases the a/d input ½ LSB from zero scale for calibration (see Table 7-4); here it provides a 3.9-μA (½ LSB of 2 mA) bias. The ½ LSB bias can also be accomplished by adjustment of the comparator offset voltage, but generally this is done for ½ LSB voltages (or the comparator offsets) of 10 millivolts or less. The ½ LSB current bias will vary in sense according to whether a straight binary or complementary scale is being used. Thus, the general (±) method is shown.

Thus generally speaking, the ½ LSB offset serves as the OFFSET (or zero scale) calibration adjustment in an a/d converter of this type. GAIN (or full scale) adjustment can be generally accomplished by either variation of the d/a reference current (see Chapter 6), or by variation of the a/d input scaling resistor. As has been stressed previously, the reference and input resistor should be *tracking* types, preferably within a common network. Calibration can then be achieved by a series trim of several percent, or alternately by variation of the reference voltage by a comparable amount.

For unipolar operation, the circuit is quite straightforward, and the DAC-08 can be wired to accommodate either straight binary or complementary binary codes. It should be noted that an advantage of the DAC-08 for this type of converter is that the unused output (Io or Io) can be tied to the analog input, which causes the total input current seen by the source to remain constant (at 2 mA). This then relaxes the requirement for a low source impedance somewhat, compared to a noncomplementary output type d/a converter. The input source voltage must of course remain constant (< ½ LSB) during the SAR bit switching steps, for accuracy consistent with resolution of the a/d converter in use. Often, general purpose op-amp types such as 741s cannot satisfy this requirement.

For bipolar applications, the circuit is operated in an offset mode, by connection of the R_{offset} resistor, as shown. In general this resistor should be of a value to supply a current to the sum point of exactly one-half scale (similar to its selection criteria for a bipolar d/a converter, see Chapter 6). This resistor should also track the reference and input resistors. It should be adjustable with respect to the reference resistor to calibrate the negative input range limit for LSB offset (see Table 7-5); for this example calibration would be performed at −4.980 volts. The input resistor is adjusted for calibration at the positive full-scale end of the scale; in this case with a +4.941-volt input. Note that with these two adjustments the 2.4 megohm, ½ LSB offset becomes redundant, and need not be used for bipolar scales.

As can be noted from the table, there is very little change required in implementing one form of converter over another. For complementary scaled converters, note that both the d/a converter and comparator

Table 7-5. Input Calibration Voltages for Popular Bipolar A/D Converter Scalings

Bits	Neg. Scale (00 . . . 00)	1 LSB (00 . . . 01)	Pos. Scale (11 . . . 11)	Negative Scale +1/2 LSB[1]	Postive Scale −1/2 LSB[1]
6	−2.5 −5 −10	0.078 0.156 0.313	2.422 4.844 9.688	−2.461 −4.922 −9.844	2.383 4.766 9.531
8	−2.5 −5 −10	0.020 0.039 0.078	2.480 4.961 9.922	−2.490 −4.980 −9.961	2.471 4.941 9.883
10	−2.5 −5 −10	0.0048 0.0098 0.0195	2.495 4.990 9.980	−2.4976 −4.9951 −9.9902	2.4927 4.9854 9.9707
12	−2.5 −5 −10	0.0012 0.0024 0.0048	2.4988 4.9976 9.9951	−2.4994 −4.9988 −9.9976	2.4982 4.9963 9.9927
14	−2.5 −5 −10	305 μV 610 μV 0.0012	2.49969 4.99939 9.99878	−2.49985 −4.99969 −9.99939	2.49954 4.99908 9.99817
16	−2.5 −5 −10	76 μV 153 μV 305 μV	2.49992 4.99985 9.99969	−2.49996 −4.99992 −9.99985	2.49989 4.99977 9.99954

Notes: Midscale = 10 . . . 00 = 0.000 V in.
Inverted scale bipolar circuits use same levels with sign reversal(s).
1 Cal. levels are 1/2 LSB from end limits.
a. Adjust OFFSET control for alternate display, 00 . . . 00 → 00 . . . 01.
b. Adjust GAIN control for alternate display, 1 . . . 10 → 11 . . . 11.

wiring is changed. Note that with the 2502 (or 2503) SAR, twos complement bipolar operation is easily achieved by using the available MSB output (pin 15).

The input-output logic connections for the circuit have generally been explained above. A 2.2-MHz TTL clock source is required, such as one of the circuits of Fig. 7-9, or a simple TTL astable as shown in the inset of Fig. 7-67.

The previous discussion has touched generally on the major design considerations to be undertaken in an a/d converter of this type, using typically available hardware. The remainder of this section illustrates other examples of SA d/a converters implemented in this fashion. For further information on the details of operation of this circuit, consult Precision Monolithics Application Note AN-16 (see references at end of section).

DAC-100 6-μs 8-Bit A/D Converter

Fig. 7-70 illustrates how the DAC-100 internal reference d/a converter can be used for 8-bit SA a/d applications. This circuit is arranged for an analog input range of 0 to +10 volts, and completes an 8-bit

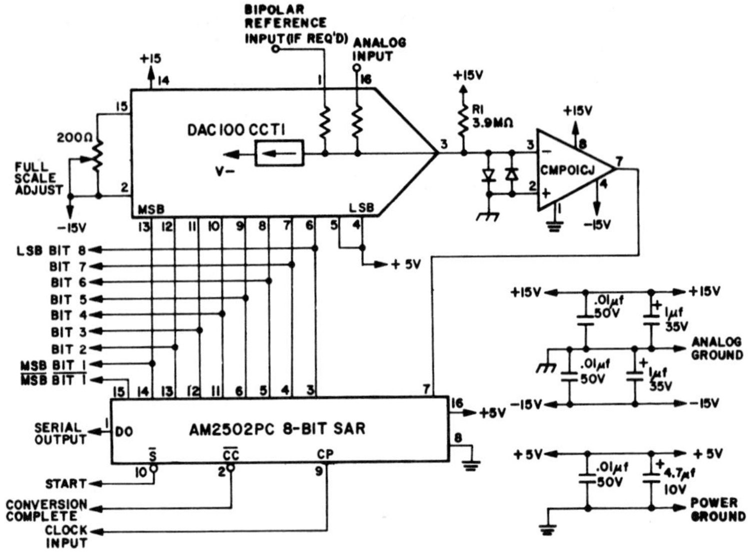

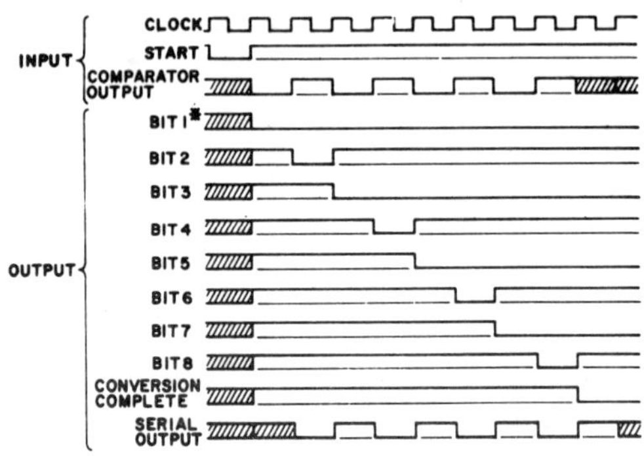

*NOTE: NEGATIVE TRUE LOGIC IS USED.
SHADED AREAS INDICATE LEVEL DEPENDS UPON PREVIOUS STATE
INPUT = +6.66 VOLTS, FULL SCALE = +10 VOLTS

Courtesy Precision Monolithics, Inc.

Fig. 7-70. DAC-100 6-μs, 8-bit complementary binary a/d converter, unipolar, 0 to 10 volts.

conversion in 6 μs, using a 1.5-MHz clock. The output is in complementary binary form (TTL compatible).

As the DAC-100 has 10-bit resolution, the two lowest order bits are disabled by tying them to the +5-volt logic supply. The 3.9-megohm resistor supplies the ½ LSB comparator offset, and IN4148 clamp diodes are used on the d/a converter output for protection. No offset calibration is necessary; gain is adjusted by altering the DAC-100 full-scale output current via the 200-ohm full-scale adjust pot.

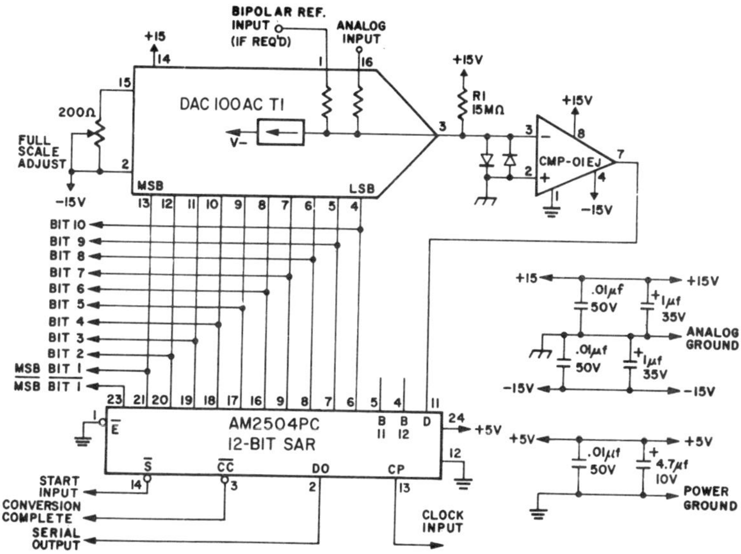

Courtesy Precision Monolithics, Inc.

Fig. 7-71. DAC-100 9-μs, 10-bit complementary binary a/d converter, unipolar, 0 to 10 volts.

The timing diagram shows the sequence of operation, and illustrates an output code for an input of +6.66 volts (01010101 in complementary binary = 170; 170/256 × 10 = 6.666) within ½ LSB.

DAC-100 9-μs 10-Bit A/D Converter

Fig. 7-71 shows the DAC-100 being utilized to the full extent of its resolution, in a 10-bit, 9-μs d/a application. The input range is 0 to 10 volts, and the circuit uses a 1.5-MHz clock source.

For 10 bits, the ten highest order bits of the 2504 are used, and as shown here, the two lowest bits are unused. The SAR will as shown step through a full 13 step (12 + 1) SAR routine; alternatively it may be short cycled as described above. It should be noted that short cycling can save conversion time, in this case two clock periods, or about 1.3 μs.

Comparator offset is provided here by the 15-megohm resistor, which yields a 1-μA current offset. With the magnitude of LSB voltage associated with 10 bits, the comparator offset is not negligible, unless a low offset device is used, or it is nulled. The CMP-01E has a 0.8-millivolt offset, which is equivalent to less than 0.1 LSB error. Therefore no offset calibration is required, and gain is adjusted as generally described before (with reference to Table 7-4 for 10-bit levels).

AD561 10-Bit A/D Converter

The AD561 is applied quite simply for either unipolar or bipolar a/d uses as shown in Fig. 7-72. This circuit provides either a straight

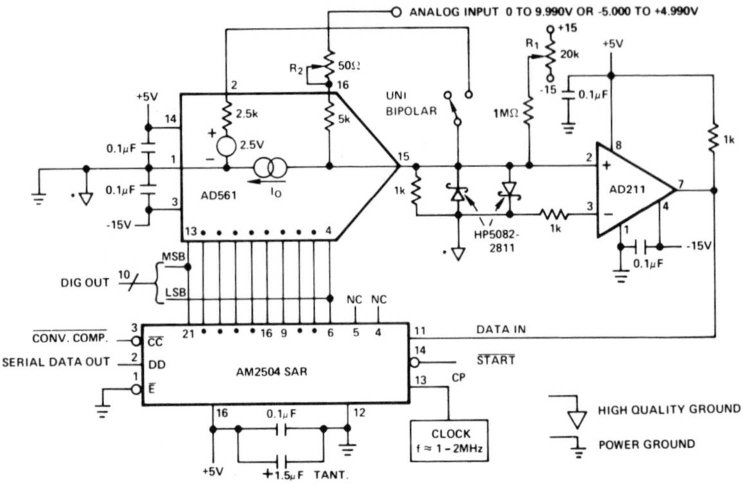

CONVERSION TIME (μs)

	UNIPOLAR	BIPOLAR
6 BITS	1.4	2.1
8 BITS	2.7	4.0
10 BITS	5.5	12

Courtesy Analog Devices, Inc.

Fig. 7-72. AD561 5.5/12-μs, 10-bit binary a/d converter, unipolar/bipolar.

binary output for unipolar inputs, or offset binary/twos complement for bipolar inputs. Conversion times for 10-bit operation are 5.5 μs (unipolar) and 12 μs (bipolar), with 2.4- and 1.1-MHz clocks respectively.

The AD561 has the required resistors for scaling integrally, and for 10-bit applications trims are used externally. R1 adjusts the camparator offset for ½ LSB, which also compensates for the residual offset of the device in use, the AD211. Gain is adjusted for full-scale calibration via R2, with an input of 9.9854 volts.

For bipolar inputs, the 2.5K internal resistor is connected to the sum point, thus providing a 1-mA offset. R1 and R2 are trimmed for offset and gain in accordance with Table 7-5, for ±5-volt scale.

For shorter bit lengths, such as 8 bits, a 2502 or 2503 SAR can be used. For these conditions no gain or offset trim of the AD561 will be necessary due to the accuracy of its pretrimmed scaling resistors, and the low offset voltage of the comparator. Fixed ½ LSB current offset should be provided as previously outlined.

For 10-bit uses, in particular with bipolar ranges, the response time of 211 (or 311) comparator types is the dominant limit to speed, and is on the order of 500 ns (this is the reason for the bipolar 12-μs conversion time). Faster comparators, such as are listed above, can yield conversion times of about 400 ns/bit with the AD561, when properly applied.

AD 562, AD563 25-μs 12-Bit A/D Converters

Both the AD562 and AD563, when applied to SA a/d converter use can yield designs with a high degree of flexibility and accuracy at moderate speeds, achieving 12-bit conversions in 25 μs. These devices are illustrated in Fig. 7-73. Both devices are applied quite simply for various ranges, unipolar and bipolar, by simple jumper options.

The AD562, shown in Fig.7-73A, requires an external 10-volt reference source, such as the AD581 or AD2700. This SA design uses the 25L04, a low-power, slightly slower version of the 2504 12-bit SAR. The AD562 requires about 1.3 μs for 12-bit settling, to which is added 500 ns for the comparator and 100 ns for logic. Maximum rate is then about 1.9 μs/bit, or 25 μs for 12-bit conversions. This circuit therefore requires a 520-kHz TTL clock source. The various input ranges possible are shown in the table.

The AD563 is shown in a comparable hookup in Fig. 7-73B. The basic attractiveness of this design is the inclusion of the reference source; otherwise, it is quite similar to the AD562. Both designs have binary-coded outputs, with serial output options from the SAR. Adjustment of the reference current provides full scale gain calibration (100-ohm pot for the 562, 20 ohm for the 563), while offset can be adjusted via the comparator terminals, or summing a variable current at the D/A converter output (see Fig. 6-42A for an example). For 12-bit unipolar or bipolar calibration levels, see Table 7-4 or 7-5.

A complete and integral a/d device which operates quite similar to both these circuits is the AD572.

DAC-80 26-μs 12-Bit A/D Converter

Another example of a d/a device which offers a great deal of flexibility to a SA a/d design is the DAC-80, which is illustrated in a general purpose 12-bit a/d converter in Fig. 7-74.

This circuit is conceptually similar to some of the others already described. A current output DAC-80 type is used, and the 2-mA full scale output current is compared against the input current developed in the internal 5K scaling resistors. As shown in the table, this network can be jumper programmed for the popular unipolar and bipolar input scales.

The DAC-80 is a d/a device which is coded in a complementary binary fashion. As can be noted, the comparator inputs are therefore reversed from an a/d converter of this type which uses a straight binary coded d/a converter. The comparator response and logic speed requires about a 2 μs/bit allowance for settling, for 12-bit accuracy. Thus the clock frequency is 500 kHz (maximum), and a 12-bit conversion requires 26 μs.

The a/d converter is trimmed for full-scale calibration via the GAIN pot, and zero scale via the OFFSET pot. For calibration levels, see Table

7-4 or 7-5. This circuit can also use the DAC-85 d/a converter (as a direct substitute for the DAC-80) and will yield greater overall accuracy and stability (with similar speed). This circuit function is also achieved in a single integral a/d device, the ADC 80, discussed later.

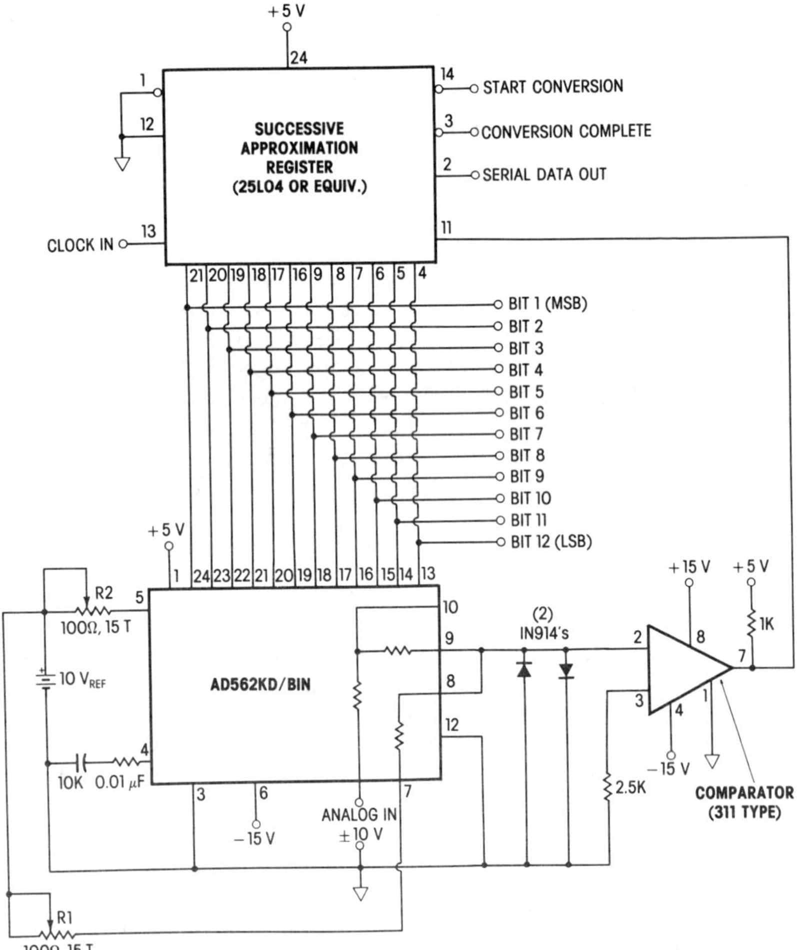

AD562 RANGE OPTIONS

UNIPOLAR			BIPOLAR		
RANGE	INPUT	JUMPER	RANGE	INPUT	JUMPER
0 to +10 V	10	—	±10 V	11	8-9, R1
0 to +5 V	10	9-11	±5 V	10	8-9, R1
			±2.5 V	10	8-9, R1, 9-11

(A) AD562.

Fig. 7-73. AD562/AD563 25-μs, 12-bit

DAC-08 1-μs 8-Bit A/D Converter

The 2502/2503/2504 series of SARs, while generally useful down to about 200 ns-per-bit, a/d speed become restrictive for appreciably faster conversion rates. For speeds of 100 ns-per-bit or less, SA opera-

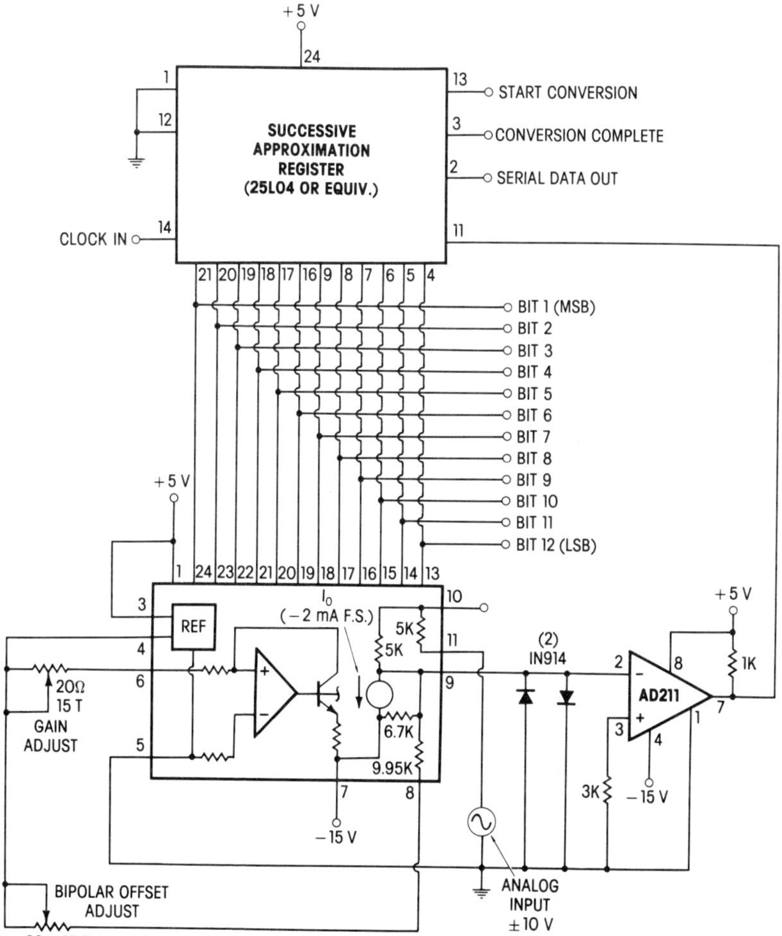

AD563 RANGE OPTIONS

UNIPOLAR			BIPOLAR		
RANGE	INPUT	JUMPER	RANGE	INPUT	JUMPER
0 to +10 V	10	–	±10 V	11	R3
0 to +5 V	10	9-11	±5 V	10	R3
			±2.5 V	10	R3, 9-11

(B) AD563.

Courtesy Analog Devices, Inc.

binary, unipolar/bipolar a/d converters.

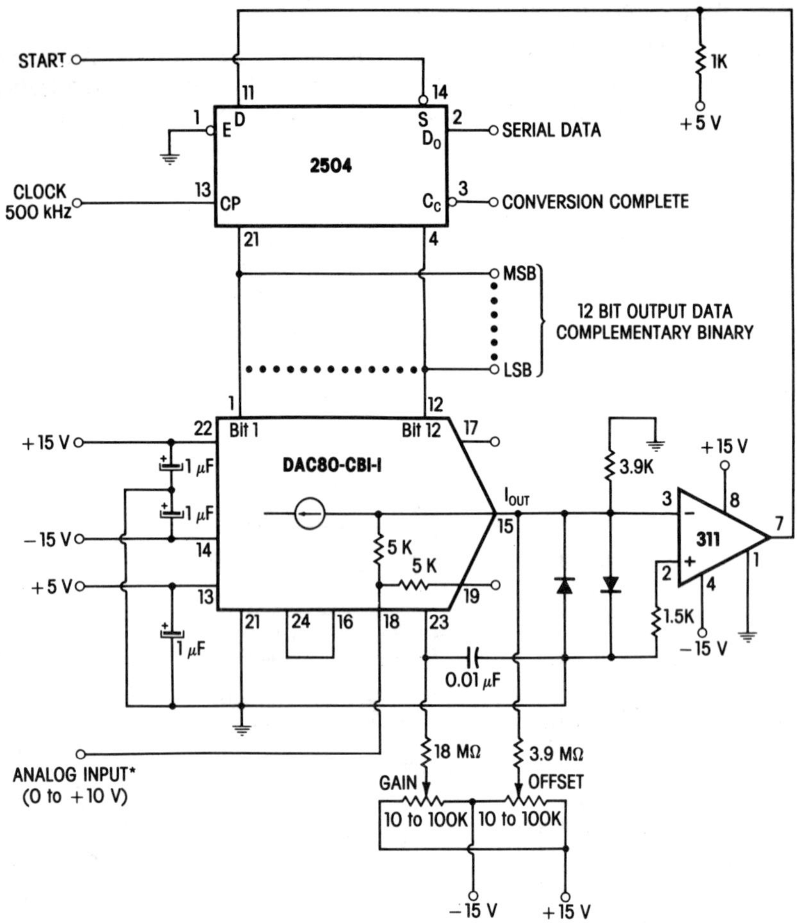

***DAC-80 RANGE OPTIONS**

UNIPOLAR			BIPOLAR		
RANGE	INPUT	JUMPER	RANGE	INPUT	JUMPER
0 to +10 V	18	17-21	±10 V	19	15-17
0 to +5 V	18	17-21, 15-19	±5 V	18	15-17
			±2.5 V	18	15-17, 15-19

Fig. 7-74. DAC-80 26μs, 12-bit complementary-binary, unipolar/bipolar a/d converter.

tion can be achieved through the use of a high speed comparator and discrete high speed logic. Such a circuit is shown in Fig. 7-75, using the DAC-08 as the d/a converter, an AM686 comparator, and a 74S series logic.

This design uses a basic 13-MHz clock, from which are derived two basic clocking periods, 160 ns and 80 ns. The first four significant d/a bits are clocked with the 160-ns clock period, while the four lesser bits

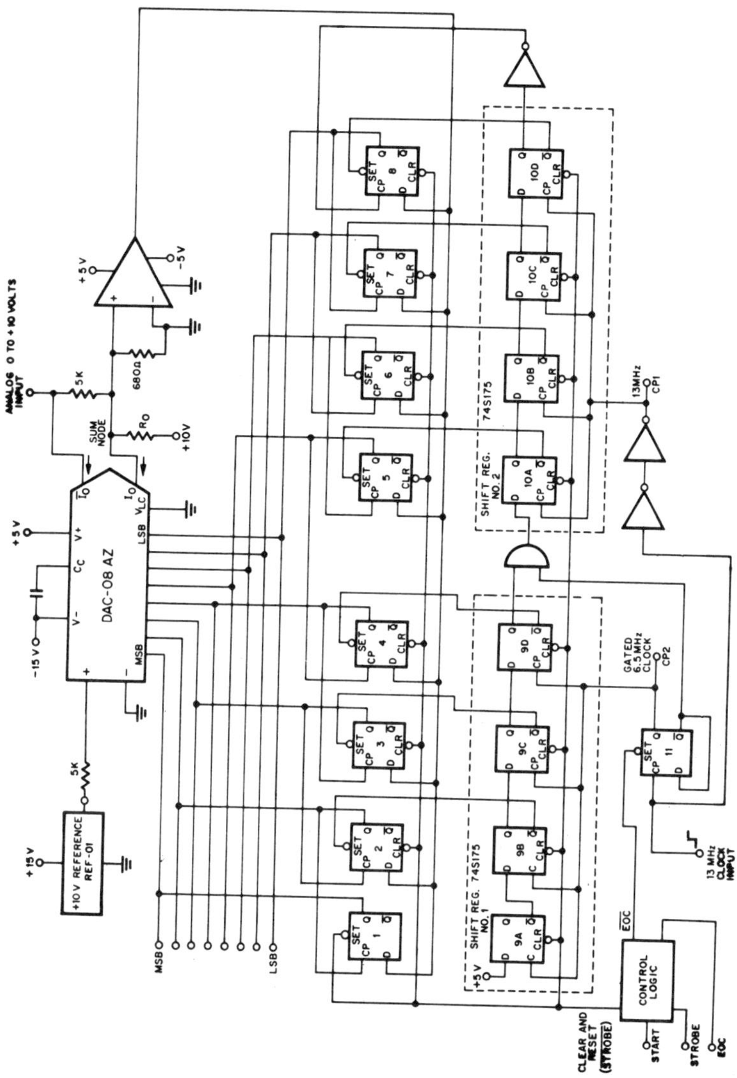

(A) Basic configuration.

Fig. 7-75. DAC-08 1-μs, 8-bit binary, unipolar a/d converter.

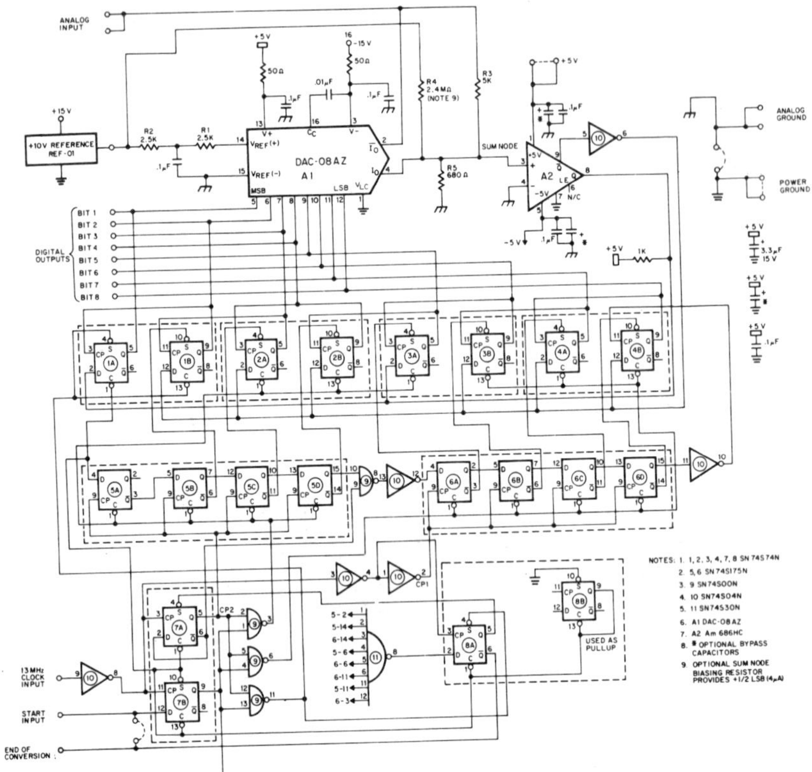

(B) Detailed diagram.

Fig. 7-75 (cont). DAC-08 1-μs, 8-bit binary,

are speeded up to 80 ns. This scheme allows the first four bits a longer time to settle, and thus achieve the required accuracy. The lower four bits, which settle faster, require less time. This concept decreases the necessary time for settling to ¾ of that which would be required via a conventional approach, and yields a conversion time of 1 μs. The timing diagram is shown in Fig. 7-75C.

Two necessary prerequisites for achieving these speeds accurately and reliably are an appropriately fast d/a converter and a comparator. The DAC-08 settling time of 85 ns is achieved by the use of a low load impedance of 680 ohms, and a minimal capacitance layout. The AM686 comparator, under the overdrive conditions as used here, has a response time of 25 ns. A detailed circuit diagram is shown in Fig. 7-75B.

The circuit is arranged for a 0 to +10-volt unipolar input, with a 5K input resistor. Details of gain and offset calibration are generally as

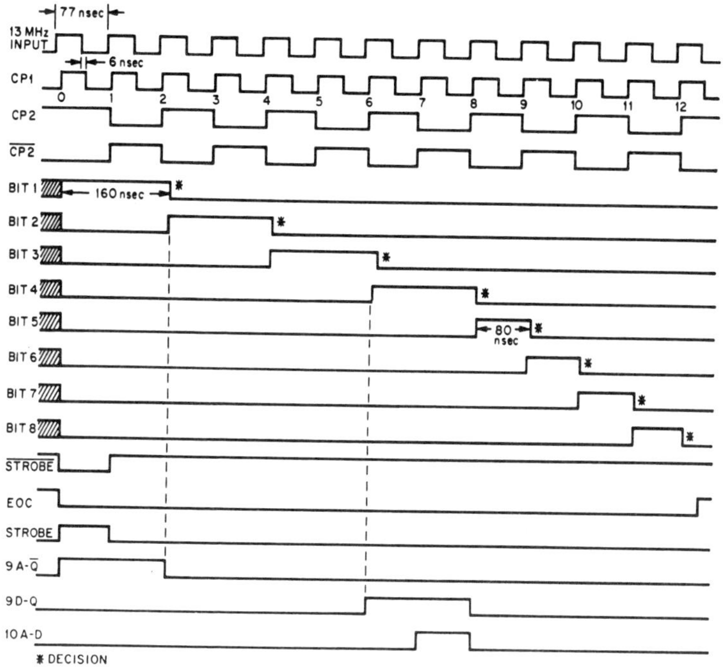

Courtesy Precision Monolithics, Inc.

(C) Timing diagram.

unipolar a/d converter.

described previously. R4, which furnishes the ½ LSB offset current to the sum mode, may require some adjustment due to comparator input errors.

The circuit requires a positive going edge at the START input to begin conversion, and EOC output goes high with a conversion completion. These two points can be jumpered for continuous conversions use. The STROBE output is used to clock a holding register on the positive edge. A suitable register would be two 74S175 quad latches. The timing diagram shown illustrates a conversion of a voltage input of zero. Further details of this circuit operation can be found in Precision Monolithics Applications Note AN-16 (see references at end of section).

AD7570

The AD7570 10-bit/8-bit a/d converter is illustrated in Fig. 7-76. This device can convert unipolar and bipolar input voltages over the

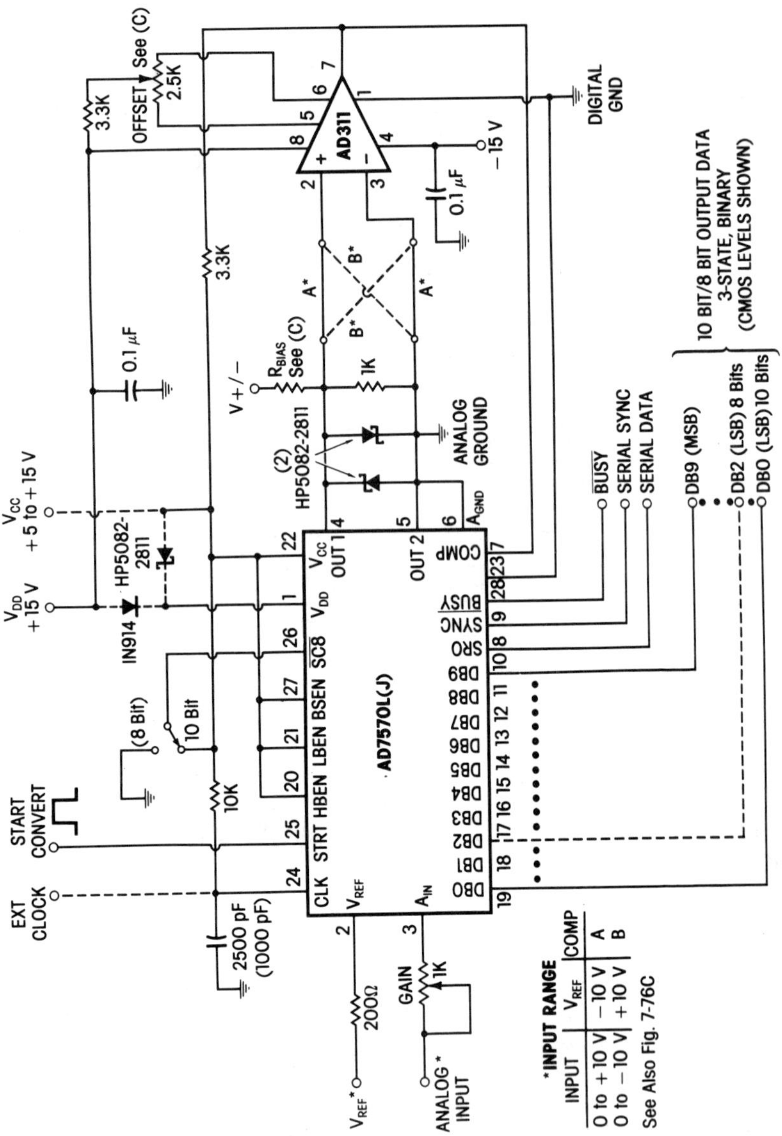

(A) Basic connections, unipolar input of 0 to +10 volts or 0 to −10 volts.

Fig. 7-76. AD7570 120-µs, 10-bit binary

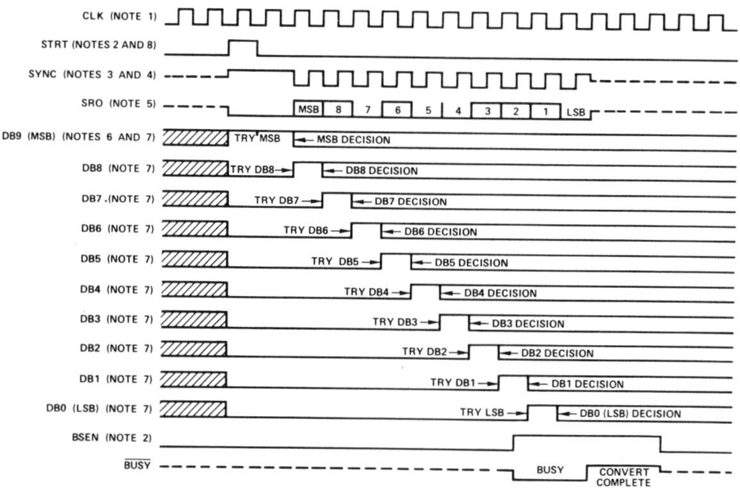

NOTES:

1. Internal Clock Runs Only During Conversion Cycle (External Clock Shown).
2. Externally Initiated.
3. Serial Sync Lags $\overline{\text{Clock}}$ by ≈ 200 ns.
4. Dotted Lines Indicate "Floating" State.
5. For Illustrative Purposes, Serial Out Shown as 1101001110.
6. Cross Hatching Indicates "Don't Care" State.
7. Set and Reset of Output Data Bits Lags Clock Positive Edge by ≈ 200 ns.
8. STRT Falling Edge Should not Coincide with CLK in Falling Edge.
9. Shown for $\overline{\text{SC8}}$ = 1.

Courtesy Analog Devices, Inc.

(B) Timing diagram.

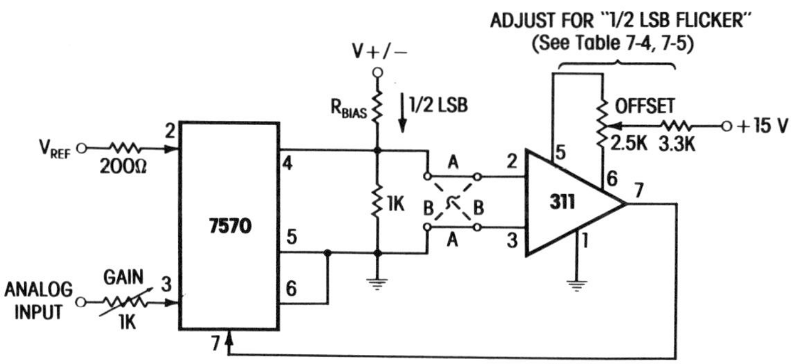

INPUT	V_{REF}	COMP CONN.	V+/−	R_{BIAS}
0 to +10 V	−10 V	A	+15 V	30 MΩ (10 Bits)
0 to −10 V	+10 V	B	−15 V	7.5 MΩ (8 Bits)

Courtesy Analog Devices, Inc.

(C) Calibration details for 10- and 8-bit operation.

unipolar/bipolar a/d converter.

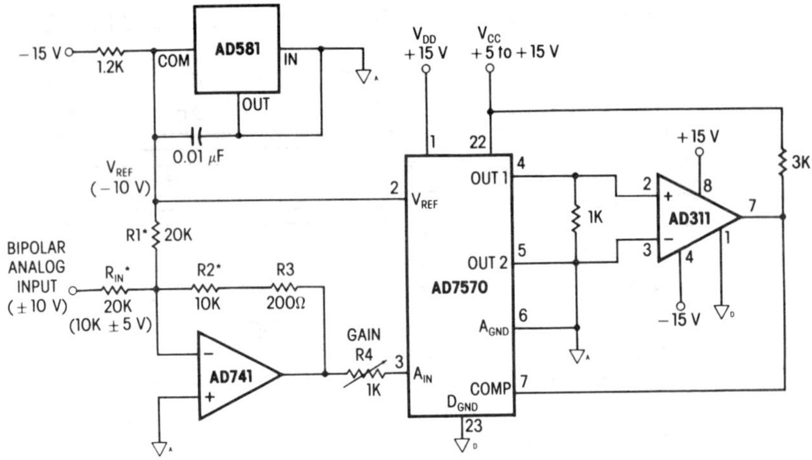

*Matched, Tracking Types Recommended

NOTE: If positive V_{REF} is used, connect minus input of comparator to OUT 1 (pin 4) of the AD7570. (See Fig. 7-76C)

ANALOG INPUT (AIN) NOTES 1, 2, 3	DIGITAL OUTPUT CODE MSB LSB
+(FS − 1LSB)	1 1 1 1 1 1 1 1 1 1
+(FS − 2LSB)	1 1 1 1 1 1 1 1 1 0
+(1/2 FS)	1 1 0 0 0 0 0 0 0 0
+(1LSB)	1 0 0 0 0 0 0 0 0 1
0	1 0 0 0 0 0 0 0 0 0
−(1LSB)	0 1 1 1 1 1 1 1 1 1
−(1/2 FS)	0 1 0 0 0 0 0 0 0 0
−(FS − 1LSB)	0 0 0 0 0 0 0 0 0 1
−FS	0 0 0 0 0 0 0 0 0 0

NOTES:

1. Analog inputs shown are nominal center values of code.
2. "FS" is full scale; i.e., (VREF).
3. For 8-bit operation, 1LSB equals (−VREF) (2^{-7}); for 10-bit operation, 1LSB equals (−VREF) (2^{-9}).

(D) Bipolar operation (simplified).

Fig. 7-76 (cont). AD7570 120-μs, 10-bit binary unipolar/bipolar a/d converter.

range of +10 to −10 volts, in speeds of 120 μs (10 bits) or 40 μs (8 bits). Output data is in 3-state binary form, and can be compatible with either TTL or CMOS levels, by choice of a power supply voltage. The AD7570 is nearly a complete a/d converter, containing all required elements of an SA a/d type with the exception of the comparator and reference, which are added externally.

The AD7570 is shown in terms of a unipolar range 0 to +10 volts (or 0 to −10 volts) application in Fig. 7-76A, which will also be used to point out other general considerations to its operation.

With an analog input applied to the AIN terminal through the 1K GAIN pot, the appropriate reference voltage is applied to the VREF terminal. In general, the full-scale analog input will be equal and opposite to that of the reference, as noted in the table. In going from one range/reference voltage to the other, the inputs to the comparator must also be reversed, as noted by jumper pairs A and B. These are the major considerations as to operation between ranges. Further details of calibration are noted in Fig. 7-76C.

The AD7570 requires two supplies, V_{DD} and V_{CC}, analog and digital respectively. V_{DD} is generally +15 volts, but V_{CC} can be adjusted between +5 and +15 volts, for TTL and CMOS respectively. If different levels are used, the optional protection components should be used. Best accuracy results however when the supplies are a common +15 volts, as shown here. This yields 15-volt CMOS digital output levels, and 15-volt compatible input levels also. Similarly if V_{CC} is 5 volts, inputs and outputs are TTL compatible.

Bipolar clamp diodes are used on the d/a output/comparator input, (pin 4), with a 1K load resistor. The AD7570 has an internal clock whose frequency is set up by the RC components at pin 24. As shown the clock frequency is 100 kHz, which results in the maximum conversion speed of 120 μs for 10-bit operation (AD7570L). For 8-bit operation (AD7570J) the frequency can be increased to 250 kHz with the alternate values, resulting in a 40-μs conversion time. The short-cycle pin, $\overline{SC8}$, is wired to V_{CC} for 10-bit operation, and grounded for 8 bits. If desired, an external clock can be used by applying it direct to pin 24.

The 3-state outputs of the AD7570 are controlled in two bytes, through inputs HBEN (high byte enable) which controls the two most significant bits (only). LBEN (low byte enable) controls the 8 lower order bits. The outputs are in the third state when these inputs are low, and active when high or logic one (here they are continuously active, being wired to V_{CC}).

Conversion in the AD7570 is started when the STRT (convert start) input is taken high, and returns low. The internal clock runs only during the conversion sequence. A timing diagram is shown in Fig. 7-76B, along with pertinent data.

Outputs of the AD7570, in addition to those discussed above, are $\overline{BUSY}$, SRO (serial output) and SYNC (serial synchronization). The BUSY output is low during conversion, and high during standby. This output is 3-state, and is addressed by BSEN (busy enable), with a sense of control similar to HBEN and LBEN. As shown here, BUSY is continuously active, with BSEN wired high.

The SRO and SYNC outputs are active only during conversion, and float at other times. These two outputs are used in conjunction with one another for serial loading of a holding register.

In any configuration using the AD7570, best linearity is realized when the d/a output terminal (pin 4) sees a voltage level of 0 to ±1 millivolt. This is best achieved by null of the comparator during calibration, as is diagrammed in simplified form in Fig. 7-76C.

Here, as can be noted, are repeated the optional connections for V_{REF} and the comparator inputs, along with a ½ LSB current-offset resistor, R_{bias}. R_{bias} is returned to either +15 volts or −15 volts, as appropriate, with a value proportional to 10 or 8 bits (see table). The offset of the 311 comparator is adjusted for the ½ LSB split quantization level flicker, as outlined in Table 7-4 or 7-5, for either 8- or 10-bit operation. The GAIN control is adjusted in a conventional fashion. The reference voltage for the AD7570 should be either +10 volts or −10 volts, which is most conveniently derived from an AD581 (see Fig. 7-76D).

For bipolar operation of the AD7570, the circuit of Fig. 7-76D is appropriate. This circuit is simplified, and shows only analog details of operation. Digital output of the AD7570 will be binary scaled for a bipolar range of −10 to +10 volts; twos complement operation can be acheived by externally inverting the MSB. Both inverted and normal bipolar scalings are possible, as described below.

In this circuit the AD581 is used in its "zener" mode to derive a stable −10-volt source. The op-amp inverter produces a nominal +5-volt output which then is modulated by the −10 to +10-volt analog input. The coding arrangement is shown in the table. With a −10-volt reference as shown all 0s result with an analog input of +10 volts (inverted bipolar scale). With a +10-volt reference, an input of −10 volts will result in an all 0s output (conventional bipolar scale). Note that different ranges can also be accommodated by reducing R_{IN} in proportion. R1, R2, and R_{IN} should be stable matched pair types, preferably on a common substrate with low tracking TCs.

Integral SA A/D Converters

This section deals with SA a/d devices which are complete and ready to connect for operation, thus the term "integral" SA a/d converters. For the most part, these devices require a minimum of user design attention, and have a wide range of flexibility. Most of them can be connected for all standard analog input ranges, and have both serial and parallel data outputs.

A variety of (maximum) device resolutions are available, from 8 through 12 bits, and most units can be truncated or short cycled for shorter bit lengths and/or faster operation. All devices have integral clock sources, but some can also be operated from external sources.

All devices utlize the basic SAR conversion sequence generally described in the first part of this section.

ADC82 8-Bit A/D Converter

The ADC82 8-bit SA a/d converter is illustrated in Fig. 7-77. This device can convert unipolar and bipolar input voltages with a speed of 2.8 μs, using the internal clock. The output data is in complemented form for either unipolar or bipolar ranges, and an $\overline{\text{MSB}}$ output is also available for complementary twos-complement bipolar operation.

As shown in the diagram of Fig. 7-77A, the ADC82 is connected for a unipolar range of 0 to +10 volts. The internal clock is activated by the jumper between CLOCK OUT and CLOCK IN.

A conversion is started by raising the CONVERT COMMAND input to a high level; the conversion then begins on the falling edge of the convert command (see timing diagram Fig. 7-77B). The CONVERT COMMAND input must remain low during the conversion cycle for proper operation. As can be noted, the STATUS output (or EOC) is high during the conversion interval, and goes low when the data is valid, or conversion complete. The internal clock runs only during the conversion cycle, thus it may conveniently be used to clock an external shift register to load the serial output data.

The various input ranges possible with the ADC82 are set up by a series of jumper options, as noted in the table.

Gain and offset trim of the ADC82 may be considered an option, as with no trim the errors due to gain and offset are 0.1% or less (½ LSB). The necessary trim components shown are in the inset, if it is desired to trim these errors to zero.

The circuit as shown is suited for convert on command use. With an external clock, it can be used for continuous conversion applications by opening the CLOCK OUT and CONVERT COMMAND terminals, and feeding the external clock to CLOCK IN. Conversion will start on every 10th clock pulse, and the frequency should be no more than 3.2 MHz (it can of course, be lower).

The circuit of Fig. 7-77C can be used for convert on command applications, with an external clock. The leading edge of the convert command must be synchronous with the leading edge of the clock. Logic shown is TTL.

For continuous conversion with the internal clock, the circuit of Fig. 7-77D is used. The gated astable formed by the 7400 gates ensures power on start up.

The ADC82 uses ±15-volt and +5-volt supplies, which should be bypassed with tantalum capacitors. The analog and digital grounds should be kept separate, connected at one point only. These general points are also applicable to other devices which follow.

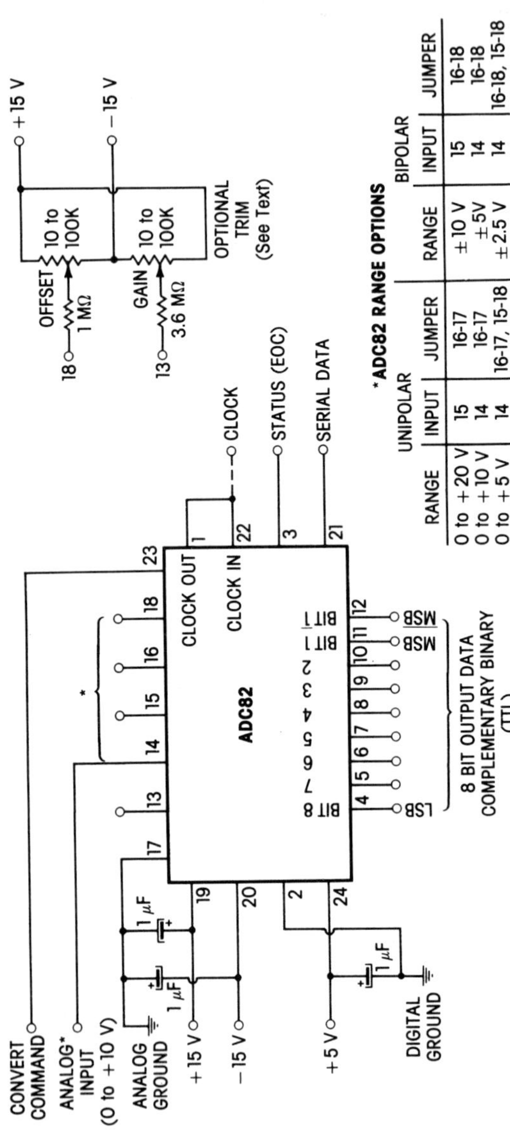

***ADC82 RANGE OPTIONS**

UNIPOLAR			BIPOLAR		
RANGE	INPUT	JUMPER	RANGE	INPUT	JUMPER
0 to +20 V	15	16-17	±10 V	15	16-18
0 to +10 V	14	16-17	±5V	14	16-18
0 to +5 V	14	16-17, 15-18	±2.5 V	14	16-18, 15-18

(A) Connection diagram.

Fig. 7-77. ADC82 2.8-μs, 8-bit complementary

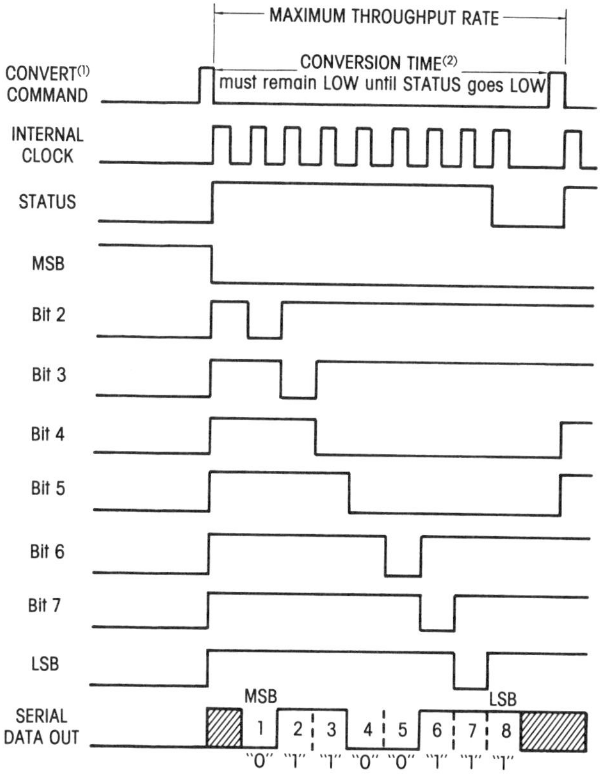

NOTE: 1. Conversion is initiated by the "falling edge" of convert command.
2. 2.8 μs.

Courtesy Burr-Brown Research Corp.

(B) Timing diagram.

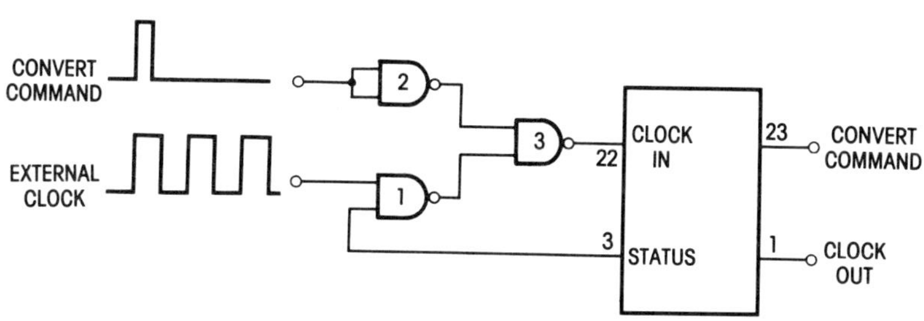

(C) Convert on command with external clock.

binary, unipolar/bipolar a/d converter.

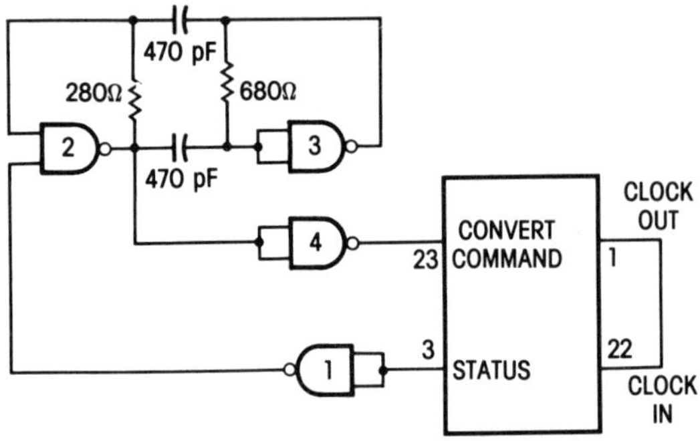

(D) Continuous conversion with internal clock.

Fig. 7-77 (cont). ADC82 2.8-μs, 8-bit complementary binary, unipolar/bipolar a/d converter.

ADC80 12- or 10-Bit A/D Converter

The ADC80 12- or 10-bit SA a/d converter is illustrated in Fig. 7-78. This device can convert unipolar and bipolar input voltages, with a speed of 25 μs (12 bits) using the internal clock. Output data is in complementary form for either unipolar or bipolar ranges, and an MSB output is also available for complementary twos-complement bipolar operation.

As shown in Fig. 7-78A, the ADC80 is connected for a unipolar range of 0 to +10 volts, with the internal clock operating. An external clock can optionally be applied to the EXT CLOCK terminal, with the CLOCK INHIBIT pin grounded.

A conversion is started in the ADC80 by the rising edge of the convert command input (see timing diagram, Fig. 7-78B). This input pulse must be of a specified width and the input must remain low during a conversion cycle. The status (or EOC) output is high during conversion, and goes low when the conversion is complete. The internal clock runs only during conversion and is available at the CLOCK OUT terminals.

Various input ranges are possible with the ADC80, and are set up by the jumper options listed in the table.

Gain and offset trim of the ADC80 are accomplished by the trim networks shown. These networks are recommended for the lowest gain and offset errors, with the fixed components physically close to the device package for minimum noise pickup. Without these trims the typical errors are 0.1% or less.

The ADC80 is available with two accuracy grades, the ADC80-12 for 12-bit uses, and the ADC80-10 for 10-bit uses, with ½ LSB linearity errors, for the respective resolutions. The ADC80-12 completes a 12-bit conversion in 25 μs, while the ADC80-10 requires 21μs for a 10-bit conversion.

As shown in the conversion diagram the ADC80 is configured for 12-bit use with conversions started by external command. Resolution of the SAR is controlled by the SHORT CYCLE pin which is connected to +5 volts for 12-bit use, or bit 11 for 10-bit use.

Alternately, the ADC80 can be operated with an external clock in addition to the short cycle feature, as shown in Fig. 7-78C. A triggered, external clock option is shown in Fig. 7-78C; the convert command must be synchronous with the clock leading edge (also shown is the 10/12 bit resolution option). For continuous conversions with an external clock, the ADC80 is connected as shown in Fig. 7-78C. Note that with the option of an external clock and short cycle, the speed of the ADC80 can be increased appreciably. Maximum (external) clock rates are 0.560, 0.910, and 1.5 mHz respectively, for 12, 10, and 8 bit uses.

For continuous conversion using the internal clock, the hookup of Fig. 7-78D is appropriate, using a 7400 quad gate. This configuration can also be short cycled, if desired.

ADC84/ADC85 12/10 Bit A/D Converter

The ADC84/ADC85 12- or 10-bit SA a/d converter is illustrated in Fig. 7-79. This device can convert unipolar and bipolar input voltages with a speed of 10 μs (12 bits) or 6 μs (10 bits), using the internal clock source. Output data is in complementary binary form, for either unipolar or bipolar ranges, and an $\overline{\text{MSB}}$ output is also available for complementary twos-complement bipolar operation. A serial output is also provided.

As shown in Fig. 7-79A, the ADC85 is connected for a unipolar range of 0 to +10 volts with the internal clock operating. This circuit may be used with either the ADC84 or ADC85; the primary difference between these two units is in the package, and the fact that the ADC85 has a lower TC (10 ppm/°C vs 20 for the ADC84).

A conversion is started in the ADC84/ADC85 by a pulse which is positive going, with conversion then started on the trailing (negative) edge (see timing diagram of Fig. 7-79B). The convert command must remain low during the conversion. The status (or EOC) output is high during conversion and goes low when conversion is complete. The internal clock is available at pin 19, and is adjustable in frequency via pin 17.

To take full advantage of the speed capability of the ADC84/ADC85, the clock frequency should be optimized for the resolution chosen, as outlined in Fig. 7-79C. For 12-bit use, the CLOCK RATE CONTROL is grounded and SHORT CYCLE is tied to +5 volts. For 10- or 8-bit uses, the appropriate short cycle connection is made (see table) and the clock frequency is adjusted by a clock rate trim network. For 10-bit use this clock should be 1.8 MHz; for 8 bits, it should be 2.25 MHz. This will result in conversion times of 6 μs and 4 μs, respectively.

Various input range and connection options are possible with the ADC84/ADC85, as indicated in Fig. 7-79A. From the table it can be noted that the analog input can either be applied directly, or passed through an internal unity gain, noninverting buffer (for high source

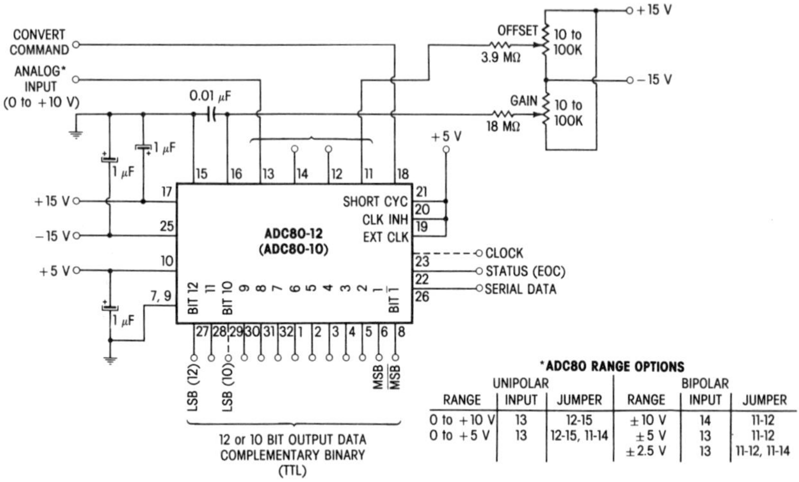

*ADC80 RANGE OPTIONS

UNIPOLAR			BIPOLAR		
RANGE	INPUT	JUMPER	RANGE	INPUT	JUMPER
0 to +10 V	13	12-15	±10 V	14	11-12
0 to +5 V	13	12-15, 11-14	±5 V	13	11-12
			±2.5 V	13	11-12, 11-14

(A) Connection diagram.

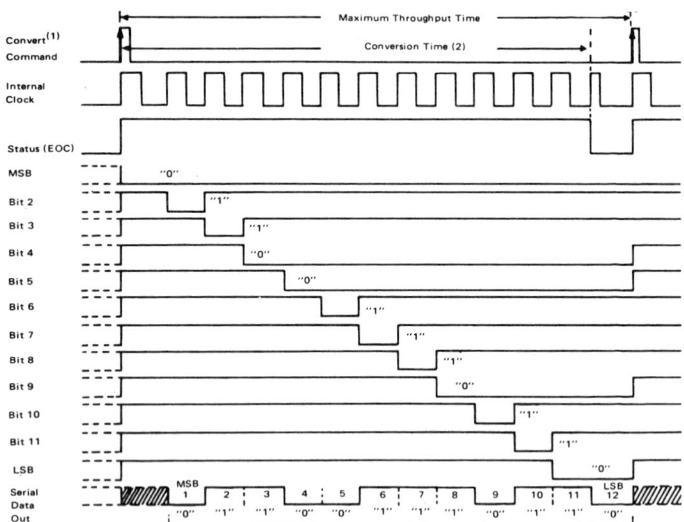

NOTES: (1) The convert command must be between 100 ns and 2 μs wide and must remain low during a conversion. The conversion is initiated by the "rising edge" of the convert command.
(2) 25 μs for 12 bits and 21 μs for 10 bits.

Courtesy Burr-Brown Research Corp.

(B) Timing diagram.

Fig. 7-78. ADC80 25(21)-μs, 12(10)-bit complementary

impedances). As noted from the schematic, the buffer is activated by the alternate (dotted) input connections. When used, this buffer's 2-μs settling time must be added to the conversion time. Gain and offset trim of the ADC84/ADC85 are accomplished by the trim networks

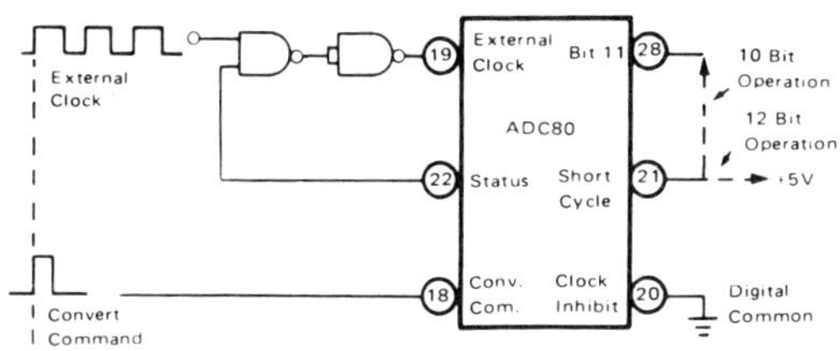

1. TRIGGERED

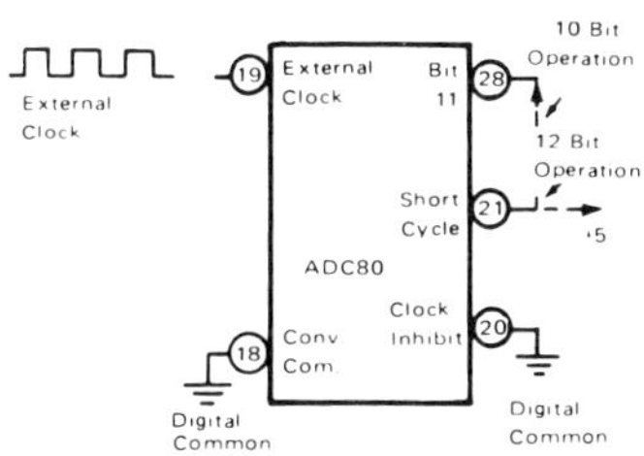

2. CONTINUOUS CONVERSION

RESOLUTION (BITS)	12	10	8
Connect Pin 21 to	Pin 9	Pin 28	Pin 30
Maximum Conversion Time[1]			
Internal Clock (μsec)	25	22	18
External Clock (μsec)	23	12	6
Maximum Nonlinearity At +25°C (% of FSR)	0.012[2]	0.048[3]	0.20[3]

NOTES: (1) Max conversion time to maintain ±½LSB Nonlinearity error.
(2) 12 Bit Models only.
(3) 10 or 12 Bit Models.

3. OPERATING GUIDELINES

Courtesy Burr-Brown Research Corp.

(C) Short cycle/external clock operation.

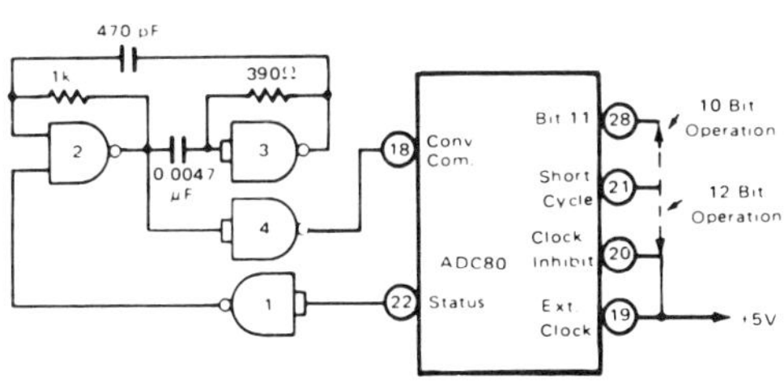

Courtesy Burr-Brown Research Corp.

(D) Continuous conversion with internal clock.

binary, unipolar/bipolar a/d converter.

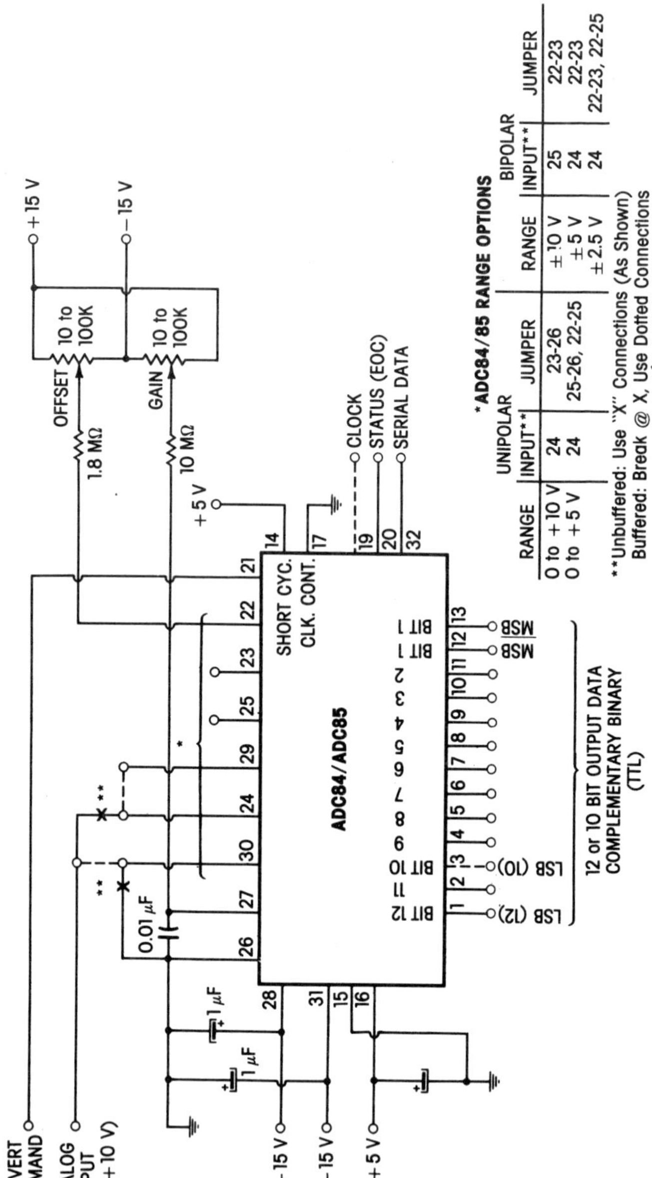

*ADC84/85 RANGE OPTIONS

UNIPOLAR			BIPOLAR		
RANGE	INPUT**	JUMPER	RANGE	INPUT**	JUMPER
0 to +10 V	24	23-26	±10 V	25	22-23
0 to +5 V	24	25-26, 22-25	±5 V	24	22-23
			±2.5 V	24	22-23, 22-25

**Unbuffered: Use "X" Connections (As Shown)
Buffered: Break @ X, Use Dotted Connections

(A) Connection diagram.

Fig. 7-79. ADC84/85 10(6)-μs, 12(10)-bit complementary

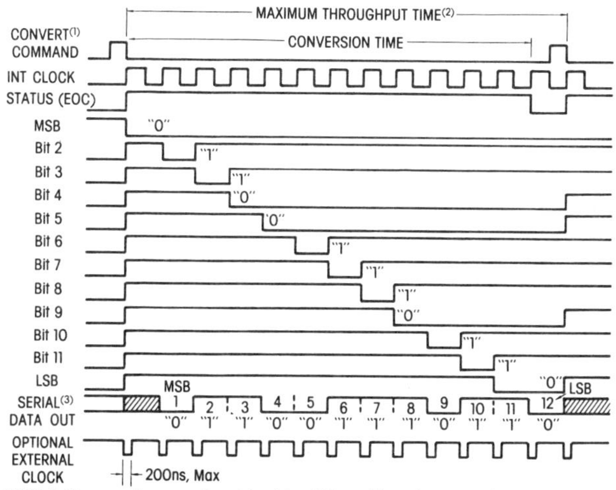

NOTES: 1. The convert command must be at least 50 ns wide and must remain low during a conversion. The conversion is initiated by the "trailing edge" of the convert command.
2. 10.5 μs for 12 bits and 6.4 μs for 10 bits.
3. Use trailing edge of clock to strobe serial output.

Courtesy Burr-Brown Research Corp.

(B) Timing diagram.

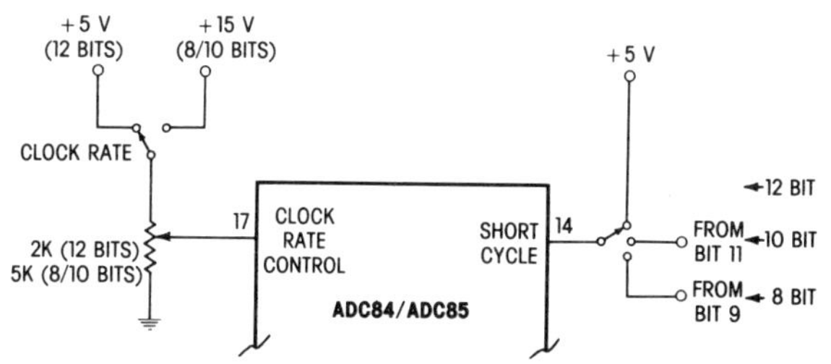

Resolution (Bits)	12	10	8
Connect Pin 17 to(1)	Pin 15	Pin 16	Pin 28
Connect Pin 14 to	Pin 16	Pin 2	Pin 4
Maximum Conversion Speed (μsec)(2)	10	6	4
Maximum Nonlinearity at 25°C (% of FSR)	0.012(3)	0.048(4)	0.20(4)

Notes: 1) Connect only if clock rate control is not used.
2) Max. conversion speeds to maintain ±1/2 LSB nonlinearity error.
3) 12 bit models only. 4) 10 or 12 bit models.

(C) Short cycle/clock rate options.

binary, unipolar/bipolar a/d converter.

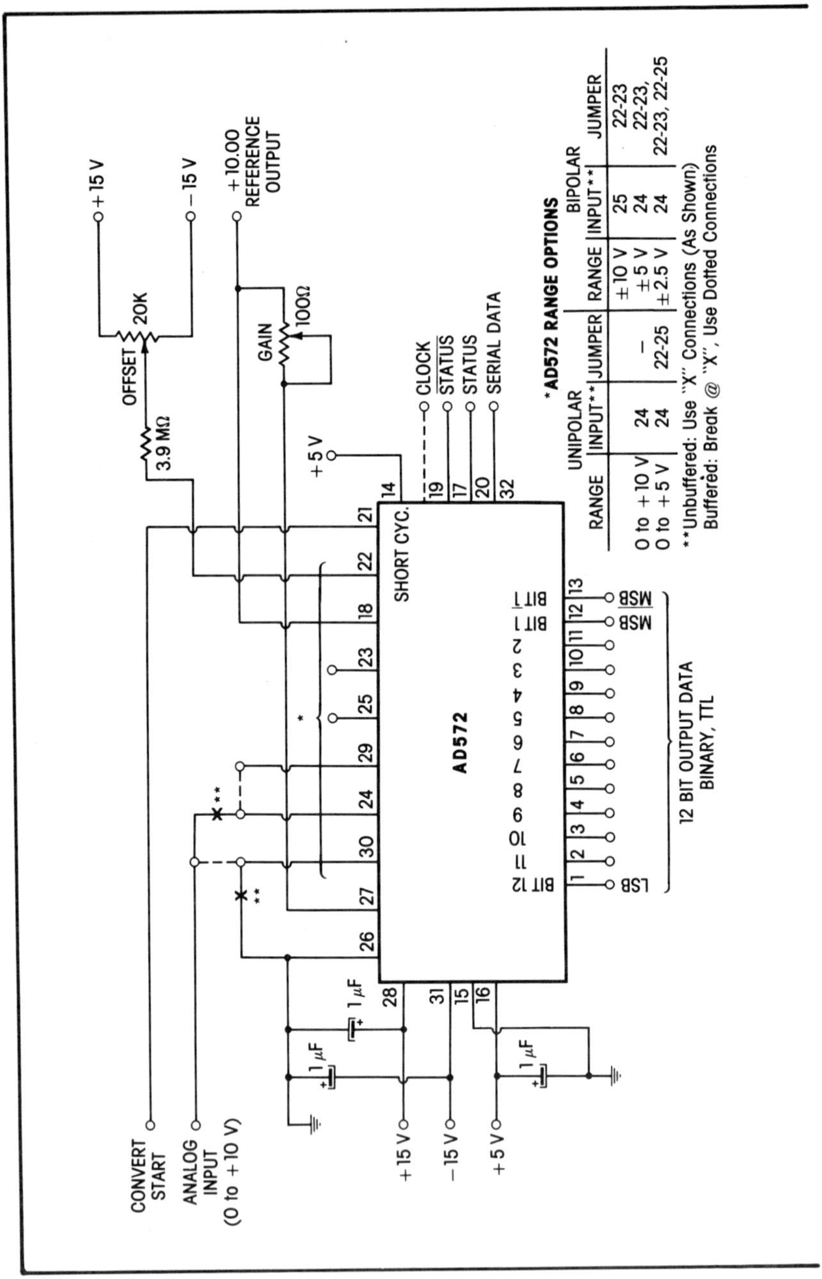

(A) Connection diagram.

Fig. 7-80. AD572 25-μs, 12-bit binary,

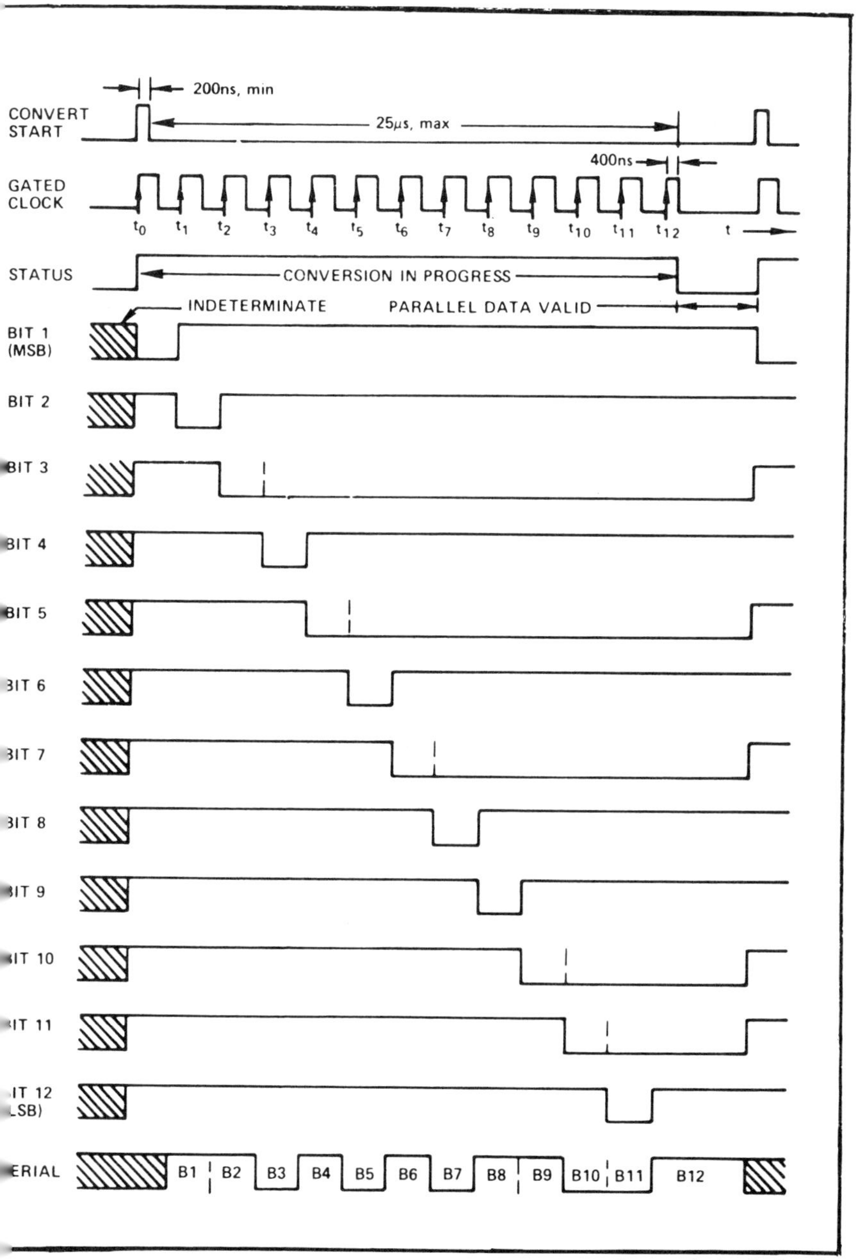

Courtesy Analog Devices, Inc.

(B) Timing diagram.

unipolar/bipolar a/d converter.

shown, which are recommended for the lowest gain and offset errors. The fixed components should be close to the device physically. Without trim, the typical errors are 0.1% or less.

AD572 12-Bit A/D Converter

The AD572 12-bit a/d converter is illustrated in Fig. 7-80. This device can convert unipolar and bipolar voltages with a speed of 25 μs, using the internal clock source. Output data is in binary form for either unipolar or bipolar ranges, and an MSB output is also available for twos-complement operation. A serial output is also provided.

As shown in Fig. 7-80A, the AD572 is connected for a unipolar input range of 0 to +10 volts. Other input ranges are accommodated by vari-

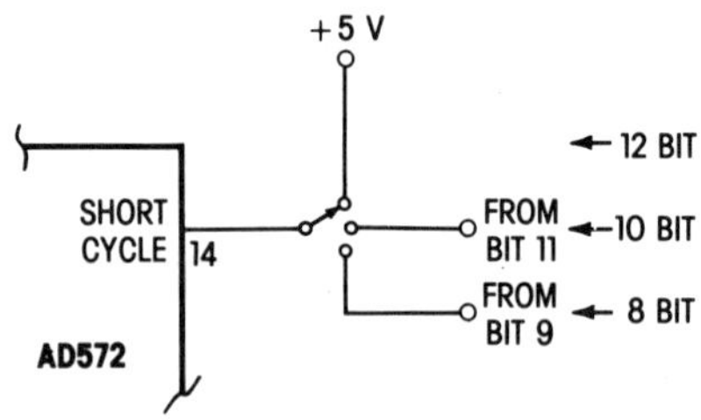

1. WIRING DETAILS

Connect Short Cycle Pin 14 to Pin:	Bits	Resolution (% FSR)	Maximum Conversion Time (μs)	Status Flag Reset at: (Figure 2)
16	12	0.024	25	t_{12} + 400ns
2	10	0.10	21	t_{10} + 400ns
4	8	0.39	17	t_8 + 400ns

2. PERFORMANCE DETAILS

(C) Short-cycle operation.

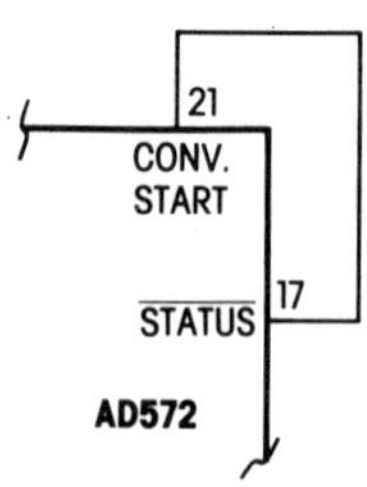

(D) Continuous conversion option.

Fig. 7-80 (cont). AD572 25-μs, 12-bit binary,

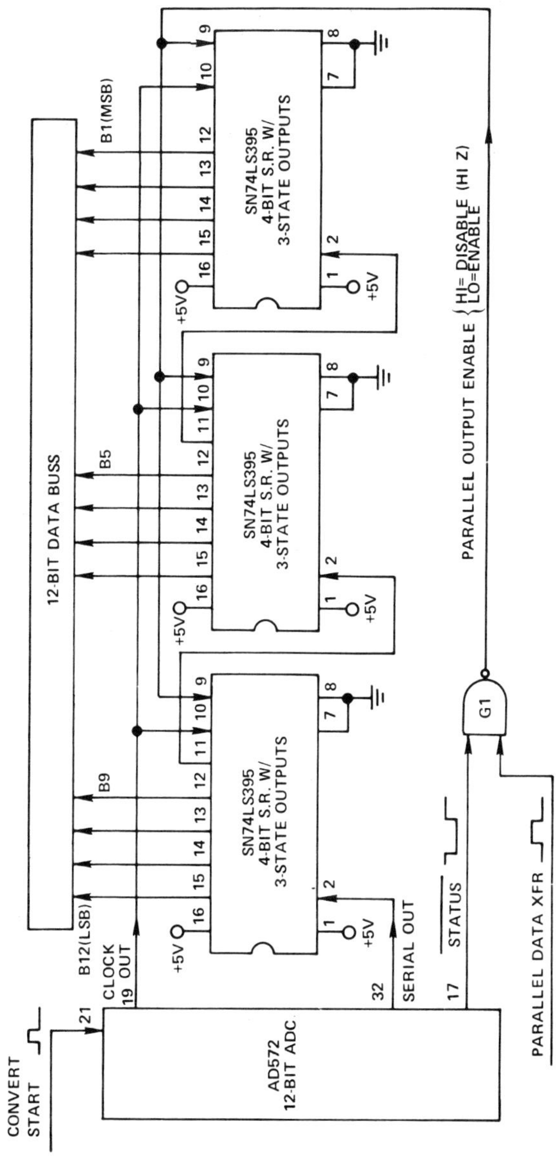

Courtesy Analog Devices, Inc.

(E) Serial output operation, with holding register.

unipolar/bipolar a/d converter.

ous jumper options (see table). An input buffer is provided which can be activated by choosing the alternate (dotted) input connections. When this buffer amp is used, its 2-μs settling time must be added to the basic conversion time.

A conversion is started in the AD572 by a CONVERT START input which rises to a 1, then the internal clock is enabled on the trailing (negative) edge (see timing diagram of Fig. 7-80B). The STATUS output is set to a high level on the leading edge of CONVERT START, is high during conversion, and goes low when the conversion is complete. The internal clock is available at pin 19, and runs only during conversion.

The AD572 may be operated in a short-cycle mode for faster conversion time as shown in Fig. 7-80C. The SHORT CYCLE input is simply wired to the appropriate node, according to the diagram and table.

Gain and offset trim of the AD572 are accomplished by the trim networks shown, which are recommended for the lowest errors. Components should preferably be located physically close to the device. Without trim, and the GAIN pot replaced by a fixed 50-ohm 1% resistor, typical errors are 0.05% or less.

The AD572 has a buffered +10.00 volt reference voltage brought out on pin 18, which can be used for external applications, such as ratiometric configurations.

As noted above output data is in parallel binary form and a serial output is available. The STATUS output is also available as a complement, or $\overline{\text{STATUS}}$. The serial output and clock signals can very simply be used to clock the data into a serial to parallel converting holding register, as shown in Fig. 7-80D. Three 74LS395 expandable 4-bit shift registers are used for this purpose. These devices have 3-state outputs, which are controlled by pin 9.

As shown, the outputs are open (third state) when the DATA XFR line is low, and become active when a conversion is complete, and this line is taken high. Note that this circuit can also be used as a 12-bit serial to parallel converter with holding register, remote from the a/d converter itself, if desired.

REFERENCES

1. Aldridge, D. *A/D Conversion Techniques With the MC6800 Microprocessor System*. Motorola Application Note AN-757, December, 1975. Motorola Semiconductor Products, Inc., Phoenix, Arizona.

2. ———. *A 3½ Digit DVM Using an IC Dual Ramp System*. Motorola Application Note AN-746, September, 1975. Motorola Semiconductor Products, Inc. Phoenix, Arizona.

3. ———. *Autopolarity Circuits for MC1405 Dual Ramp System*. Motorola Engigineering Bulletin EB-35, June, 1975. Motorola Semiconductor Products, Inc., Phoenix, Arizona.

4. ———. *Autoranging Digital Multimeter Using the MC14433 CMOS A/D Converter.* Motorola Application Note AN-769, January 1977. Motorola Semiconductor Products, Inc., Phoenix, Arizona.

5. ———; Kelley, S. *Input Buffers for the MC1505.* Motorola Engineering Bulletin EB-24A, January 1976. Motorola Semiconductor Products, Inc., Phoenix, Arizona.

6. Alvsten, B. "Calculate With a V/F Converter." *Electronic Design,* June 7, 1974.

7. ———. "Linearize Your V/F Converter." *Electronic Design,* November 8, 1973.

8. Bradshaw, P.; Osgood, S. *Do's and Dont's of Applying A/D Converters.* Intersil Application Bulletin A018, March 1977. Interrsil Inc., Cupertino, California.

9. Brown, L. *Designing With High Speed Comparators.* Advanced Micro Devices Application Note, December 1975.

10. Buurma, G. *CMOS Schmitt Trigger—A Uniquely Versatile Design Component.* Applications Note AN-140. National Semiconductor Corp., Santa Clara, California.

11. Cate, T. "Advances in Voltage-to-Frequency Converters." *EE Times IC Applications Conference,* 1977.

12. ———. "Automotive Applications of Monolithic V-F Converters." *SAE International Automotive Congress & Exposition,* February 1977.

13. ———. "IC V/F Converters Readily Handle Other Functions Such as F/V, A/D." *EDN,* January 5, 1977.

14. Connors, S. "V/F Converters: A/D's With Advantages." *EDN,* June 5, 1974.

15. Dean, J. A.; Rupley, J. P. *Astable and Monostable Oscillators Using COS/MOS Digital Integrated Circuits.* RCA Application Note ICAN-6267, November 1973. RCA Solid State Div., Somerville, New Jersey.

16. Evans, L. *The Integrating A/D Converter.* Intersil Application Bulletin A017, February 1977. Intersil, Inc., Cupertino, California.

17. ———; Fullagar, D. "A Low Cost 4½ Digit A/D Converter." *Proceedings EE Times IC Applications Conference,* 1976.

18. Fishbeck, J. "10-Bit Binary to 3-Digit BCD Conversion Using the ADC-102 as a 3-Digit BCD Converter." *Analog Dialogue,* Vol. 6, No. 2.

19. Fullagar, D. "Interfacing Data Converter and Microprocessors." *Electronics,* December 9, 1976.

20. ———. *Selecting A/D Converters.* Intersil Application Bulletin AD16, March 1977. Intersil, Inc., Cupertino, California.

21. ———; Dufort, M. *Low-Cost Digital Panel Meter Designs.* Intersil Application Bulletin A023, May 1977. Intersil Inc., Cupertino, California.

22. Ghest, R. C. *A Successive Approximation Register.* Advanced Micro Devices Application Note, 1973.

23. ———. "Successive Approximation Registers Fast, Powerful Building Blocks." *EDN,* February 20, 1973.

24. Gilbert, B. "A Versatile Monolithic Voltage-to-Frequency Converter." *IEEE JSSC,* Vol. SC-11, No. 6, December 1976.

25. Goodenough, F. "Solve Your Measurement Problems With a V/F Converter." *Electronic Design,* July 19, 1975.

26. ———. "Modular V/F and F/V's: Simple Solutions to Everyday Conversion Problems." *Electronic Products,* February, 1977.

27. Gordon, B. "What's Wrong With A/D Converter Specs." *EEE* Magazine (out of print). Reprinted by Analogic, Wakefield, Massachusetts.

28. Graehme, J. "Use V/F Converters for Analog Data Transmission." *Electronic Design,* April 1, 1975.

29. Grandbois, G. *Applying the LD130 ±3 Digit A/D Converter.* Application Note AN76-5, January 1977. Siliconix, Incorporated.

30. ———. *Build an Autoranging DMM With the LD130 A/D Converter.* Design Aid DA76-3, January 1977. Siliconix, Incorporated.

31. ———. *Function and Application of 3½ Digit A/D Converter Set.* Application Note AN74-1, January 1977. Siliconix, Incorporated.

32. ———. *LD130 3 Digit DVM Demonstrator Board.* Design Aid DA76-2, May 1976. Siliconix, Incorporated.

33. Guzeman, D. *Interfacing the 8700 A/D Converter With the 8080 μP System.* Application Note No. 8, August 1976. Teledyne Semiconductor.

34. Hadley, M. *CMOS Schmitt Triggers.* Motorola Application Note AN-768, January 1977. Motorola Semiconductor Products, Inc., Phoenix, Arizona.

35. Hoeschele, D. F., Jr. *Analog-to-Digital/Digital-to-Analog Conversion Techniques.* John Wiley & Sons, Inc., New York, 1968.

36. Jung, W. G. *IC Op-Amp Cookbook.* Howard W. Sams & Co., Indianapolis, 1974.

37. ———. *IC Timer Cookbook.* Howard W. Sams & Co., Inc., Indianapolis, 1977.

38. ———. "Wide Frequency-Range Clock Source Needs Few Discrete Components." *Electronic Design,* April 26, 1977.

39. Kelley, S. *Applications of MC1405/MC14435 in Digital Meters.* Motorola Application Note AN-748, July 1975. Motorola Semiconductor Products, Inc., Phoenix, Arizona.

40. ———. *4½Digit DVM System.* Motorola Engineering Bulletin EB-36, January 1976. Motorola Semiconductor Products, Inc., Phoenix, Arizona.

41. ———; Fletcher, P. *Battery Powered Digital Multimeter.* Motorola Engineering Bulletin EB-55, September 1975. Motorola Semiconductor Products, Inc., Phoenix, Arizona.

42. Kime, R. C., Jr. "The Charge Balancing AD Converter: An Alternative to Dual Slope Integration." *Electronics,* May 24, 1973.

43. Kress, D. "Highest-Performance 10-Bit IC DAC." *Analog Dialogue,* Vol. 11, No. 1, 1977.

44. ———; Gilbert, B. "Versatile Monolithic V/F or I/F Converter." *Analog Dialogue,* Vol. 10, No. 2, 1976.

45. Marosi, G. "Design a Precision A/D Converter." *Electronics Design,* April 12, 1975.

46. National Semiconductor Application Note AN-148. *Driver Update New Drivers for LED Displays.* National Semiconductor Corp., Santa Clara, California.

47. National Semiconductor Application Note AN-84. *Driving 7-Segment Gas Discharge Display Tubes With National Semiconductor Circuits.* National Semiconductor Corp., Santa Clara, California.

48. National Semiconductor Application Note AN-99. *Driving 7-Segment LED Displays With National Semiconductor Circuits.* National Semiconductor Corp., Santa Clara, California.

49. Osgood, S. "CMOS A/D Converter IC Applications Contest." *EE Times,* June 21, 1976.

50. ———. "Using the 8700 Series CMOS A/D Converter." *Proceedings EE Times IC Applications Conferences,* 1976.

51. Paiva, M. *8750 3½ Digit A/D Converter.* Application Note No. 9, June 1977. Teledyne Semiconductor.

52. Pease, R.; Teledyne Inc. *Amplitude to Frequency Converter.* US Patent No. 3, 746,968, Filed September, 1972.

53. Pauliot, F. "Have You Considered V/F Converters?" *Analog Dialogue,* Vol. 9, No. 3, 1975.

54. Ritmanich, W. et al. "The A07550-A 13-Bit Quad-Slope, A/D Converter." *Proceedings EE Times IC Applications Conference,* 1976.

55. ———. "13-Bit Monolithic CMOS A/D Converter." *Analog Dialogue,* Vol. 10, No. 1, 1976.

56. Schmid, H. *Electronic Analog/Digital Conversions.* Van Nostrand Reinhold, New York, 1970. P. 240.

57. ———. "Practical Guide to A/D Conversion, Part I." *Electronic Design,* December 5, 1968.

58. Schmoock, J. C. "The Monolithic Voltage to Frequency Converter." *Proceedings EE Times IC Applications Conference,* 1976.

59. Schoeff, J. A. "A Monolithic Analog Subsystem for High Accuracy A/D Conversion." *IEEE ISSCC Digest of Technical Papers,* February 1973.

60. ———; Soderquist, D. *DAC-08 Applications Collection.* Precision Monolithics Application Note AN-17, September 1976. Precision Monolithics, Inc., Santa Clara, California.

61. ———. *Low-Cost, High Speed A/D Conversion With the DAC-08.* Precision Monolithics Application Note AN-16, December 1975. Precision Monolithics, Inc., Santa Clara, California.

62. Sheingold, D. H. *Analog-Digital Conversion Handbook.* Analog Devices, Norwood, Massachusetts, 1972, P. II-46, P. II-80, P. 11-86.

63. ———. *Analog-Digital Conversion Notes.* Analog Devices, Norwood, Massachusetts, 1977. P. 121, 162, 166.

64. Soderquist, D. *A Low-Cost Easy-to-Build SA A/D Converter*. Precision Monolithics Application Note AN-11, December 1975. Precision Monolithics, Inc., Santa Clara, California.

65. Tuma, W. *Digital Voltmeters and the MM5330*. National Semiconductor Application Note AN-155. National Semiconductor Corp., Santa Clara, California.

66. Vanderkool, M. K. "3½ Digit A/D Converter Chip Set." *Proceedings of EE Times IC Applications Conference,* 1975.

67. Watts, M. *CMOS Oscillators*. National Semiconductor Application Note AN-118. National Semiconductor Corp., Santa Clara, California.

68. Wheelock, N. "A/D Converter Chip Sets Do More Than Just Cut DVM Parts Count." *EDN,* February 5, 1977.

69. ———. *More Applications of the LD110/LD114 3½ Digit A/D Converter*. Application Note AN76-4, January 1977, Siliconix, Incorporated.

70. ———. *3½ Digit DVM Demonstrator Board*. Design Aid DA74-1, January 1977, Siliconix, Incorporated.

71. Whitmore, J.; VanAken, R. "10-Bit Monolithic CMOS A/D Converter." *Analog Dialogue,* Vol. 9, No. 2, 1975.

72. Zuch, E. L. "Voltage-to-Frequency Converters: Versatility Now at a Low Cost." *Electronics,* May 15, 1975.

73. Manufacturers' Data Sheets:

Advanced Micro Devices	AM2502/2503/2504, AM685, AM686
Analog Devices	AD561, AD571, AD537, AD7550 AD562, AD572, AD536 AD563, AD7570
Burr-Brown	ADC80 ADC82 ADC84/85, VFC32 DAC80 DAC85
Fairchild Semiconductor	μA710
Intech	A8400
Intersil Inc.	ICL7103, 7104 ICL7106 ICL7107 ICL8052, 8053
Motorola Semiconductor Products	MC1408 MC14433 MC3410
National Semiconductor	LM311, LM331
Precision Monolithics	CMP-01 DAC-08 DAC-100
Raytheon	RC4151, RC4152
Siliconix	LD110/111 LD111/114 LD130/131
Teledyne Semiconductor	4400, 8700 series

8

IC Converter System Applications

This chapter of the applications part of the book focuses on uses of d/a and a/d devices which are more specialized than their basic operation covered in Chapters 6 and 7. Most of these uses are system oriented, that is the d/a or a/d converter is employed as a solution to some larger system type of problem.

To some extent, a few d/a and a/d devices can even be regarded as low cost components, and used as such. Some of the application examples for multiplying d/a converters, and those of some V/F ICs fall into this category of use.

Due to the nature of this chapter, it builds upon already established device operating characteristics. Logically then, previous chapters should be digested prior to this one, as basic details are not repeated.

Examples of specific applications for d/a and a/d converters are definitely boundless, and those shown here can be best regarded as starting points or suggestions for more imaginative adaptations. The use of the reference material cited at the end of the chapter is encouraged for it will be helpful as further background information.

PROGRAMMABLE VOLTAGE AND CURRENT SOURCES

A very useful and basic class of application for which d/a converters are highly applicable is as programmable voltage and current sources. Such circuits are not very far removed from the basic d/a function, but are often specialized in light of the intended application. Examples of such types of use are as programmable power supplies, test stimulus sources, and other instrumentation related uses.

Although a voltage output d/a converter is by definition a programmable source of voltage, it usually cannot furnish appreciable *power,* typically being limited to a few milliamperes of output current. Of course, a direct and straightforward remedy to this is to simply substitute a power op amp for the current-to-voltage converter stage—a solution which is amenable to many of the d/a circuits in Chapter 6. However, since power op amps can be expensive and difficult to obtain, alternate solutions are desirable.

The circuit of Fig. 8-1 is a relatively straightforward approach to this application, as it implements the power op-amp function by using a standard op amp buffered by a "pass transistor" that can output over 1 ampere of current. Actually A3 is a three-terminal regulator, which is highly attractive as it affords high current gain, integral safe area protection, and is a single low cost component. Since it is controlled by the op amp (inside the loop) its own output voltage and/or drift is not material, and accuracy of the output voltage is determined by A1, A2, and the reference voltage.

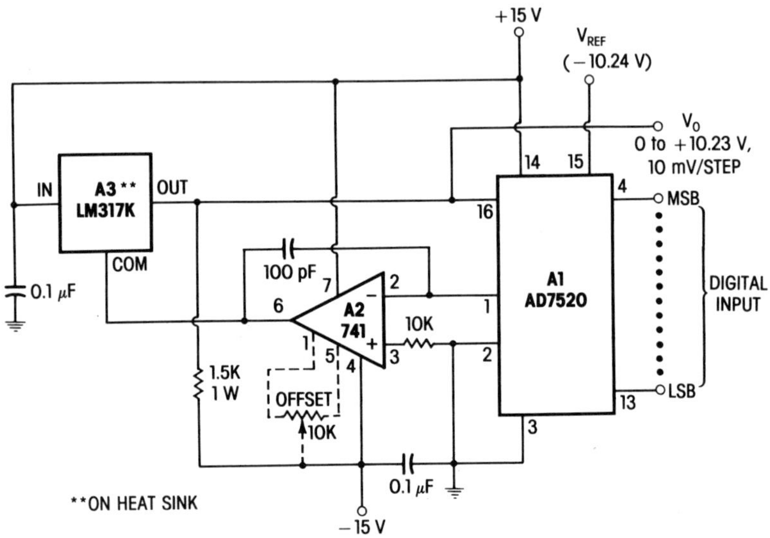

Fig. 8-1. Programmable voltage source, 10 V/1 A.

As shown here, a −10-volt reference is used, which provides a positive output range of 0 to +10 volts, with 10 bit control resolution. Linearity is determined essentially by the grade of AD7520 used. For best performance A2 should be offset nulled.

The circuit is calibrated for zero scale by the OFFSET control, and for full scale output by trim of the reference voltage. Note that if V_{REF} = −10.24 volts, the output will increment in 10 mV/LSB steps, which gives a very useful resolution/accuracy combination for a lab or system supply.

The LM317K should be adequately heat sinked for reliable operation. If desired, an LM350K can be directly substituted for the LM317K, to provide 3 amperes of output capability.

Although this circuit uses a negative reference and outputs positive voltage as shown, it can be adapted for negative outputs (with a positive reference) as well. This would involve use of an LM337 (a negative regulator analogous to the LM317) as the pass transistor.

With the flexibility of reference voltage range afforded by the AD7520, various other reference voltages and/or output levels can be accommodated also. Lower control resolution is possible, of course, by simply grounding the undesired lower order bits, or substituting an AD7523.

Fig. 8-2 illustrates a programmable voltage source with a 30-volt, 1-ampere capability, and the option of either a positive or negative output voltage range.

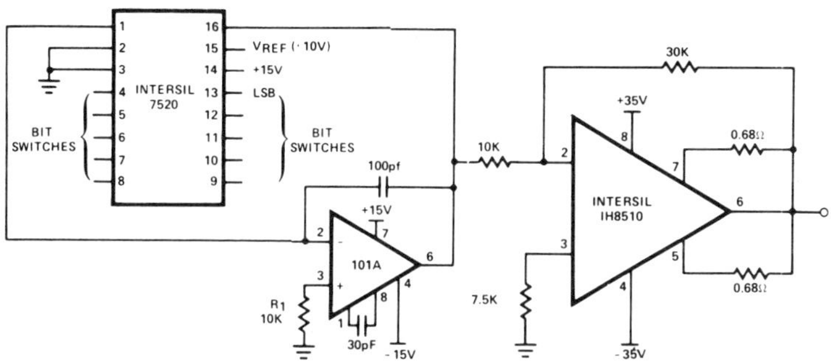

Courtesy Intersil, Inc.

Fig. 8-2. Programmable voltage source, 30 V/1 A.

This circuit is in basic function a 7520 d/a converter, followed by a power amp with a gain of three. The gain of 3 scales the 10-volt maximum op-amp output up to a higher range, within the capability of the power op amp. The power op amp used here is an Intersil IH8510, a device which operates from ± 35-volt supplies and can output up to 30 volts and 1 ampere (it should of course be heat sinked for such operation).

Inasmuch as the IH8510 is a bipolar output device, output voltage polarity in this circuit is determined only by the reference applied to the 7520. Since there are two inverting amplifiers cascaded, a positive reference results in a positive output. Further details on the operation of this circuit can be found in Intersil Application Bulletin A021.

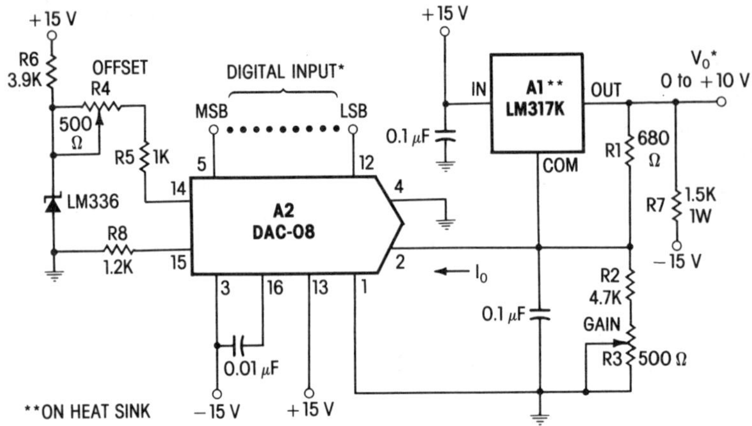

***OUTPUT TABLE**

DIGITAL INPUT	V_O As Shown	V_O General
11111111	10.000 V	$1.25 \frac{R1 + R2 + R3}{R1}$
⋮	⋮	⋮
10000000	5.000 V	⋮
⋮	⋮	⋮
00000000	0.039 V	$1.25 \frac{R1 + R2 + R3}{R1} - I_0 (R2 + R3)$

Fig. 8-3. Programmable 1-A source, with adjustable range.

Fig. 8-3* illustrates a programmable voltage source with some quite interesting properties, yet with a relative minimum of complexity. The previously described voltage source circuits have been basically unipolar, 0 to some V_{max}. The circuit of Fig. 8-3 is different in that it can be set up for range limits which are relatively arbitrary, and can be set by the user over a wide range of limits between 0 and 30 volts. The circuit can furnish 1 A of output current while operating over this range.

This circuit may be best understood by considering it in two components—A1 that comprises a voltage regulator circuit, in conjunction with R1-R3, and A2 that is a variable current source d/a converter which produces the variation in output voltage.

The upper range limit of output is set by the 1.25-volt output voltage of A1, and R1-R3, as shown in the table. This limit of output will appear with an all 1s digital input. As the digital input to A2 is reduced, the output I_o increases from zero and subtracts from the maximum output. The lower voltage limit is defined by the expression in the table. As

* W. G. Jung. "Current Inverters Make Current Sources Out of Monolithic IC D/A Sink Outputs." *Electronic Design,* June 16, 1978.

shown, R3 and R4 serve as GAIN and OFFSET controls respectively and trim the circuit for a 10-volt (maximum), 39-millivolt (minimum) range. Maximum range can be up to +20 volts, and minimum can be any voltage from 0 to this limit, achieved by selecting I_o (maximum). The output current of 1 A can be increased to a current of 3 A by using an LM350K for A1.

Inasmuch as a great many IC d/a converters are natural current output devices, they can be very effectively and efficiently used as programmable current sinks (and sources). Two current generator circuits with a very wide range of applicability are illustrated in Fig. 8-4.

In Fig. 8-4A* is shown a circuit which "mirrors" or inverts the normal current sinking output from an IC d/a converter and delivers a current

* W. G. Jung. "IC D/A Plus 3 Terminal Regulator = Power D/A." *Electronic Design,* July 1978.

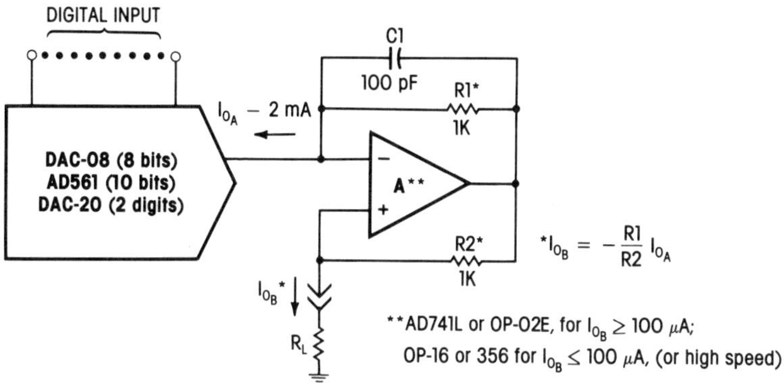

(A) Basic circuit.

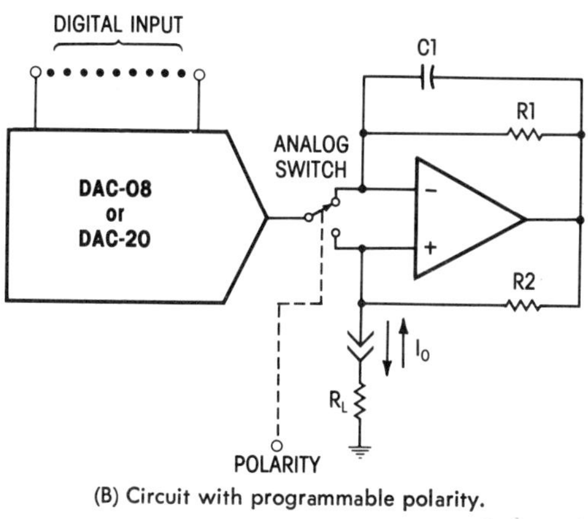

(B) Circuit with programmable polarity.

Courtesy Electronic Design

Fig. 8-4. General purpose programmable current source.

source to a load, R_L. The output current I_{oB} is related to the input current I_{oA}, according to the relation shown. As can be noted, the relative values of R1 and R2 can be adjusted (over a wide range) to either increase or decrease the current. If the amplifier used is a FET type unity such as a 356 or OP-16, output currents can extend down to the nanoampere range with high accuracy, or up to ampere range with a power op amp (or buffered standard type).

A virtue of this circuit is the fact that the current can be delivered to R_L accurately, even if it is referred to voltages other than ground. The general restraint is that the voltage across R_L (or inputs of A) must be within the voltage compliance limits of the d/a converter used. The d/a converters listed are high-compliance, high-impedance output types, and will work well with this circuit. Suitable op-amp types are listed in the figure.

In Fig. 8-4B the circuit is shown with a programmable current polarity option. The addition of an analog switch steers the d/a converter output current to the op-amp current mirror, or directly to the load. Thus the current in the load can be either current sourcing or sinking. As long as the d/a converter driving the switch is a true current source, the analog switch will not materially contribute dc errors. A suitable analog switch that could be used here was previously shown in Fig. 6-16B.

For both of these circuits, the choice of d/a converter can be from a variety of device types and resolutions; DAC-08 (8 bit) AD561 (10 bit), for binary, DAC-20 for 2 digit BCD, or DAC-76 for exponential coding.

Process Control Current Regulators

A common industrial circuit requirement is the process control current generator used to remotely control various types of electronic circuitry. The transmission of the control signal in a current mode form removes sensitivity to ground potential differences and noise pickup in general. Two forms of digitally controlled current transmitters for this application are shown in Fig. 8-5. The standard for this type of data transmission is a 4- to 20-mA current range, with 4 mA corresponding to zero and 20 mA to full scale.

In Fig. 8-5A an 8-bit circuit is shown which provides a 4- to 20-mA source of current to load R_L. With all 0s to the DAC-08, it produces no output current and the fixed 0.5-mA current from A3-Q1 drives R5. The voltage drop across R5 is duplicated across R6 by feedback from A3-Q2. With R6 being a value ⅛ that of R5, this creates a 4-mA current in Q2, and thus R_L.

As the digital input to A2 increases, the current to R5 increases to 2.5 mA which when multiplied by the current gain of 8 produces the 4- to 20-mA scale. Output voltage range for the specified current range

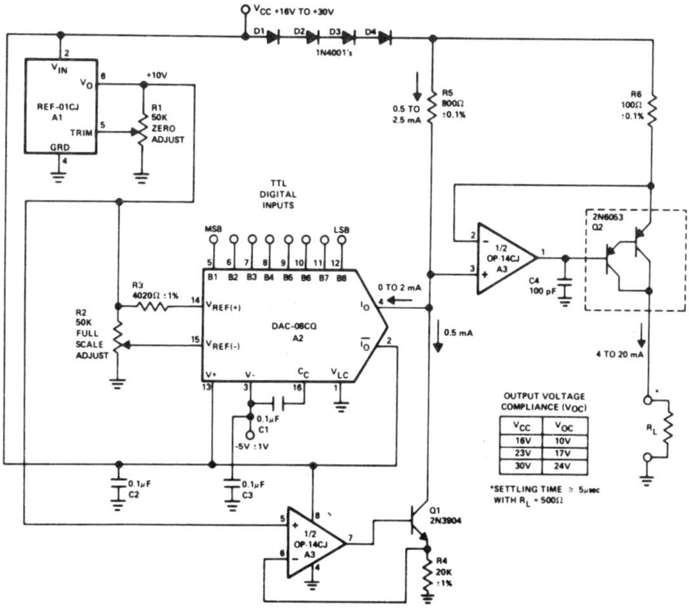

Courtesy Precision Monolithics, Inc.

(A) DAC-08, 8-bit control.

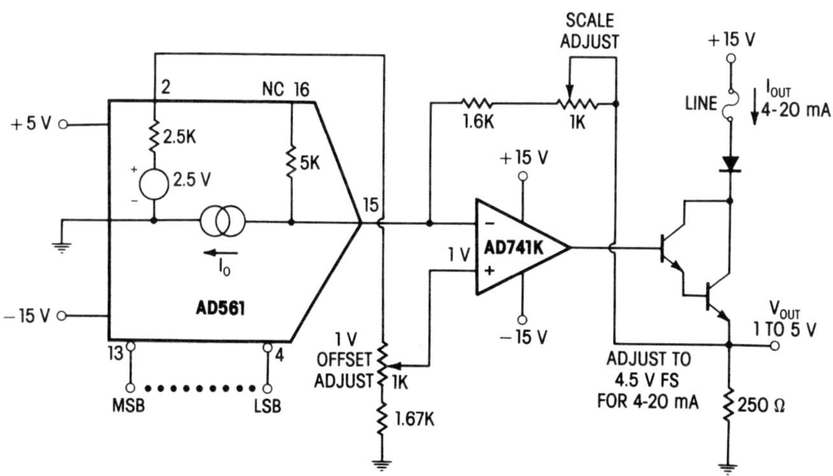

Courtesy Analog Devices, Inc.

(B) AD561, 10-bit control.

Fig. 8-5. Process control 4 to 20 mA current generators.

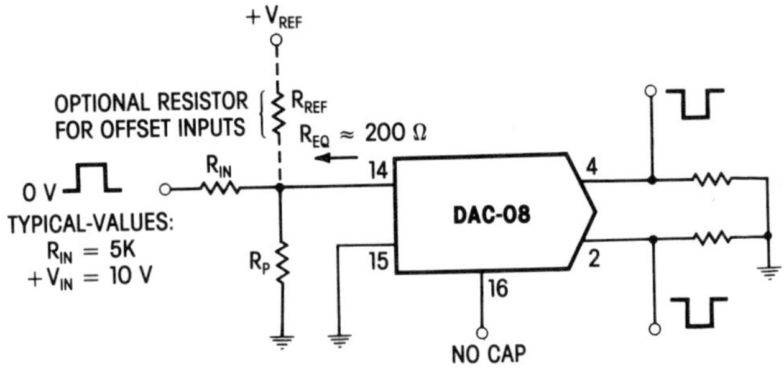

Courtesy Precision Monolithics, Inc.

Fig. 8-6. Pulsed reference d/a converter.

is shown in the figure, and is dependent upon the positive supply voltage used. Calibration is achieved by R1 (for 4 mA) and R2 (20 mA).

A 10-bit control circuit which provides a 4- to 20-mA current sinking output is shown in Fig. 8-5B. In this circuit an AD561 is used in conjunction with an op amp and npn Darlington transistor. With all 0s in, the 1K OFFSET pot is adjusted for 4 mA of output current. With all 1s in, the 1K SCALE ADJUST pot is set for 20 mA of output current.

Although the load here is shown as being referred to a +15-volt supply, it may in general be returned to any positive supply within the breakdown rating of the transistor used.

PROGRAMMABLE PULSE SOURCES

With the relatively high-speed operation inherent to a number of the available multiplying d/a types, their application in programmable pulse sources is quite logical and useful. Such circuits find application in instrumentation and video control applications.

Fig. 8-6 illustrates a basic scheme for extending the reference channel bandwidth of a DAC-08 or DAC-20 type d/a converter, allowing digital control of pulses applied to the reference input. When the equivalent resistance (R_{EQ}) is on the order of 200 ohms, no reference compensation is necessary for stability. As a result, the reference current slew rate can be raised to 16 mA/μs by removing the capacitor normally connected to pin 16. Into a 1K load, such as would be used for fastest speed, the voltage output slew rate will then be 16 V/μs, allowing pulse rise times on the order of 125 ns (assuming minimal load capacitance).

R_{IN} is a series input resistance chosen to drop the desired full scale (peak) reference current for the input pulse "1" level, such as the 5K/10 V combination shown. Either or both pulse outputs can be used with this circuit. The DAC-08 will give 8 bit, 2 binary pulse amplitude control; the DAC-20 will give 2 digit bcd control.

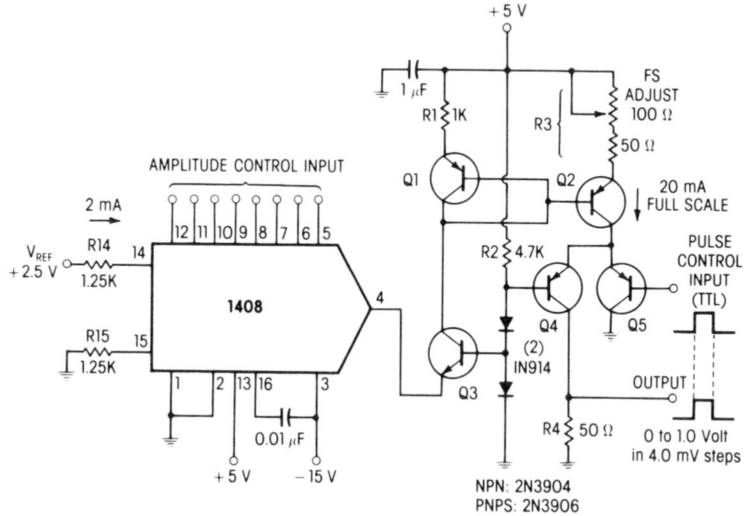

(A) Single ended.

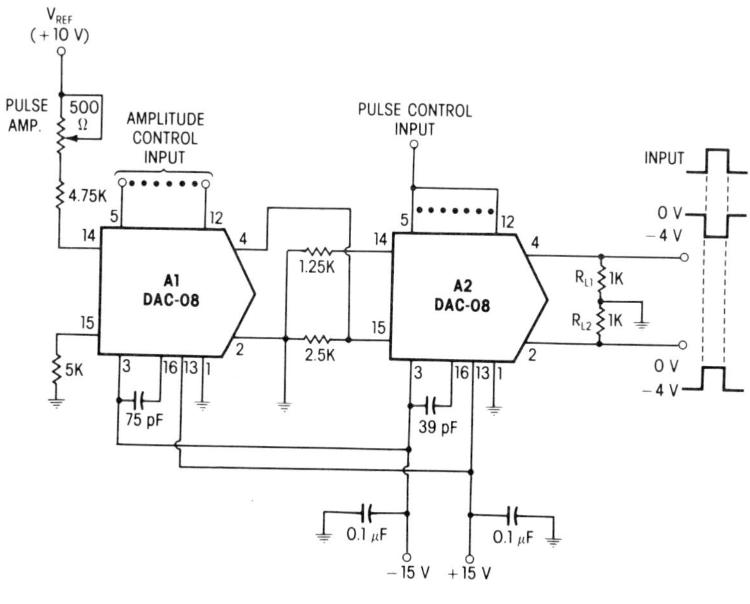

(B) Differential.

Fig. 8-7. Programmable-amplitude pulse sources.

Programmable pulse sources which are even faster in speed capability are illustrated in Fig. 8-7. Fig. 8-7A shows a circuit which delivers an 8-bit programmable positive pulse over a range of 0 to 1 volt into a 50-ohm load. Response time is quite fast, and well below 100 ns, being limited mainly by the switching transistor used for Q4-Q5. The pulse-control input is TTL compatible, as is the amplitude input to the 1408 d/a converter.

As can be noted, the extra speed is achieved by operating the d/a converter at dc only, to control the current fed to Q4-Q5. R3 trims the circuit to the correct full-scale pulse amplitude, with an all 1s input. Finer resolution of control can be achieved by using a 10-bit converter such as the MC3410 or AD561, with appropriate pin changes.

Fig. 8-7B shows a scheme to control pulse amplitude with two d/a converters. Here A1 is operated in a more or less conventional fashion, delivering a 0 to −5-volt programmable (dc) output to the reference input of A2. Also A2 operates as a pulsed current switch, delivering the output current programmed by A1 to either R_{L1} or R_{L2}.

Note that A2 is set up for a full-scale current of 4 mA, which results when an all 1s input is applied to A1. Maximum output-pulse amplitude is 4 volts, and is trimmed by the PULSE AMPLITUDE control.

Fig. 8-8 illustrates two circuits that are useful in the interfacing of fast-pulsed logic levels between portions of a system operating at different ground potential(s). In large systems such techniques are often desirable to minimize spurious interference and maximize noise immunity.

Fig. 8-8A shows a system that capitalizes on the high output impedance of a DAC-08 type d/a converter, in driving a single-ended logic output line. The line is pulsed to a 2-mA source level for a "1" logic input; to a 2-mA sink level for a "0" input. Bidirectional current output capability is provided by the 2N3906 current mirror transistor pair.

At the receiver end of this transmission link is either a TTL or CMOS inverter, as appropriate. These receivers are simple inverter stages operated in a quasi-linear fashion. The resistive feedback biases the input node in the center of its active region. The ±2-mA (200-μA CMOS) logic input is sufficient to saturate either stage, and allow fully reconstituted output logic levels. Throughput delay is 100 ns or less for a TTL system, and less than 2 μs for a CMOS system. The 100-ohm input and output resistors are not necessary for operation, but advisable for fault protection.

Fig. 8-8B shows a scheme which offers even higher noise immunity, by using differential data transmission from the DAC-08 outputs. Here the 200-μA current pulses in the 200-ohm loads at the receiver are sensed and converted into CMOS output levels, with a swing of V+ volts (p-p). As shown the main feature of this circuit is the low power

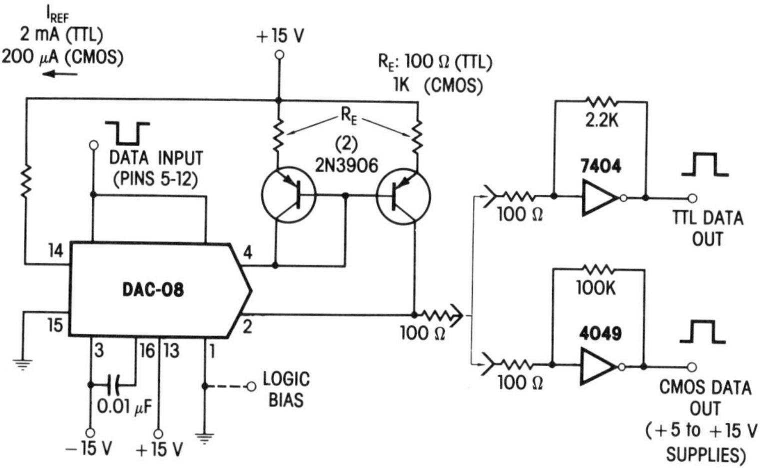

(A) Single ended.

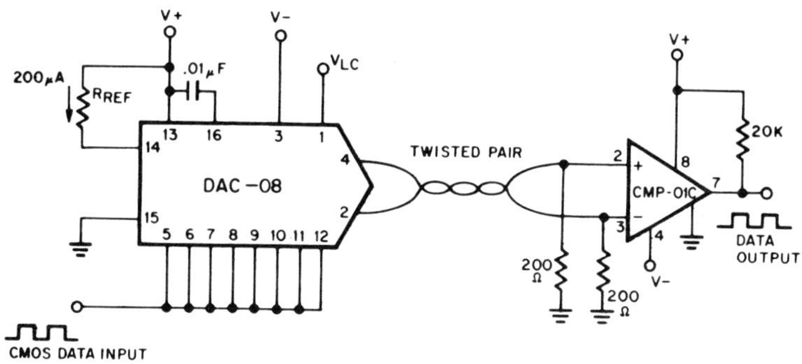

Courtesy Precision Monolithics, Inc.

(B) Differential.

Fig. 8-8. D/A line driver interface.

consumption—it can be operated at higher speeds by raising the DAC-08 reference current (as done in Fig. 8-8A). Without a pullup to V+, the CMP-01 comparator output is TTL compatible.

DIGITALLY PROGRAMMABLE OSCILLATORS

A circuit type which is of increasing importance with the advent of the low cost microprocessor is the digitally programmable oscilllator, or frequency source. Such circuits are useful in instrumentation, control applications, and frequency synthesis. Some of the advantages of a digitally programmable oscillator are frequency agility, fast settling between two discrete frequencies, choice of output waveform (pulse, square, or ramp) and compatibility with a variety of logic families. A sampling of a few circuits which can effectively serve this function are described in the following section.

In block diagram form, the basic elements of a digitally controlled oscillator are shown in Fig. 8-9. As can be noted, a unipolar d/a converter with the appropriate range, accuracy and resolution comprises the input block, and produces an output voltage proportional to the digital input. This voltage is used as the input control voltage for a V/F converter, chosen for the appropriate frequency range, output waveform and level, and control sensitivity.

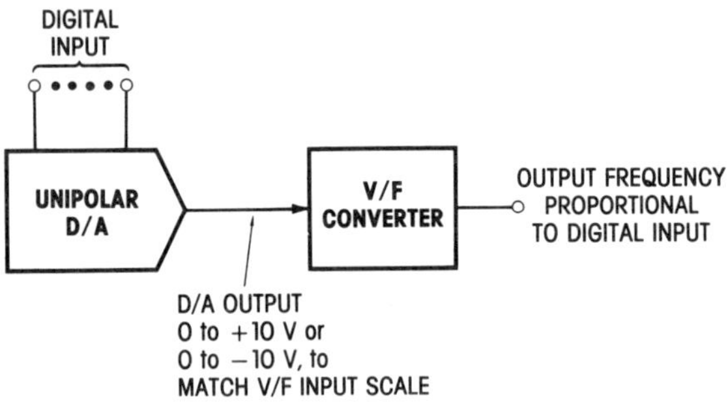

Fig. 8-9. Block diagram of digitally controlled oscillator.

As may be appreciated this is a quite straightforward approach and can yield a workable oscillator system using a variety of device combinations. The highest performance V/F circuits have linearities at or approaching 12 bits; for example, such as the VFC-32 or 8400. These devices could be directly driven by a DAC80 (12-bit) or AD561 (10-bit) d/a converter with minimal attention to interfacing details. If available design time is limited and only a few systems are to be built, this method of approach may be favored.

An alternate approach, given more time to explore details of device operation, is to design the digitally programmed oscillator as a system in itself. In this manner the relative strengths of different device characteristics can be optimized for the system.

In Fig. 8-10 is illustrated an example of this method of approach, an 8-bit linearly programmable oscillator with a pulse-type output waveform. The design uses only two ICs and features several error-minimization advantages.

If the circuit is examined closely it will be noted that the right portion (exclusive of the DAC-08) is a charge-balancing type of current-to-frequency (I/F) converter (see Fig. 7-12).

The reference current in this case is I2, the input current is I1, the output current of the DAC-08. Since I1 is proportional to the digital input, so also is the operating frequency of the I/F converter.

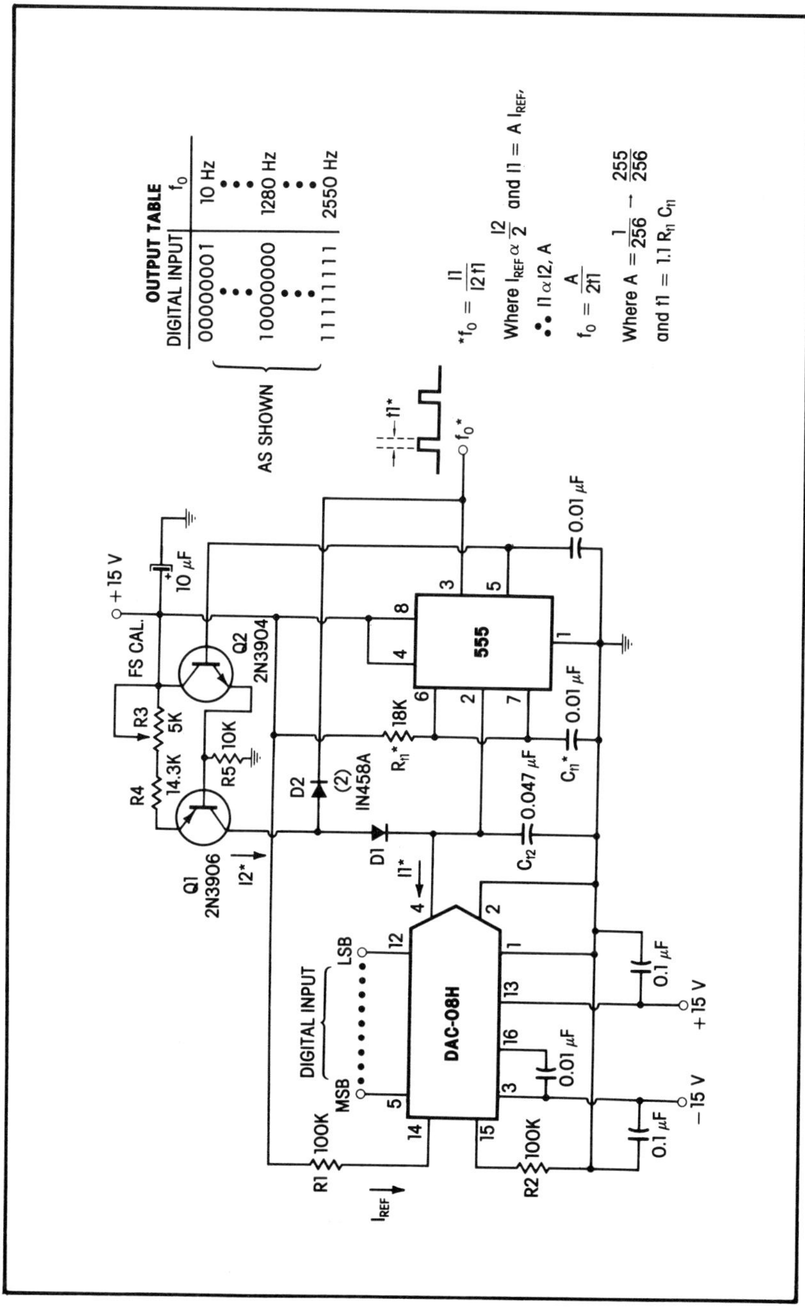

Fig. 8-10. An 8-bit programmable oscillator with a pulse output.

Error cancellation techniques in this circuit include the fact that both I1 and I2 are proportional to V+, therefore supply induced errors cancel and no separate reference source is required.

The circuit as shown operates at a full-scale frequency of 2550 Hz, with a 10 Hz/step sensitivity. Full-scale frequency is adjusted by varying t1, by adjustment of C_{t1} (and C_{t2}, in proportion). For full-scale calibration R3 trims the circuit.

The circuit is capable of better than 8-bit linearity, therefore a DAC-08H device is specified. A higher resolution d/a converter can be used, but it must have a true current source output to function properly, with at least a 5-volt compliance rating. Note that the output is a positive pulse of a width t1, which occurs at the f_o rate. As shown, the levels are 15-volt compatible.

A version of an 8-bit programmable oscillator with a square wave output is shown in Fig. 8-16. This circuit is composed of a d/a converter which is interfaced (somewhat similarly to Fig. 8-9) with a V/F converter, the AD537. However, it also uses some error minimization.

The AD1408-9 (9-bit linear 1408 type) d/a converter is scaled for a 1-mA full-scale current output, which matches the AD537K, by setting up a 1-mA I_{REF}. The 1-volt reference from the AD537 is used for this purpose after being buffered by the 741 op amp. Since this reference is common to both D/A and V/F converters, drift will tend to cancel.

A polystyrene capacitor is used for C_t, with R3 selected to cancel its TC by an opposite AD537 TC. To minimize net system TC, R3 can be trimmed for minimum overall TC if desired. As shown the full-scale frequency is 9961 Hz, with $C_t = 0.01\ \mu F$; this can readily be adapted as desired. Worst case linearity of the D/A-V/F combination is 0.17%; typically, however, it will be better than this. The output is a square wave, and is TTL compatible as shown.

A very interesting type of oscillator is shown in Fig. 8-12. This circuit has an exponential digital-frequency characteristic, as inherent to the DAC-76 d/a converter. This allows it to achieve either a 72-dB or 78-dB frequency span, the latter being the case here.

The principle of operation here is one of two equal and opposite current generators, which charge and discharge a timing capacitor, C_t. The voltage across the capacitor is sensed by a bistable comparator, the 555.

The logic output of the 555 is fed back to the SB input of the DAC-76, to control the selection of outputs. The I_o (+) outputs when active, feed a precision current inverter, which produces a source current to C_t. The I_o (−) outputs, when active, provide a sink current to C_t.

With appropriate component selection, this circuit is capable of operating over the full range of the DAC-76, a range of nearly 10,000/1.

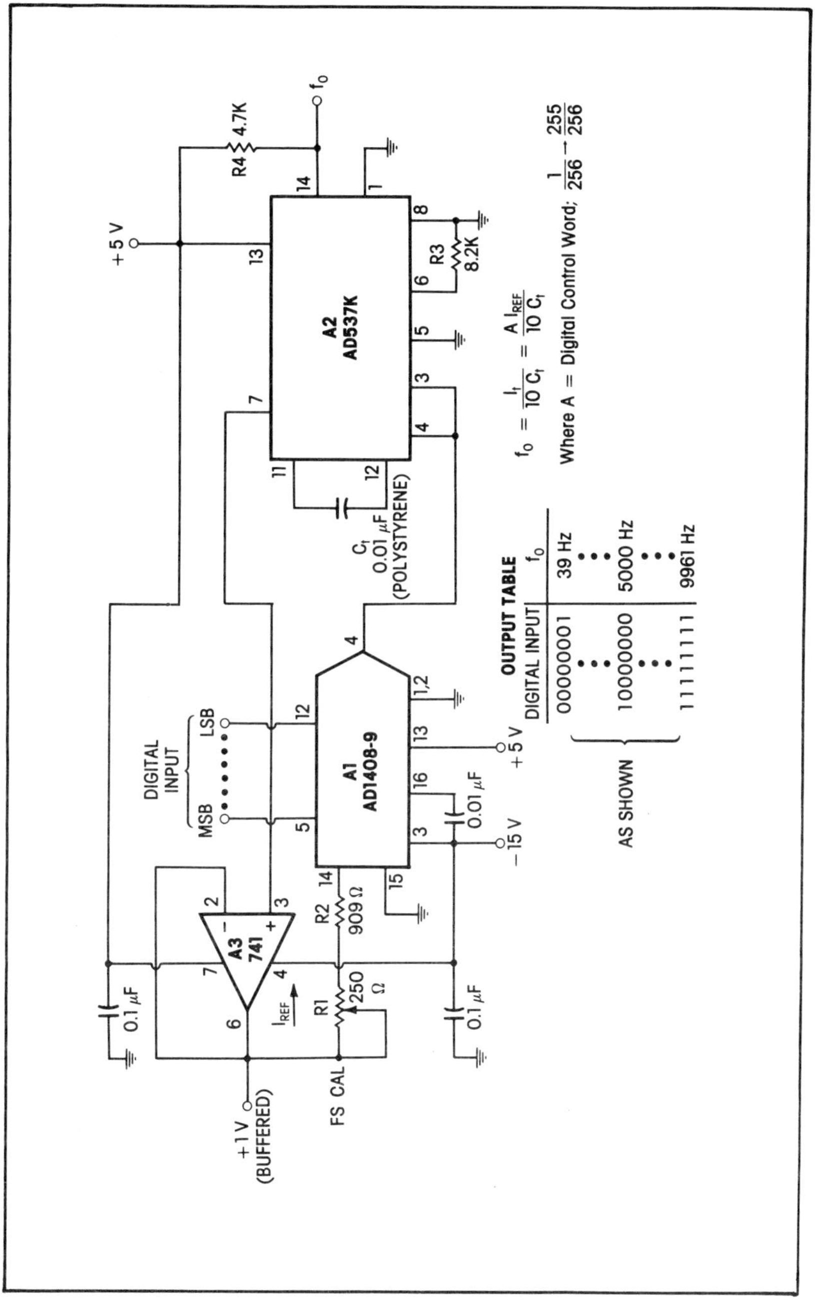

Fig. 8-11. An 8-bit programmable oscillator with a square-wave output.

For correct tracking at the lower scale range, A2 must either be a low offset-voltage type, or be offset nulled for best waveform symmetry. It should also be a FET type preferably, and have a high slew rate.

The 1K mirror resistors R4-R5 should be matched (or trimmed) for best waveform symmetry at full scale. R1 is trimmed for the correct full-scale output frequency; as shown here it is 20 kHz with an all 1s digital input.

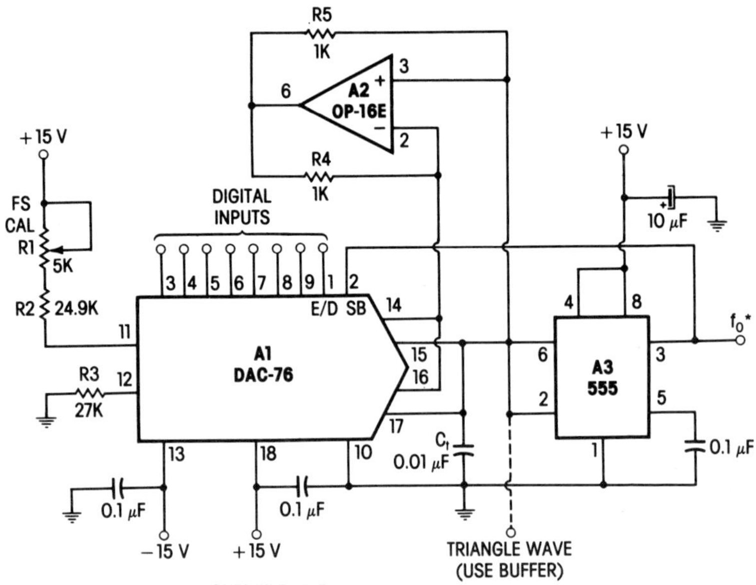

***OUTPUT TABLE**

DIGITAL INPUT	f_o
00000001	2.45 Hz
⋮	⋮
10000000	1213 Hz
⋮	⋮
11111111	20000 Hz

Fig. 8-12. An 8-bit programmable oscillator with exponential control and a square-wave output.

Power-supply sensitivity is quite low in this circuit, on the order of 0.01%/V, allowing it to be used over a 10 to 15-volt V+ range.

Although the basic output is a square wave, a triangular wave is optionally available across C_t if a noninverting buffer is added, using an FET op amp. Operating range can be readily adjusted for other frequencies simply by choosing C_t. The 555 and slew rate of A2 will limit operation to about 50 kHz or less, for best accuracy.

WAVEFORM GENERATORS

One of the more powerful tools which is available through d/a conversion is waveform synthesis, or the generation of precise signal waveforms. Compared to analog techniques, digital waveform synthesis can be accurate in amplitude to 1 part in 2^n, with crystal-referenced time-base stability. And arbitrary waveforms can also be generated through the use of ROMs, with virtually any waveshape and accuracy.

Fig. 8-13 illustrates a basic block diagram for the simplest form of waveform generator. This circuit is composed of a clock, an "n" bit

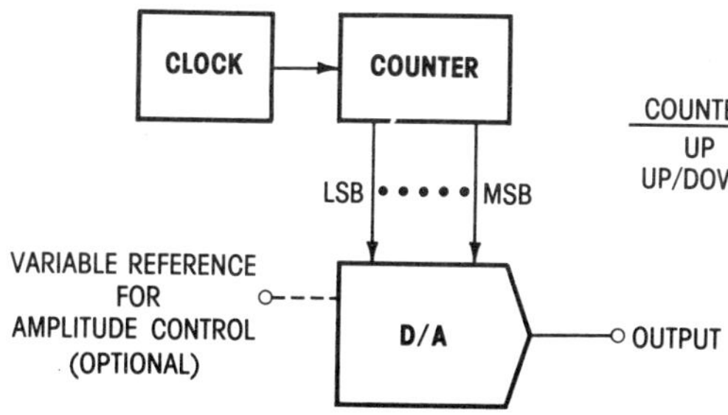

COUNTER	OUTPUT WAVEFORM
UP	SAWTOOTH
UP/DOWN	TRIANGLE

Fig. 8-13. Basic d/a waveform generator.

counter, and an "n" bit d/a converter. With a simple binary up counter, this circuit will exercise the d/a converter over its entire transfer characteristic sequentially, generating a linear ramp or sawtooth output waveform in the process. Amplitude limits are determined by the d/a converter in use, and can be either unipolar or bipolar, as desired, without regard to logic input. Amplitude control is as determined by reference input and/or feedback scaling, and a second d/a converter may be used to modulate the d/a reference for waveform amplitude control, if desired.

For a sawtooth waveform, the circuit will complete one output cycle in 2^n clock periods. The settling time of the d/a converter in use will determine the upper speed limit, which will be in general

$$f_{max} = \frac{1}{2^n t_s}$$

where,

f_{max} is the sawtooth frequency,
t_s is the d/a converter settling time.

A 1-μs, 8-bit d/a converter will allow a 3.9-kHz maximum output frequency, for example.

The selection criteria for a d/a converter should include settling time, linearity, and in particular glitch-free or low-glitch operation.

Fast switching devices are useful here, along with synchronous counters.

An up/down counter allows the circuit to increment and decrement the d/a scale, and so to produce a symmetrical triangular waveform output. The period in this case is one half that of the sawtooth (for a given d/a converter settling time).

Options not shown in the figure, but which would be added with relative ease are two-speed clock operation between the up and down cycles, which would yield asymmetric sawtooth waveforms. The digitally controlled oscillator and V/F techniques described earlier can be useful in this regard.

Quite often this type of waveform generator is used to generate nonlinear waveforms, as for example, the sine function. Modification to the basic generator to allow this are shown in Fig. 8-14.

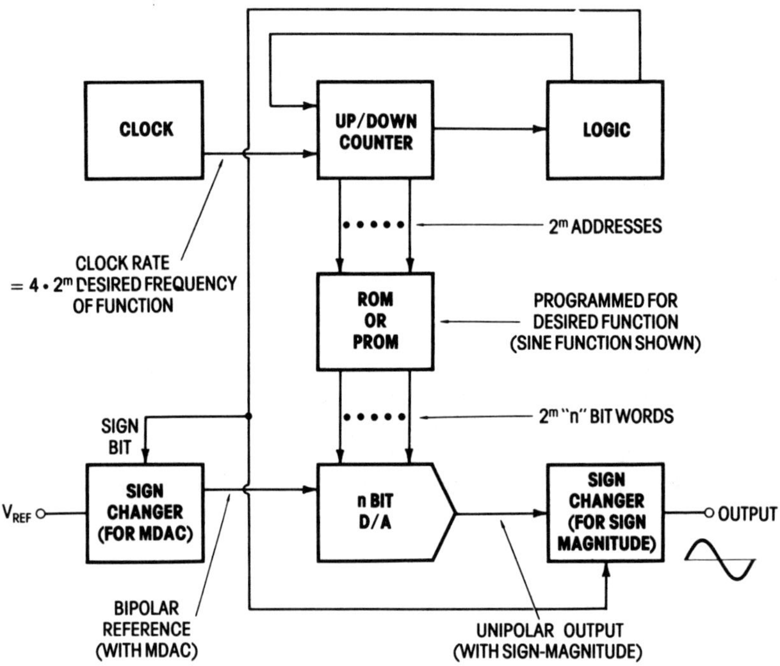

Fig. 8-14. Sine function (or arbitrary) waveform generator.

Here a ROM (or PROM) is inserted between the up/down counter and the d/a converter input. This ROM is programmed for the desired output function, as for example a sine table. For waveforms which have quadrant-to-quadrant symmetry, this will be a 90° table. For the first 90° of the sine period the up/down counter increments; the second 90° it decrements back to 0 count (180°). Over the last 180° of the period this up/down count repeats, but logic is used to switch the d/a converter output sign for the third and fourth (negative) quadrants.

In this circuit the d/a converter can be either a bipolar reference unit (such as the AD7520 for instance), or the sign control can be accomplished after the d/a converter output with a sign-change circuit.

As previously, amplitude control is optionally available by control of the reference. Note that in this circuit the final frequency is ¼ times $1/2^m$ times the clock rate, since the up/down counter must be cycled twice per output cycle, and it is of a length of 2^m input periods.

An 8-bit circuit example of a relatively simple but yet flexible waveform generator is shown in Fig. 8-15. This circuit can generate both positive and negative sawtooth and triangular waves which are amplitude controlled. Fastest operation can be achieved with a high-speed

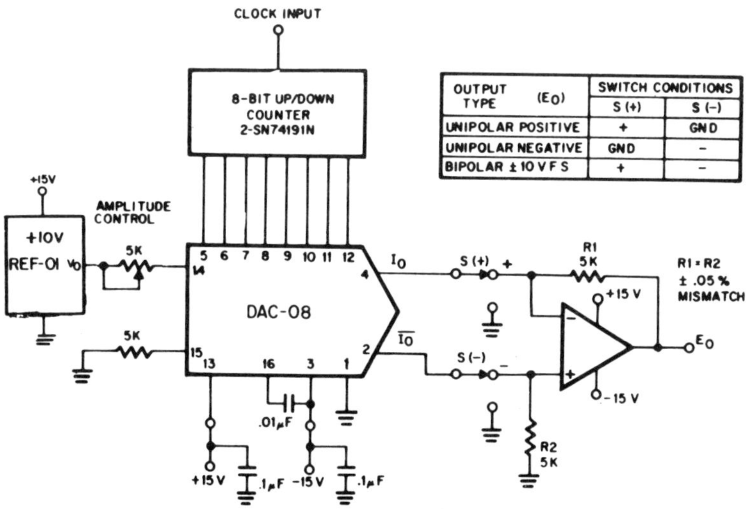

OUTPUT TYPE (Eo)	SWITCH CONDITIONS S (+)	SWITCH CONDITIONS S (-)
UNIPOLAR POSITIVE	+	GND
UNIPOLAR NEGATIVE	GND	−
BIPOLAR ± 10 V F S	+	−

NOTES: 1. BIPOLAR OUTPUT IS SYMMETRICAL AROUND ZERO, ADJUSTABLE PEAK TO PEAK AMPLITUDE.
2. FOR TRIANGLE WAVE, COUNT UP TO FULL, REVERSE AND COUNT DOWN.
3. FOR POSITIVE-GOING SAWTOOTH, COUNT UP TO FULL, CLEAR, REPEAT.
4. FOR NEGATIVE-GOING SAWTOOTH, COUNT DOWN, CLEAR, REPEAT.

Courtesy Precision Monolithics, Inc.

Fig. 8-15. DAC-08 8-bit waveform generator.

amplifier such as the OP-16, with its less than 1-μs settling time. Note that the function mode switches can be analog types, making the circuit electronically programmable.

For unipolar sine waveform requirements, where waveform fidelity and low distortion are not critical, there is an alternate synthesis technique which is simple, efficient, and highly economical. This is illustrated in Fig. 8-16.

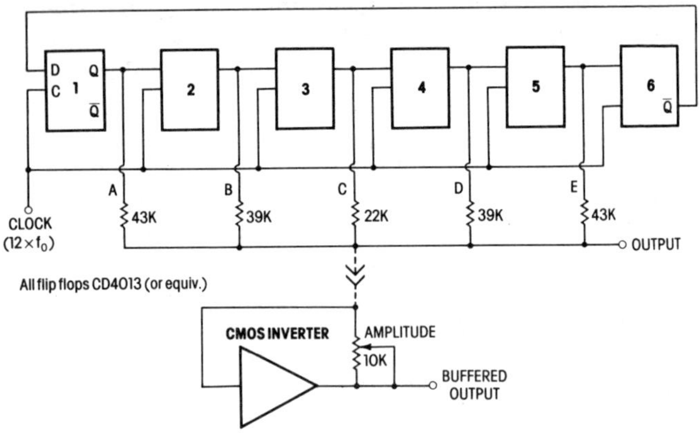

(A) Basic technique.

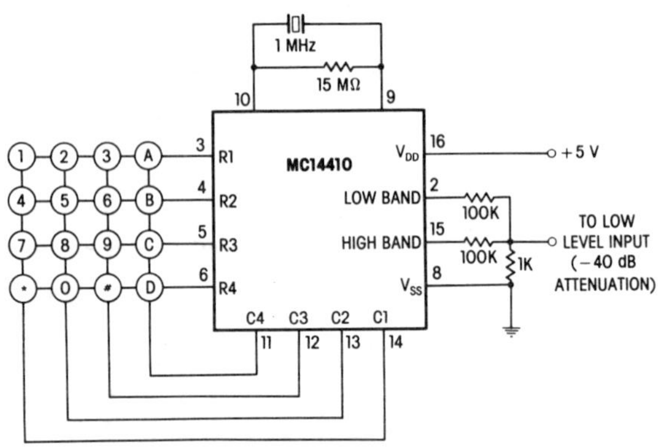

Note: Application using Chomerics Keypads ER21623 (4 rows x 3 columns) or ER21611 (4 rows x 4 columns)

1. LOW LEVEL TONE GENERATOR CIRCUIT

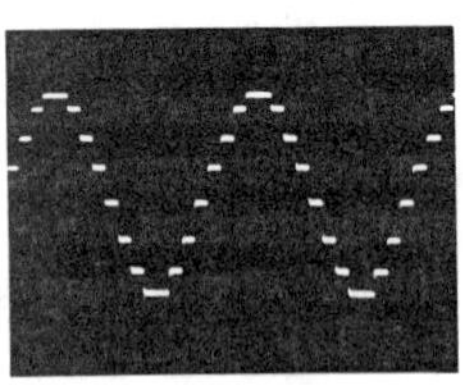

2. WAVEFORM

INPUT LINE ACTIVATED (LOW)	FREQUENCY GENERATED** f_L (Hz)	f_H (Hz)
R1	697	–
R2	770	–
R3	852	–
R4	941	–
C1	–	1209
C2	–	1336
C3	–	1477
C4	–	1633

**All frequencies are accurate to ±0.2% (crystal tolerance not included)

3. OUTPUT FREQUENCIES

Courtesy Motorola Semiconductor Products, Inc.

(B) Example using MC14410 2-of-8 tone encoder.

Fig. 8-16. Simplified sine-wave synthesis.

The basic idea of this synthesis technique is a multistep approximation to a sine waveform, using phase-shifted square waves. The square waves are generated by a multistage shift register in a "walking-ring" configuration, as shown in basic form in Fig. 8-16A.

For an "n" step sine-wave approximation, the register length is "n" bits. As can be noted n − 1 outputs are used, with the complemented nth output fed back to the D input of the register. The resistor matrix forms a specialized type of d/a converter, with the resistor ratios chosen to provide a close match to the sine function. Output frequency is, in general, the input frequency divided by 2n.

In the example shown, three 4013 dual D flip-flops are used to show the general concept. Longer register lengths would more efficiently use such devices as the 4018 or 74C164. The shift register must have "n" parallel outputs available for connection of the matrix.

The output is a pseudosine wave with a p-p level of V+ volts, and a quiescent (sine-wave zero crossing) level of V+/2 volts. It can be used either directly or buffered. Amplitude can be controlled either by variation of the supply voltage to the register, or if a buffer is used, in the feedback circuit as shown. In this example of a 6-level synthesis the THD is 15% (unfiltered); this can be lowered easily if desired by a capacitor to ground to remove the higher harmonics.

Obviously, one of the more attractive virtues of this technique is the fact that it can allow crystal stability sine waves to be generated at low cost. In fact, several dedicated synthesizer chips are now available which employ this principle. One of these, the Motorola MC14410 two-of-eight tone encoder, is shown in Fig. 8-16B. The MC14410 generates 2 sine wave tones by programmable division from a common 1-MHz crystal reference. A typical tone generator application circuit and the sine wave which is produced are shown in Fig. 8-16B. As can be noted from the table, the frequencies are accurate to 0.2%, or less.

This device and others with related topology are intended for telephone and data-communications applications.

COUNTER-TYPE A/D CONVERTERS

A class of very simple feedback-type a/d converter is the counter type, based as the name implies, on a counting process. Counter a/d converters are feedback types, somewhat akin to the SA type in that they use a d/a converter within a feedback loop, and seek a closed loop null condition to determine the analog input. They are distinctly different however, in that they are not as efficient timewise as the SA type, and are therefore inherently much slower.

An SA a/d converter completes a conversion in n + 1 clock cycles, while a counter type requires 2^n-1 clock cycles—an obvious difference. Nevertheless counter type a/d converters can often offer certain advantages where speed is not the prime criteria, namely low cost and simplicity. Two closely allied counter type a/d converters are discussed in this section.

A basic example of a counter a/d converter is shown in Fig. 8-17. The three fundamental components are the d/a converter (A1), a comparator (A2), and a counter (A3-A4). In this example an 8-bit binary counter and d/a converter are used; conceptually, a design could employ a fewer or greater number of binary bits, or be bcd-coded as well (if desired).

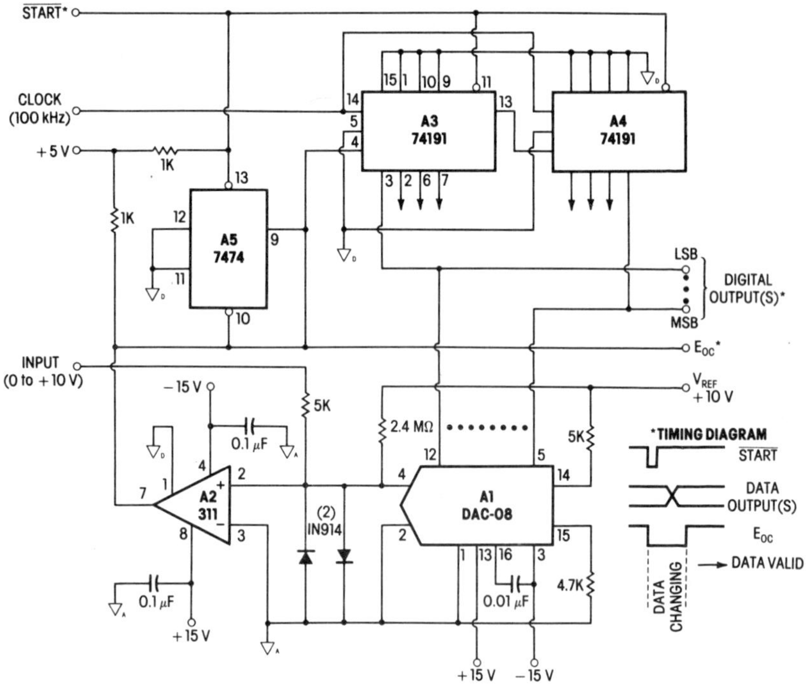

Fig. 8-17. Counter-ramp a/d converter.

This basis of operation is that during each conversion interval the counter is reset to zero, which causes the d/a converter to go to an analog zero scale. The counter then is incremented step by step until the comparator signals that the d/a converter output is equal to the analog input. The counter is then stopped, and the stored count represents the digital equivalent of the analog input. Since the counter must traverse its entire scale of 2^n counts to reach a full scale input, the maximum conversion time takes 2^n-1 counts; lower analog inputs take correspondingly less time. Note that the data is changing continuously during the conversion interval (as the counter ramps) and will be valid only when the comparator signals the conversion complete.

Although a design of one of these a/d converter types is fairly straightforward, several pitfalls should be avoided. High speed synchronous counters are preferred over ripple counters, to minimize glitching in the d/a converter. Similarly, a low-glitch d/a converter is preferred. The comparator used should be laid out as appropriate for high speed, using techniques as previously described in the SA a/d converter section (Chapter 7). Analog and digital grounds should be separated, and good bypassing used.

In the design here, 74191 counters are used, hardwired for the upcount mode. They are synchronously clocked in common, and enabled by A5, a control flip-flop. To start a cycle, A3-A4 are reset to zero by a $\overline{\text{START}}$ pulse to the load input (which loads all 0s). When comparator output goes low, this resets A5 and stops the count. It also locks out any counter state changes, should the analog input change before the next $\overline{\text{START}}$ pulse. The positive edge of A5's output signals a complete conversion, EOC.

As with SA a/d types, a data buffer will often be useful to hold the valid data between conversions. One of the designs discussed in the SA section can be used for this purpose, with clocked loading accomplished by EOC.

As may be obvious, many of the basic design criteria of an SA feedback a/d converter apply to this circuit as well. The input and feedback resistors set up the analog scale, and are preferably tracking types. A ½ LSB current bias is also used, provided by the 2.4-megohm resistor to +10 volts.

Although a DAC-08 type is shown here, a 1408 type is a pin for pin substitute, if a +5-volt supply is used. Coding is for a unipolar 0 to +10-volt range. A 100-kHz clock is shown, allowing a conversion to be completed in 2.55 ms. Faster (or slower) conversion times are made possible, simply by adjusting the clock rate.

The basic scheme shown in Fig. 8-17 has a low input impedance, essentially the value of the input resistor (5K). Two alternate connections which offer high input impedance are shown in Fig. 8-18. Fig. 8-18A shows a unipolar positive range circuit, with the d/a output connected to a voltage before comparison by the op-amp A. Shown in Fig. 8-18B is a negative unipolar range circuit. Note that this version must use a DAC-08 for a −10-volt scale; a 0 to −5-volt scale can use a 1408 also.

The tracking a/d converter is a form of counter-type a/d converter which continuously tracks the analog input, as the name implies. It is similar to the basic counter (ramp) type in that the d/a converter is driven by a counter, but the counter is an up/down type. Therefore, the circuit requires no convert command, as it inherently is outputting a

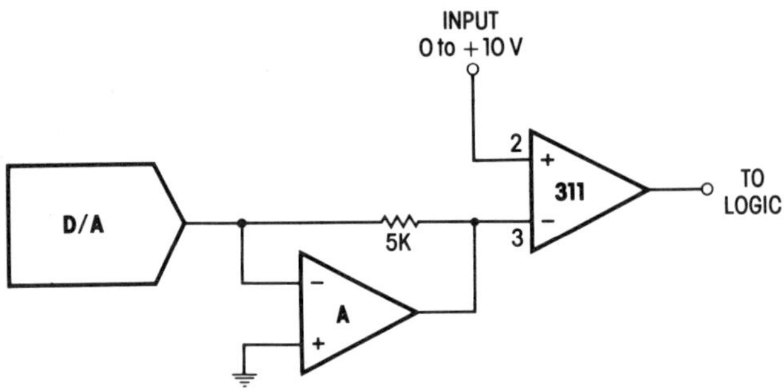

(A) Positive input range.

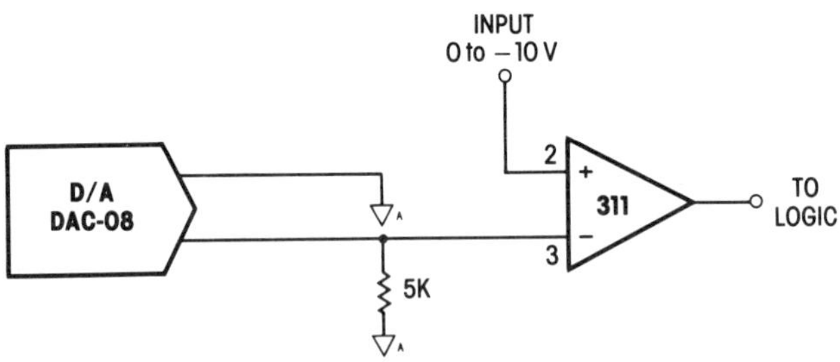

(B) Negative input range.

Fig. 8-18. Alternate analog input connections.

digital code corresponding to the analog input at that particular instant. In this regard the tracking a/d converter is somewhat unique; very few a/d types offer this capability.

The tracking a/d converter, however is like the simple counter/ramp type in that it has an inherent slew limit which will determine the maximum input rate of change to which it can respond, and track faithfully. This will be proportional to the clock rate used, for a given full-scale input range (this will be evident from the example which follows). Also, the virtue of a constant and correct output is valid only when the loop is in a locked condition. If the input is switched abruptly between different levels (such as in a multiplexer), the loop will be momentarily broken while the a/d converter acquires the new input level. During the acquisition (or slewing) interval to the new code, the output data will not be valid.

A basic 8-bit tracking a/d converter design is illustrated in Fig. 8-19. As can be noted, the circuit is similar to the previously described counter/ramp a/d converter, and both circuits can be built with essentially the same parts.

Operational differences here are the fact that the 74191s are operated in their up/down mode, as controlled by the comparator output.

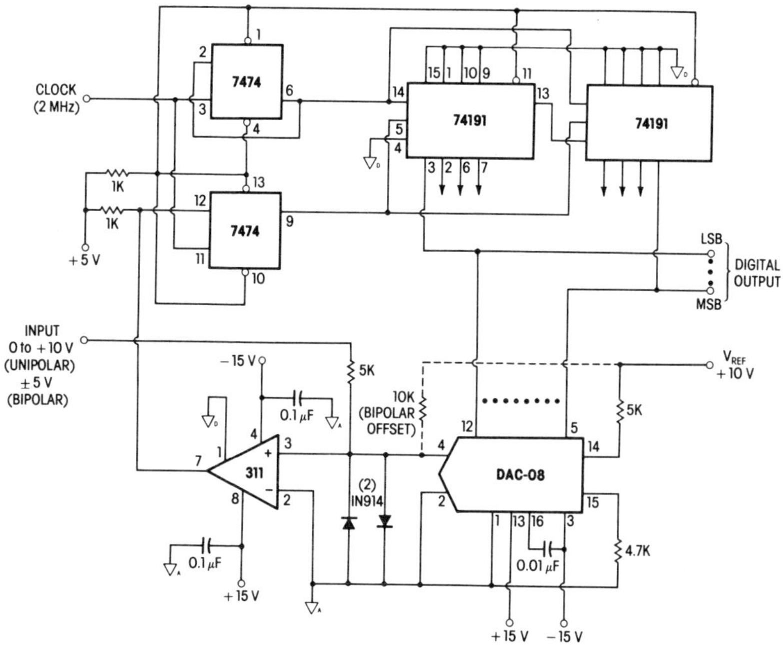

Fig. 8-19. A tracking a/d converter.

If the analog input is increasing, the counters must count up to track the signal. Conversely, if it is decreasing, they must count down to track. For a constant analog input, the loop will "dither" up and down, or ±1 LSB about the ideal code level. This is an important fundamental difference for this type of converter; the fact that the output is not accurate within the conventional ±½ LSB, but rather ±1 LSB. Thus in a tracking converter of this type, there will always be a 1-LSB "bobble"—evidence that the loop is tracking correctly.

The design example shown is relatively straightforward, with perhaps the exception of the clocking scheme used. A D flip-flop buffer is used on the comparator output to satisfy clock timing restrictions of the 74191 devices. As noted previously, proper layout techniques are of paramount importance to correct operation.

The circuit can be used as shown in either a 0 to +10-volt unipolar mode, or ±5-volt bipolar with the addition of a bipolar offset resistor. This circuit can also use a 1408 type d/a converter, with a +5-volt supply. A variety of other different coding and analog range/polarity options are also possible using the DAC-08.

The maximum tracking rate of an a/d converter of this type will be determined by both the d/a 1-LSB current level and clock rate, and is expressed in μA/μs (if voltage output mode d/a converter were used, V/μs would be the appropriate figure of merit). For the design as shown, the counters are clocked at a 1-MHz rate (1-μs period), and the d/a current for 1 LSB is 7.8 μA, and therefore the tracking rate is 7.8 μA/μs. This can be expressed in terms of voltage input by simply multiplying by the value of R_{IN} (5K), which yields an input SR of 39 mV/μs. For a 10-volt range input, this then corresponds to a 1.2-kHz maximum input frequency. Higher input rates of change can be accommodated by raising either the d/a output or clock rate, or both. At present, device limitations put the maximum input rates of change between approximately 0.2 and 0.4 V/μs for tracking type a/d converters.

Applications which can use the tracking a/d converter characteristics to advantage are those which represent relatively slowly changing data, self-sufficient (or noninterrogative) transducers (no conversion command), the requirement for continuous real time output data, and lastly any applications where relatively low cost/complexity is a requisite.

Some features of a tracking a/d converter can provide rather unique capabilities. For example the zero droop digital sample/hold system of Fig. 8-20 is one of these. Here a tracking a/d converter outputs continuous digital data. A simple D-type register can be used to selectively freeze a single "n" bit word when clocked by the sample/hold control line. Since the storage is done in a logic register, the sampled

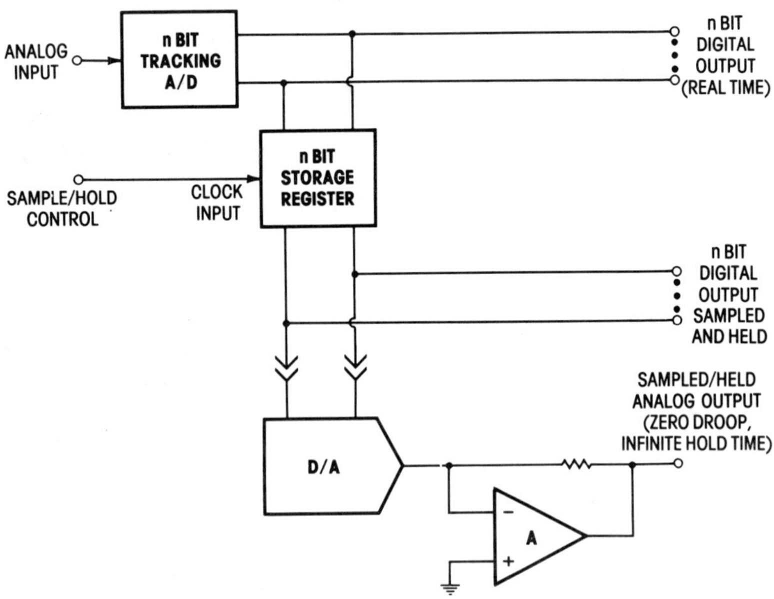

Fig. 8-20. Digital sample-hold system with zero drop.

output is available indefinitely. It may be used either directly in digital form, or a d/a converter can be used to reconvert it into an analog format.

Another useful application of a tracking a/d converter is in a bridge transducer digital nulling system, shown in Fig. 8-21. Here a resistive bridge (sensing temperature, pressure, etc.) is simultaneously sensed by a different comparator and driven by a DAC-08. As the bridge becomes unbalanced due to the input being sensed, the d/a converter

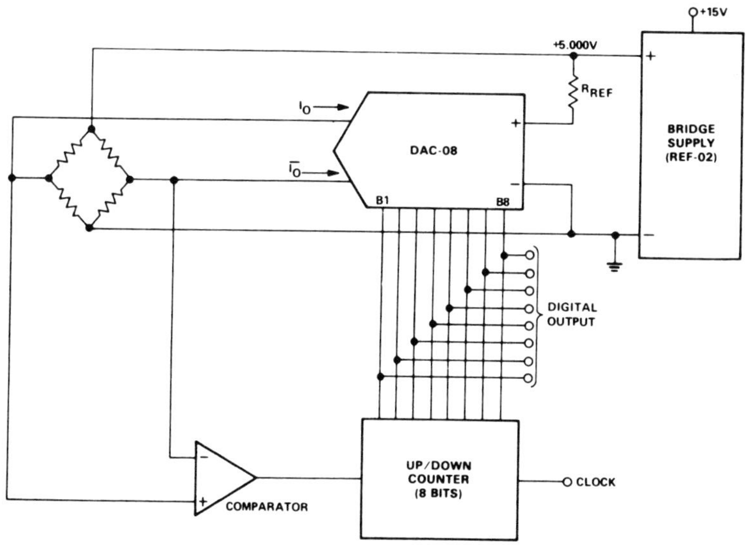

Courtesy Precision Monolithics, Inc.

Fig. 8-21. Bridge transducer digital null system.

supplies a compensating feedback. The d/a converter drive then represents a measure of the unknown input to the bridge. Note that both the bridge and d/a converter can utilize the same reference voltage—a contribution to system efficiency and error reduction.

ALGEBRAIC OPERATIONS

D/A conversion techniques, particularly those involving multiplying d/a converters, can be very useful in situations where two digital words are to be algebraically manipulated and subsequently converted to analog form. For example, 8-bit digital multipliers can involve several logic packages in addition to the final d/a converter. An alternate approach to the solution of such algebraic computations is to do the operation directly in the d/a process, using two multiplying d/a con-

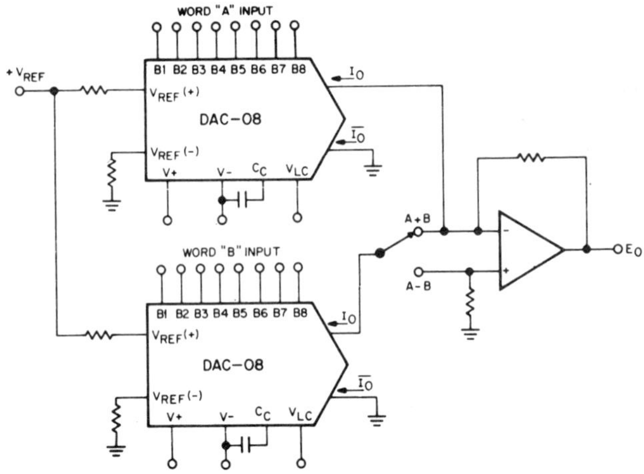

(A) Unipolar words.

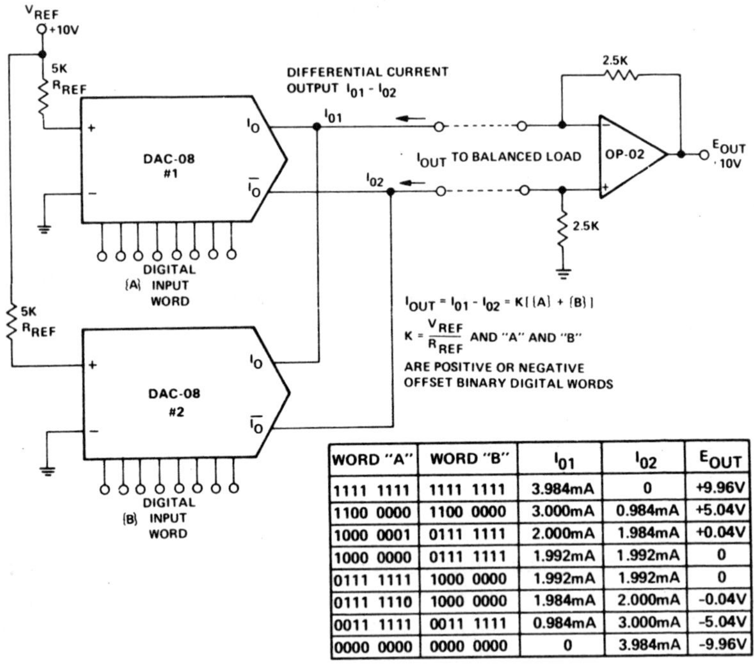

WORD "A"	WORD "B"	I_{01}	I_{02}	E_{OUT}
1111 1111	1111 1111	3.984mA	0	+9.96V
1100 0000	1100 0000	3.000mA	0.984mA	+5.04V
1000 0001	0111 1111	2.000mA	1.984mA	+0.04V
1000 0000	0111 1111	1.992mA	1.992mA	0
0111 1111	1000 0000	1.992mA	1.992mA	0
0111 1110	1000 0000	1.984mA	2.000mA	-0.04V
0011 1111	0011 1111	0.984mA	3.000mA	-5.04V
0000 0000	0000 0000	0	3.984mA	-9.96V

(B) Bipolar words.

Courtesy Precision Monolithics, Inc.

Fig. 8-22. Analog addition/subtraction of digital words.

verters. Several examples of such uses with 8- and 10-bit d/a converter types follow.

A simple example of an efficient algebraic operation is the addition or subtraction of two digital words, shown in Fig. 8-22.

Here two current-output d/a converters generate analog currents corresponding to digital words "A" and "B." The currents are either added or subtracted in an output summing/difference amplifier. Since the maximum output current (A + B) is twice that of a conventionally scaled d/a converter of this type, the op-amp scaling resistors should be one-half normal value (2.5 K for a 2-mA output d/a converter).

As shown this circuit is used with two unipolar-coded input digital words (Fig. 8-22A). Note that with the DAC-08 programmable logic threshold the input words used need not necessarily be from the same logic family.

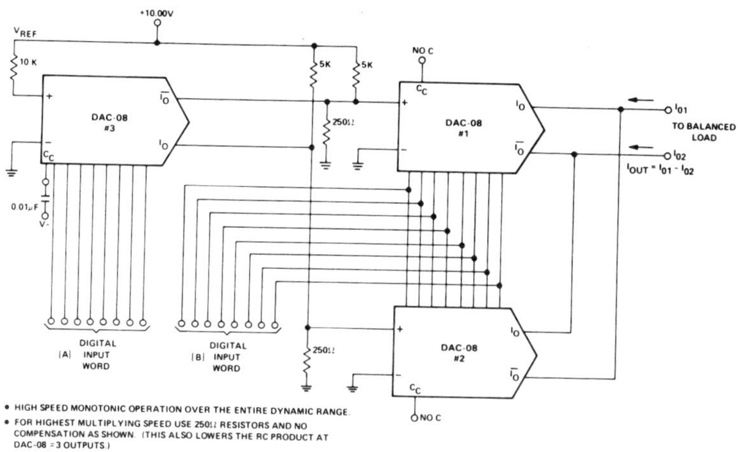

Courtesy Precision Monolithics, Inc.

Fig. 8-23. Analog multiplication of digital words.

A circuit which responds to two bipolar coded digital words is shown in Fig. 8-22B. As can be noted this circuit is no more complex, but offers the bipolar or four-quadrant capability by the use of the DAC-08 differential outputs. A similar balanced op-amp configuration is used, and a partial code table is listed with the figure.

A high speed multiplication of two digital words is shown in Fig. 8-23. Here DAC-08s #1 and #2 are differentially connected, and device #3 modulates their reference inputs. Reference shunt resistance and no compensation is used on devices #1 and #2 for maximum bandwidth.

This circuit is four-quadrant in operation, and can handle two bipolar input words. For maximum speed the output load should be low value resistors, 1K or less. Alternately a differential current-to-voltage op-amp connection (such as in Fig. 8-22B) can be used with a high speed op amp such as the OP-16.

The algebraic operation of division is the inverse of multiplication; therefore it is electrically accomplished by a multiplier in a feedback path. An example is shown in Fig. 8-24.

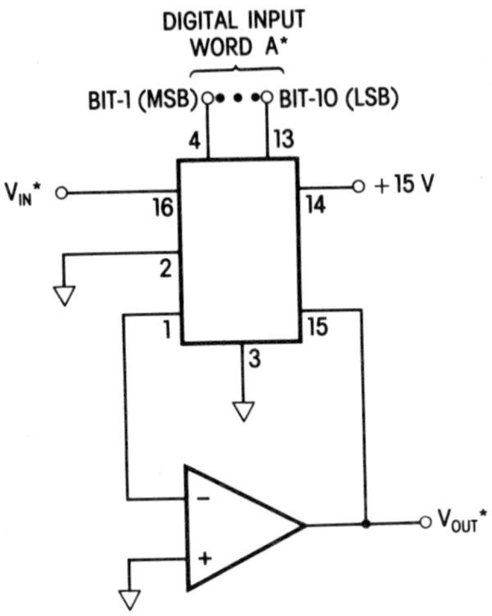

$$^{*}V_{OUT} = \frac{V_{IN}}{A}; \quad \frac{1}{1024} \le A \le \frac{1023}{1024}$$

DIGITAL INPUT	1/A
1 1 . . . 1 1	1024/1023
⋮	
1 0 . . . 0 0	1024/512
⋮	
0 0 . . . 0 1	1024/1
0 0 . . . 0 0	∞ (disallowed)

Courtesy Analog Devices, Inc.

Fig. 8-24. Analog division by digital word.

This circuit divides an analog input voltage by the 10-bit digital word, A. The AD7520 device series is very flexible in this regard as V_{IN} can range over ±10 volts. The connections to the ladder and feedback paths are reversed from the normal state.

Two application points of interest here are that the all zero state of A is disallowed, as it defines division by zero. Also, for accurate division at low values of A, the op amp used should be a high gain precision type such as the AD504L or AD510L. Of course, for operation at low values of A, V_{IN} must of necessity be low in level to prevent saturation of the op amp. Note that other "7520" type devices can be used as well in this circuit; the AD7522 and AD7523 are examples.

An extension of this basic technique is applied in Fig. 8-25, a circuit which produces an analog quotient of two digital words A and B, using AD7520s.

In this circuit A2 is a divider operating essentially as just described dividing the analog input at pin 16 by the word A (with similar restrictions). Device A2 is a conventional multiplier connection, producing an analog product of V_{REF} and word B. Overall, the circuit produces an output voltage V_o, here

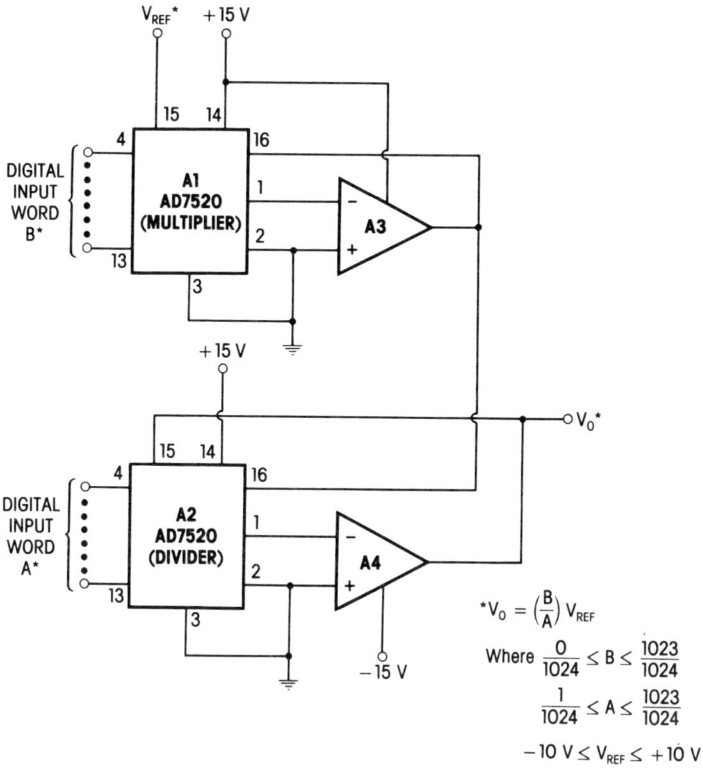

Fig. 8-25. Analog quotient of two digital words.

$$V_o = \frac{B}{A} V_{REF}.$$

Points to be noted are that V_{REF} may range up to ±10 volts as desired (or convenient), and B and A are limited as noted. Digital words A and B control basically unipolar scales, but the output sign can be controlled by the choice of V_{REF}, as desired. For general use, dual op-amp types are suitable; if high precision is desired selected single devices are more appropriate.

The function of an analog square of two digital words is also quite simply generated with a pair of AD7520s, as shown in Fig. 8-26.

The only difference between this figure and Fig. 8-25 is that both A1 and A2 are connected as multipliers, and are driven by a common digital word A. The circuit obviously can be used as an analog multiplier of two words (A and B) by driving the devices separately. Higher order functions (cubes, squares etc.) are possible as well. Again, flexibility is inherent in that V_{REF} can be up to ±10 volts.

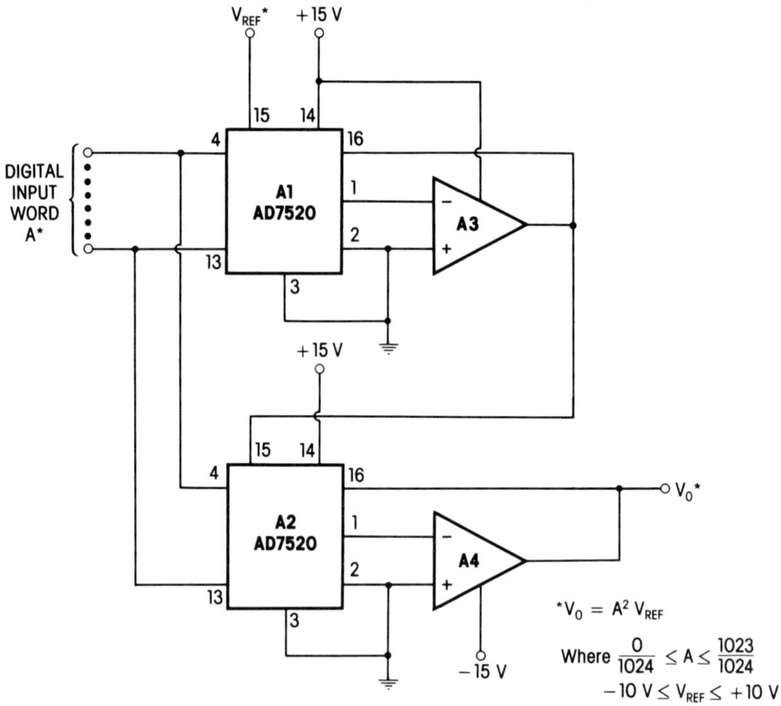

Fig. 8-26. Analog square of two digital words.

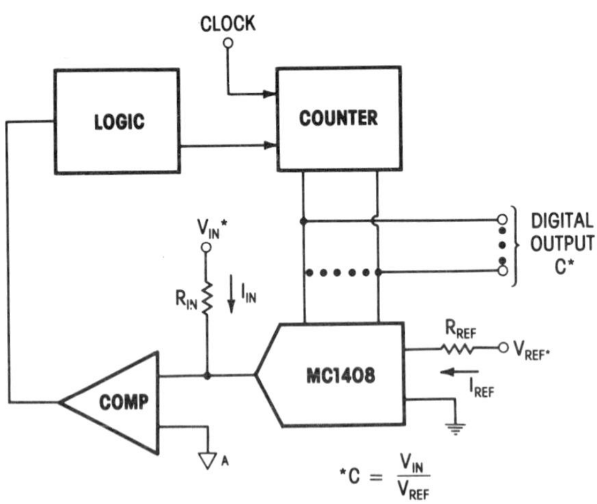

Fig. 8-27. Digital quotient of two analog variables.

The circuit of Fig. 8-27 provides a digital output word C, where C is the quotient of two analog variables V_{IN} and V_{REF}. As should be obvious this circuit is a variation on the previously described counter type a/d converter.

With a multiplying d/a converter used in such an a/d converter, ratiometric operation is inherent. It is then possible (within limits) to vary either V_{IN} or V_{REF}, with the digital output an indicator of this ratio. The circuit of either Fig. 8-17 or Fig. 8-19 could be adapted for this purpose.

It should be noted that there are also existing devices which operate in such a fashion inherently. The AD7570 is one example, performing the conversion in SA fashion, with 8- or 10-bit accuracy.

A final useful function which can be easily implemented with d/a converters is a relative magnitude indicator, shown in Fig. 8-28. In this circuit two similar d/a converters produce outputs across matched load resistors of 2000 ohms. The resulting output voltages are examined for relative magnitude by a comparator, which furnishes a logic output indicative of the magnitude between input words A and B.

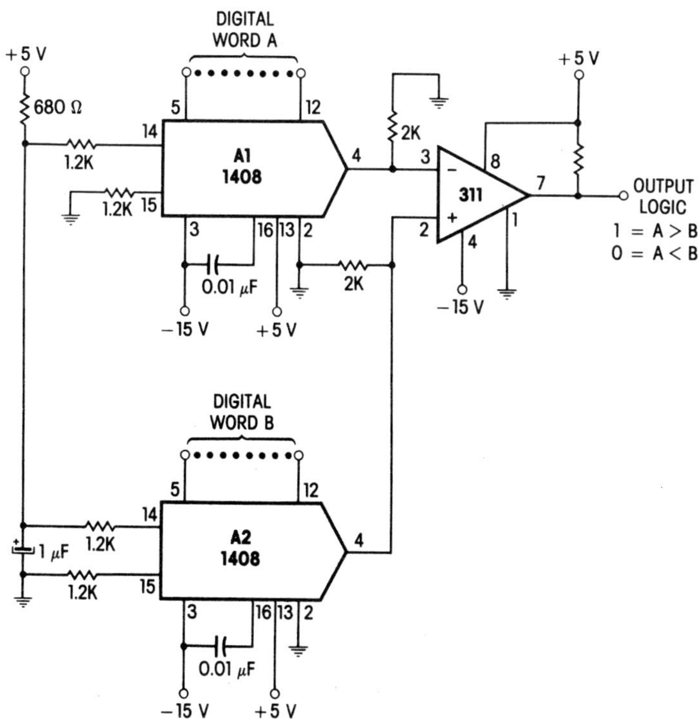

Fig. 8-28. Logical relative-magnitude indicator between two digital words.

Note that since only *relative* magnitude is being examined here, the reference input to the d/a converters is not critical and the logic supply is used.

The circuit as shown does not detect the A-B condition. This can be implemented by some additional comparison logic to provide a third output state, and is suggested as a challenge for the reader.

COMPONENT TEST/MEASUREMENT

The testing of standard passive circuit components such as resistors and capacitors can be readily accomplished with a/d conversion techniques, and in many instances quite accurately. Some examples follow using a variety of a/d devices, and are applicable to both automated instrumentation systems and lab/bench uses.

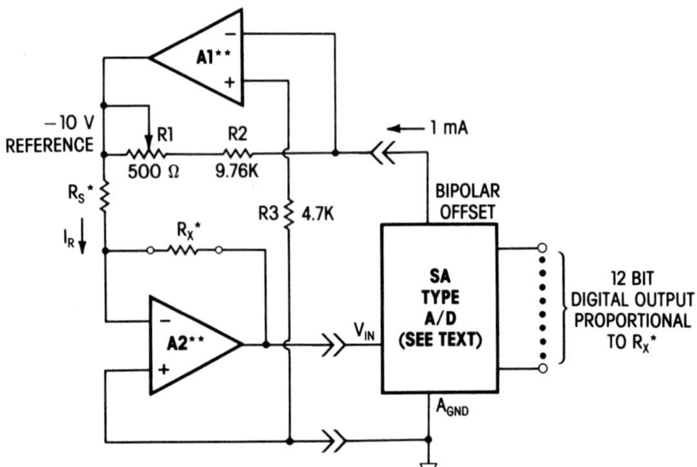

**A1 = OP-02C or AD741K
A2 = FET input, low (or nulled) offset voltage, low drift— OP-16E or 356A.

***SCALING TABLE (12 BITS)**

I_R	R_S	$R_{X\,(MAX)}$	Ω/LSB
2.44 mA	4,096 Ω	4,095 Ω	1
244 μA	40,960 Ω	40,950 Ω	10
24.4 μA	409,600 Ω	409,500 Ω	100
2.44 μA	4.096 MΩ	4.095 MΩ	1K
244 nA	40.96 MΩ	40.95 MΩ	10K

Fig. 8-29. General purpose digital resistance measurement.

A general scheme for measuring the value of an unknown resistance is illustrated in Fig. 8-29. This technique is applicable to any of the SA type a/d converters described in Chapter 7, and with slight alteration, many integrating a/d converters as well.

As noted in the figure, the internal reference of the a/d converter is tapped by feeding the 1-mA (typical) bipolar offset current to an external inverting op amp. Since this pin would normally be grounded for a unipolar a/d mode operation, this correctly loads the reference diode and does not alter the a/d converter calibration (or drift).

A1 produces a new reference voltage of −10 volts. This voltage, being slaved to the converter's reference, is then used to develop reference current I_R. This current, as will be shown, can be scaled over a wide range to produce various resistance measurement sensitivities.

A2 is a precision FET input op amp, which is used to force current I_R in R_x, and to hold it accurately over a wide range. The voltage drop across R_x is directly proportional to its value, and the buffered output of A2 feeds the V_{IN} terminal of the a/d converter, operating on a 0 to +10-volt scale.

If R_s (and I_R) are appropriately chosen, the output code of the a/d converter can be made to be integrally weighted in ohms per LSB. For example, if R_s is 4096 ohms, I_R is 2.44 mA and the a/d input will be 1 LSB of voltage for each ohm of R_x resistance. The output will then read in terms of 1 LSB per ohm (see table). Of course, R_s can be chosen to accommodate various full scale ranges, as noted. If the R_s values used are ratio matched as shown, the −10-volt reference trim is the only calibration necessary for the circuit.

Although a 12-bit example is shown, other a/d scales are possible, using 10- or 8-bit a/d devices. A similar ratiometric arrangement can be made for bcd-coded converters, as well. Also, many integrating a/d converter types can be used for such a test configuration by adding an input resistor for A1 which will drive A1 with 1 mA of current, at the particular reference voltage for the a/d type in use.

It should be noted that if overrange logic is added, R_s can be programmed to automatically adjust I_R for various R_x ranges.

A bit more specific example of this general technique is shown in Fig. 8-30, and is suitable for use as an ohms-to-frequency adapter for a lab counter.

This circuit uses the AD537 V/F converter to convert a voltage which is proportional to R_x into a corresponding frequency, which is simply read on a counter. The typically used lab counter has resolution/accuracy much greater than necessary to this use, and models with variable gate times can conveniently move the decimal place of the displayed "resistance" value, as appropriate. For example, a gate time of 1 second will read in hertz, and the most usable range of this circuit converts ohms as 1 ohm = 1 Hz, or 100K ohm = 100 kHz (full-scale range).

The interesting feature of a V/F converter for this type of use is that the typically high dynamic range can allow the most often used resistance values to be displayed on a single range, which spans four decades.

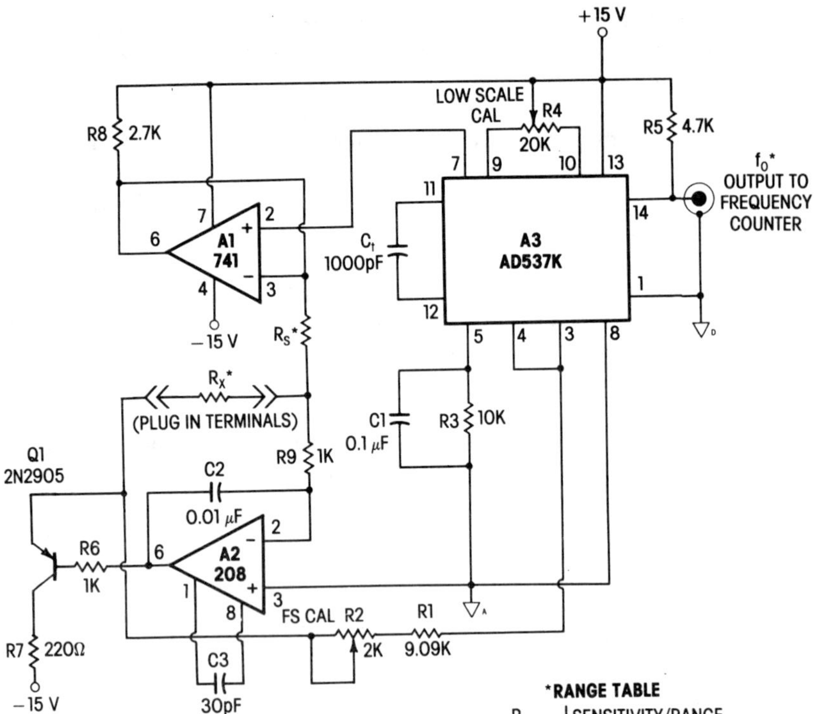

*RANGE TABLE

R_S	SENSITIVITY/RANGE
100 Ω	1 Ω/100 Hz, 1 K
10 K	1 Ω/Hz, 100 K
10 MΩ	1 K/Hz, 100 MΩ

Fig. 8-30. Ohms-to-frequency converter.

On the 1 ohm / Hz range for example, values can be read as low as 10 ohms with 1 Hz resolution. After calibration at 100 kHz full scale by R2, and low scale by R4 (with 100K and 100 ohm standards, respectively) linearity error will typically be no more than 0.06%.

R_S, R1, and R2 should be stable precision types for best stability and repeatability, and C_t should be an npo ceramic. C1 and C2 serve as noise bypasses; C2 should be an ultralow-leakage unit, such as a polystyrene.

As can be noted from the chart, two additional ranges are suggested. The 1 ohm/100 Hz range allows greater resolution and accuracy for resistances below 1000 ohms, with a 1K = 100 kHz full-scale limit. This scale can resolve resistances below 0.1 ohm. With this range in use, R8 or a high output current op amp for A1 and A2 should be used, as it drives R_x with 10 mA of current.

The higher scale range allows resistances up in the megohms to be used. If a low bias current input unit such as the 208 is used for A2, this will allow excellent accuracy to well above 10 megohms. Other bipolar input devices should not be used on this range simply because of their high bias currents. An effective solution to this problem, should one want to use all three listed ranges, is to use the devices just as indicated, as they solve the dynamic range considerations.

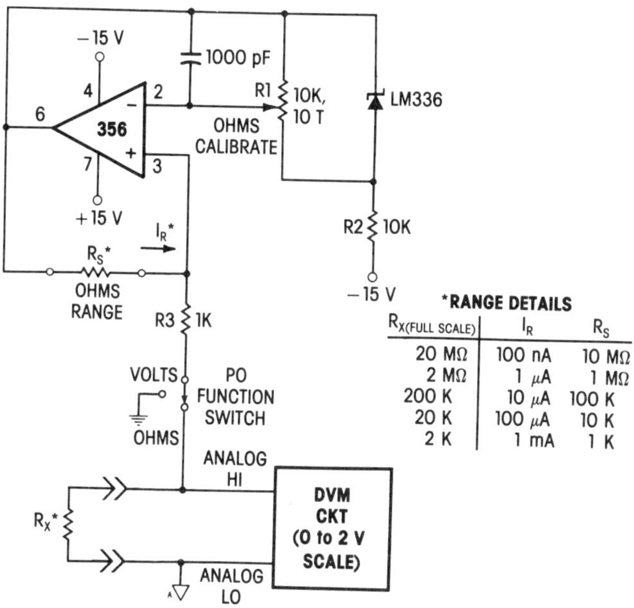

*RANGE DETAILS

$R_{X(FULL\ SCALE)}$	I_R	R_S
20 MΩ	100 nA	10 MΩ
2 MΩ	1 μA	1 MΩ
200 K	10 μA	100 K
20 K	100 μA	10 K
2 K	1 mA	1 K

Fig. 8-31. Ohms-to-volts converter.

Fig. 8-31 illustrates another variation for an ohms measurement circuit—a general purpose ohms-to-volts converter which can be added to any 0 to 2-volt range integrating a/d circuit. It uses a self-contained reference and accuracy is such that it can be used with up to 4½ digit a/d dvms with meaningful results.

This circuit also operates to apply a constant current to R_x, with the resulting voltage proportional to R_x being displayed on the dvm. Scaling of I_R is such that the displayed voltage is read as a value of 0 to .999 (to 1.999 overrange) for 3½ digits, and multiplied times a range scaling factor. For example, a 1500-ohm resistor reads as 1.500 (1.5 × 1K = 1.5K).

The table gives the appropriate current scaling values for R_s, for ranges of up to 20 megohms full scale. A simple reference voltage trim, R1, adjusts ohms calibration; it will be correct for all ranges if the R_s values are of the correct ratio. There is no offset control necessary—this is provided by the a/d converter auto-zero function.

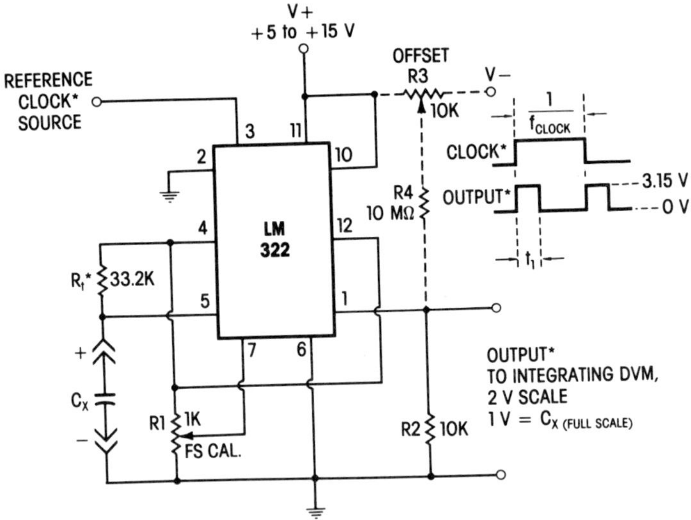

*$V_0 = (t_1)(f_{CLOCK})(K)\ V_{REF}$
Where $t_1 = R_t\ C_X$, $V_{REF} = 3.15$ And
$0.9 \leq K \leq 1.1$ (Calibration Constant)

$C_{X(FULL\ SCALE)}$	f_{CLOCK}
10000 pF	1000 Hz
10 μF	1 Hz

Fig. 8-32. Capacitance-to-volts converter.

Capacitance can also be measured with a/d techniques, as illustrated in Fig. 8-32. This circuit is a capacitance-to-volts converter, and is also intended to be used with a 0 to 2-volt scale integrating dvm.

This circuit functions basically as a capacitance-to-pulse width converter. It is driven at a fixed frequency by a reference clock source which produces a basic time interval of $1/f_{clock}$. The pulse period t1 is some fraction of this period, and the output pulse amplitude is regulated to a 3.15-volt reference level by the LM322 timer. The maximum pulse width for a given range is set by R_t, and is designed to produce 1 volt = full scale capacitance, with overrange to 1.999 volts (3½ digits). The capacitance is read as 0 to 0.999 (or 0 to 1.999) times the range scale factor. For example, 8200 pF would read as 0.820 on the 10,000-pF scale.

Because of the very high resolution of this converter, it can resolve only a few picofarads on the 10,000-pF scale, which allows only two basic ranges to cover most practical values. For best low-scale accuracy on the 10,000-pF range, an optional offset trim can be added (shown dotted). This allows "cancellation" of the 10 to 20 pF of stray circuit capacitance.

R1 serves as a full scale trim, and is adjusted for a reading of 1.000, with a 10,000-pF reference capacitor connected. This calibration will suffice for both ranges if the clock is exactly ratioed 1000/1 by counting. The circuit applies a 0 to +2-volt voltage to C_x, so polarity should be observed for polarized types.

Active components are also readily instrumented for test by a/d and d/a techniques. A block diagram (simplified) of a representative system is shown in Fig. 8-33.

This diagram shows how a family of I_D versus V_{GS} curves might be generated for an n channel FET. A voltage output d/a converter generates a rising staircase of voltage, which is applied to the U.U.T. as V_{GS}. A second voltage-output d/a converter generates a staircase, V_{DS}, and the drain current of the U.U.T. produces a drop across resistor R_D. The output across R_D is referred to ground by the IA, and may be either displayed in analog form in curve-tracer fashion, or digitized by an a/d converter.

This is of course a much simplified example, as the myriad of semiconductor types generates a diverse set of test requirements. Bipolar transistors require current drive, for instance, which can be readily supplied by current output d/a converters, scaled and polarized as necessary. The programmable current source and voltage source sections are adaptable to these needs, as are the waveform generators.

A useful general point to be made here is that the resolution and accuracy of even inexpensive 8-bit d/a converters are usually more than sufficient to meet these test requirements, as most semiconductor parameters need not be characterized to much better than 1% accuracy.

More complex active components, such as op amps, can also be thoroughly exercised using conversion techniques. A detailed treatment of op-amp test techniques is beyond the scope of this book, but a useful example is shown in Fig. 8-34.

While op amps are characterized by a host of performance parameters, perhaps the two most often important are offset voltage and bias

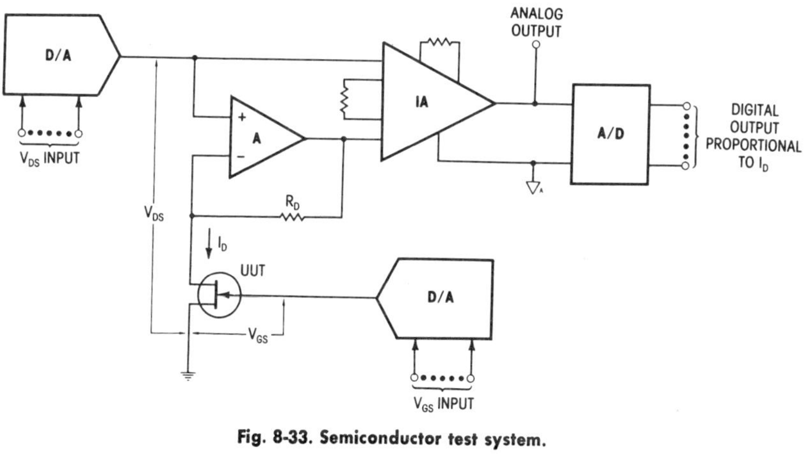

Fig. 8-33. Semiconductor test system.

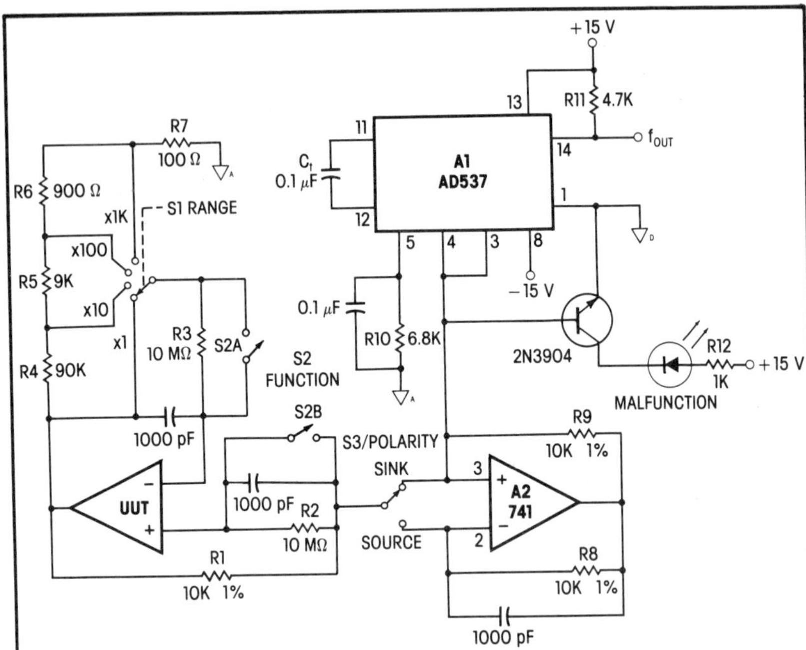

TABLE I
TESTS/CONDITIONS

TEST	SWITCH	STATE	MEASURE
A	S1	AS. REQ.(1)	V_{IO}
	S3	AS. REQ.(2)	
	S2A, S2B	CLOSED	
B	S1	AS. REQ.(1)	I_{IB} (+)/NPN
	S3	SINK	
	S2A	CLOSED	
	S2B	OPEN	
C	S1	AS. REQ.(1)	I_{IB} (−)/NPN
	S3	SOURCE	
	S2A	OPEN	
	S2B	CLOSED	
D	S1	AS. REQ.(1)	I_{IB} (+)/PNP
	S3	SOURCE	
	S2A	CLOSED	
	S2B	OPEN	
E	S1	AS. REQ.(1)	I_{IB} (−)/PNP
	S3	SINK	
	S2A	OPEN	
	S2B	CLOSED	

TABLE II
SCALE FACTORS

S1/RANGE	SCALE FACTOR (x1000 = FS)
x1	1 nA/Hz 10 mV/Hz
x10	100 pA/Hz 1 mV/Hz
x100	10 pA/Hz 100 μV/Hz
x1K	1 pA/Hz 10 μV/Hz

Notes: 1. Adjust S1/RANGE as required to obtain full scale display (see Table II).
2. Since V_{IO} can be + or −, adjust S3 as required to extinguish MALFUNCTION LED.

Fig. 8-34. Op-amp tester.

current. This circuit tests these two parameters over a wide range, and displays the result converted to frequency.

The heart of this circuit is a V/F converter, the AD537, operating at a full-scale frequency of 1 kHz. The parameter being measured is scaled so that the full scale range equals 1 kHz, such as the 1 nA/Hz bias current range, where 1000 nA of bias current = 1000 Hz. This system can, for most ranges, read the parameter to much more than typically required accuracy, and is an effective low cost tool for amplifier screening and selection.

There are a number of switches shown; their operation is summarized in Tables I and II. The notes in the table describe their operational sequence. A current inverter A2, is necessary to adapt source currents, which will be generated by positive outputs from the U.U.T. Note that if very low ($\leq$ 1 nA) currents are to be measured by this circuit, low-leakage sockets, components, and construction techniques should be used.

PROGRAMMABLE OFFSET AND GAIN TECHNIQUES

With the advent of microprocessor-controlled data systems, a class of d/a applications which becomes increasingly useful is the programmable signal processing block. Typically, such a stage is configured to digitally program signal offset and gain parameters. This section discusses this type of d/a converter use in the light of several possible circuit types.

The circuit of Fig. 8-35 is a general purpose scheme which can be used to digitally null the residual offset of an op amp. This is achieved by applying differential correction currents to the amplifier offset null terminals.

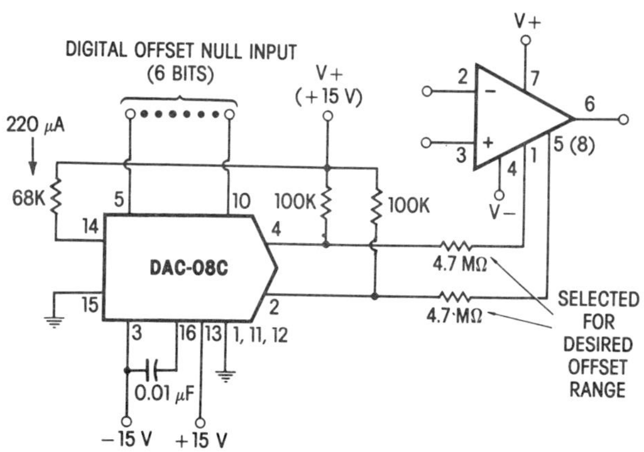

Fig. 8-35. General purpose op-amp offset null technique.

Functionally, the circuit replaces the manual offset pot. The practical advantages are that this scheme offers very high resolution adjustment, can be automated for closed loop auto zero, will work with practically any amplifier type, and can be remotely operated.

The circuit works by generating a voltage of +22 volts across the two 100K resistors. This is converted to currents of $\pm 5\ \mu A$ by the 4.7-megohm resistors, which drive the amplifier null pins (1 and 5 or 1 and 8, dependent upon amplifier type).

Applied to the appropriate amplifier pins, this scheme will work with most IC op amps, but sensitivity will vary between types. General purpose op amps such as 741s or 301As have a correction sensitivity of 1 to 2 millivolts of offset change per microampere of correction current applied. The series resistors should be tailored for the IC in use to provide the desired range.

It should be noted that this scheme can be applied to any op-amp circuit configuration, as it does not require access to the input terminals. Used in conjunction with an a/d converter in an auto-zero system, it can effectively eliminate offset drift, in a manner similar to integrating a/d converter types. The offset voltage in this case is stored in a d/a input latch (or memory location), instead of an auto-zero capacitor.

It is sometimes desirable to purposely introduce a defined offset into a signal channel, for example in the case of an analog quantity to be algebraically added to an unknown signal. Fig. 8-36 illustrates a general method which can implement a 10-bit programmable signal offset of ±2.5 volts.

This circuit is a relatively straightforward marriage of an IA (AD-521) and a bipolar-scaled d/a converter, the AD561. The programmable ±2.5-volt output of A2 is applied to the reference input of the IA, causing the output to be predictably offset, as commanded by the digital input. Smaller offset ranges are possible by loading the AD561 via R_x; larger ranges by adding gain in the IA (See Chapter 5 section on IAs).

This offset system does of course function independent of the IA gain programmed by R_s and R_g; however, dynamic range limitations and error contributions of the AD521 should still be considered. Note that inasmuch as IAs typically change offset as gain is programmed, this circuit can cancel this effect.

A very interesting and useful scheme for programmable gain calibration of a d/a converter (and also many a/d converters) is shown in Fig. 8-37.

For multiplying d/a converter types, such as the AD562 shown here, accessibility is usually provided to the direct reference input (pin 4). Thus current may be drawn away from this node, as well as sourced into it. If the reference current is initially forced to be high by a fixed increment, current drawn away from the sum point can be used for

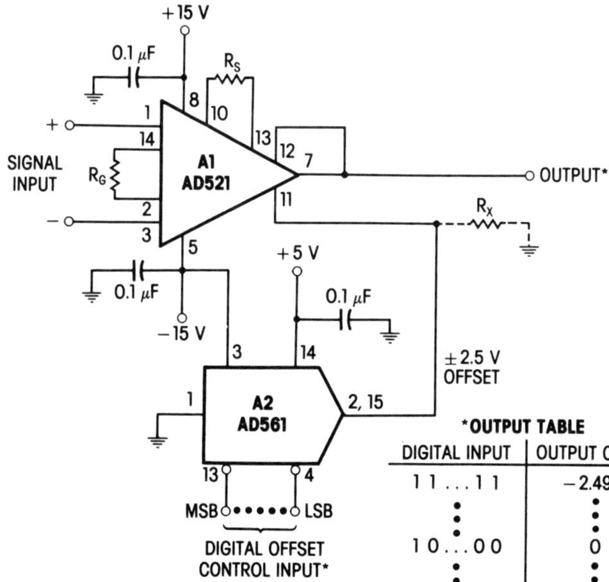

***OUTPUT TABLE**

DIGITAL INPUT	OUTPUT OFFSET
1 1 . . . 1 1	−2.495
⋮	⋮
1 0 . . . 0 0	0
⋮	⋮
0 0 . . . 0 0	+2.5

Fig. 8-36. Programmable offset instrumentation amplifier.

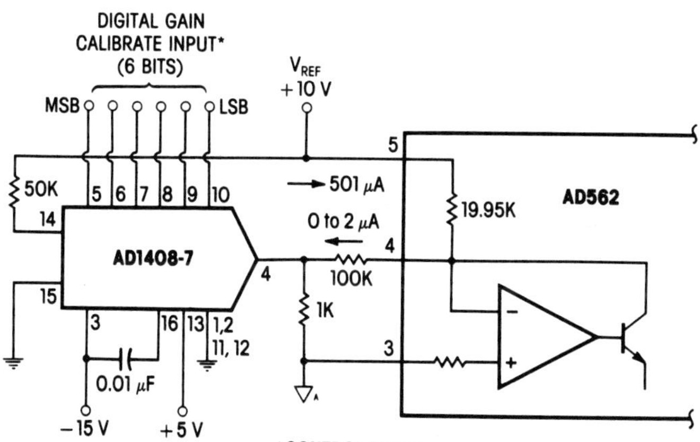

***CONTROL TABLE**

DIGITAL INPUT	GAIN
111111	+0.2%
⋮	⋮
100000	nominal
⋮	⋮
000000	−0.2%

Fig. 8-37. D/A gain calibration.

calibration. As shown here, the AD562 is programmed for a gain 0.2% high by connecting pin 5 to a +10-volt reference; and a variable 0 to −2 μA current from pin 4 will cause the gain to vary from 0.2%, through nominal, to −0.2%.

The 0 to −2 μA programmable current is supplied by a 1408 d/a converter, with 6-bit programming resolution. This will yield a basic gain change increment of 0.00625% per LSB, which is ¼ the basic resolution of a 12-bit a/d converter.

Note that the performance of the components used to achieve this function is not highly critical, as their contribution to overall gain error is reduced by the small percentage weighting.

Although this scheme is shown as being applied to an AD562 d/a converter, it is also generally applicable to other multiplying d/a converter types, and a/d converters which utilize them.

A circuit which in a system sense is closely allied to the gain calibration method just described is shown in Fig. 8-38. This circuit accomplishes the system function of gain calibration by programming the reference voltage, ±1% about the nominal level. It can be used to self-calibrate itself, as well as to gain calibrate a d/a or a/d converter system.

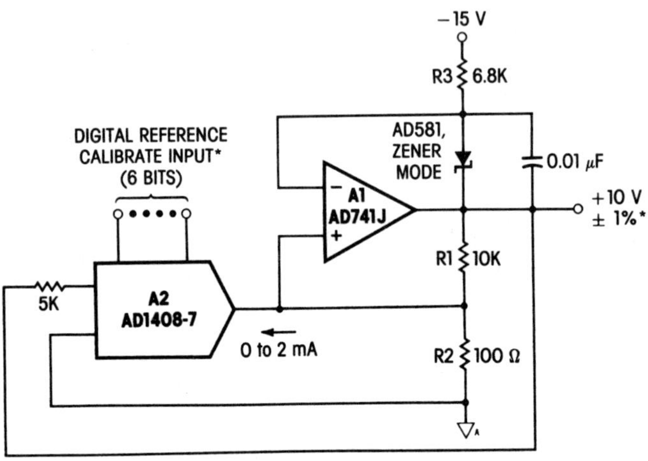

*CONTROL TABLE

DIGITAL INPUT	REFERENCE
111111	−1%
•	•
•	•
•	•
100000	nominal (+10 V)
•	•
•	•
•	•
000000	+1%

Fig. 8-38. Programmable reference source.

The circuit consists of two portions—A1 and the reference diode, and the d/a converter. A1 and R1-R2 form a bootstrapped buffer circuit for the reference, providing an output which is 1% high, as set by the R2/R1 + R2 ratio. A2 is connected to drive a 0 to −2 mA current in R2, which swings the output 0 to −0.2 volt from nominal. Since the output is initially 1% high (or 0.1 volt) this is effectively a ±1% range of adjustment.

With the 6-bit resolution shown, the reference can be incremented in 0.03%/LSB steps (3 mV/LSB for a 10-volt scale). Higher resolution trim can be accommodated by using the full 8-bit d/a resolution, if desired.

A circuit which can be used for signal path gain calibration is shown in Fig. 8-39. This circuit, as may be recognized, is basically a unity gain inverter, with gain variable ±1% about the nominal unity value.

The circuit is actually equivalent to an inverter with a variable input resistance. Gain as set up by (R2 ∥ R3)/R1 is 0.99. However, the A2 multiplying d/a converter forms a shunt input path, which lowers the effective value of R1. Resistors R5-R6 form a preattenuation network which reduces the input level applied to A2 to 2%. From the R5-R6 junction to A1, the d/a converter effectively appears as a 500K-to-infinity variable resistor, as programmed by the input code. With all

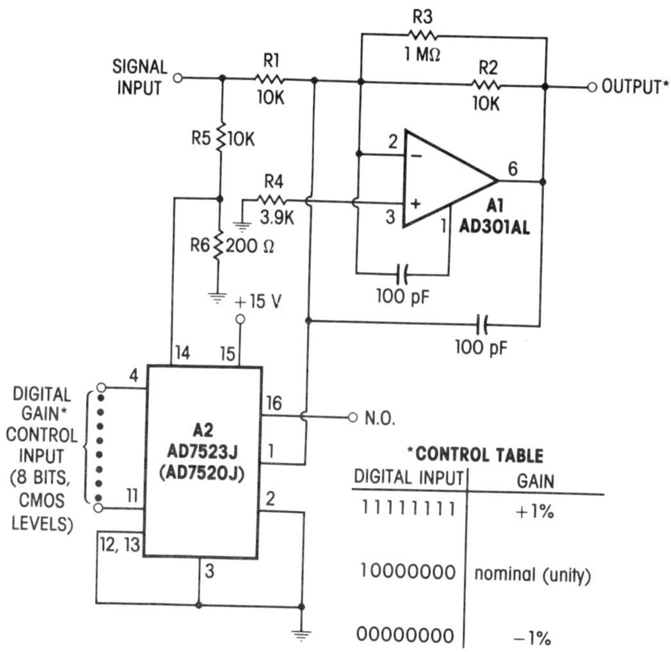

*CONTROL TABLE

DIGITAL INPUT	GAIN
11111111	+1%
10000000	nominal (unity)
00000000	−1%

Fig. 8-39. Signal path programmable gain.

1s the gain is +1%; with all 0s it is −1%. Gain is variable in 0.008%/LSB increments.

With the 8-bit resolution use as shown here the AD7523 is an efficient choice for the d/a converter. Note that with the bipolar reference range, this circuit can accommodate inputs of ±10 volts. The AD7523 requires a 15-volt CMOS drive; the AD7520 (pin-for-pin) can use either TTL or CMOS.

With many of these programmable circuits it is often highly desirable to be able to store the d/a data within a local register. These can be either one of the commonly available logic packages (see Chapter 6), or register input d/a converters can be used for the functions. Examples are the AD7522 and the AM6080.

V/F-F/V CONVERTER APPLICATIONS COLLECTION

This section of the chapter is devoted to systems applications of V/F devices, operated in both their V/F and F/V modes. Because of the overwhelming variety of uses for which these devices are suited, many of the applications discussions which follow will be necessarily brief. However, they will all as a minimum illustrate a concept, but may or may not have detailed design data for a specific IC. Inasmuch as the basic operation of all IC devices has been covered in some detail in Chapters 6 and 7, the conceptual block diagrams should be sufficient in most cases to get the reader "on board" as to how to implement the circuit with actual hardware.

Frequency Multiplier/Divider

One of the most unusual capabilities of a F/V-V/F combination is to multiply or divide a frequency (or range of frequencies), as shown in Fig. 8-40. The circuit consists of an F/V converter which converts frequency f1 into a voltage V1. V1 drives a V/F converter, reconverting to a new frequency f2. The "gain" of this system (K) can be equal to, greater than, or less than unity, allowing it to scale an input frequency either up or down.

Either the gain of the F/V or V/F converter can accomplish the scaling, and it should be noted that the scalor K will hold essentially constant over the dynamic range of the devices used. A pair of like devices such as 9400s, A-8400s or VFC-32s will work well in this application. Ripple must be minimized on the F/V output, otherwise undesirable fm will appear on f2.

Frequency Difference Monitor

A common application is the comparison of two frequencies, for monitoring purposes or phase detection. The circuit of Fig. 8-41 accomplishes this task in an interesting fashion.

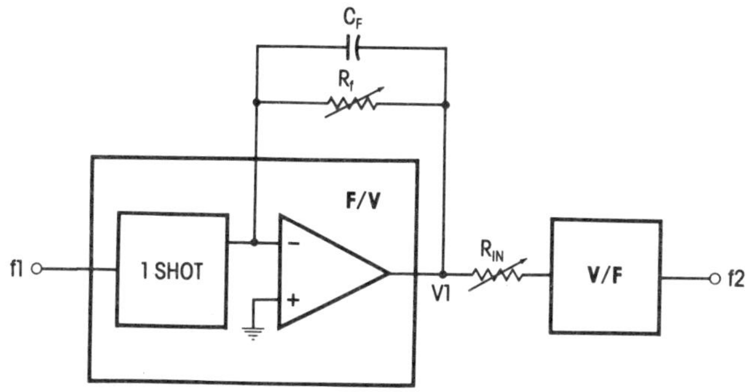

Notes: f2 = Kf1, Where K can be Greater or Less Than 1.
f2 tracks f1 with K a constant of proportionality.
Scale factor of either F/V or V/F can manipulate K as desired.

Fig. 8-40. A frequency multiplier/divider.

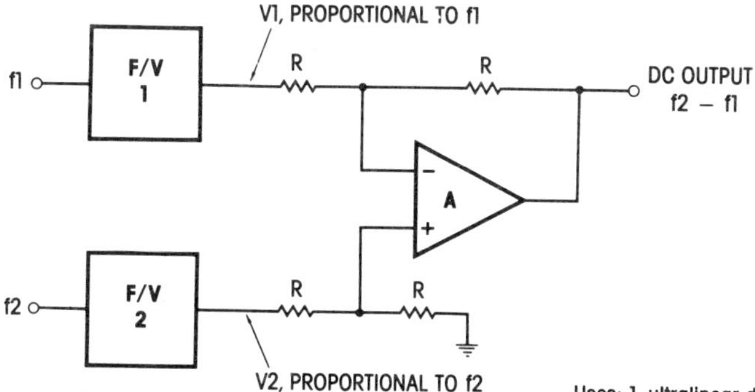

Notes: Circuit produces dc output voltage with sign and magnitude proportional to f2 − f1 relation.

Uses: 1. ultralinear detector
2. pll
3. phase detector

Fig. 8-41. Frequency difference monitor/phase detector.

Here two F/V converters convert frequencies f1 and f2 to voltages V1 and V2. The op amp subtracts the two voltages, producing an output proportional to the difference of f2 and f1. The circuit will respond linearly to both sign and magnitude of the relative frequencies. A few of the possible applications are listed in the figure. A notable feature of this circuit is the extremely good linearity over wide ranges—essentially as good as the F/V devices used to implement it.

Relative Frequency Detector

If detection of only the *relative* difference in a pair of frequencies is the prime requirement, the circuit of Fig. 8-42 will suffice. Here a comparator simply compares the voltages V1 and V2, which are proportional to the two frequencies. The comparator output indicates the relative frequency state, as shown.

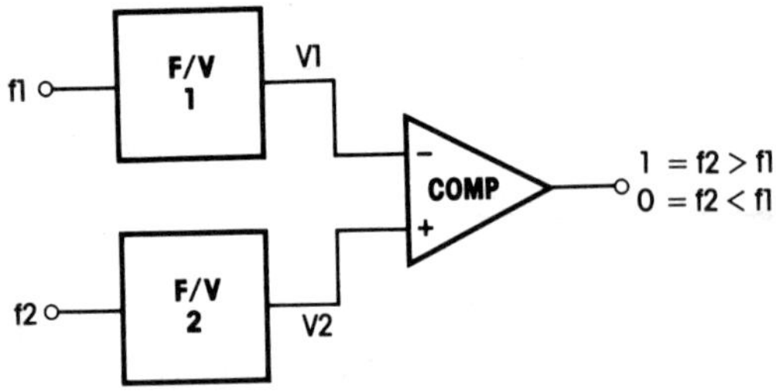

Fig. 8-42. Relative frequency detector.

Tone Detector

The circuit of Fig. 8-43 indicates the presence or absence of a tone within a specified frequency band. Here a single F/V produces an output voltage which is examined by a window comparator. If the tone were 3.5 kHz for example, a true indication would result; if it were lower than 3 or more than 4 kHz, a false indication results.

To improve noise immunity, particularly when a number of tones are being sensed, input prefiltering may be desirable.

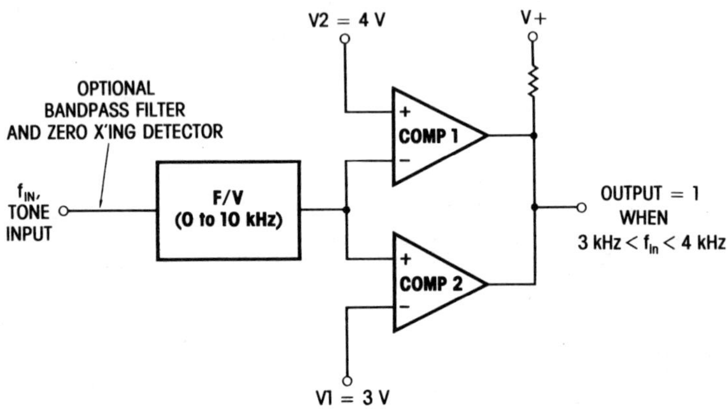

Fig. 8-43. Tone detector.

Speedometer

Since F/V converters produce an output which is directly proportional to input pulse rate, they serve as excellent speedometers or tachometers, as shown in Fig. 8-44.

Typical use of such a type is the sensing of the speed of rotational machinery. Common pickup transducers are variable reluctance magnetic pickups, and optical slot detectors. Magnetic transducers will usually require a comparator for amplification, but LED-phototransistor detectors will usually drive an F/V directly.

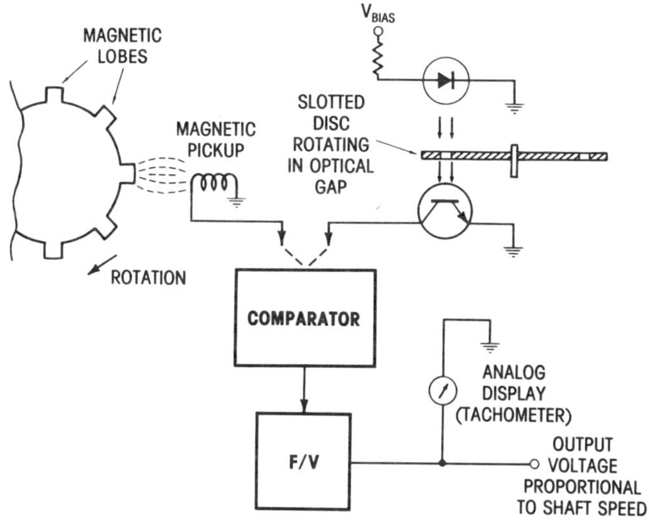

Fig 8-44. Speedometer or tachometer.

The analog output of the F/V converter can be displayed on an analog meter for a speedometer or tachometer function. Or, it may be used in voltage form as part of a control loop (next circuit).

Speed Control System

Fig. 8-45 shows how an F/V tachometer can be used in a closed loop servo system to control the speed of a rotating motor shaft.

Using a pickup as just described, the F/V converter produces an output voltage proportional to the shaft speed. This voltage is compared

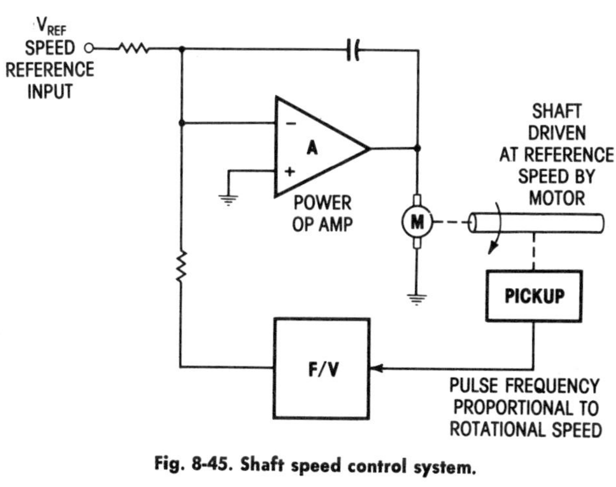

Fig. 8-45. Shaft speed control system.

to a speed reference input V_{REF}, and the amplified error drives the servo motor.

This is one example of a velocity servo system; another would be a flowmeter, where a F/V converter can control fluid rate via a control valve.

Precision FM Modulator

A very useful application for a V/F converter is as a precision fm modulator for use in data transmission, telemetry, and communications. The wide dynamic range, excellent linearity and easily transmittable output of a V/F converter are assets for this type of use.

In Fig. 8-46 is sketched a simple fm modulator, which will frequency modulate a fixed carrier frequency. The carrier is fixed by the reference current summed into the V/F converter; the modulation input adds to

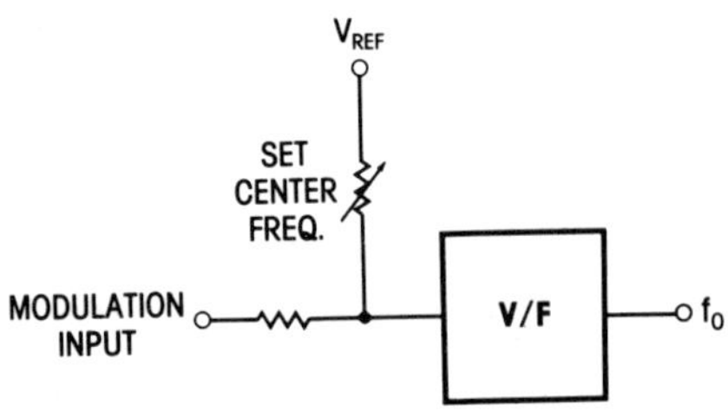

Fig. 8-46. Precision fm modulator.

or subtracts from this reference current, and linearly modulates the carrier. The center frequency stability will be as good as the V/F converter and reference current, and for small percentages of deviation, the modulation linearity can be extremely good, 0.01% or less.

Frequency-Stabilized FM Modulator

Fig. 8-47 shows a variation on this modulator, adding a phase detector and reference control loop to slave the output center frequency to a crystal reference frequency. The carrier stability then becomes that of the reference oscillator, but with the modulation linearity still that of the V/F converter. Note that the final carrier frequency is easily made programmable, using a divide by "n" counter.

Precision FM Demodulator

An F/V converter makes a natural precision, wide bandwidth, fm demodulator, which is a fitting complement to the V/F converter's capabilities as an fm modulator. With V/F and F/V devices so operated in concert, precision fm data systems can be economically built.

The fm demodulator shown in Fig. 8-48 is very straightforward, and is really a direct connection of a F/V converter. If demodulation of a

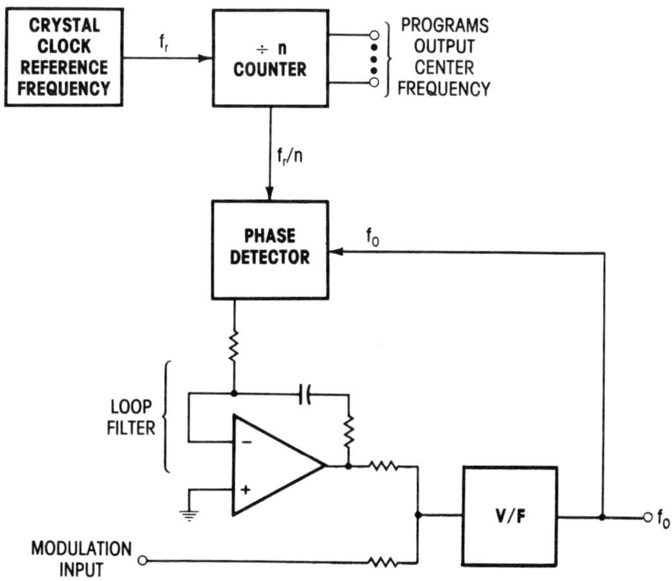

Fig. 8-47. Frequency-stabilized fm modulator.

fixed center frequency is desired, the F/V output scale can be offset with a resistor, as shown.

Data bandwidth will be limited by the response time of the F/V integrator network, which should be optimized to suit the particular task.

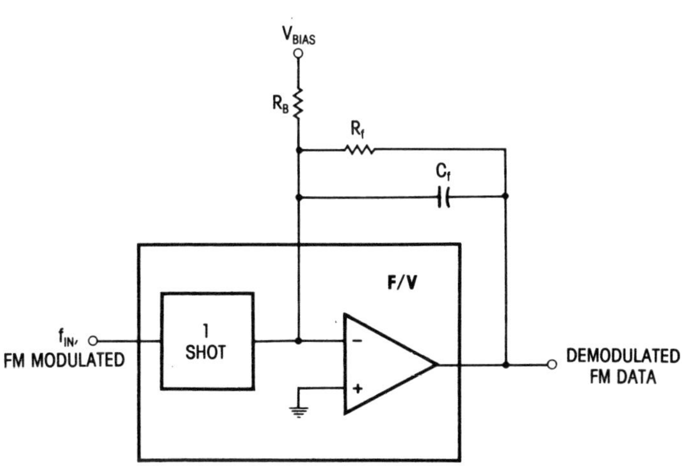

Fig. 8-48. Precision fm demodulator.

Phase-Locked Loop FM Demodulator

A more sophisticated form of fm demodulator is illustrated in Fig. 8-49, and is based on phase-locked loop (PLL) principles. It can be used either to demodulate a carrier to recover the modulation, or to reconstitute the carrier (with the modulation removed).

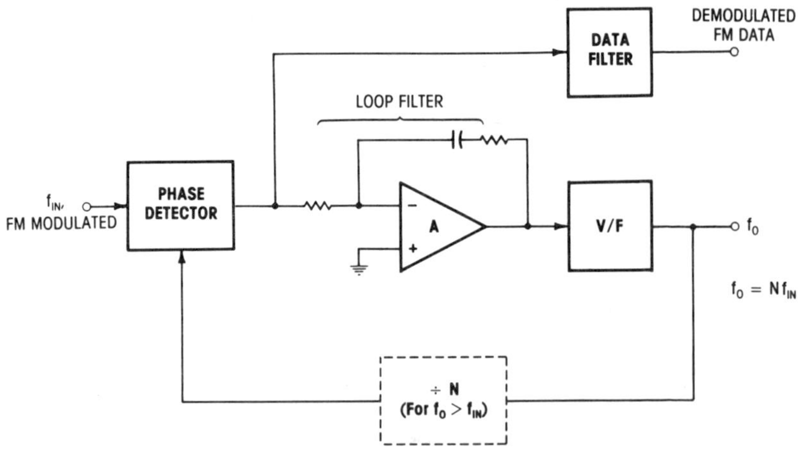

Fig. 8-49. Phase-locked loop fm demodulator.

The system uses a V/F converter, phase detector, and loop filter as the basic components of the PLL. Frequency scaling can be accomplished with respect to the carrier by appropriate division of the V/F converter output. In this system the PLL filter and data filters are separated, therefore each may be optimized separately.

FSK Modulator

The frequency-shift key (fsk) modulator is widely used in the transmission of digital data over narrow band systems. Basically, it encodes the 0 to 1 logic states as two separate tones. The tones need not be highly separated in percentage of frequency, which minimizes the bandwidth required.

A V/F system for fsk modulation is shown in Fig. 8-50. As may be noted it produces two different output frequencies for a 0 and 1 logic state by causing two distinct V/F input currents to flow for the respective logic levels. Although the details are not shown in this simplified system, the logic input should be buffered so that the typically inexact logic 0 to 1 levels do not cause frequency drift errors.

The output of the V/F converter should be a square wave if it is to be transmitted on phone lines or recorded. This suggests a 9400 or AD537 as the V/F device, or one of the other devices with a divide-by-two

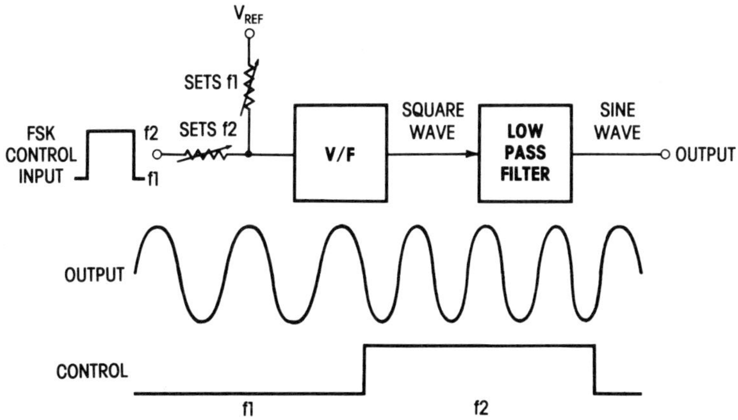

Fig. 8-50. FSK modulator.

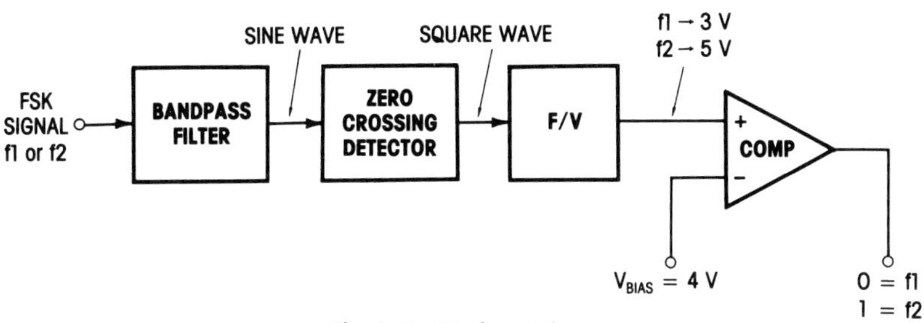

Fig. 8-51. FSK demodulator.

stage buffer. It should further be low-pass filtered to remove higher harmonics.

FSK Demodulator

A companion demodulator for an fsk system is shown in Fig. 8-51. It is composed of an F/V converter preceded by a bandpass filter which brackets the f1-f2 frequencies, a zero crossing detector to drive the F/V converter, and an output comparator.

The tones f1 and f2 will result in two corresponding voltages, for example, 3 volts and 5 volts. The comparator is biased between the two voltages; the output will then be a 1 for f2, a 0 for f1.

It should be noted that with relatively close f1-f2 tone pair spacing, a single data transmission channel can accommodate a number of digital bits, 1 bit per tone pair.

Cassette Data Storage Systems

It is desirable to be able to store analog and digital data by low cost means. One of these is the inexpensive cassette recorder, widely avail-

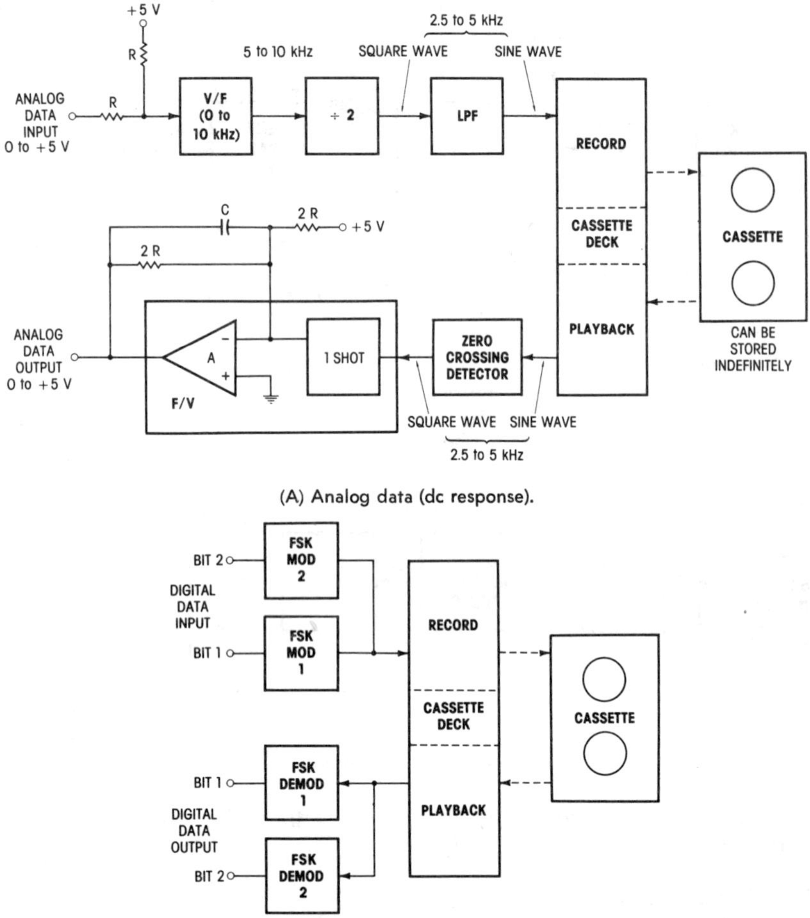

(A) Analog data (dc response).

(B) Digital data.

Fig. 8-52. Cassette data storage tape recording systems.

able and convenient to use. Two systems for implementing cassette data storage are shown in Fig. 8-52.

In Fig. 8-52A is shown an analog data system, which accepts dc analog inputs in the range of 0 to +5 volts. This data modulates an offset biased V/F converter which deviates from 5 to 10 kHz, producing a square wave, which is then low-pass filtered prior to recording. Note that this offset technique narrows the effective percentage of recording bandwidth considerably, and is compatible with the low-cost type medium.

On playback, a zero crossing detector restores the 2.5- to 5-kHz square waves which drive an F/V converter. This F/V is offset biased to shift the voltage corresponding to 2.5 kHz back to zero. It also has

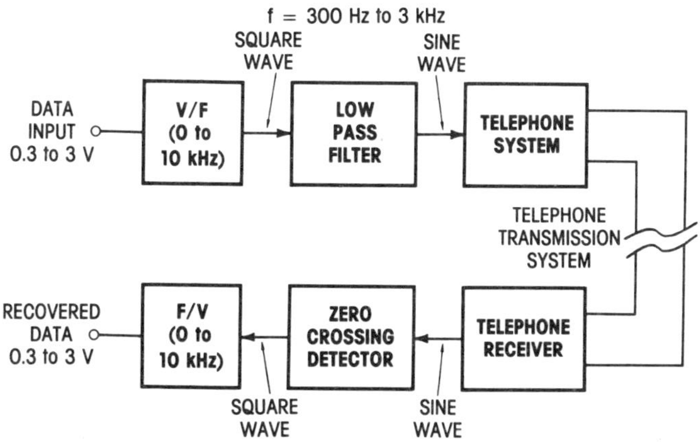

Fig. 8-53. Dc data transmission on telephone lines.

a twice normal gain to restore the 2.5-kHz span to that which originally corresponded to 5 kHz. The output is the original 0 to +5-volt analog range.

A digital data system is shown in Fig. 8-52B. It appears a great deal more simple, in block form, but the fsk modulator and demodulator blocks are composed of the elements described above. Note that if f1/f2 tone pairs are spaced appropriately over the recorder's bandwidth, several bits can be accommodated using this technique, using only one record/play channel. Both channels of a stereo format recorder will double the data density.

Telephone Data Transmission System

From the previous discussion, it can be seen that one of the more attractive virtues of V/F-F/V system is the ability to transmit dc and analog data in a frequency format. This allows it to be squeezed through not only tape recorders, but telephone systems as well, as shown in Fig. 8-53.

This system is a relatively simple 1/1 data repeater, with direct V/F and F/V conversion. The only addition to provide compatibility with the telephone system is the low-pass filtering of the V/F output square wave. The off-line sine wave at the other end of the system is shaped before reconversion by the F/V converter. Some filtering of line noise may be desirable before the comparator.

It should be noted that this system allows data transmission over distances of hundreds of miles, at relatively low cost.

Optoisolated Data Amplifier

A very useful means of transmitting the serial digital mode output data from a V/F converter is via an optoisolator, as shown in Fig.

8-54. In this circuit the "front end" is composed of a V/F converter powered by an isolated power supply. This portion of the system converts the input analog data into a proportional frequency, and drives the LED of the optoisolator. Virtually all available V/F converters can drive LEDs, with currents up to 15 mA (or more).

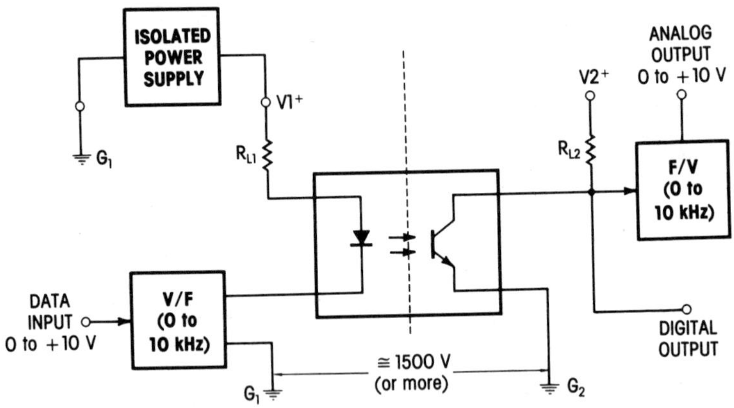

Fig. 8-54. Optoisolated data amplifier.

The optoisolator allows a ground potential difference of up to 1500 volts or more, dependent upon the ratings of the device in use. At its output is delivered the pulse stream produced by the V/F converter. The digital output can be used directly from the isolator, or it can be reconverted to analog by an F/V converter.

The optoisolated V/F-F/V pair constitute an isolation amplifier, a data processing function of great use in medical instrumentation. Note that V/F and F/V scaling can be manipulated as desired to yield virtually any overall input-output gain/scale factor. Note also that the scaling can conveniently be done with the data in the digital domain, by use of programmable counters between the isolator and F/V.

V/F Counter DVM (or A/D)

Fig. 8-55 illustrates the hardware necessary to implement an entire a/d function using a V/F converter, including all functions necessary. This circuit concept could be used either to implement a data acquisition type a/d converter with parallel digital outputs, or to build a lab dvm with a V/F "front end."

Basically the circuit is a combination of a V/F converter producing a pulse rate proportional to an input voltage, and counter to convert the pulse rate to parallel, stored data format. A display allows dvm use, or the latched data outputs can be used directly. An absolute value circuit

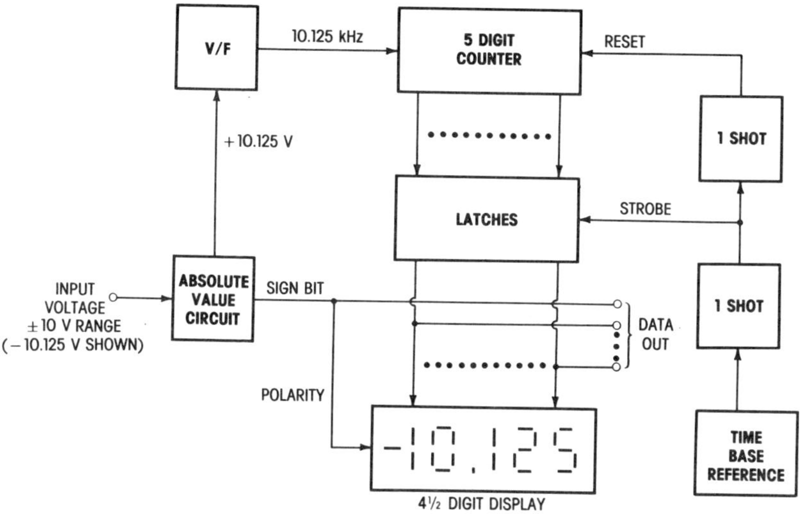

Fig. 8-55. V/F counter dvm.

converts the bipolar input to a unipolar range for the V/F converter, and sends the resulting sign bit to the display or data output line.

A 4½ digit format is indicated here, and is realistic assuming a 10-kHz range, high performance V/F converter such as the VFC-32 is used. The ½ digit overrange in this instance will not be "double scale" as in the case of most integrating a/d converters, but will be the rated overrange of the V/F converter in use—typically 110 to 120% of FS. An interesting feature of this dvm concept is that the V/F front end can be completely isolated (using optoisolators), which allows the equivalent of battery operation, and the resulting total elimination of ground loop errors, and/or safety problems.

Long Term Integration

A valuable feature of the V/F converter concept is the capability of integrating an analog signal by means of digital storage. For periods of long integration such as minutes/hours/or days, this is superior to the often used op-amp integrator, as it has no inherent drift rate (as the op amp does).

The idea is shown in Fig. 8-56, and consists basically of just a V/F converter and a *very* long digital counter. The counter will accumulate data for as long as the input gate is held open, at a rate determined by the V/F converter scale factor. The simple table shown illustrates the maximum (nonoverflow) integration periods with a 10-kHz input.

The practical advantage of this technique is that it vastly reduces sensitivity to noise, at the expense, of course, of the time interval necessary.

Note that when the gate is closed at the end of the measurement interval, the stored count can be held indefinitely.

FSK Modulator

As a representative and popular application example of V/F-F/V devices, a fsk modulator and demodulator pair are illustrated in Fig. 8-57.

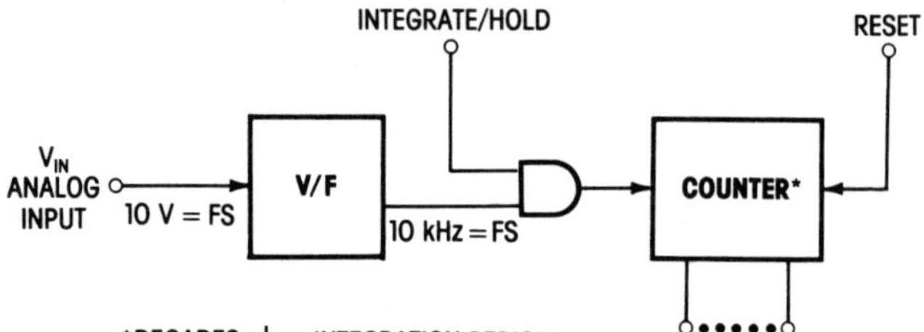

*DECADES	INTEGRATION PERIOD
5	1 SEC
6	10 SEC
7	100 SEC (1.67 MIN)
8	1,000 SEC (16.7 MIN)
9	10,000 SEC (2.7 HOURS)
10	100,000 SEC (1.15 DAYS)

Fig. 8-56. Long term integration.

In Fig. 8-57A is shown an fsk modulator using the AD537 V/F converter. As the AD537 has a square wave output and built in reference voltage, it is well suited to this application. It can be easily programmed to shift between two discrete frequencies by paralleling a fixed timing resistor during the input logic "1" state. Here the switching is accomplished by an inverted mode operated npn switching transistor (2N2222 or similar).

The ratio of the frequencies produced is determined by the ratio of the two effective timing resistances; here it is 2.2/1.2. Overall frequency calibration is set by the 2K pot. The values shown provide 1.2- and 2.2-kHz output frequencies, suitable for Bell Systems 202 data encoding. A post filter is required before transmission.

A complementary FSK demodulator is shown in Fig. 8-57B, using the 4151 or 331 in an F/V mode. If the input is an a square wave, 5-volt p-p form, it can be applied directly to the 4151 differentiator input. The 4151 functions here as a simple, single supply F/V converter, producing a dc output across R_B. For an f1 frequency of 1.2 kHz a dc voltage of about 4 volts is developed; for an f2 frequency of 2.2 kHz, 6 volts are developed. The comparator is biased to about 5 volts, and reshapes the demodulated output from the F/V converter into valid logic levels. Hysteresis is provided by the 330K feedback resistor, to eliminate chatter due to F/V ripple.

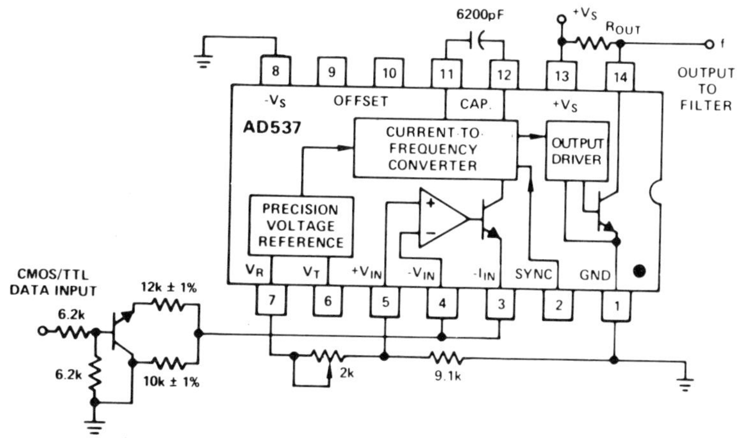

Courtesy Analog Devices, Inc.

(A) Modulator.

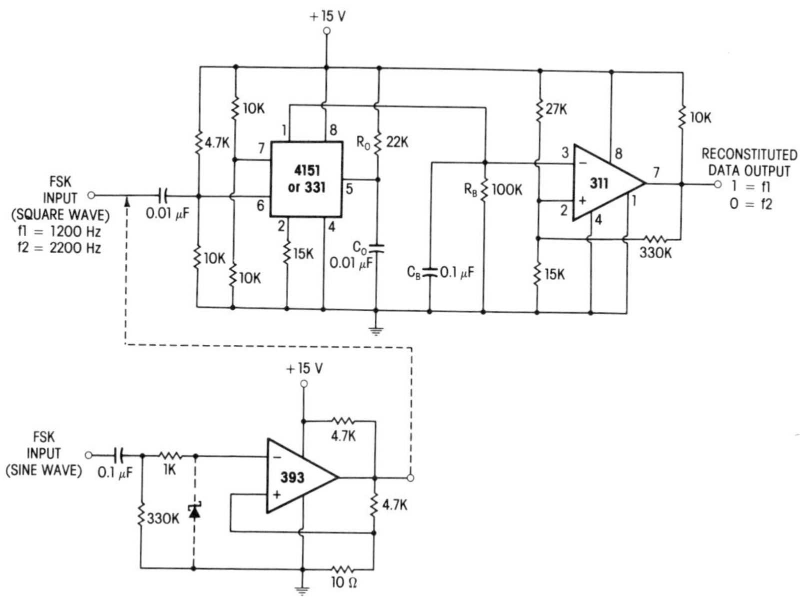

(B) Demodulator.

Fig. 8-57. FSK modulator and demodulator.

The operating speed of this demodulator is limited by the F/V converter response time, which is determined by R_B-C_B. C_B is reduced to as low as practical to speed response, without excessive ripple. For reliable operation, the highest f1-f2 keying rate should allow several R_B-C_B time constants. With the values shown this yields a maximum data rate of about 30-50 Hz.

If the data is to be demodulated from low-level sine waves, a zero crossing detector is required. The circuit shown in the inset can be used for this purpose and will operate from levels from 20 millivolts p-p to about 1 volt. Higher inputs are possible, if a series input resistor and Schottky clamp diode are added (shown dotted). Note that a dual comparator device can be used for both input and output functions (a 393 or ½ of a 339).

Two-Wire V/F Operation

Some V/F devices can be operated in a *two-wire* mode, that can be used to efficiently and simultaneously feed power to the device, and receive the output data. An example of this is shown in Fig. 8-58.

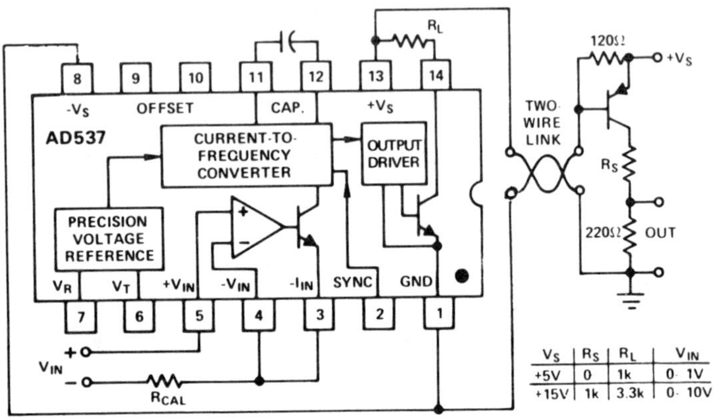

Vs	Rs	RL	VIN
+5V	0	1k	0-1V
+15V	1k	3.3k	0-10V

Courtesy Analog Devices, Inc.

Fig. 8-58. Two-wire V/F operation.

This circuit uses an AD537 V/F converter which is operated in a single supply mode. The output frequency is transmitted on the supply line as current modulation, and recovered by the pnp switching transistor (a 2N2907 or similar type). Note that it is compatible with various supply voltages and logic formats.

Transducer Excitation/Readout

A very useful system function which is available with internal reference type V/F converters is the ability to simultaneously provide excitation to a resistive transducer, and convert its output voltage (or current) to a proportional frequency. An example of this is shown using the AD537 in Fig. 8-59.

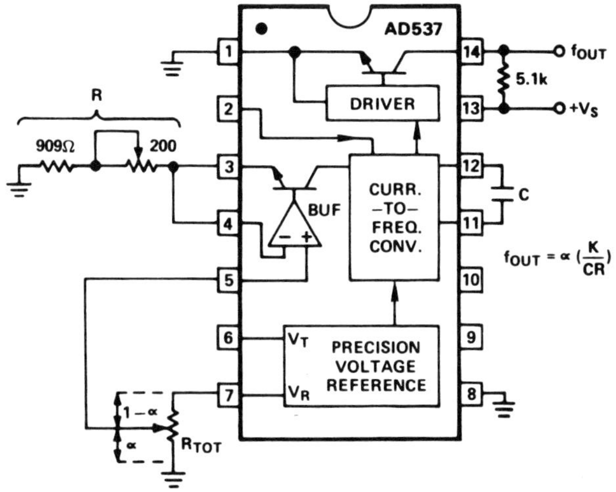

Courtesy Analog Devices, Inc.

Fig. 8-59. Transducer excitation/readout.

Here the 1-volt reference voltage of the AD537 is applied to a potentiometer from which it is desired to read the ratio, ∝. The arm of this potentiometric transducer is fed directly to the nonloading input of the AD537 input amplifier (pin 5).

The output frequency is of the form shown, where ∝ varies from 0 to unity, and K is a constant. K will vary from 0.65 to 0.98, for potentiometer values from 3K to 100K. Lower values of resistance can be driven (allowing K = 1), if a buffer amp is used with the reference output (see Fig. 8-11 for such an example).

Note that this circuit can efficiently drive a remote resistance deviation type sensor, with the AD537 also remotely located, when combined with the circuit of Fig. 8-58.

TEMPERATURE-SENSING APPLICATIONS

This section of the chapter discusses several classes of devices which are useful for the measurement of temperature. By and large these applications are in the nature of sensors or transducers which convert temperature in either °C or °F to a proportional voltage, which is then processed by a/d circuitry.

The three general classes of sensor devices which will be covered are thermocouples, thermistors and semiconductors. These types of course are not all sensors which are applicable to temperature measurement, but they happen to cover a broad range of requirements.

Thermocouples span the widest temperature range, allowing ranges up to over 1000°C. Fairly complex instrumentation is required to implement a themocouple (accurately), which adds to the cost.

Thermistors are reasonable in cost and cover several hundred degrees of range, with only modest instrumentation. Resistance temperature devices (RTDs) cover a wider range, up to about 500°C, with moderate instrumentation complexity. Semiconductor devices are varied in format, but generally require minimal instrumentation and are low in cost.

The following discussions of each class of devices illustrate the key application highlights of each. More detailed information can be found in the references.

Thermocouples

The thermocouple is one of the simplest temperature sensors in principle, as is illustrated in Fig. 8-60. In this much simplified circuit a type-J thermocouple (*iron-constantan*) is used to monitor the temperature of a source in the vicinity of 600°C.

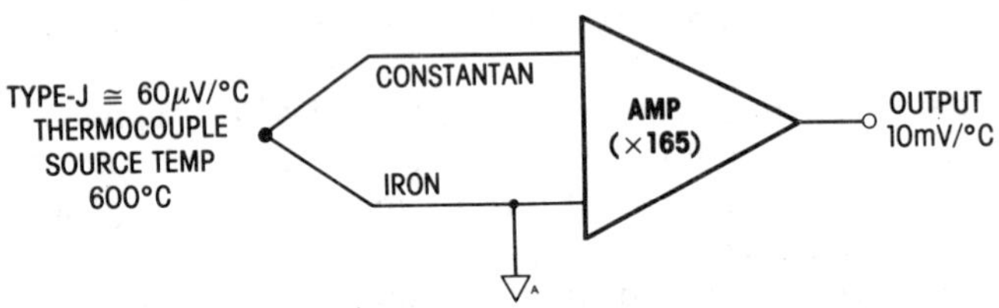

Fig. 8-60. Simple thermocouple instrumentation.

Since a thermocouple consists of only a pair of dissimilar metals which produce a predictable voltage difference over a temperature range, it follows that optimum selection matches thermocouple output and range to that of the requirements. The 600°C source here is compatible with the range of the type-J thermocouple.

An obvious disadvantage of thermocouple is its low output, which requires high amplification to reach useful voltages. The relatively low voltage coefficients also place stringent requirements on the performance of the amplifier to be used in terms of drift and general stability—a drift of 60 μV/°C in the amplifier, for example, will be nondistinguishable from that of the source.

A second drawback to thermocouples is due to their basic nature—dissimilar metals produce voltages which are a function of temperature. The mere connection of copper lead wires to the two metal conductors of a thermocouple pair produces two *additional* thermocouples, as seen in

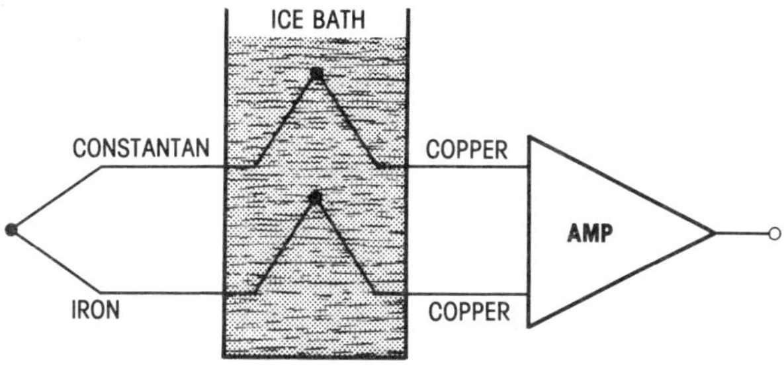

Fig. 8-61. Ice bath cold junction compensation

Fig. 8-61. To minimize the potential errors which can arise from the lead wire connection, these two parasitic thermocouples can be placed in an ice bath to cancel the error. This general process is termed "cold junction compensation," as it provides error compensation and serves as a reference for the sensor thermocouple junction.

Since ice baths are obviously unwieldy and inconvenient to general use, alternate means of cold junction compensation are often sought. A general form of this is the active cold junction compensation of Fig. 8-62. This implementation replaces the constant temperature ice bath

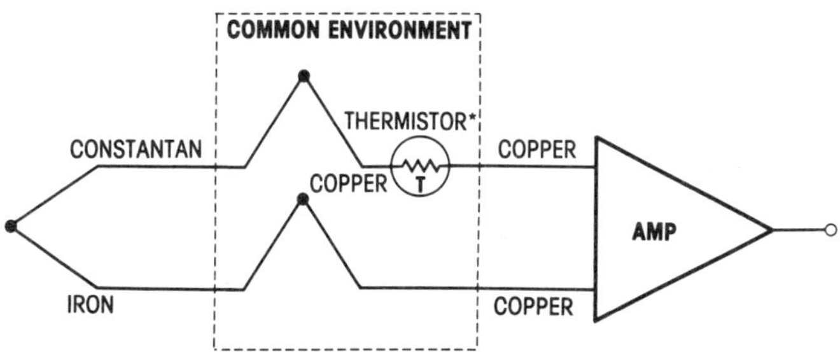

*Thermistor, or other temperature dependent, reference quality, local transducer

Fig 8-62. Active cold junction compensation.

with a temperature-dependent voltage which tracks the temperature seen by the parasitic thermocouples. The temperature dependent voltage bucks out the parasitic thermocouple voltage errors, providing a flat (nontemperature dependent) output with minimum error due to the cold junctions. The thermistor or other type of local temperature sensor must, of course, be held at the same temperature as the two parasitic junctions, so they are in a common environment.

A circuit example of cold junction compensation combined with a preamplifier is illustrated in Fig. 8-63. In this example a type-K thermocouple is used, which has a nominal 40 μV/°C output, which is at-

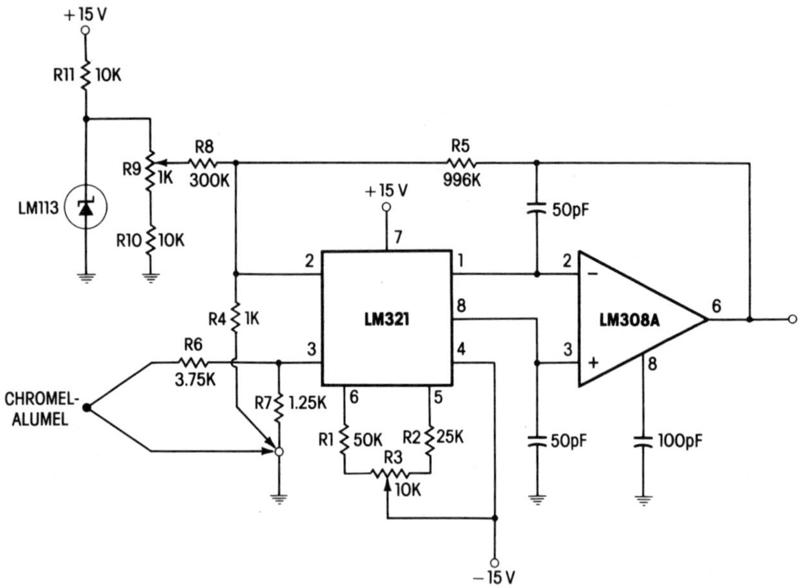

Courtesy National Semiconductor Corp.

Fig. 8-63. Circuit used for active cold junction compensation.

tenuated by a factor of 4 by R6-R7 to a 10 μV/°C scale factor at pin 3 of the LM321.

The LM321 and LM308A form a precision preamp with a predictable temperature drift characteristic, which is adjustable by R3. Circuit constants are chosen such that the preamp's drift will exactly compensate that of the reference junction when the two are at the same temperature.

Overall the circuit provides not only cold junction compensation but offset adjustment and gain scaling to a 10 mV/°C output as well. For optimum performance R4-R7 should be wirewound types, and the entire circuit enclosed in a small box to minimize gradients. The thermocouple leads should be terminated close to the LM321.

The circuit is calibrated by shorting the reference diode and R6 to ground, and adjusting R3 for a 10 mV/°K scaled room temperature output (3.00V @ 27 °C); then with the short removed from the LM113, R9 is adjusted for this voltage. With the thermocouple short then removed the circuit is ready for operation. Other thermocouple types can be used with this circuit if they are scaled for a 10 μV/°C output, by appropriate R6-R7 values.

A final application point which should be noted concerning thermocouples is that their output voltage sensitivities are *not* constant over their operating ranges; i.e., they are nonlinear. Linearity (and accuracy) can be optimized by confining operation to small portions of a total range, where linearity is within allowable errors (by choice of types).

Thermistor Temperature Sensors

A very popular class of device used for temperature-sensing applications is the thermistor. Thermistors are available in a wide variety of sizes, shapes, resistance ranges, and accuracies. General purpose types are quite nonlinear, and not well suited as linear temperature transducers, being more suited for control applications or set point maintenance.

A more sophisticated type of thermistor sensor is the *linearized network,* which is actually a composite assembly of thermistors and standard resistors designed for linear response for a specific temperature range. This class of thermistor can be connected for either a linear voltage versus temperature output (voltage mode), or for linear resistance versus temperature output (resistance mode). These two types of use are illustrated in Fig. 8-64.

Here are illustrated Yellow Springs Instrument Thermilinear®[1] linearized thermistor networks. These devices consist of a thermistor composite such as the 44018, and a fixed precision resistor composite such as the 44301. The 44018 and 44301 when combined as shown form the linearized thermistor network; together they are specified as a 44201 Thermilinear® component.

A bias voltage E_{IN} is applied across this network, and output can be taken from the center terminal to either outside terminal as common. Voltage output E_{o1} has a positive slope versus temperature, while E_{o2} has a negative slope. The voltage output equations describe the sensitivity to temperature changes.

For a specific temperature span and accuracy, a given network is selected to be compatible with the design requirements. For the 44201 network shown here, the applicable temperature range is 0 to 100°C with a nonlinearity of 0.216% or less. Voltage output sensitivity per degree is adjusted by choice of the bias voltage; here a voltage of 1.86975 volts yields a 10 mV/°C output. Since this is a high impedance network, output loading should be minimized.

Fig. 8-64B illustrates how a 44201 network might in practice be operated for a voltage mode output, using output E_{o1}. Op amp A and the reference diode produce a −1.86975-volt bias, which drives the network and bias string R1-R3. The output of R2 is set to a voltage which corresponds to 0°C, or −0.2523 volt in this case. The output is taken differentially, with polarity as noted, and should feed a high impedance load such as a dvm input. R2 is adjusted for 0°C = 0 volt, and R7 is adjusted for 100°C = 1 volt to calibrate. All fixed resistors should be pre-

---

[1] Registered trademark, Yellow Springs Instruments.

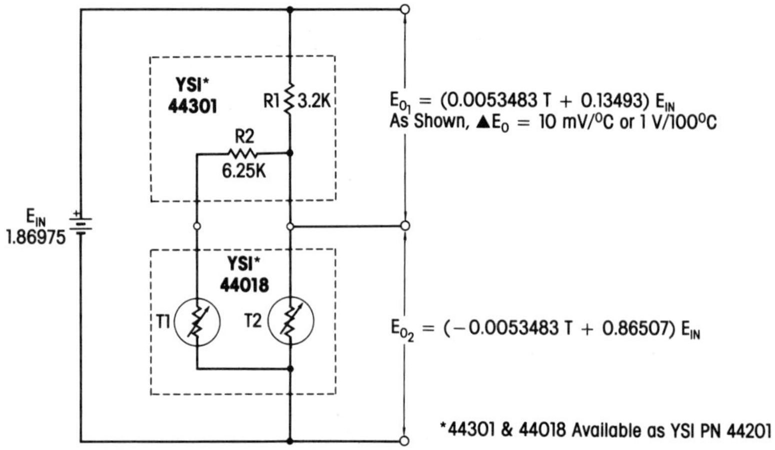

Notes: 1. R1, R2 = 0.1% Tolerance Metal Film, 30 ppm/°C (MAX) TC.
2. E_{O_1} = Positive Slope vs T
E_{O_2} = Negative Slope vs T
3. E_{IN} (MAX) = 2 V
4. R_L (MIN) = 3 Megohms

(A) Basic circuit.

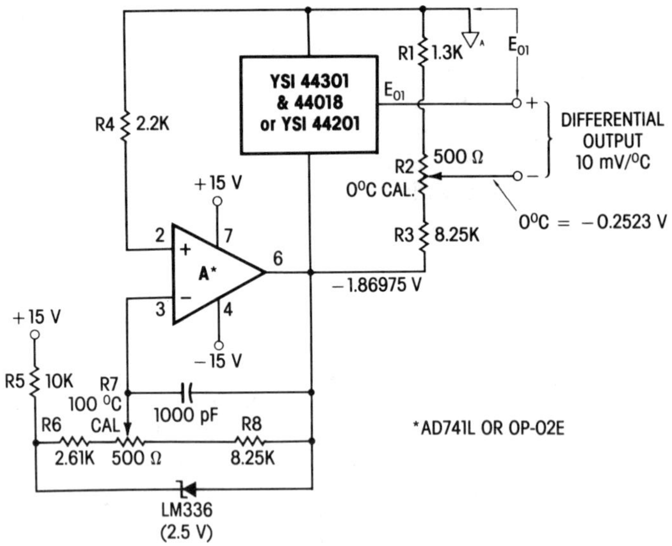

(B) Practical example.

Courtesy National Semiconductor Corp.

Fig. 8-64. Linearized thermistor network sensors in a voltage output mode.

cision types (except for R5), and trimmers should be multiturn cermet units.

To operate the same linearized thermistor network in its resistance mode, it is recommended as shown in Fig. 8-65A. The same resistor pair

(A) Basic circuit.

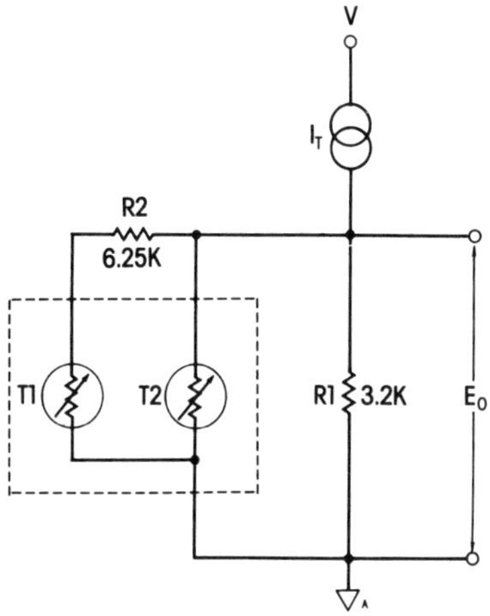

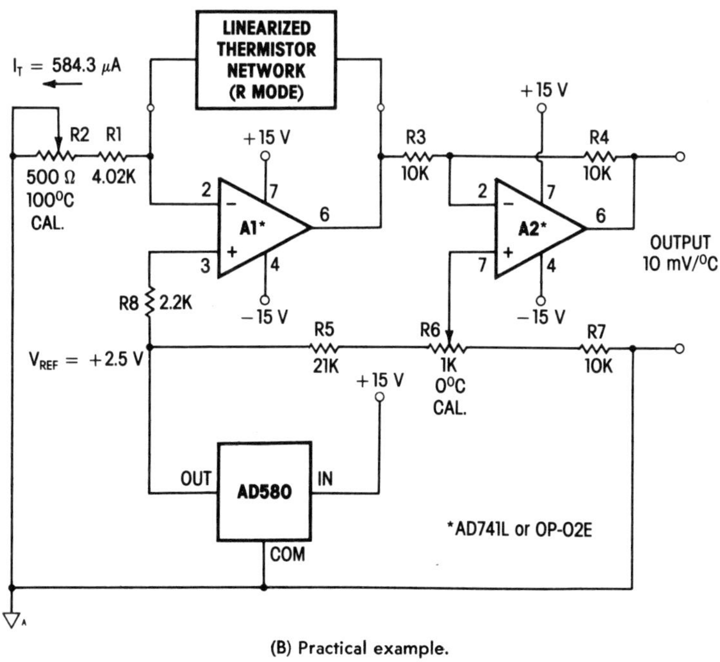

(B) Practical example.

Fig. 8-65. Linearized thermistor network sensors in a resistance mode.

values are used for R1-R2, but bias is applied in the form of a current source, I_T, and output read as the voltage across R1.

A practical circuit to employ the network in the resistance mode is shown in Fig. 8-65B. Here the two terminal connected linearized thermistor network is connected in the feedback path of A1, to which is applied a 2.5-volt reference. The current I_T which flows in the network is set up by R1-R2. This current is chosen to provide the desired output scale factor in volts/degree, after considering the resistance per degree temperature coefficient of the network used. For example, the 44201 network used here has a 17.115 ohms/°C coefficient, and a 10 mV/°C voltage coefficient is desired. The current I_T is then set as 0.01V/17.115 ohms, or 584.3 μA.

The output of the A1 amplifier stage has the required scale factor of 10 mV/°C; however, the slope is negative and the scale is also offset. The differential stage A2 level shifts the signal to remove the zero degree offset, and inverts the range, giving a positive going 10 mV/°C. Output calibration is set by R6 for 0°C = 0 volt, and R2 for 100°C = 1 volt. Since the output is buffered by A2, it can feed a wider variety of loads, compared to the simple bridge circuit.

Resistance-Temperature Devices

The resistance-temperature device, or RTD as it is more often known, is a thermal sensor which operates on the basic temperature coefficient of a wire. Most often this wire is platinum, due to its excellent accuracy and repeatability. The platinum resistance temperature device is in fact generally considered to be a standard for accuracy in temperature measurement.

Typically RTDs have 25°C resistance of 100-500 ohms, and positive temperature coefficients of about 0.4%/°C. They are applicable over ranges of about −200 to 600°C.

A circuit configuration to exploit the properties of an RTD is shown in Fig. 8-66. In Fig. 8-66A is shown the basic concept, which as can be noted, is an amplified, buffered form of bridge. The RTD composes one arm of the bridge and is excited with a constant current I_{REF}, supplied by the voltage reference, V_{REF}.

Amplifier A1 converts the RTD resistance changes to voltage changes, and drives the right side of the bridge. R_{OFFSET} and R_{IN} form the right arm of the bridge, which is at null when there is no current flow into the sum point of A2. This condition will occur when $R_{REF}/R_{TD} = R_{OFFSET}/R_{IN}$, and R_{OFFSET} can be selected to balance the output for zero volts at a specific R_{TD} resistance (temperature).

The output slope (voltage change/°C) is set by the current I_{REF}, the R_{TD} sensitivity (ΔR) and R_f/R_{IN}, as shown. All of the bridge components must of course be stable, temperature-independent types, and amplifiers A1 and A2 should be low-drift units.

(A) Basic circuit.

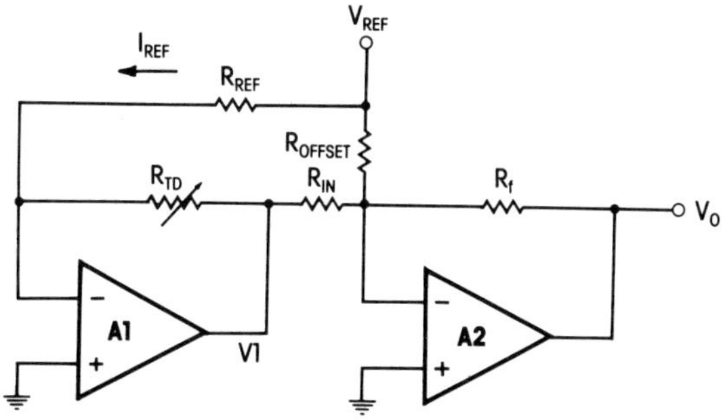

$$\text{Output Balance When } \frac{R_{REF}}{R_{TD}} = \frac{R_{OFFSET}}{R_{IN}}$$

$$\text{Output Slope} = I_{REF}\, \blacktriangle R\, \frac{R_f}{R_{IN}}$$

Where, $\blacktriangle R$ is R_{TD} Sensitivity

$$I_{REF} = \frac{V_{REF}}{R_{REF}}$$

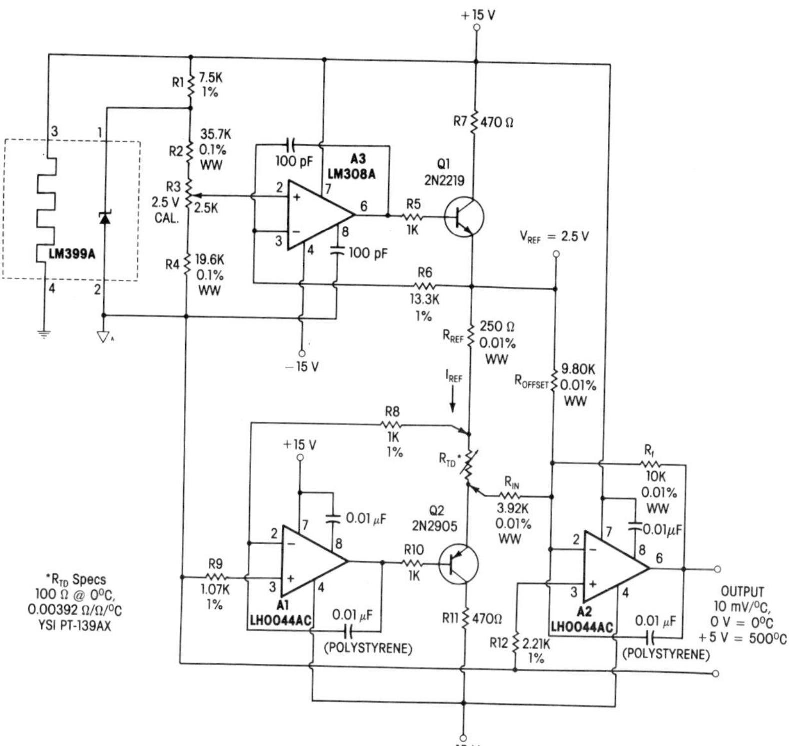

(B) Practical example.

Fig. 8-66. RTD temperature sensors.

Fig. 8-66B shows an example of how this circuit principle can be carried out, to a high degree of precision. This circuit is designed for a 100 ohm @ 25°C, 0.392%/°C RTD device, such as the Yellow Springs Instrument PT-139AX. It measures temperature over a span of 0 to 500°C, with accuracy limited essentially by the RTD element actually used.

A reference voltage of 2.5000 volts is developed by the LM399A diode and the A3-Q1 control loop, and is applied to the top of the bridge, R_{REF} and R_{OFFSET}. With the value of R_{REF} shown, this sets up a reference current of 10 mA in the RTD which keeps its self-heating errors to 0.2°C or less, and provides a scaling of 3.92 mV/°C at the RTD output. R_f and R_{IN} scale the A2 output to 10 mV/°C, and R_{OFFSET} causes the output to be 0V for 0°C.

Due to the highly precise resistances and ICs specified, the only calibration required is for scale factor, by adjustment of R3.

A four-wire lead arrangement is employed on the RTD for maximum accuracy and minimal sensitivity to wire length. The leads to the RTD should preferably be shielded to minimize noise pickup.

Semiconductor Temperature Sensors

The concept of a temperature sensor based on the properties of a transistor is not a new one, but in recent years it has become more refined, with several ICs currently available which are specifically dedicated as temperature transducers.

One of the simplest and least expensive methods which can be employed to sense temperature is the silicon transitor base-emitter junction voltage. For a fixed collector current, V_{BE} will vary at about −2 mV/°C, which is a relatively high sensitivity in terms of percent (0.3%/°C). A not insignificant virtue of this type of sensor is its rock bottom cost, with general purpose transistors available for 10-20 cents, and little additional hardware being required for implementation. Unfortunately, drawbacks of this technique are that V_{BE} is highly process dependent, and as a result direct interchangeability of sensors is not possible. Still, it is attractive for modest accuracy applications if screening and calibration of transistors is not considered disadvantageous.

Implementation of a transistor V_{BE} temperature sensor can be carried out as diagrammed in Fig. 8-67.

In Fig. 8-67A is shown a conceptual diagram of how the technique would be accomplished, on an optimum basis. The diode-connected transistor is fed a fixed reference current, and the resulting V_{BE} is read out and amplified by an op amp, configured to provide the desired gain and scale factor.

A simple example of how such a sensor might be configured in practice is shown in Fig. 8-67B. Here the sensor Q1 is used to amplify its own V_{BE}, thus serving as both the transducer *and* op amp. The collector

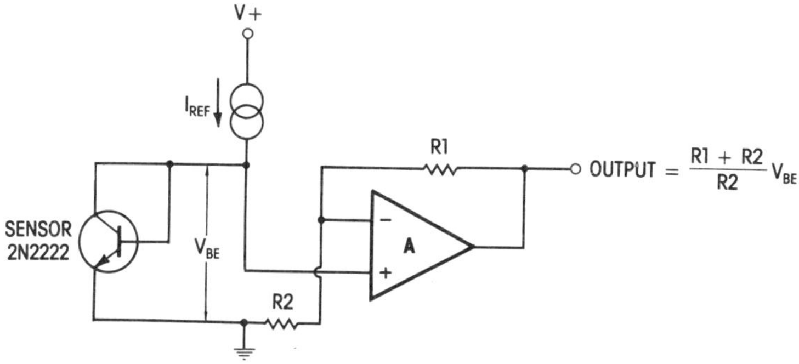

(A) Conceptual operation.

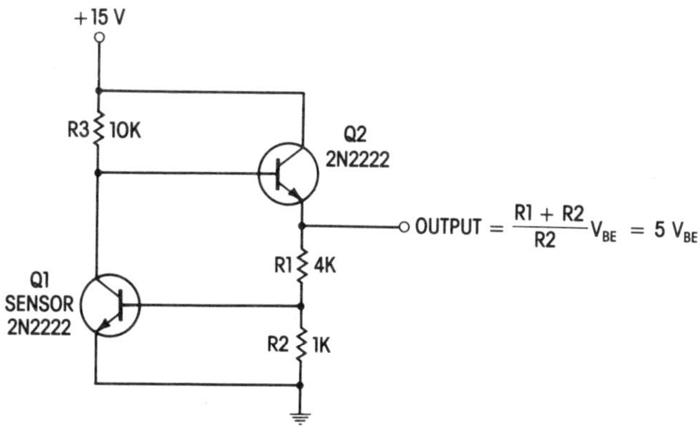

(B) Simple practical example.

Fig. 8-67. Transistor V_{BE} temperature sensors.

current of Q1 is "semiregulated" by returning R3 to a relatively high voltage. Divider resistors R1-R2 scale up the V_{BE} of Q1, and Q2 provides a low impedance output.

Although performance of this circuit is modest, and accuracy is only about 10% considering a number of error sources, it may be all that is justified for simple applications. It can be calibrated for low error at a specific temperature, and will be repeatable, for a given sensing device.

IC Temperature Transducers

More accurate and predictable performance using semiconductor temperature sensors may be obtained using monolithic IC temperature transducers. A number of ICs dedicated to this task are now available, and they offer useful combinations of cost, performance, and flexibility. A number of these types are discussed in this section.

The first IC temperature transducer generally available was the National Semiconductor LX5600 and LX5700. These devices operate over a −55 to +125°C temperature range, within TO-5 and TO-46 packages. They produce an output voltage which is calibrated directly as 10 mV/°K. The LX5600/LX5700 have ±4.4% of span errors, while their respecive "A" suffix versions have ±2.2% errors. A more recently introduced device is the LM3911, which operates similarly over a −25 to +85°C range.

Two basic circuits which illustrate the use of the LX5600/LX5700 (or LM3911) are shown in Fig. 8-68. Fig. 8-68A shows the device as operated from a negative supply, producing a 10 mV/°K output. An internal 6.8-volt zener diode stabilizes the LX5600 supply voltage (pins 3-4) and zener current is set up by the choice of series resistor R_s, selected as shown. The LX5600 and LX5700, being contained in small metal packages are well suited to point contact probes. A positive supply voltage circuit using the devices is shown in Fig. 8-68B.

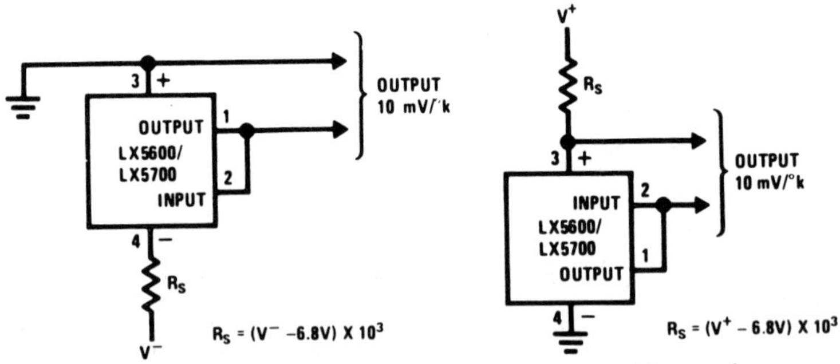

(A) Negative supply.

(B) Positive supply.

Courtesy National Samiconductor Corp.

Fig. 8-68. Basic LX5600/LX5700 Kelvin thermometers.

Another device which produces a voltage output proportional to temperature is the Precision Monolithics REF-02 voltage reference IC. The voltage proportional to absolute temperature is in fact a natural byproduct of the bandgap type voltage circuit, thus all ICs using bandgap voltage references are also temperature sensors (although all are not necessarily specified as such).

A general circuit hookup using the REF-02 is illustrated in Fig. 8-69. This circuit can be used to produce either centigrade or Fahrenheit scaled thermometers by appropriate component selection. The REF-02 provides signals for both temperature and offset, therefore the circuit is efficient in this regard. It can be remotely located from the op-amp buffers if desired, using shielded 4-wire cable.

In calibrating the circuit, R_p adjusts the offset for the scale; centigrade or Fahrenheit. R_{b1} then trims the scale factor for the desired slope.

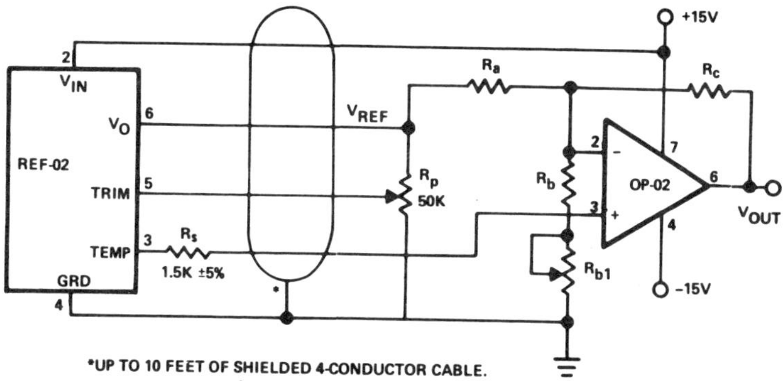

(A) Schematic.

TCVOUT SLOPE (S)	10mV/°C	100mV/°C	10mV/°F
TEMPERATURE RANGE	−55° to +125°C	−55° to +125°C	−67°F to +257°F
OUTPUT VOLTAGE RANGE	−0.55V to +1.25V	−5.5V to +12.5V*	−0.67V to +2.57V
ZERO SCALE	0V @ 0°C	0V @ 0°C	0V @ 0°F
R_a (±1% resistor)	9.09K	15K	7.5K
R_{b1} (±1% resistor)	1.5K	1.82K	1.21K
R_{bp} (Potentiometer)	200Ω	500Ω	200Ω
R_c (±1% resistor)	5.11K	84.5K	8.25K

*For 125°C operation, the op amp output must be able to swing to +12.5V; increase V_{IN} to +18V from +15V if this is a problem.

(B) Value/performance table.

Courtesy Precision Monolithics, Inc.

Fig. 8-69. REF-02 thermometer circuit.

Overall accuracy is dependent upon the grade of REF-02 used, and ranges from ±0.9% (REF-02C to ±0.4% for the REF-02E, over a 0-70°C range. Further details on this circuit can be found in PMI Applications Note AN-18, "Thermometer Applications of the REF-02."

More recently, IC temperature transducers have been introduced that produce output *currents* proportional to absolute temperature. One of these is the National Semiconductor LM234/LM334 series devices, the other is the Analog Devices AD590 series. The key specifications and operational features of these devices are discussed in representative applications below.

One of the virtues of a current operating-mode transducer is the fact that currents can more easily be transmitted in analog form without noise pickup. Another is the fact that current-mode operation eliminates the need for regulated supplies. Both of these points are useful advantages for a temperature transducer, particularly when it is to be remotely operated.

A simple circuit which uses the LM234 (LM334) as a remote sensing centigrade thermometer is shown in Fig. 8-70. This circuit produces an output of 10 mV/°C, and is suitable for use with a 1-volt scale dvm (or dpm) as a digital thermometer.

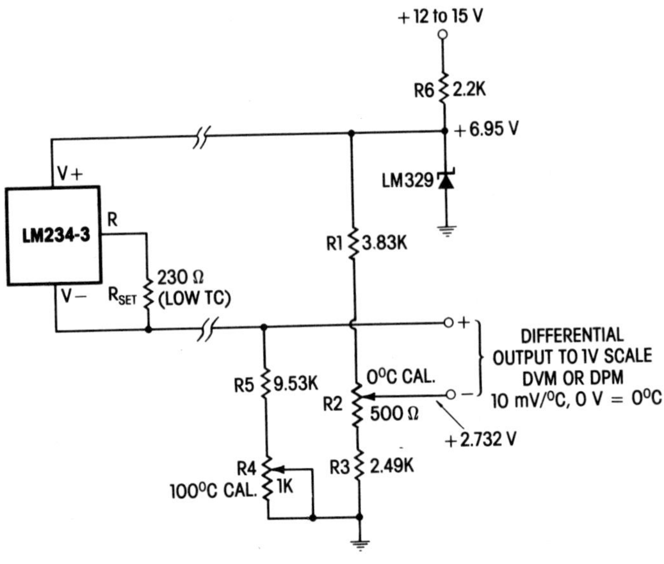

Fig. 8-70. LM234(334) centigrade-reading thermometer with remote sensor.

The LM234 (334) is a three-terminal current source, and is programmed for the desired sensitivity by a resistor (R_{SET}) connected between the R and V− pins. With R_{SET} equal to 230 ohms (as shown), the sensitivity is 1 μA/°K. This coefficient will be within ±1% (or ±3°C) for the LM234-3, and within ±2% (or ±6°C) for the LM234-6. The devices operate over a 1-volt to 30-volt supply voltage range.

In the circuit as shown, R_{SET} is a low TC resistor (metal film or wirewound) located at the remote sensor location with the LM234-3. A bias voltage is applied to the device's V+ terminal; in this circuit it is +6.95 volts from an LM329 reference. This voltage need not be tightly regulated to attain rated accuracy. The reference diode is used in this circuit primarily to provide a fixed offset voltage for the centigrade scaling.

Output current from the LM234-3 is converted to a voltage by the load resistor R4-R5, and is 10K for a 10 mV/°C scale. R4 trims out resistor and sensor calibration errors and is adjusted for a 1-volt differential output at 100°C. R2 trims the (−) output line for a bias level of +2.732 volts at 0°C, or a differential output of zero. The output is read on a differential input dvm or dpm, and spans a temperature range of −25° to +100°C. An LM134-3 sensor can be used for a range of −55 to +125°C, if desired.

Fig. 8-71 shows how a current mode temperature transduced can be used to implement a buffered, single ended output, centigrade thermometer. This circuit uses the Analog Devices AD590 two-terminal temperature transducer.

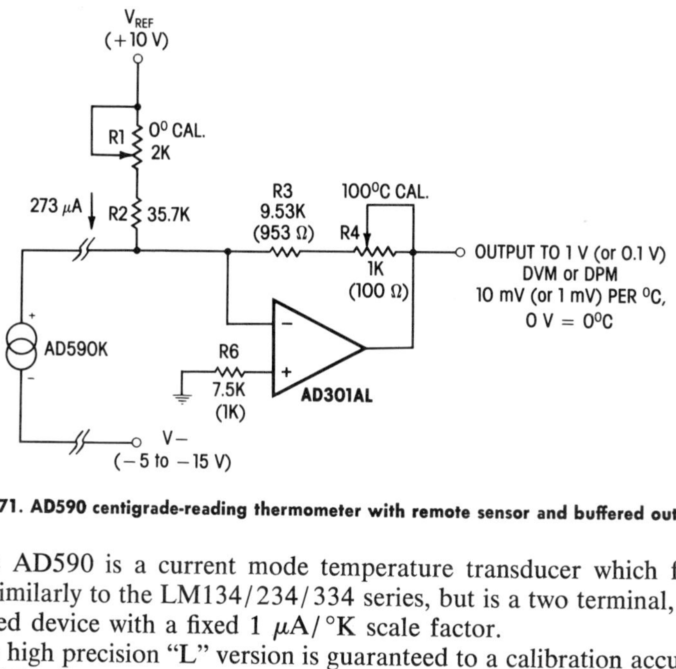

Fig. 8-71. AD590 centigrade-reading thermometer with remote sensor and buffered output.

The AD590 is a current mode temperature transducer which functions similarly to the LM134/234/334 series, but is a two terminal, pretrimmed device with a fixed 1 μA/°K scale factor.

The high precision "L" version is guaranteed to a calibration accuracy of ±1°C at +25°C and ±1°C linearity over the −55°C to +150°C range; the "K" version is ±2°C accurate with ±1°C linearity; and the "J" version is ±5°C accurate and a ±2°C linearity. The devices operate over a 4 to 30-volt supply range.

In this circuit the AD590K is remotely located at the temperature sensing point, and biased from a −5 to −15-volt potential (which need not be tightly regulated). The 1μA/°K output current is fed to a current-to-voltage converter op amp, which provides a buffered voltage output. A fixed 273.2-μA current provides the offset required for centigrade scaling, and is derived from a stable +10-volt source.

Scaling for this circuit can be either 10 mV/°C or 1 mV/°C, dependent upon the values of the feedback resistors R3-R4. R4 provides a trim for 100°C calibration, and R1 a 0°C trim. After trimming for zero error at these two reference temperatures, this circuit will have a maximum error due to sensor nonlinearity of ±0.5°C over the −55 to 150°C range. Alternately, if trimming is considered disadvantageous, an AD590L can be used with precision resistors and reference voltage and can achieve a ±2.5°C error over this range.

A low-drift op amp should be used for best results, such as the AD-301AL (compensated) or the AD741L. The output of this circuit can also directly drive a dvm or dpm to read temperature over a −55 to +150°C range.

Temperature-to-Frequency Converters

An interesting form of temperature transducer is the temperature-to-frequency converter. Using V/F (or I/F) techniques, the temperature-to-frequency converter can convert temperature directly to a corresponding frequency, at the sensing point in many cases, allowing efficient and accurate transmission in a digital format. With the relatively low cost of IC V/F converters and IC sensors, this scheme can allow data from many sensing points to be economically processed.

An example of a Kelvin scaled temperature-to-frequency converter using an LM234-3 sensor and an LM331 V/F is illustrated in Fig. 8-72.

This circuit has a number of useful features and can be used in a number of ways, dependent upon the application requirements. The

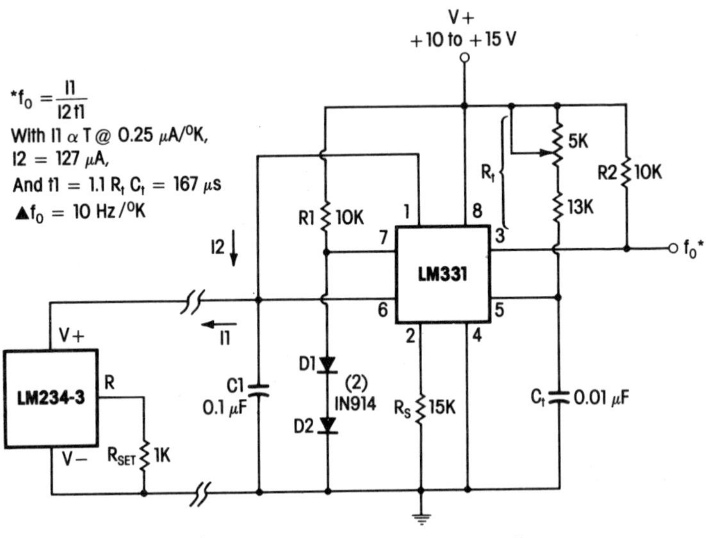

Fig. 8-72. LM234(334)/LM331 Kelvin scaled temperature-to-frequency converter.

heart of the circuit is the LM331 V/F converter, which is used here as an I/F converter, converting the temperature-dependent output current of the LM234-3 to a proportional frequency. The output is Kelvin scaled, with a sensitivity of 10 Hz/°K.

This circuit can be recognized as a current based charge balancing converter, with C1 the capacitor which integrates the charge. D1 and D2 provide a +1.2-volt threshold bias for the LM331's comparator, which in turn indirectly determines the voltage bias across the LM234, which will be in the range of 1 to 2 volts. This low voltage and the low set current result in negligible transducer self-heating effects.

The sensor output current is impressed directly across C1 to drive the LM331. Due to the relatively large value of C1 and the fact that it does not appear in the frequency expression means that the sensor can be remoted at the end of a long line (of high capacitance) without error. And due to current mode operation of the sensor, the circuit is relatively immune to electrostatic pickup.

Calibration of the circuit is achieved by trim of the LM331 pulse width (t1), via the timing resistance R_t. This adjusts out all circuit scaling errors, and should be done at a reference temperature such as 0°C (273.2°K) for a frequency of 2732 Hz. Note that if the sensor is remoted, R_t serves as a convenient local trim.

The circuit can be used with an input analog multiplexer if desired, to interrogate a number of sensors. In such a case it may then be desirable to individually trim the sensor R_{SET}. It can be used over a power supply range of +5 to +15 volts, with minimal change in callibration due to supply variations.

REFERENCES

1. Brugler, J. S. "Silicon Transistor Biasing for Linear Collector Current Temperature Dependence." *IEEE JSSC,* June 1977.

2. Dobkin, R. C. *Applications for an Adjustable IC Power Regulator.* National Semiconductor Application Note AN-178, January 1977. National Semiconductor Corp., Santa Clara, California.

3. ———. "Monolithic Temperature Transducer." *ISSCC Digest of Tech Papers,* February 1974.

4. ———. *3 Terminal Regulator Is Adjustable.* National Semiconductor Application Note AN-181, March 1977. National Semiconductor Corp., Santa Clara, California.

5. ———. *Versatile IC Preamp Makes Thermocouple Amplifier With Cold Junction Compensation.* National Semiconductor LB-24, June 1973. National Semiconductor Corp., Santa Clara, California.

6. Erdi, G. *Thermometer Applications of the REF-02.* Precision Monolithics Application Note AN-18, August 1976. Precision Monolithics, Inc. Santa Clara, California.

7. "Intersil Application Bulletin A021, 1977. Intersil, Inc., Cupertino, California.

8. Jung, W. G. "Current Inverters Make Current Sources Out of Monolithic IC D/A Sink Outputs." *Electronic Design,* January 18, 1978.

9. ———. "IC D/A + 3 Terminal Voltage Regulator = Power DAC." *Electronic Design,* 1978.

10. ———. *IC Timer Cookbook.* Howard W. Sams & Co., Inc., Indianapolis, 1977.

11. McDermott, J. "Knowing Component Temps Can Improve Your Design." *Electronic Design,* November 22, 1977.

12. Nelson, C. *Super Matched Transistor Pair.* Application Note, National Semiconductor Corp., Santa Clara, California.

13. Schoef, J.; Soderquist, D. *DAC-08 Applications Collection.* Precision Monolithics Application Note AN-17, December 1975. Precision Monolithics, Inc., Santa Clara, California.

14. ———; ———. *Differential and Multiplying Digital to Analog Converter Applications.* Precision Monolithics Application Note AN-19, December 1976. Precision Monolithics, Inc., Santa Clara, California.

15. Sheingold, D. H. "Analog-Digital Conversion Notes." Analog Devices, Inc., Norwood, Massachusetts, 1977.

16. Soderquist, D. *"3 IC 8 Bit Binary Digital to Process Current Converter With 4-20 mA Output."* Precision Monolithics Application Note AN-21, January 1977. Precision Monolithics, Inc., Santa Clara, California.

17. Timko, M. P. "A Two Terminal IC Temperature Transducer." *IEEE JSSC,* Vol. SC-11 No. 6, December 1976.

18. Tuma, W. *Low Cost LED Thermometer.* National Semiconductor LB-30, July 1975. National Semiconductor Corp., Santa Clara, California.

19. Williams, J. "Designer's Guide to Temperature Sensing." Parts 1-3, *EDN,* May 5, 1977.

20. Yates, W. W. "Thermistors." *Design News,* November 22, 1976.

21. ———. *YSI-Sostman Platinum Temperature Measurement Products.* Product Brochure, Yellow Springs Instrument Company, Inc., Yellow Springs, Ohio.

22. ———. *YSI Thermilinear Components.* Product Brochure, January, 1977, Yellow Springs Instrument Company, Inc., Yellow Springs, Ohio.

23. *Temperature Measurement Handbook.* Omega Engineering Inc., Stamford, Connecticut.

24. Manufacturers' Data Sheets

Analog Devices	AD301AL	AD590
	AD521	AD741
	AD537	AD1408
	AD561	AD7520
	AD562	AD7523
	AD580	
Motorola	MC1408	MC14410

National Semiconductor	LF356	LM336
	LH0044	LM337
	LM234/LM334	LM350
	LM308A-1	LM358
	LM311	LM393
	LM317	LM339A
	LM322	LX3911
	LM331	LX5600
Precision Monolithics	DAC-08	OP-02
	DAC-76	REF-02
Raytheon	RC4151	

PART III

Appendices

A

Manufacturers' Data Sheets

This appendix contains catalog specification sheets for a selected group of IC d/a and a/d devices from those discussed in this book. Because of the broad range of devices being covered, it is not feasible to provide specification data for all of them within this section. The devices which are presented here are chosen as being representative of each device type and category, as evidenced by strong industry popularity and/or multiple source availability. The data sheets reproduced here are those of the original manufacturers.

Due to space limitations, it is not possible to reproduce these data sheets in their entirety, therefore presentation here is confined to the essential "electrical characteristics" information, for devices performing in a standard commercial environment (0-70°C). It is suggested that the reader avail himself/herself of more complete and most recent data, by writing directly to the manufacturer in question.

DAC-08

8 BIT HIGH SPEED MULTIPLYING D/A CONVERTER

UNIVERSAL DIGITAL LOGIC INTERFACE

GENERAL DESCRIPTION

The DAC-08 series of 8 bit monolithic multiplying Digital-to-Analog Converters provide very high speed performance coupled with low cost and outstanding applications flexibility.

Advanced circuit design* achieves 85 nsec settling times with very low "glitch" and at low power consumption. Monotonic multiplying performance is attained over a wide 40 to 1 reference current range. Matching to within 1 LSB between reference and full scale currents eliminates the need for full scale trimming in most applications. Direct interface to all popular logic families with full noise immunity is provided by the high swing, adjustable threshold logic inputs.

High voltage compliance dual complementary current outputs are provided, increasing versatility and enabling differential operation to effectively double the peak-to-peak output swing. In many applications, the outputs can be directly converted to voltage without the need for an external op amp.

All DAC-08 series models guarantee full 8 bit monotonicity, and nonlinearities as tight as ±0.1% over the entire operating temperature range are available. Device performance is essentially unchanged over the ±4.5V to ±18V power supply range, with 33 mW power consumption attainable at ±5V supplies.

FEATURES

- Fast Settling Output Current . . . . 85 nsec
- Full Scale Current Prematched to ±1 LSB
- Direct Interface to TTL, CMOS, ECL, HTL, PMOS
- Nonlinearity to ±0.1% Max Over Temp Range
- High Output Impedance and Compliance . . −10V to +18V
- Differential Current Outputs
- Wide Range Multiplying Capability . . . . 1 MHz Bandwidth
- Low FS Current Drift . . . . ±10ppm/°C
- Wide Power Supply Range . . . . ±4.5V to ±18V
- Low Power Consumption . . . . 33 mW @ ±5V
- Low Cost

The compact size and low power consumption make the DAC-08 attractive for portable and military/aerospace applications; devices processed to MIL-STD-883A, Level B are available.

DAC-08 applications include 8 bit, 1 μsec A/D converters, servo-motor and pen drivers, waveform generators, audio encoders and attenuators, analog meter drivers, programmable power supplies, CRT display drivers, high speed modems and other applications where low cost, high speed and complete input/output versatility are required.

EQUIVALENT CIRCUIT

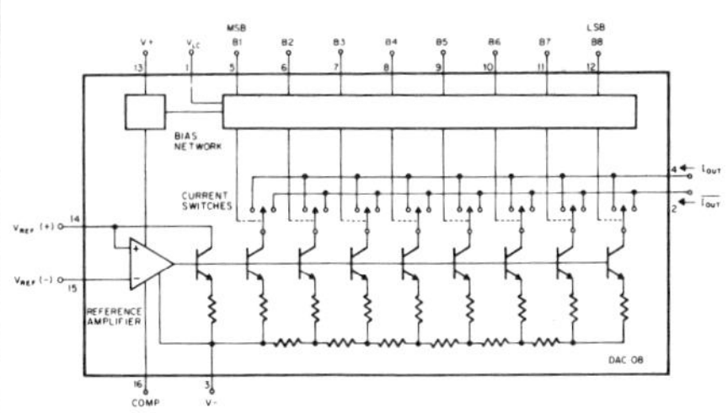

FIGURE 1

ORDERING INFORMATION AND PIN CONNECTION

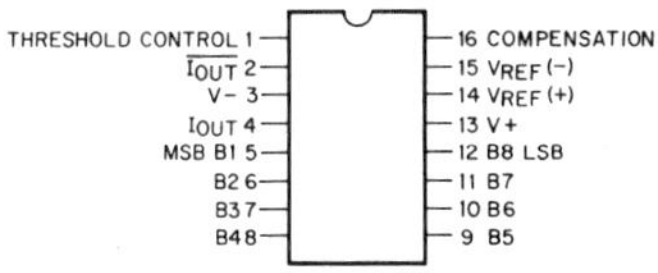

TOP VIEW

16 PIN HERMETIC DUAL-IN-LINE (Q-Suffix)

MODEL	TEMP RANGE	NONLINEARITY
DAC-08AQ	-55°/+125°C	±0.1%
DAC-08Q	-55°/+125°C	±0.19%
DAC-08HQ	0°/70°C	±0.1%
DAC-08EQ	0°/70°C	±0.19%
DAC-08CQ	0°/70°C	±0.39%

Military Temperature Range Devices With MIL-STD-883A Class B Processing:

ORDER: DAC-08-883-AQ
DAC-08-883-Q

The premium performance of this product is achieved through an advanced processing technology. All Precision Monolithics products are guaranteed to meet or exceed published specifications.

TECHNICAL SPECIFICATIONS DECEMBER 1976

*PATENTS APPLIED FOR

Courtesy Precision Monolithics, Inc.

ELECTRICAL CHARACTERISTICS

These specifications apply for $V_S = \pm 15V$, $I_{REF} = 2.0$ mA, $T_A = 0°C$ to $70°C$ unless otherwise specified. Output characteristics refer to both I_{OUT} and $I_{\overline{OUT}}$.

			DAC-08H			DAC-08E			DAC-08C			
Parameter	**Symbol**	**Conditions**	**Min**	**Typ**	**Max**	**Min**	**Typ**	**Max**	**Min**	**Typ**	**Max**	**Units**
Resolution			8	8	8	8	8	8	8	8	8	bits
Monotonicity			8	8	8	8	8	8	8	8	8	bits
Nonlinearity		$T_A = 0°C$ to $70°C$	–	–	±0.1	–	–	±0.19	–	–	±0.39	% FS
Settling Time	t_s	To ±½ LSB, all bits switched ON or OFF $T_A = 25°C$	–	85	135	–	85	150	–	85	150	nsec
Propagation Delay Each bit	t_{PLH}	$T_A = 25°C$	–	35	60	–	35	60	–	35	60	nsec
All bits switched	t_{PHL}		–	35	60	–	35	60	–	35	60	nsec
Full Scale Tempco	TCI_{FS}		–	±10	±50	–	±10	±50	–	±10	±80	ppm/°C
Output Voltage Compliance	V_{OC}	Full scale current change < ½ LSB R_{OUT} > 20 Megohm typ.	-10	–	+18	-10	–	+18	-10	–	+18	Volts
Full Scale Current	I_{FS4}	$V_{REF} = 10.000V$ $R_{14}, R_{15} = 5.000k\ \Omega$ $T_A = 25°C$	1.984	1.992	2.000	1.94	1.99	2.04	1.94	1.99	2.04	mA
Full Scale Symmetry	I_{FSS}	$I_{FS4} - I_{FS2}$	–	±0.5	±4.0	–	±1.0	±8.0	–	±2.0	±16	μA
Zero Scale Current	I_{ZS}		–	0.1	1.0	–	0.2	2.0	–	0.2	4.0	μA
Output Current Range	I_{FSR}	V– = -5.0V	0	2.0	2.1	0	2.0	2.1	0	2.0	2.1	mA
		V– = -7.0V to -18V	0	2.0	4.2	0	2.0	4.2	0	2.0	4.2	mA
Logic Input Levels Logic "0"	V_{IL}	$V_{LC} = 0V$	–	–	0.8	–	–	0.8	–	–	0.8	Volts
Logic "1"	V_{IH}		2.0	–	–	2.0	–	–	2.0	–	–	Volts
Logic Input Current Logic "0"	I_{IL}	$V_{LC} = 0V$ $V_{IN} = -10V$ to $+0.8V$	–	-2.0	-10	–	-2.0	-10	–	-2.0	-10	μA
Logic "1"	I_{IH}	$V_{IN} = 2.0V$ to $18V$	–	0.002	10	–	0.002	10	–	0.002	10	μA
Logic Input Swing	V_{IS}	V– = -15V	-10	–	+18	-10	–	+18	-10	–	+18	Volts
Logic Threshold Range	V_{THR}	$V_S = \pm 15V$	-10	–	+13.5	-10	–	+13.5	-10	–	+13.5	Volts
Reference Bias Current	I_{15}		–	-1.0	-3.0	–	-1.0	-3.0	–	-1.0	-3.0	μA
Reference Input Slew Rate	dI/dt	See Figs. 5, 27	4.0	8.0	–	4.0	8.0	–	4.0	8.0	–	mA/μsec
Power Supply Sensitivity	$PSSI_{FS+}$	V+ = 4.5V to 18V	–	±0.0003	±0.01	–	±0.0003	±0.01	–	±0.0003	±0.01	%/%
	$PSSI_{FS-}$	V– = -4.5V to -18V $I_{REF} = 1.0$ mA	–	±0.002	±0.01	–	±0.002	±0.01	–	±0.002	±0.01	%/%
Power Supply Current	I+	$V_S = \pm 5V$, $I_{REF} = 1.0$ mA	–	2.3	3.8	–	2.3	3.8	–	2.3	3.8	mA
	I–		–	-4.3	-5.8	–	-4.3	-5.8	–	-4.3	-5.8	mA
	I+	$V_S = +5V, -15V$ $I_{REF} = 2.0$ mA	–	2.4	3.8	–	2.4	3.8	–	2.4	3.8	mA
	I–		–	-6.4	-7.8	–	-6.4	-7.8	–	-6.4	-7.8	mA
	I+	$V_S = \pm 15V$ $I_{REF} = 2.0$ mA	–	2.5	3.8	–	2.5	3.8	–	2.5	3.8	mA
	I–		–	-6.5	-7.8	–	-6.5	-7.8	–	-6.5	-7.8	mA
Power Dissipation	P_D	±5V, $I_{REF} = 1.0$ mA	–	33	48	–	33	48	–	33	48	mW
		+5V, -15V, $I_{REF} = 2.0$ mA	–	108	136	–	103	136	–	108	136	mW
		±15V, $I_{REF} = 2.0$ mA	–	135	174	–	135	174	–	135	174	mW

Courtesy Precision Monolithics, Inc.

ORDERING INFORMATION

Device	Temperature Range	Package
MC1408L6	0°C to +75°C	Ceramic DIP
MC1408L7	0°C to +75°C	Ceramic DIP
MC1408L8	0°C to +75°C	Ceramic DIP
MC1408P6	0°C to +75°C	Plastic DIP
MC1408P7	0°C to +75°C	Plastic DIP
MC1408P8	0°C to +75°C	Plastic DIP
MC1508L8	–55°C to +125°C	Ceramic DIP

MC1408
MC1508

Specifications and Applications Information

EIGHT-BIT MULTIPLYING DIGITAL-TO-ANALOG CONVERTER

. . . designed for use where the output current is a linear product of an eight-bit digital word and an analog input voltage.

- Eight-Bit Accuracy Available in Both Temperature Ranges
  Relative Accuracy: ±0.19% Error maximum
  (MC1408L8, MC1408P8, MC1508L8)
- Seven and Six-Bit Accuracy Available with MC1408 Designated by 7 or 6 Suffix after Package Suffix
- Fast Settling Time – 300 ns typical
- Noninverting Digital Inputs are MTTL and CMOS Compatible
- Output Voltage Swing – +0.4 V to -5.0 V
- High-Speed Multiplying Input
  Slew Rate 4.0 mA/μs
- Standard Supply Voltages: +5.0 V and -5.0 V to -15 V

EIGHT-BIT MULTIPLYING DIGITAL-TO-ANALOG CONVERTER

SILICON MONOLITHIC INTEGRATED CIRCUIT

L SUFFIX
CERAMIC PACKAGE
CASE 620

P SUFFIX
PLASTIC PACKAGE
CASE 648

FIGURE 1 – D-to-A TRANSFER CHARACTERISTICS

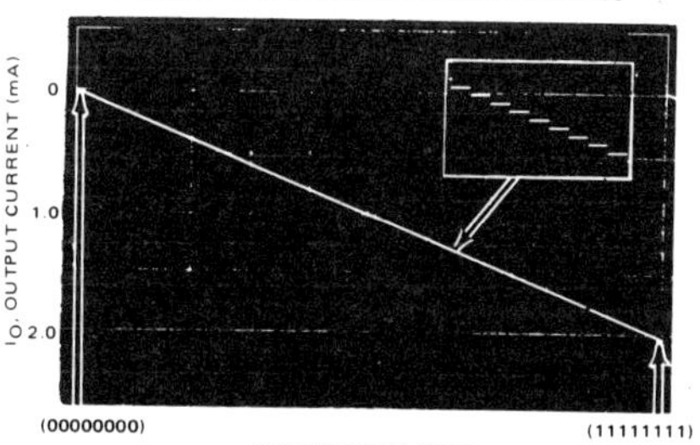

FIGURE 2 – BLOCK DIAGRAM

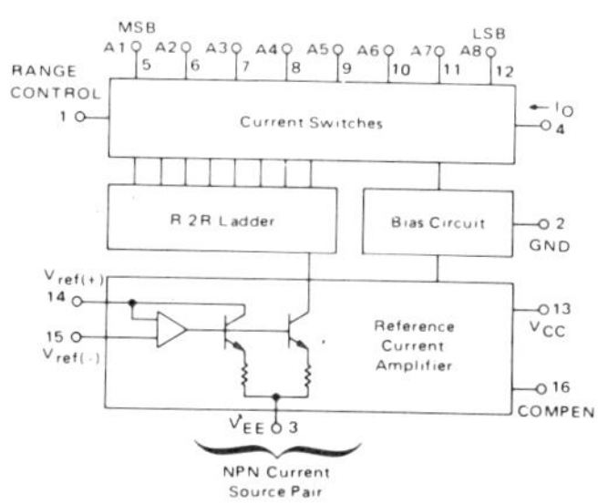

TYPICAL APPLICATIONS

- Tracking A-to-D Converters
- Successive Approximation A-to-D Converters
- 2 1/2 Digit Panel Meters and DVM's
- Waveform Synthesis
- Sample and Hold
- Peak Detector
- Programmable Gain and Attenuation
- CRT Character Generation
- Audio Digitizing and Decoding
- Programmable Power Supplies
- Analog-Digital Multiplication
- Digital-Digital Multiplication
- Analog-Digital Division
- Digital Addition and Subtraction
- Speech Compression and Expansion
- Stepping Motor Drive

MAXIMUM RATINGS (T_A = +25°C unless otherwise noted.)

Rating	Symbol	Value	Unit
Power Supply Voltage	V_{CC} V_{EE}	+5.5 -16.5	Vdc
Digital Input Voltage	V_5 thru V_{12}	0 to +5.5	Vdc
Applied Output Voltage	V_O	+0.5, -5.2	Vdc
Reference Current	I_{14}	5.0	mA
Reference Amplifier Inputs	V_{14}, V_{15}	V_{CC}, V_{EE}	Vdc
Operating Temperature Range MC1508 MC1408 Series	T_A	 -55 to +125 0 to +75	°C
Storage Temperature Range	T_{stg}	-65 to +150	°C

ELECTRICAL CHARACTERISTICS (V_{CC} = +5.0 Vdc, V_{EE} = -15 Vdc, $\frac{V_{ref}}{R14}$ = 2.0 mA, MC1508L8: T_A = -55°C to +125°C. MC1408L Series: T_A = 0 to +75°C unless otherwise noted. All digital inputs at high logic level.)

Characteristic	Figure	Symbol	Min	Typ	Max	Unit
Relative Accuracy (Error relative to full scale I_O) MC1508L8, MC1408L8, MC1408P8 MC1408P7, MC1408L7, See Note 1 MC1408P6, MC1408L6, See Note 1	4	E_r	 – – –	 – – –	 ±0.19 ±0.39 ±0.78	%
Settling Time to within ±1/2 LSB [includes t_{PLH}] (T_A = +25°C) See Note 2	5	t_S	–	300	–	ns
Propagation Delay Time T_A = +25°C	5	t_{PLH}, t_{PHL}	–	30	100	ns
Output Full Scale Current Drift		TCI_O	–	-20	–	PPM/°C
Digital Input Logic Levels (MSB) High Level, Logic "1" Low Level, Logic "0"	3	 V_{IH} V_{IL}	 2.0 –	 – –	 – 0.8	Vdc
Digital Input Current (MSB) High Level, V_{IH} = 5.0 V Low Level, V_{IL} = 0.8 V	3	 I_{IH} I_{IL}	 – –	 0 -0.4	 0.04 -0.8	mA
Reference Input Bias Current (Pin 15)	3	I_{15}	–	-1.0	-5.0	µA
Output Current Range V_{EE} = -5.0 V V_{EE} = -15 V, T_A = 25°C	3	I_{OR}	 0 0	 2.0 2.0	 2.1 4.2	mA
Output Current V_{ref} = 2.000 V, R14 = 1000 Ω	3	I_O	 1.9	 1.99	 2.1	mA
Output Current (All bits low)	3	$I_{O(min)}$	–	0	4.0	µA
Output Voltage Compliance ($E_r \leq$ 0.19% at T_A = +25°C) Pin 1 grounded Pin 1 open, V_{EE} below -10 V	3	V_O	 – –	 – –	 -0.55, +0.4 -5.0, +0.4	Vdc
Reference Current Slew Rate	6	$SR\ I_{ref}$	–	4.0	–	mA/µs
Output Current Power Supply Sensitivity		PSRR(–)	–	0.5	2.7	µA/V
Power Supply Current (All bits low)	3	I_{CC} I_{EE}	– –	+13.5 -7.5	+22 -13	mA
Power Supply Voltage Range (T_A = +25°C)	3	V_{CCR} V_{EER}	+4.5 -4.5	+5.0 -15	+5.5 -16.5	Vdc
Power Dissipation All bits low V_{EE} = -5.0 Vdc V_{EE} = -15 Vdc All bits high V_{EE} = -5.0 Vdc V_{EE} = -15 Vdc	3	P_D	 – – – –	 105 190 90 160	 170 305 – –	mW

Note 1. All current switches are tested to guarantee at least 50% of rated output current.
Note 2. All bits switched.

MOTOROLA *Semiconductor Products Inc.*

Courtesy Motorola Semiconductor Products, Inc.

OPERATING TEMPERATURE RANGE:
C SUFFIX 0 to +70°C

A/D CONVERTERS

PRELIMINARY

LD110/LD111A HIGH PERFORMANCE 3½ DIGIT A/D CONVERTER SET

LD110 DIGITAL A/D PROCESSOR / LD111A ANALOG A/D PROCESSOR

Features

- 0.02% Accuracy (of Reading) ±1 Count
- Two Voltage Ranges – 1.999 V and 199.9 mV
- Sampling Rates up to 40 Samples/Second
- Buffered Analog Input ($Z_{IN} > 1000M\,\Omega$)
- Buffered Reference Input ($Z_{IN} > 1M\,\Omega$)
- Auto-Zero
- Auto-Polarity
- 10 μV Resolution (A Useable 20 mV Scale)
- Differential Input Capability
- LD111A Directly Replaces LD111
- Typical T.C. of 5 ppm/°C
- Overrange and Underrange Signals Available for Auto-Ranging Capability
- TTL Compatible Outputs

Applications

- DVM/DPM
- A/D Converters
- Controllers
- Digital Weighing Systems
- Digital Thermometers
- Process Control Instrumentation
- Scientific Instrumentation

Description

The "Quantized Feedback" conversion scheme used in the LD110/LD111A set provides an Auto-Zeroing, Auto-polarity A/D system requiring only a single reference voltage.

The monolithic LD111A high performance analog processor contains a bipolar comparator, a bipolar integrating amplifier, a bipolar reference amplifier, two MOSFET input unity gain amplifiers, several P-channel enhnacement mode analog switches and the necessary level shifting drivers to allow the analog and digital processors to be directly interfaced. The high impedance input and reference buffer amplifiers eliminate source loading errors and provide the outstanding temperature coefficient inherent in this system. Break-before-make switch action insures that neither the analog input nor the reference voltage will be shorted to ground at any time.

The PMOS LD110 synchronous digital processor combines the counting, storage and data multiplexing functions with the random logic necessary to control the quantized charge-balancing function of the analog processor. Seventeen static latches store the 3½ digits of BCD data as well as overrange, underrange and polarity information. Nine push-pull output buffers (capable of driving one standard TTL load each) provide the sign, digit strobe and multiplexed BCD data outputs, all of which are active high. The digit scan is an interlaced format of digits 1, 3, 2 and 4.

FUNCTIONAL DIAGRAM

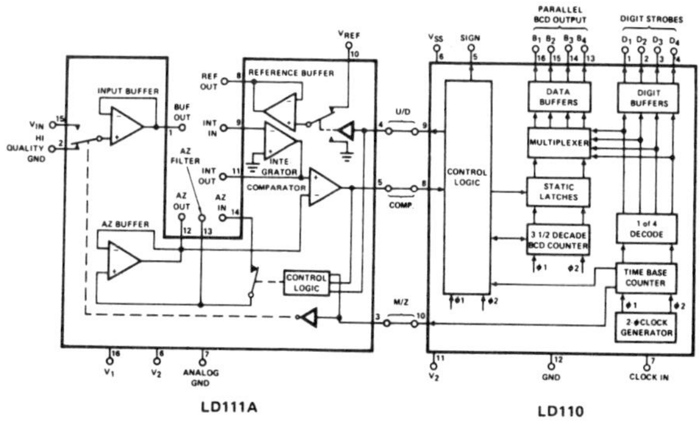

SWITCH STATES ARE FOR A LOGIC "0" AT U/D AND M/Z INPUTS.

PIN CONFIGURATIONS

Dual-In-Line Packages

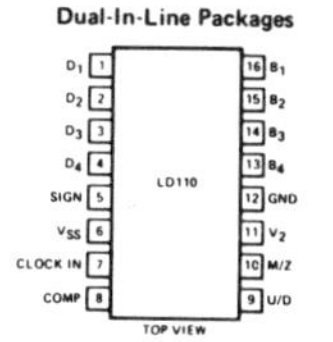

ORDER NUMBER: LD110CJ
SEE PACKAGE 8

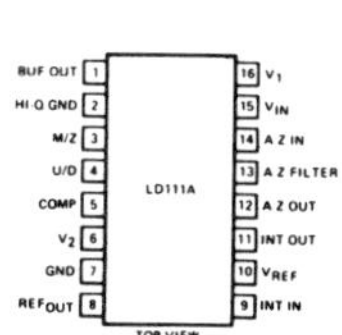

ORDER NUMBER: LD111ACJ
SEE PACKAGE 8

Courtesy Siliconix, Inc.

ABSOLUTE MAXIMUM RATINGS

I_{IN} (Pin 15, LD111A) . . . . 1 mA
$V_1 - V_2$ (LD111A) . . . . 30 V
V_{SS} . . . . 6 V
$V_{SS} - V_2$ (LD110) . . . . 20 V
Voltage On Any Pin Relative
to V_{SS} (LD110) . . . . 0.3 V to −20 V
V_{REF} . . . . V_1

Operating Temperature . . . . 0 to 70°C
Storage Temperature . . . . −65 to +125°C
Power Dissipation (Package)* . . . . 750 mW

*Device mounted with all leads welded or soldered to PC board. Derate 6.3 mW/°C above 25°C.

ELECTRICAL CHARACTERISTICS

All DC parameters are 100% tested at 25°C. Lots are sample tested for AC parameters to assure conformance with specifications.

			Characteristics	Min	Typ	Max	Unit	Test Conditions, Unless Noted: V_1 = +12 V, V_2 = −12 V, V_{SS} = 5 V, V_{REF} = 8.2 V, T_A = 25°C, R_1 = 100K Ω
1	GENERAL		Linearity		0.02		% rdg	
2			Noise		0.1		LSB	Peak-to-Peak Noise Apparent When Going From One Steady Reading to Another
3			Gain T.C.		5		ppm/°C	
4		NMR	Normal Mode Rejection		40		dB	f_{noise} = 60 Hz
5		f_{IN}	Clock Frequency	2	30.7	250	kHz	50% Duty Cycle
6		$r_{DS(on)}$	ON Resistance, Auto Zero Switch		6	20	K Ω	$V_{AZ(in)}$ = −4.0 V, I_S = −30 µA
7	INPUT	I_{CL}	Clock Input Current, Low			−500	µA	V_{CLOCK} in = 0.4 V
8		I_{IN}	Input Bias Current		4		pA	T_A = 25°C
9					40			T_A = 70°C
10	AMPLIFIERS	I_{source}	Reference Buffer	−400	−800		µA	V_{INL} (U/D) = 0.8 V, V_{OUT} = 0
11		I_{sink}	AZ Buffer		800			V_{AZ} = −4 V, V_{OUT} = 0 V
12		I_{sink}	Input Buffer	400	800			V_{IN} = −2 V, V_{OUT} = 0 V
13		I_{source}	Input Buffer	−50	−100			V_{IN} = 2 V, V_{OUT} = 0 V
14		V_{offset}	AZ Buffer	−100		100	mV	V_{OUT} = 0V
15	OUTPUT	V_{OL1}	Measure/Zero Voltage, Low			0.4	V	I_{OL} = 150 µA
16		V_{OH1}	Measure/Zero Voltage, High	2.4				I_{OH} = −200 µA
17		V_{OL2}	Up/Down Logic Voltage, Low			0.4		I_{OL} = 250 µA
18		V_{OH2}	Up/Down Logic Voltage, High	2.4				I_{OH} = −200 µA
19		V_{OH3}	Analog Comparator Voltage	2.4				I_{OH} = −100 µA
20		V_{OL3}	Digits, Bits, Voltage, Low			0.4		I_{OL} = 1.6 mA
21		V_{OL4}	Sign Voltage, Low			0.6		I_{OL} = 1.6 mA
22		V_{OH4}	Data Bits Voltage, High	2.4				I_{OH} = −200 µA
23		V_{OH5}	Digits, Sign Voltage, High	2.4				I_{OH} = −800 µA
24	SUPPLY	I_1	Supply Current		2.2	4.0	mA	
25		I_{2A}	Supply Current, LD111A		−1.8	−3.5		
26		I_{2D}	Supply Current, LD110		−17	−23		
27		I_{SS}	Supply Current		17.4	24		
28		$PSRR_1$	Power Supply Rejection Ratio, V_1	80	85		dB	
29		$PSRR_2$	Power Supply Rejection Ratio, V_2	60	65			
30			Reference Voltage Rejection		1		% rdg/ % Change	R_{REF} = R_2 = 100K Ω, V_{IN} = 2 V

Typical values are for DESIGN AID ONLY, not guaranteed and not subject to production testing LD110 IPANII LD111A CMAMI-A

Courtesy Siliconix, Inc.

RAYTHEON

VOLTAGE-TO-FREQUENCY CONVERTER

4151

DESCRIPTION

The RC4151 and RM4151 provide a simple low-cost method of analog-to-digital conversion. They have all the inherent advantages of the voltage-to-frequency conversion technique. The output of the 4151 is a series of pulses of constant duration. The frequency of the pulses is proportional to the applied input voltage. These converters are designed for use in a wide range of data conversion and remote sensing applications.

FEATURES

- Single supply operation (+8V to +22V)
- Pulse output compatible with all logic forms
- Programmable scale factor (K)
- Linearity ±0.05% typical - precision mode
- Temperature stability ±100% ppm/°C typical
- High noise rejection
- Inherent monotonicity
- Easily transmittable output
- Simple full scale trim
- Single-ended input, referenced to ground
- Also provides frequency-to-voltage conversion

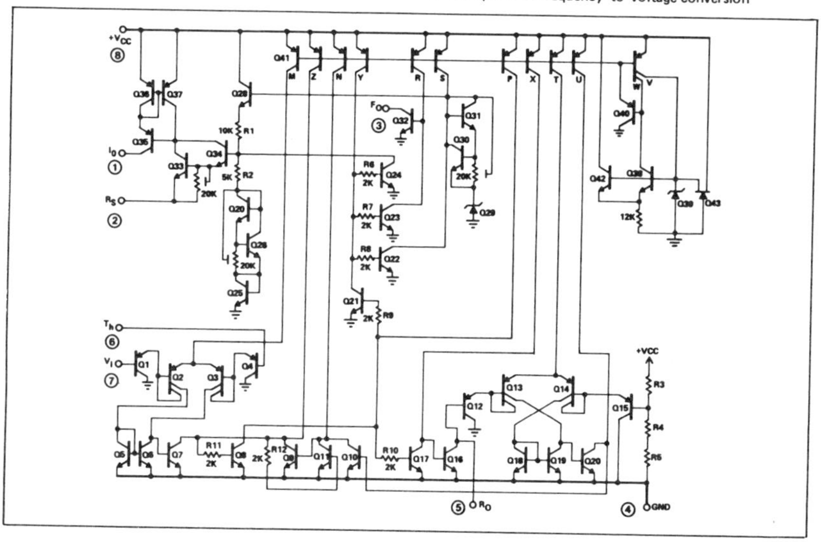

Figure 1. 4151 Schematic Diagram

TE (TO-99) METAL CAN

(Top View)

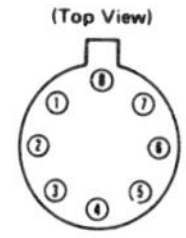

Order Part Nos.: RC4151T, RM4151T

NOTE: PIN 4 CONNECTED TO CASE

NB MINIATURE DUAL-IN-LINE

(Top View)

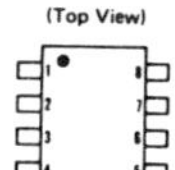

Order Part Nos.: RC4151NB, RV4151NB

PIN	FUNCTION
1	CURRENT SOURCE
2	SCALE FACTOR
3	LOGIC OUTPUT
4	GROUND
5	ONE-SHOT R, C
6	THRESHOLD
7	INPUT VOLTAGE
8	V_{CC}

Courtesy Raytheon Semiconductor Division

RAYTHEON

ABSOLUTE MAXIMUM RATINGS

Supply voltages		+8.0 to +22V
Output sink current		20mA
Internal power dissipation		500mW
Input voltage		-0.2V to $+V_{CC}$
Output short circuit to ground		Continuous
Temperature Range	Storage	Operating
RM4151:	-65°C to +150°C	-55°C to +125°C
RV4151:	-55°C to +125°C	-40°C to + 85°C
RC4151:	-55°C to +125°C	0°C to + 70°C

ELECTRICAL CHARACTERISTICS V_{CC} = +15V, T_A = +25°C, unless otherwise specified

PARAMETER	CONDITIONS	UNITS	MIN	TYP	MAX
SUPPLY CURRENT	$8V < V_{cc} < 15V$	mA	2.0	3.5	6.0
	$15V < V_{cc} < 22V$	mA	2.0	4.5	7.5
CONVERSION ACCURACY					
Scale Factor	Circuit Figure 3, V_I = 10V R_S = 14.0K	kHz/V	0.90	1.0	1.10
Drift with Temperature	Circuit Figure 3, V_I = 10V	ppM/°C	–	± 100	–
Drift with V_{cc}	Circuit Figure 3, V_I = 1.0V $8V < V_{cc} < 18V$	%/V	–	0.2	1.0
INPUT COMPARATOR					
Offset Voltage		mV	–	5.0	10.0
Offset Current		nA	–	± 50.0	±100.0
Input Bias Current		nA	–	-100.0	-300
Common Mode Range (Note 1)		V	0	0 to V_{cc}-2	V_{cc} -3.0
ONE-SHOT					
Threshold Voltage, Pin 5		xV_{cc}	0.63	0.667	0.70
Input Bias Current, Pin 5		nA	–	-100.0	-500
Reset V_{SAT}	Pin 5, 1 = 2.2 mA	V	–	0.15	0.50
CURRENT SOURCE					
Output Current (R_S = 14.0kΩ)	Pin 1, Figure 3, V = 0	μA	–	138.7	–
Change with Voltage	Pin 1, V = 0V to V = 10V	μA	–	1.0	2.5
Off Leakage	Pin 1, V = 0V	nA	–	1.0	50.0
REFERENCE VOLTAGE	Pin 2, Figure 3	V	1.70	1.9	2.08
LOGIC OUTPUT					
V_{SAT}	Pin 3, I = 3.0 mA	V	–	0.15	0.50
V_{SAT}	I = 2.0 mA	V	–	0.10	0.35
Off Leakage		μA	–	0.1	1.0

Note 1: Input Common Mode Range includes ground.

Courtesy Raytheon Semiconductor Division

ANALOG DEVICES

12-Bit Successive Approximation Integrated Circuit A/D Converter

AD572

FEATURES

Performance

True 12-Bit Operation: Max Nonlinearity < ±0.012%
Low Gain T.C.: < ±15ppm/°C (AD572B)
Low Power: 900mW
Fast Conversion Time: < 25μsec
Monotonic Feedback DAC Guarantees No Missing Codes
Hermetically-Sealed, Electrostatically-Shielded DIP

Versatility

Military/Aerospace Temperature Range:
-55 to +125°C (AD572S)
MIL-STD-883B Processing Available
Positive-True Serial or Parallel Logic Outputs
Short-Cycle Capability

Value

Precision +10V Reference for External Application
Internal Buffer Amplifier
High Reliability Welded Metal Package
Low Cost: $95.00 (1-24 pcs, AD572A)

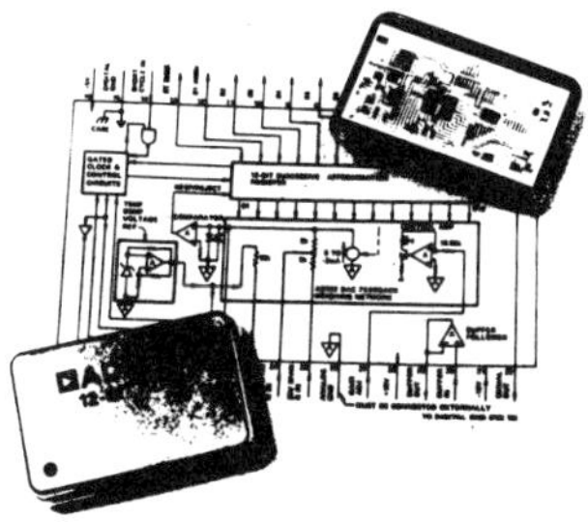

GENERAL DESCRIPTION

The AD572 is a complete 12-bit successive approximation analog-to-digital converter that includes an internal clock, reference, comparator, and buffer amplifier. Its hybrid IC design utilizes MSI digital and linear monolithic chips and active laser trimming of high-stability thin-film resistors to provide modular performance, flexibility, and ease of use, combined with IC size, price, and reliability.

Important performance characteristics of the AD572 include a maximum linearity error at 25°C of ±0.012%, gain T.C. below 15ppm/°C, typical power dissipation of 900mW, and conversion time of less than 25μsec. Of considerable significance in military and aerospace applications is the guaranteed performance from -55 to +125°C of the AD572S, and the availability of units processed to MIL-STD-883B. Monotonic operation of the feedback D/A converter guarantees no missing output codes over temperature ranges of 0 to +70°C, -25 to +85°C, and -55 to +125°C.

The design of the AD572 includes scaling resistors that provide analog input signal ranges of ±2.5, ±5.0, ±10, 0 to +5, or 0 to +10 volts. Adding flexibility and value are the +10V precision reference, which also can be used for external applications, and the input buffer amplifier. All digital signals are fully DTL and TTL compatible, and the data output is positive-true and available in either serial or parallel form.

The AD572 is packaged in a hermetically-sealed, all-metal DIP. Welding...rather than solder sealing...eliminates any possibility of contamination from flux and solder particles. The metal construction provides excellent shielding from random electrostatic and/or electromagnetic radiation which could cause incorrect output codes. To insure a level of reliability consistent with its performance, each AD572 receives a stringent pre-cap visual inspection, high temperature storage and temperature cycling, acceleration testing, fine and gross leak testing, and operating burn-in.

The AD572 is available in three versions with differing guaranteed performance characteristics and operating temperature ranges; the "A" and "B" are specified from -25 to +85°C, and the "S" from -55 to +125°C.

PRODUCT DESCRIPTION

The AD572 functional diagram and pin-out are shown in Figure 1. The device consists of the following monolithic bipolar transistor and thin-film resistor circuit elements:

1. 12-bit successive-approximation register
2. 12-bit feedback DAC weighing network
3. low-drift comparator
4. temperature-compensated precision +10V reference
5. high-impedance buffer follower
6. gated clock and digital control circuits

Route 1 Industrial Park; P.O. Box 280; Norwood, Mass. 02062
Tel: 617/329-4700 TWX: 710/394-6577
West Coast 213/595-1783 | Mid-West 312/894-3300 | Texas 214/231-5094

Courtesy Analog Devices, Inc.

SPECIFICATIONS (typical @ +25°C, ±15V and +5V unless otherwise noted)

MODEL	AD572AD	AD572BD	AD572SD
RESOLUTION	12 Bits	•	•
ANALOG INPUTS			
Voltage Ranges			
Bipolar	±2.5, ±5.0, ±10.0V	•	•
Unipolar	0 to +5, 0 to +10V	•	•
Impedance (Direct Input)			
0 to +5V, ±2.5V	2.5kΩ	•	•
0 to +10V, ±5V	5.0kΩ	•	•
±10V	10kΩ	•	•
Buffer Amplifier			
Impedance (min)	100MΩ	•	•
Bias Current	50nA	•	•
Settling Time			
to 0.01% of FSR for 20V step	2μsec	•	•
DIGITAL INPUTS			
Convert Command	Note 1	•	•
Logic Loading	1 TTL Load	•	•
TRANSFER CHARACTERISTICS			
Gain Error (Note 2)	±0.05% FSR (Adj to Zero)	•	•
Unipolar Offset Error	±0.05% FSR (Adj to Zero)	•	•
Bipolar Offset Error	±0.1% FSR (Adj to Zero)	•	•
Linearity Error (max)	0.012% FSR	•	•
Inherent Quantization Error	±½ LSB	•	•
Differential Linearity Error	±½ LSB	•	•
No Missing Codes	Guaranteed: 0 to +70°C	Guaranteed: -25 to +85°C	Guaranteed: -55 to +125°C
Power Supply Sensitivity			
±15V	±0.002% FSR/%ΔV_S	•	•
+5V	±0.001% FSR/%ΔV_S	•	•
TEMPERATURE COEFFICIENTS			
Gain (max)	±30ppm/°C (-25 to +85°C)	±15ppm/°C (-25 to +85°C)	±15ppm/°C (-25 to +85°C) ±25ppm/°C (-55 to +125°C)
Unipolar Offset	±3ppm FSR/°C	±3ppm FSR/°C (max)	••
Bipolar Offset (max)	±15ppm FSR/°C	±7ppm FSR/°C	••
Linearity	±3ppm FSR/°C	±2ppm FSR/°C	••
CONVERSION TIME (max)	25μsec	•	•
DIGITAL OUTPUTS (All Codes Positive-True)			
Parallel Data			
Unipolar Code	Binary	•	•
Bipolar Code	Offset Binary/Two's Complement	•	•
Output Drive	2 TTL Loads	•	•
Serial Data (NRZ format)			
Unipolar Code	Binary	•	•
Bipolar Code	Offset Binary	•	•
Output Drive	2 TTL Loads	•	•
Status	Logic "1" during Conversion	•	•
$\overline{\text{Status}}$	Logic "0" during Conversion	•	•
Output Drive	2 TTL Loads	•	•
Internal Clock			
Output Drive	2 TTL Loads	•	•
Frequency	500kHz	•	•
INTERNAL REFERENCE VOLTAGE	+10.00V, ±5mV	•	•
Max External Current	±4mA	•	•
Voltage Temperature Coefficient (max)	±20ppm/°C	±10ppm/°C	••
POWER REQUIREMENTS			
Supply Voltages/Currents	+15V, ±5% @ +25mA	•	•
	-15V, ±5% @ -20mA	•	•
	+5V, ±5% @ +50mA	•	•
Total Power Dissipation	925mW	•	•
TEMPERATURE RANGE			
Specification	-25 to +85°C	•	-55 to +125°C
Operating	-55 to +125°C	•	•
Storage	-55 to +150°C	•	•

*Same specification as AD572AD
**Same specification as AD572BD

Specifications subject to change without notice.

Note 1 Positive pulse 200nsec wide (min). Leading edge ("0" to "1") resets registers. Trailing edge ("1" to "0") initiates conversion.

Note 2 With 50Ω, 1% fixed resistor in place of Gain Adjust pot; see Figures 4 and 5.

Courtesy Analog Devices, Inc.

ANALOG DEVICES

Low Cost 10-Bit Monolithic D/A Converter

AD561

PRELIMINARY TECHNICAL DATA

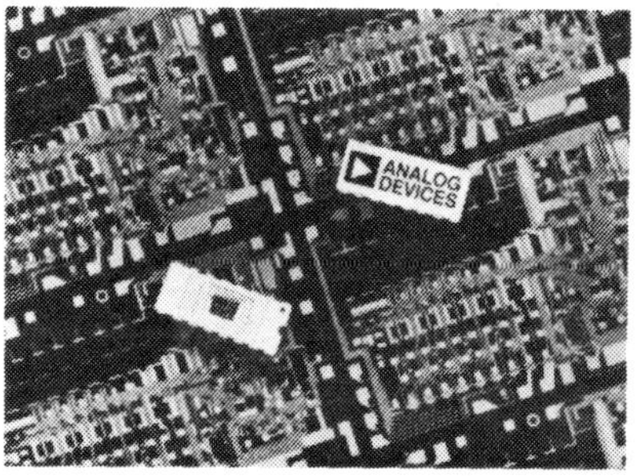

FEATURES

Low Cost ($9.95 in 100's, AD561J)
Complete Current Output Converter
High Stability Buried Zener Reference
Laser Trimmed to High Accuracy (1/4LSB Max Error, AD561K, T)
Trimmed Output Application Resistors for 0 to +10, ±5 Volt Ranges
Fast Settling – 250ns to 1/2LSB
Guaranteed Monotonicity Over Full Operating Temperature Range
TTL/DTL and CMOS Compatible (Positive True Logic)
Single Chip Monolithic Construction
Hermetically-Sealed Ceramic DIP (All Grades)

PRODUCT DESCRIPTION

The AD561 is an integrated circuit 10-bit digital-to-analog converter combined with a high stability voltage reference fabricated on a single monolithic chip. Using 10 precision high-speed current-steering switches, a control amplifier, voltage reference, and laser-trimmed thin-film SiCr resistor network, the device produces a fast, accurate analog output current. Laser trimmed output application resistors are also included to facilitate accurate, stable current-to-voltage conversion; they are trimmed to 0.1% accuracy, thus eliminating external trimmers in many situations.

Several important technologies combine to make the AD561 the most accurate and most stable 10-bit DAC available. The low temperature coefficient, high stability thin-film network is trimmed at the wafer level by a fine resolution laser system to 0.01% typical linearity. This results in an accuracy specification of ±1/4LSB max for the K and T versions, and 1/2LSB max for the J and S versions.

The AD561 also incorporates a low noise, high stability subsurface zener diode to produce a reference voltage with excellent long term stability and temperature cycle characteristics which challenge the best discrete zener references. A temperature compensation circuit is laser-trimmed to allow custom correction of the temperature coefficient of each device. This results in a typical full-scale temperature coefficient of 15ppm/°C; the T.C. is tested and guaranteed to 30ppm/°C max for the K and T versions, 60ppm/°C max for the S, and 80ppm/°C for the J.

All grades are packaged in a 16-pin hermetically-sealed ceramic dual-in-line package. The AD561J and K versions are specified for operation over the 0 to +70°C temperature range, the AD561S and T for operation over the full military temperature range from -55°C to +125°C.

PRODUCT HIGHLIGHTS

1. Advanced monolithic processing and laser trimming at the wafer level have made the AD561 the most accurate 10-bit converter available while keeping costs consistent with large volume integrated circuit production. The AD561K and T have 1/4LSB max relative accuracy and 1/2LSB max differential nonlinearity. The low T.C. R-2R ladder guarantees that all AD561 units will be monotonic over the entire operating temperature range.
2. Digital system interfacing is simplified by the use of a positive true straight binary code. The digital input voltage threshold is a function of the positive supply level; connecting V_{CC} to the digital logic supply automatically sets the threshold to the proper level for the logic family being used. Logic sink current requirement is only 25μA.
3. The high speed current steering switches are designed to settle in less than 250ns for the worst case digital code transition. This allows construction of successive-approximation A/D converters in the 3 to 5μs range.
4. The AD561 has an output voltage compliance range from -2 to +10 volts, thus allowing direct current-to-voltage conversion with just an output resistor, omitting the op amp. The 40MΩ open collector output impedance results in negligible errors due to output leakage currents.
5. Every AD561 is subjected to long term stabilization bakes and temperature cycled ten times from -65°C to +150°C prior to final test to insure reliability and long-term stability.

Route 1 Industrial Park; P.O. Box 280; Norwood, Mass. 02062
Tel: 617/329-4700 TWX: 710/394-6577
West Coast 213/595-1783 Mid-West 312/894-3300 Texas 214/231-5094

Courtesy Analog Devices, Inc.

SPECIFICATIONS (T_A = +25°C, V_{cc} = +5V, V_{ee} = -15V, unless otherwise specified)

MODEL	AD561J MIN	AD561J TYP	AD561J MAX	AD561K MIN	AD561K TYP	AD561K MAX	UNITS
RESOLUTION		10 Bits			10 Bits		
ACCURACY (Error Relative to Full Scale)		±1/4 (0.025)	±1/2 (0.05)		±1/8 (0.012)	±1/4 (0.025)	LSB % of F.S.
DIFFERENTIAL NONLINEARITY		±1/2			±1/4	±1/2	LSB
DATA INPUTS							
TTL, V_{CC} = +5V							
Bit ON Logic "1"	+2.0			•			V
Bit OFF Logic "0"			+0.8			•	V
CMOS, 10V ≤ V_{CC} ≤ 16.5V (See Figure 1)							
Bit ON Logic "1"	70% V_{CC}			•			V
Bit OFF Logic "0"			30% V_{CC}			•	V
Logic Current (Each Bit)(T_{MIN} to T_{MAX})							
Bit ON Logic "1"		+5	+100		•	•	nA
Bit OFF Logic "0"		-5	-25		•	•	μA
OUTPUT							
Current							
Unipolar	1.5	2.0	2.4	•	•	•	mA
Bipolar	±0.75	±1.0	±1.2	•	•	•	mA
Resistance (Exclusive of Application Resistors)		40M			•		Ω
Unipolar Zero (All Bits OFF)		0.01	0.05		•	•	% of F.S.
Capacitance		25			•		pF
Compliance Voltage	-2	-3	+10	•	•	•	V
SETTLING TIME TO 1/2LSB All Bits ON-to-OFF or OFF-to-ON		250			•		ns
POWER REQUIREMENTS							
V_{CC}, +4.5VDC to +16.5VDC		8	10		•	•	mA
V_{EE}, -10.8 to -16.5VDC		12	16		•	•	mA
POWER SUPPLY GAIN SENSITIVITY							
V_{CC}, +4.5VDC to +16.5VDC		2	10		•	•	ppm of F.S./%
V_{EE}, -10.8VDC to -16.5VDC		4	25		•	•	ppm of F.S./%
TEMPERATURE RANGE							
Operating		0 to +70			•	•	°C
Storage		-65 to +150			•	•	°C
TEMPERATURE COEFFICIENTS							
With Internal Reference							
Unipolar Zero		1	10		1	5	ppm of F.S./°C
Bipolar Zero		2	20		2	10	ppm of F.S./°C
Full Scale		15	80		15	30	ppm of F.S./°C
Differential Nonlinearity		2.5			2.5		ppm of F.S./°C
MONOTONICITY		Guaranteed over full operating temp. range			Guaranteed over full operating temp. range		
PROGRAMMABLE OUTPUT RANGES (See Figs. 5, 6)		0 to +10 -5 to +5			• •		V V
CALIBRATION ACCURACY							
Full Scale Error with Fixed 25Ω Resistor		±0.1			•		% of F.S.
Bipolar Zero Error with Fixed 10Ω Resistor		±0.1			•		% of F.S.
CALIBRATION ADJUSTMENT RANGE							
Full Scale (With 50Ω Trimmer)		±0.5			•		% of F.S.
Bipolar Zero (With 20Ω Trimmer)		±0.2			•		% of F.S.

•Specifications same as AD561J specs.
Specifications subject to change without notice.

Courtesy Analog Devices, Inc.

amplifiers / DAC-ADC / multipliers-dividers
analog functions/modular power supplies
active filters/data conversion products

DAC80

IC D/A CONVERTER

FEATURES

- LOW COST – $18.50 in 100 QUANTITIES
- 12 BIT RESOLUTION
- ±1/2 LSB MAXIMUM NONLINEARITY
- COMPLETE WITH INTERNAL REFERENCE
- FAST SETTLING – 300 nsec to ±.01% – (Current Output Model)
- CERAMIC DUAL-IN-LINE PACKAGE

DESCRIPTION

The DAC80 12 bit D/A converter offers quality performance in a 24 pin dual-in-line ceramic case. This D/A converter is complete with internal reference and optional output amplifier and is engineered to preserve the performance normally found only in much larger, high cost modular units.

Highly stable laser trimmed thin film resistors and fast settling current switches provide maximum nonlinearity error of ±0.012% of Full Scale Range, ±30 ppm/°C maximum gain drift and monotonicity over a 0 to +70°C operating temperature range. Offered with a choice of TTL compatible complementary 12 bit binary (CBI) or 3 digit BCD (CCD) input codes, these D/A converters contain low drift matched and trimmed internal resistors that are used to provide user selectable output voltage ranges of ±2.5, ±5, ±10, 0 to +5, 0 to +10 volts (V models) or output current ranges of 1 mA or 0 to -2 mA.

Voltage output models settle to ±0.01% of Full Scale Range in just 3 microseconds for a 10 volt step change, and current output models settle to ±0.01% in only 300 nanoseconds.

A closer look at the detailed specifications will reveal that the DAC80 is truly the best quality 12 bit IC D/A converter in its class on the market.

BURR-BROWN

International Airport Industrial Park•Tucson, Arizona 85734
Tel 602-294-1431•Twx 910-952-1111•Cable BBRCORP•Telex 66-6491

Courtesy Burr-Brown Research Corp.

SPECIFICATIONS

ELECTRICAL

Typical at 25°C and rated power supplies <u>unless</u> otherwise noted.

MODEL	DAC80		Units
Binary	CBI		
Decimal		CCD	
INPUT			
DIGITAL INPUT			
Resolution	12		Bits
		3	Digits
Logic levels (TTL/CMOS Compatible)(1)			
Logical "1"	$+2 < e_d < +5.5$ at +40 μA		Volts
Logical "0"	$0 < e_d < +0.8$ at -1.0 mA		Volts
TRANSFER CHARACTERISTICS			
ACCURACY			
Linearity Error @ 25°C, max	±½	±¼	LSB
Differential Linearity Error	±½		LSB
Gain Error(2)	±0.1		%
Offset Error(2)	±0.05		% of FSR(3)
Minimum Temperature Range for Guaranteed Monotonicity	0 to +70		°C
DRIFT(4) (0°C to +70°C)			
Gain, max	±30		ppm/°C
Exclusive of internal reference, max	±15		ppm/°C
Offset			
Unipolar	±1		ppm of FSR/°C
Bipolar, max	±15		ppm of FSR/°C
Differential Linearity	±2		ppm of FSR/°C
Linearity Error, max 0°C to +70°C	±½	±½	LSB
CONVERSION SPEED			
Voltage Models			
Settling time to ±0.01% of FSR			
For FSR change			
with 10 kΩ Feedback	5		μsec
with 5 kΩ Feedback	3		μsec
For 1 LSB change	1.5		μsec
Slew Rate	20		V/μsec
Current Models			
Settling time to ±0.01%			
For FSR change			
10 to 100 Ω load	300		nsec
1 kΩ load	1		μsec
OUTPUT			
ANALOG OUTPUT			
Voltage Models			
Ranges – CBI Units	±2.5, +5, ±10,+5,+10		Volts
CCD Units	+10		Volts
Output Current, min	±5		mA
Output Impedance (DC)	0.05		Ω
Short Circuit Duration	Indefinite to Common or Power Supplies		
Current Models			
Ranges	±1, -2		mA
Output Impedance - Bipolar	4.4		kΩ
Unipolar	15		kΩ
Compliance	±2.5		Volts
INTERNAL REFERENCE VOLTAGE	6.3		Volts
Maximum External Current(5)	200		μA
Tempco of Drift, max	±20		ppm/°C
POWER SUPPLY SENSITIVITY			
+15 V Supply	±0.02		% of FSR/%Vs
-15 and +5 V Supplies	±0.002		% of FSR/%Vs
POWER SUPPLY REQUIREMENTS			
Rated Voltage	±15 and +5		Volts
Range	±14.0 to ±16.0 and +4.75 to +16.0		Volts
Supply Drain			
±15 V (including 5 mA load)	±25		mA
+5 V	+20		mA
TEMPERATURE RANGE			
Specification	0 to +70		°C
Operating (double above specs)	-25 to +85		°C
Storage	-55 to +100		°C
PRICES	See page 4 for prices.		

1. Adding external CMOS hex buffers CD 4009A or CD 4049A will provide CMOS input compatibility. Approximate cost is $2.25 (1-24) to $1.50 (100's).
2. Adjustable to zero with external trim potentiometer.
3. FSR means "Full Scale Range" and is 20V for ±10V range, 10V for ±5 range, etc.
4. To maintain drift spec internal feedback resistors must be used for current output models.
5. Maximum with no degradation of specifications.

Prices and Specifications subject to change without notice.

MECHANICAL

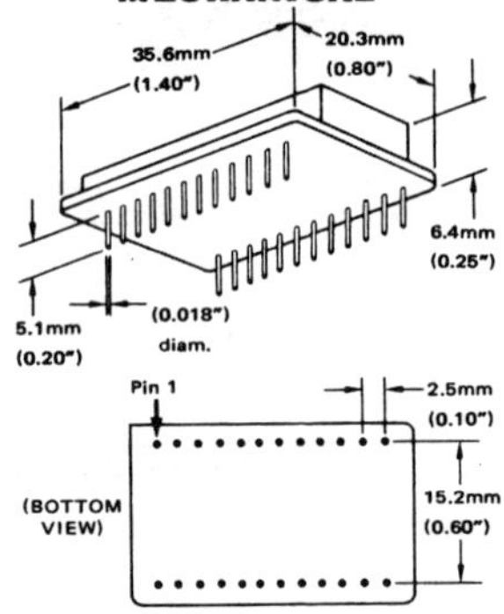

CASE: Black Ceramic
Mating Connector 245MC $4.25 ea.

PIN: Pin material and plating composition conform to method 2003 (solderability) of Mil-Std-883 (except paragraph 3.2).

WEIGHT: 2 oz.

HERMETICITY: Conforms to Method 1014 Condition C Step 1 (fluorocarbon) of Mil-Std-883 (Gross Leak)

CONNECTION DIAGRAM

(VOLTAGE MODELS)

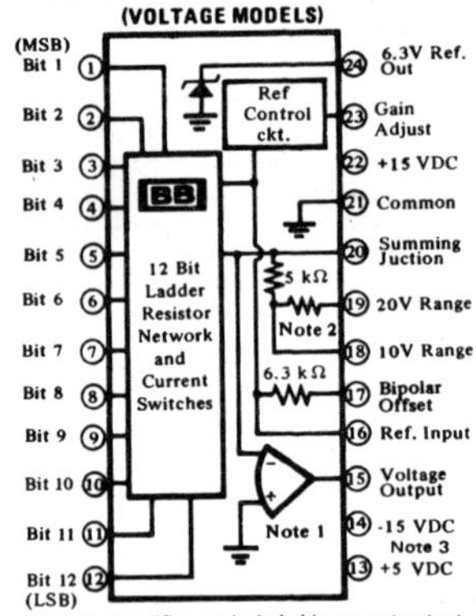

Note 1: Amplifier not included in current output models.

Note 2: 3 kΩ for CCD models
5 kΩ for CBI models

Note 3: +5V supply input may be connected to +15V supply if +5V supply is not available. This will increase internal power dissipation by 200 mW.

Courtesy Burr-Brown Research Corp.

ANALOG DEVICES

CMOS 10 & 12 BIT MONOLITHIC MULTIPLYING D/A CONVERTERS

AD7520, AD7521

FEATURES

AD7520:	10 Bit Resolution
AD7521:	12 Bit Resolution
Linearity:	8, 9 and 10 Bit
Nonlinearity Tempco:	2 ppM of FSR/°C
Low Power Dissipation:	20 mW
Current Settling Time:	500 ns
Feedthrough Error:	1/2 LSB @ 100 kHz
TTL/DTL/CMOS Compatible	

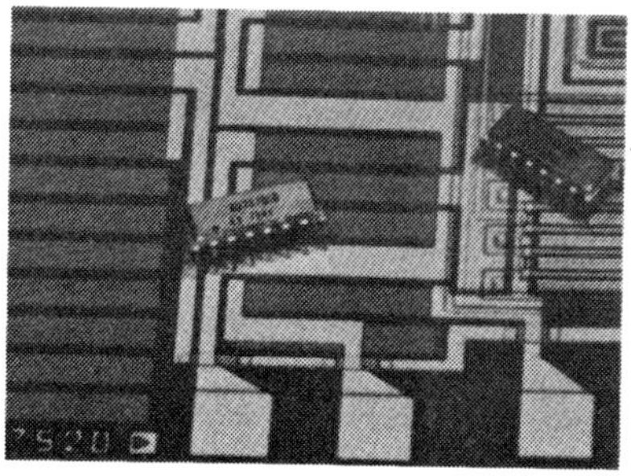

GENERAL DESCRIPTION

The AD7520 (AD7521) is a low cost, monolithic 10-bit (12-bit) multiplying digital-to-analog converter packaged in a 16-pin (18-pin) DIP. The devices use advanced CMOS and thin film technologies providing up to 10-bit accuracy with TTL/DTL/CMOS compatibility.

The AD7520 (AD7521) operates from +5V to +15V supply and dissipates only 20 mW, including the ladder network.

Typical AD7520 (AD7521) applications include: digital/analog multiplication, CRT character generation, programmable power supplies, digitally controlled gain circuits, etc.

FUNCTIONAL DIAGRAM

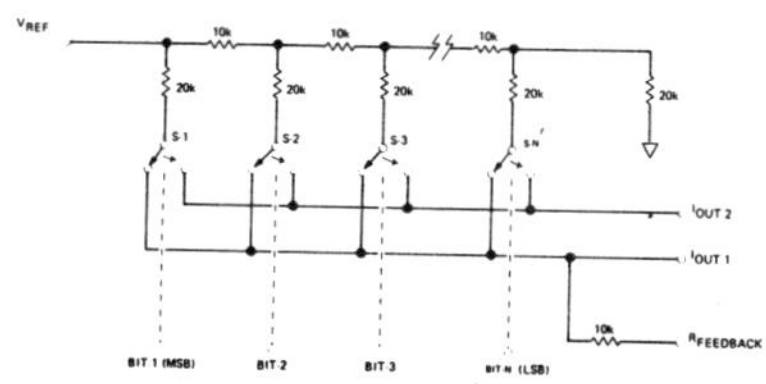

AD7520: N=10
AD7521: N=12
Logic: A switch is closed to $I_{OUT\ 1}$ for its digital input in a "HIGH" state.

ORDERING INFORMATION

Nonlinearity	Temperature Range 0 to +75°C	Temperature Range −55°C to +125°C
0.2% (8-BIT)	AD7520JD AD7520JN AD7521JD AD7521JN	AD7520SD AD7521SD
0.1% (9-BIT)	AD7520KD AD7520KN AD7521KD AD7521KN	AD7520TD AD7521TD
0.05% (10-BIT)	AD7520LD AD7520LN AD7521LD AD7521LN	AD7520UD AD7521UD

PIN CONFIGURATION

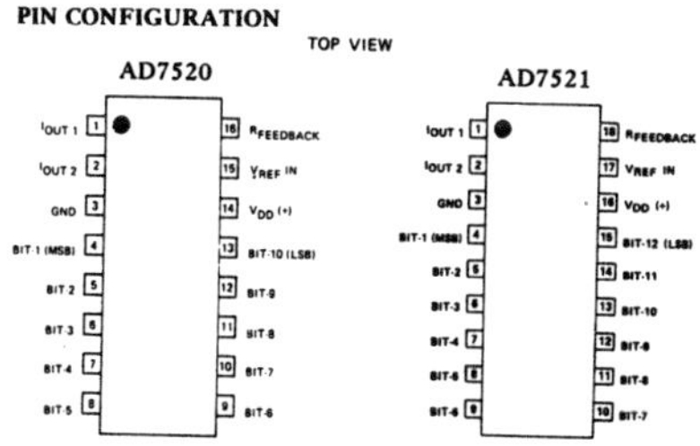

PACKAGE IDENTIFICATION

Suffix D: Ceramic DIP package
Suffix N: Plastic DIP package

Route 1 Industrial Park; P.O. Box 280; Norwood, Mass. 02062
Tel: (617) 329-4700 Telex: 924491
Cable: ANALOG NORWOODMASS

Courtesy Analog Devices, Inc.

SPECIFICATIONS (V_{DD} = +15, V_{REF} = +10 V, T_A = 25° C unless otherwise noted)

PARAMETER		AD7520	AD7521	UNITS	LIMIT	TEST CONDITIONS	FIG.
DC ACCURACY (Note 1) Resolution		10	12	Bits			
Nonlinearity	J S	0.2 (8 Bit)		% of FSR	Max	S, T, U: over −55°C to +125°C	5
	K T	0.1 (9 Bit)		% of FSR	Max	$-10V \leqslant V_{REF} \leqslant +10V$	5
	L U	0.05 (10 Bit)		% of FSR	Max		5
Nonlinearity Tempco		2		PPM of FSR/°C	Max		
Gain Error (Note 2)		0.3		% of FSR	Typ	$-10V \leqslant V_{REF} \leqslant +10V$	
Gain Error Tempco (Note 2)		10		PPM of FSR/°C	Max		
Output Leakage Current (either output)		200		nA	Max	Over specified temperature range	
Power Supply Rejection		50		PPM of FSR/%C	Typ		6
AC ACCURACY Output Current Settling Time		500		ns	Typ	To 0.05% of FSR All digital inputs low to high and high to low	10
Feedthrough Error		10		mV pp	Max	V_{REF} = 20 V pp, 100 kHz All digital inputs low	9
REFERENCE INPUT Input Resistance		10 k		Ω	Typ		
ANALOG OUTPUT Voltage Compliance (both outputs)		See absolute max. ratings					
Output Capacitance	$I_{OUT\ 1}$	120		pF	Typ	All digital inputs high	8
	$I_{OUT\ 2}$	37		pF	Typ		
	$I_{OUT\ 1}$	37		pF	Typ	All digital inputs low	8
	$I_{OUT\ 2}$	120		pF	Typ		
Output Noise (both outputs)		Equivalent to 10 kΩ Johnson noise					7
DIGITAL INPUTS (Note 3) Low State Threshold		0.8		V	Max	Over specified temperature range	
High State Threshold		2.4		V	Min		
Input Current (low to high state)		1		μA	Typ		
Input Coding		Binary				See Tables 1 & 2 on page 6	
POWER REQUIREMENTS Power Supply Voltage Range		+5 to +15		V			
I_{DD}		5		nA	Typ	All digital inputs at GND	
		2		mA	Max	All digital inputs high or low	
Total Dissipation (Including ladder)		20		mW	Typ		

NOTES:
1. Full scale range (FSR) is 10V for unipolar mode and ±10V for bipolar mode.
2. Using the internal $R_{FEEDBACK}$
3. Digital input levels should not go below ground or exceed the positive supply voltage, otherwise damage may occur.

ABSOLUTE MAXIMUM RATINGS
(T_A = 25°C unless otherwise noted)

V_{DD} (to Gnd) . . . . +17 V
V_{REF} (to Gnd) . . . . ±25 V
Digital Input Voltage Range . . . . V_{DD} to Gnd
Output Voltage Compliance . . . . –100mV to V_{DD}

Power Dissipation (package)
up to +75°C . . . . 450 mW
derates above +75°C by . . . . .6 mW/°C
Operating Temperature . . . . –55°C to +125°C
Storage Temperature . . . . –65°C to +190°C

CAUTION:

1. Do not apply voltages higher than V_{DD} or less than GND potential on any terminal except V_{REF}.
2. The digital control inputs are zener protected; however, permanent damage may occur on unconnected units under high energy electrostatic fields. Keep unused units in conductive foam at all times.

Specifications subject to change without notice.

Courtesy Analog Devices, Inc.

ICL7106, 7107
3½ Digit Single Chip A/D Converter

FEATURES

- **Guaranteed zero reading for 0 volts input on all scales.**
- **True polarity at zero for precise null detection.**
- **1 pA input current typical.**
- **True differential input and reference.**
- **Direct display drive - no external components required. — LCD ICL7106**
  — LED ICL7107
- **Low noise - less than 15μV pk-pk.**
- **On-chip clock and reference.**
- **Low power dissipation - typically less than 10mW.**
- **No additional active circuits required.**
- **Evaluation Kit available.**

GENERAL DESCRIPTION

The Intersil ICL7106 and 7107 are high performance, low power 3-1/2 digit A/D converters. All the necessary active devices are contained on a single CMOS I.C., including seven segment decoders, display drivers, reference, and a clock. The 7106 is designed to interface with a liquid crystal display (LCD) and includes a backplane drive; the 7107 will directly drive an instrument-size light emitting diode (LED) display.

The 7106 and 7107 bring together an unprecedented combination of high accuracy, versatility, and true economy. High accuracy like auto-zero to less than 10μV, zero drift of less than 1μV/°C, input bias current of 10 pA max., and roll-over error of less than one count. The versatility of true differential input and reference is useful in all systems, but gives the designer an uncommon advantage when measuring load cells, strain gauges and other bridge-type transducers. And finally the true economy of single power supply operation (7106), enabling a high performance panel meter to be built with the addition of only 7 passive components and a display.

TYPICAL CONNECTION DIAGRAMS

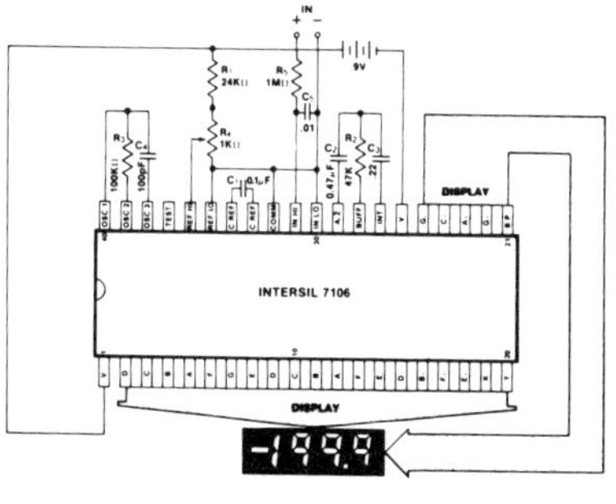

ICL7106 with Liquid Crystal Display

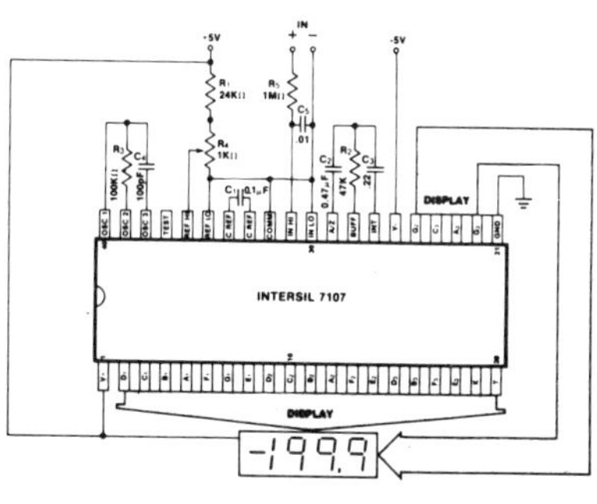

ICL7107 with LED Display

ORDERING INFORMATION

Part	Package	Temp. Range	Order Part #
7106	40 pin ceramic DIP	0°C to +70°C	ICL7106CDL
7106	40 pin plastic DIP	0°C to +70°C	ICL7106CPL
7107	40 pin ceramic DIP	0°C to +70°C	ICL7107CDL
7107	40 pin plastic DIP	0°C to +70°C	ICL7107CPL
7106 Kit 7107 Kit	Evaluation kits contain IC, display, circuit board, passive components and hardware. See page 10.		ICL7106EV/Kit ICL7107EV/Kit

PIN CONFIGURATION

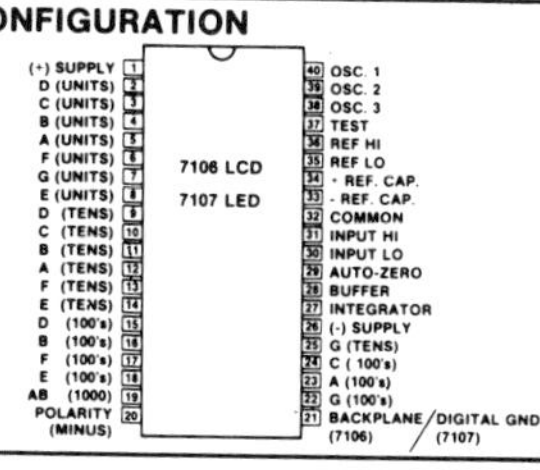

INTERSIL, INC., 10710 N. TANTAU AVE., CUPERTINO, CA 95014
(408) 996-5000 TWX: 910-338-0171

Printed in U.S.A.

Courtesy Intersil, Inc.

ICL7106/ICL7107

ABSOLUTE MAXIMUM RATINGS

ICL 7106

Supply Voltage (V+ to V−) 15V
Analog Input Voltage (either input) (Note 1) ...... V+ to V−
Reference Input Voltage (either input) .... V+ to V−
Clock Input .... Test to V+
Power Dissipation (Note 2)
  Ceramic Package .... 1000 mW
  Plastic Package .... 800 mW
Operating Temperature .... 0° C to +70° C
Storage Temperature .... −65° C to +160° C
Lead Temperature (Soldering, 60 sec) .... 300° C

ICL 7107

Supply Voltage V+ .... +6V
V− .... −9V
Analog Input Voltage (either input) (Note 1) ...... V+ to V−
Reference Input Voltage (either input) .... V+ to V−
Clock Input .... Gnd to V+
Power Dissipation (Note 1)
  Ceramic Package .... 1000 mW
  Plastic Package .... 800 mW
Operating Temperature .... 0° C to +70° C
Storage Temperature .... −65° C to +160° C
Lead Temperature (Soldering, 60 sec) .... 300° C

Note 1: Input voltages may exceed the supply voltages provided the input current is limited to ±100μA.

Note 2: Dissipation rating assumes device is mounted with all leads soldered to printed circuit board.

ELECTRICAL CHARACTERISTICS (Note 3)

CHARACTERISTICS	CONDITIONS	MIN	TYP	MAX	UNITS
Zero Input Reading	Vin = 0.0V Full Scale = 200.0 mV	−000.0	±000.0	+000.0	Digital Reading
Ratiometric Reading	Vin = Vref Vref = 100 mV	999	999/1000	1000	Digital Reading
Rollover Error (Difference in reading for equal positive and negative reading near Full Scale)	−Vin = +Vin ≃ 200.0mV	−1	±.2	+1	Counts
Linearity (Max. deviation from best straight line fit)	Full scale = 200mV or full scale = 2.000V	-1	±.2	+1	Counts
Common Mode Rejection Ratio (Note 4)	Vcm = ±1V, Vin = 0V. Full Scale = 200.0mV.		50		μV/V
Noise (Pk - Pk value not exceeded 95% of time)	Vin = 0V Full Scale = 200.0mV		15		μV
Leakage Current @ Input	Vin = 0V		1	10	pA
Zero Reading Drift	Vin = 0 0°< T_A <70° C		0.2	1	μV/° C
Scale Factor Temperature Coefficient	Vin = 199.0mV 0< T_A < 70° C (Ext.Ref.0ppm/° C)		1	5	ppm/° C
Supply Current (Does not include LED current for 7107)	Vin = 0		0.8	1.8	mA
Analog Common Voltage (With respect to pos. supply)	25KΩ between Common & pos. Supply	2.4	2.8	3.2	Volts
Temp. Coeff. of Analog Common (with respect to pos. Supply)	25KΩ between Common & pos. Supply		80		ppm/° C
7106 ONLY Pk-Pk Segment Drive Voltage (Note 5)	V Supply = 9V	4	5	6	Volts
7106 ONLY Pk-Pk Backplane Drive Voltage (Note 5)	V Supply = 9V	4	5	6	Volts
7107 ONLY Segment Sinking Current (Except Pin 19)	+Supply = 5.0V Segment voltage = 3V	5	8.0		mA
7107 ONLY Segment Sinking Current (Pin 19 only)	+Supply = 5.0V Segment voltage = 3V	10	16		mA

Note 3: Unless otherwise noted, specifications apply to both the 7106 and 7107 at $T_A = 25°$ C, $f_{clock} = 48$kHz. 7106 is tested in the circuit of Figure 1. 7107 is tested in the circuit of Figure 2.

Note 4: Refer to "Differential Input" discussion on page 4.

Note 5: Back plane drive is in phase with segment drive for 'off' segment, 180° out of phase for 'on' segment. Frequency is 20 times conversion rate. Average DC component is less than 50mV.

Courtesy Intersil, Inc.

ADC80

signal conditioning components/amplifiers
data acquisition systems/data converters
microcomputer peripherals/power supplies
signal conditioning components/amplifiers
data acquisition systems/data converters
microcomputer peripherals/power supplies
signal conditioning components/amplifiers
data acquisition systems/data converters
microcomputer peripherals/power supplies

IC A/D CONVERTERS

- **COMPACT DESIGN - Self Contained with Internal Clock, Comparator, and Reference**
  32 Pin Ceramic Package
- **FAST CONVERSION SPEEDS**
  Provide Fast Signal Sampling Rates
  12 Bits - 25μsec, 10 Bits - 21μsec
  Faster Conversion Speeds Obtainable with "Short Cycling" and Optional External Clock
- **LOW COST - ADC80AG-12 $47.50 in 100's**

DESCRIPTION

The Model ADC80AG-10 and ADC80AG-12 are 10 and 12 bit successive approximation A/D converters. They utilize state-of-the-art IC and laser trimmed thin film components, and are packaged in a compact 32-pin ceramic package.

Complete with internal reference, the ADC80 offers versatility and performance formerly offered only in larger modular or rack-mount packages.

Thin film internal scaling resistors are provided for the selection of analog input signal ranges of ±2.5, ±5, ±10, 0 to +5 or 0 to +10 volts.

Gain and offset errors may be externally trimmed to zero, offering initial accuracies of better than ±0.0122% (±½LSB). The model ADC80 is specified for -25°C to +85°C operation.

The fast conversion speeds of 25 microseconds for 12 bit and 21 microseconds for 10 bit resolution make the ADC80 excellent for a wide range of applications where system throughput sampling rates from 40 kHz to 47 kHz are required. In addition, the ADC80 may be short cycled and an external clock may be used to obtain faster conversion speeds at lower resolutions.

Data is available in parallel and serial form with corresponding clock and status signals. All digital input and output signals are DTL/TTL compatible. Power supply voltages are ±15 and +5VDC.

signal conditioning components/amplifiers
data acquisition systems/data converters
microcomputer peripherals/power supplies
signal conditioning components/amplifiers
data acquisition systems/data converters
microcomputer peripherals/power supplies
signal conditioning components/amplifiers
data acquisition systems/data converters
microcomputer peripherals/power supplies
signal conditioning components/amplifiers
data acquisition systems/data converters
microcomputer peripherals/power supplies
signal conditioning components/amplifiers
data acquisition systems/data converters
microcomputer peripherals/power supplies
signal conditioning components/amplifiers
data acquisition systems/data converters
microcomputer peripherals/power supplies
signal conditioning components/amplifiers
data acquisition systems/data converters
microcomputer peripherals/power supplies
signal conditioning components/amplifiers
data acquisition systems/data converters
microcomputer peripherals/power supplies
signal conditioning components/amplifiers
data acquisition systems/data converters
microcomputer peripherals/power supplies
signal conditioning components/amplifiers
data acquisition systems/data converters
microcomputer peripherals/power supplies
signal conditioning components/amplifiers

BURR-BROWN

International Airport Industrial Park•Tucson, Arizona 85734
TEL: 602-294-1431•TWX: 910-952-1111•CABLE: BBRCORP

Courtesy Burr-Brown Research Corp.

SPECIFICATIONS

Typical at 25°C and rated power supplies unless otherwise noted.

ELECTRICAL

MODEL	ADC80AG-12	ADC80AG-10	UNITS
RESOLUTION	12	10	Bits
INPUT			
ANALOG INPUTS			
Voltage Ranges			
Bipolar	±2.5, ±5, ±10		V
Unipolar	0 to +5, 0 to +10		V
Impedance (Direct Input)			
0 to +5V, ±2.5V	2.5		kΩ
0 to +10V, ±5V	5		kΩ
±10V	10		kΩ
DIGITAL INPUTS (1)			
Convert Command	Positive Pulse 100 ns Wide (min) 2 μsec Wide (max). ("0" to "1" Initiates Conversion)		
Logic Loading	1		TTL Load
External Clock	1		TTL Load
TRANSFER CHARACTERISTICS			
ERROR			
Gain Error (3)	±0.1		%
Offset Error (3)			
Unipolar	±0.05		% of FSR (2)
Bipolar	±0.1		% of FSR
Linearity Error (max) (4)	±0.012	±0.048	% of FSR
Inherent Quantization Error	±1/2		LSB
Differential Linearity Error	±1/2		LSB
No Missing Codes Temp. Range	0 to +50	0 to +70	°C
Power Supply Sensitivity			
±15V	±0.0030		% of FSR/%Vs
+5V	±0.0015		% of FSR/%Vs
DRIFT			
Specification Temperature Range	-25 to +85		°C
Gain, (max)	±30		ppm/°C
Offset			
Unipolar	±3		ppm of FSR/°C
Bipolar, (max)	±15		ppm of FSR/°C
Linearity, (max)	±3		ppm of FSR/°C
Monotonicity	GUARANTEED		
CONVERSION SPEED (max) (5)	25	21	μsec
OUTPUT			
DIGITAL DATA			
(All codes complementary)			
Parallel			
Output Codes (6)			
Unipolar	CSB		
Bipolar	COB, CTC		
Output Drive	2		TTL Loads
Serial Data Codes (NRZ)	CSB, COB		
Output Drive	2		TTL Loads
Status	Logic "1" during conversion		
Status Output Drive	2		TTL Loads
Internal Clock			
Clock Output Drive	2		TTL Loads
Frequency (7)	500		kHz
INTERNAL REF. VOLTAGE	6.3		V
Max. External Current (with no degradation of specifications)	200		μA
Tempco of Drift (max)	±20		ppm/°C
POWER REQUIREMENTS			
Rated Voltages	±15, +5		V
Range for Rated Accuracy	4.75 to 5.25 and ±14.0 to ±16.0		V
Supply Drain +15V	+20		mA
-15V	-20		mA
+ 5V	+70		mA
TEMPERATURE RANGE			
Specification	-25 to +85		°C
Operating (derated specs)	-55 to +100		°C
Storage	-55 to +125		°C
PRICES (1 - 24)	$82.50	$77.50	

1) DTL/TTL compatible i.e., Logic "0" = 0.8V max, Logic "1" = 2.0V min for inputs and for digital outputs, Logic "0" = +0.4V max and "1" = 2.4V min.
2) FSR means Full Scale Range - for example, unit connected for ±10V range has 20V FSR.
3) Adjustable to zero with external trimpots.
4) Error shown is the same as ±½LSB max for resolution of A/D converter.
5) Conversion time with internal clock.
6) See Table I. CSB - Complementary Straight Binary.
COB - Complementary Offset Binary.
CTC - Complementary Two's Complementary.
7) For conversion speeds specified.

MECHANICAL

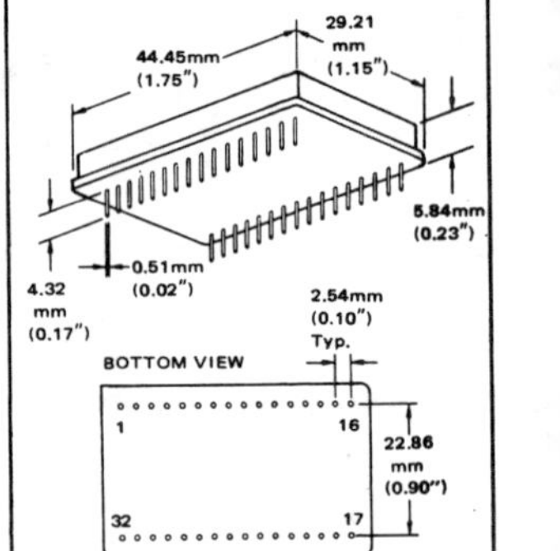

PINS: Pin material and plating composition conform to method 2003 (Solderability) of Mil-Std-883 (except paragraph 3.2).
CASE: Ceramic
MATING CONNECTOR: 2302MC - Set of two 16 pin strips $6 per set.
WEIGHT: 13 grams (0.46 oz).

Connection Diagram

TOP VIEW

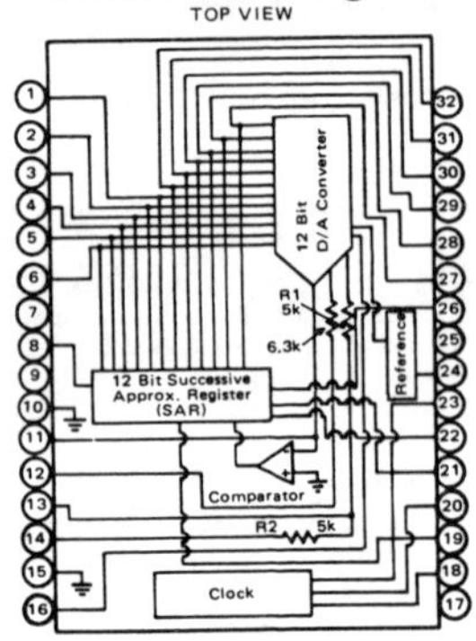

PIN CONNECTIONS

1 Bit 6	32 Bit 7
2 Bit 5	31 Bit 8
3 Bit 4	30 Bit 9
4 Bit 3	29 Bit 10 (LSB-10 Bits)
5 Bit 2	28 Bit 11
6 Bit 1 (MSB)	27 Bit 12 (LSB-12 Bits)
7 +5V Analog Supply	26 Serial Out
8 Bit 1 (MSB)	25 -15V
9 +5V Digital Supply	24 Ref. Out (+6.3V)
10 Digital Common	23 Clock Out
11 Comparator IN	22 Status
12 Bipolar Offset	21 Short Cycle
13 R1 10V Range	20 Clock Inhibit
14 R2 20V Range	19 External Clock
15 Analog Common	18 Convert Command
16 Gain Adjust	17 +15V

Courtesy Burr-Brown Research Corp.

B

IC Converter Source Guide

This appendix contains addresses and telephone numbers for major manufacturers of data conversion devices. In general all converter devices discussed within this book are listed here, according to their (original) manufacturer. In addition, related data conversion products are denoted as to source, by a general listing.

No attempt is made here to cross index alternate source part numbers, or to list in depth all products of all manufacturers. The reader is encouraged to assemble his or her own file of complete (and current) catalog data.

Manufacturer	Devices
Advanced Micro Devices 901 Thompson Place Sunnyvale, California 94086 (408) 732-2400	Am 2502/2503/2504 Am6070, Am6071, Am6072 Am6073, Am6080, Am6081 Sample/Holds
Analog Devices P.O. Box 280 Norwood, Massachusetts 02062 (617) 329-4700	AD561, AD562, AD563 AD537, AD571, AD572 AD7520, AD7521, AD7522, AD7523 AD7550, AD7570 Analog Switches, References, IAs Sample/Holds
Analogic Audubon Road Wakefield, Massachusetts 01880 (617) 246-0300	
Beckman Instruments 2500 Harbor Blvd.	

Manufacturer	Devices
Fullerton, California 92634 (714) 871-4848	
Burr-Brown International Airport Industrial Park Tucson, Arizona, 85734 (602) 294-1431	DAC80, DAC85 ADC80, ADC82, ADC84/85 VFC-32
Data Device Corp. Airport Industrial Plaza Bohemia, New York, 11716 (516) 567-5600	
Datel Systems 1020 Turnpike St. Canton, Massachusetts 02021 (617) 828-8000	
Fairchild Semiconductor 464 Ellis Street Mountain View, California 94042 (415) 962-5011	μA7151
Harris Semiconductor P.O. Box 883 Melbourne, Florida 32901 (305) 724-7430	HI-562 Analog Switches, References
Hybrid Systems Corp. Crosby Drive Bedford, Massachusetts 01730 (617) 275-1570	
Intech/Function Modules 282 Brokaw Rd. Santa Clara, California 95050 (408) 244-0500	A-8400
Intersil, Inc. 10700 Tantau Avenue Cupertino, California 95014 (408) 996-5000	ICL8052/8053(A) ICL7101, ICL7103, ICL7104 ICL7106/ICL7107 Analog Switches, References
Micro Networks 324 Clark Street Worcester, Massachusetts 01606 (617) 852-5400	
Micro Power Systems 3100 Alfred Street Santa Clara, California 95050 (408) 247-5350	

Manufacturer	Devices
Motorola Semiconductor Products 5005 East McDowell Road Phoenix, Arizona 85008 (602) 244-6900	MC1406, MC1408 MC3410 MC14433 References
National Semiconductor Corp. 2900 Semiconductor Drive Santa Clara, California 95051 (408) 732-5000	LM331 Analog Switches, References, IAs Sample/Holds
Precision Monolithics 1500 Space Park Drive Santa Clara, California 95050 (408) 242-9222	DAC-02, DAC-03, DAC-04 DAC-08, DAC-12, DAC-100 DAC-76, DAC-86 Analog Switches, References, Sample/Holds
Raytheon Semiconductor 350 Ellis St. Mountain View, California 94042 (415) 968-9211	RC4151 RC4152
RCA Solid State Division Box 3200 Somerville, New Jersey 08876 (201) 722-3200	
Signetics Corp. 811 East Arques Avenue Sunnyvale, California 94086 (408) 739-7700	LD110/LD111 (A) LD111/LD114 LD130, LD131 Analog Switches
Siliconix Inc. 2201 Laurelwoowd Rd. Santa Clara, California 95054 (408) 246-8000	
Teledyne Philbrick Allied Drive at Route 128 Dedham, Massachusetts 02026 (617) 329-1600	9400 8700 Series
Teledyne Semiconductor 1300 Terra Bella Avenue Mountain View, California 94043 (415) 968-9341	
Texas Instruments P.O. Box 5012 Dallas, Texas, 75222 (214) 238-2011	

Index

E

F

G

H

I

T

U

V